T0231443

Regularization, Optimization, Kernels, and Support Vector Machines

Chapman & Hall/CRC
Machine Learning & Pattern Recognition Series

Regularization, Optimization, Kernels, and Support Vector Machines

Edited by

Johan A. K. Suykens
KU LEUVEN, BELGIUM

Marco Signoretto
KU LEUVEN, BELGIUM

Andreas Argyriou
ECOLE CENTRALE PARIS, FRANCE

CRC Press
Taylor & Francis Group
Boca Raton London New York

CRC Press is an imprint of the
Taylor & Francis Group, an **informa** business
A CHAPMAN & HALL BOOK

ress
& Francis Group

Broken Sound Parkway NW, Suite 300
Raton, FL 33487-2742

5 by Taylor & Francis Group, LLC
ress is an imprint of Taylor & Francis Group, an Informa business

im to original U.S. Government works

13: 978-1-4822-4139-6 (hbk)

Library of Congress Cataloging-in-Publication Data

ROKS (Workshop) (2013 : Leuven, Belgium)
    Regularization, optimization, kernels, and support vector machines / editors, Johan
A.K. Suykens, Marco Signoretto, Andreas Argyriou.
        pages cm
    Includes bibliographical references and index.
    ISBN 978-1-4822-4139-6 (hardback)
    1. Mathematical models--Congresses. 2. Mathematical statistics--Congresses. I.
Suykens, Johan A. K. II. Signoretto, Marco. III. Argyriou, Andreas. IV. Title.

QA401R56 2014
511'.8--dc23
                                                                    2014034076

he Taylor & Francis Web site at
/www.taylorandfrancis.com

e CRC Press Web site at
/www.crcpress.com

Contents

Preface

Scope

Obtaining reliable models from given data is becoming increasingly important in a wide range of different applications fields including the prediction of energy consumption, complex networks, environmental modelling, biomedicine, bioinformatics, finance, process modelling, image and signal processing, brain-computer interfaces, and others. In data-driven modelling approaches one has witnessed considerable progress in the understanding of estimating flexible nonlinear models, learning and generalization aspects, optimization methods, and structured modelling. One area of high impact both in theory and applications is kernel methods and support vector machines. Optimization problems, learning, and representations of models are key ingredients in these methods. On the other hand, considerable progress has also been made on regularization of parametric models, including methods for compressed sensing and sparsity, where convex optimization plays an important role.

At the international workshop ROKS 2013 Leuven,[1] July 8–10, 2013, researchers from diverse fields were meeting on the theory and applications of regularization, optimization, kernels, and support vector machines. At this occasion the present book has been edited as a follow-up to this event, with a variety of invited contributions from presenters and scientific committee members. It is a collection of recent progress and advanced contributions on these topics, addressing methods including:

- **Regularization:** ridge regression, Lasso, group Lasso, graph-guided group lasso, (weighted) total variation, ℓ_1-ℓ_2 mixed norms, elastic net, fused Lasso, ℓ_p, ℓ_0 regularization, sparse coding, hierarchical sparse coding, nuclear norm, trace norm, K-SVD, basis pursuit denoising, quadratic basis pursuit denoising, sparse portfolio optimization, low-rank approximation, overlapped Schatten 1-norm, latent Schatten 1-norm, tensor completion, tensor denoising, composite penalties, sparse multicomposite problems, sparse PCA, structured sparsity;

- **Optimization:** forward-backward splitting, alternating directions method of multipliers (ADMM), alternating proximal algorithms, gradient projection, stochastic generalized gradient, bilevel optimization,

---

[1] http://www.esat.kuleuven.be/stadius/ROKS2013

non-convex proximal splitting, block coordinate descent algorithm, parallel coordinate descent algorithm, stochastic gradient descent, semidefinite programming, robust iterative hard thresholding, alternating non-negative least squares, hierarchical alternating least squares, conditional gradient algorithm, Frank-Wolfe algorithm, hybrid conditional gradient-smoothing algorithm;

- **Kernels and support vector machines:** support estimation, kernel PCA, subspace learning, reproducing kernel Hilbert spaces, output kernel learning, multi-task learning, support vector machines, least squares support vector machines, Lagrange duality, primal and dual representations, feature map, Mercer kernel, Nyström approximation, Fenchel duality, kernel ridge regression, kernel mean embedding, causality, importance reweighting, domain adaptation, multilayer support vector machine, Gaussian process regression, kernel recursive least-squares, dictionary learning, quantized kernel LMS;

with application examples in genome-wide data understanding and association studies, brain-computer interfaces, optical character recognition, sub-wavelength imaging, robust portfolio estimation, collaborative filtering, pharmacology, system identification, image denoising, remote sensing image classification, wireless communications, and robotic radiosurgery.

The book has intentionally not been divided into different parts because there exist many aspects in common between the main subjects (see Figure).

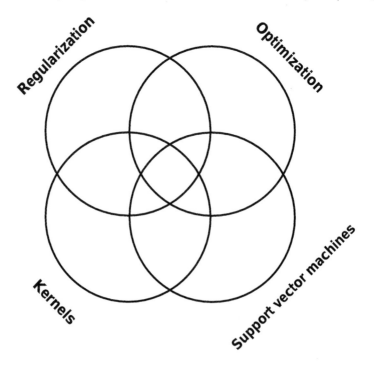

In fact it is our hope that the interested reader may discover new unexpected links and synergies. Therefore we rather aim at presenting the chapters with a natural flow of themes going from regularization, optimization, towards kernels and support vector machines.

Chapter-by-Chapter Overview

In Chapter 1, *Martin Jaggi* studies the relation between support vector machine classifiers and the Lasso technique as used in regression. An equivalence between the two methods is pointed out, together with its consequences for sparsity and regularization paths. For a given Lasso instance it is explained how an equivalent SVM instance can be constructed. Also vice versa, for a given SVM instance, how an equivalent Lasso instance can be constructed. A kernelized version of the Lasso is also discussed.

In Chapter 2, *Annalisa Barla, Saverio Salzo, and Alessandro Verri* present an alternating proximal algorithm suitable for a general dictionary learning framework with composite convex penalization terms, with stronger convergence properties than the simpler alternating minimization scheme. An analysis of the problem of computing the proximity operators is made for general sums of composite functions. The proposed algorithm is applied with mixed norms in the context of genome-wide data understanding, with identifying latent features (atoms) in array-based comparative genomic hybridization data, to reveal a genotype-phenotype relationship.

In Chapter 3, *Andreas Argyriou, Marco Signoretto, and Johan Suykens* present hybrid conditional gradient-smoothing algorithms with applications to sparse and low rank regularization for solving composite convex optimization problems. The method extends conditional gradient methods for problems with multiple nonsmooth terms. Examples and comparisons are given on regularization of matrices with combined ℓ_1 and trace norm penalties and a convex relaxation of sparse PCA.

In Chapter 4, *Suvrit Sra* presents a framework for solving a broad class of nonconvex composite objective (regularized) problems, called NIPS, which handles nonsmooth components by proximity operators. It includes forward-backward splitting with convex costs, incremental forward-backward splitting (convex), gradient projection (both convex and nonconvex), and the proximal-point algorithm. Both batch and incremental versions are discussed. The method is illustrated for sparsity regularized low-rank matrix factorization and compared with stochastic generalized gradient.

In Chapter 5, *Rémi Flamary, Alain Rakotomamonjy, and Gilles Gasso* address the problem of learning constrained task relatedness in a framework of graph-regularized multi-task learning. These similarities are learnt by optimizing a proxy on the generalization errors of all tasks. A bilevel optimization approach is taken with optimization of the generalization error and the task parameters. The global optimization is solved by means of a nonconvex proximal splitting algorithm. The interpretability of the task similarity matrices is illustrated in applications on a brain-computer interface problem of recognizing the presence of an event-related potential, and optical character recognition.

Motivated by genome-wide association studies, in Chapter 6 *Zi Wang and Giovanni Montana* propose a penalized linear regression model, the graph-guided group Lasso in two versions, which can select functionally related genes and influential SNPs within these genes that explain the variability in the trait. The method uses graph and grouping structure on hierarchical biological variants to drive variable selection at multiple levels. Theoretical properties of the smoothing effect in the proposed models are presented.

In Chapter 7, *Cheng Tang and Claire Monteleoni* survey recent work on the non-asymptotic analysis of stochastic gradient descent on strongly convex functions. They discuss how the degree of strong convexity and the degree of smoothness of a function influence the convergence rate of stochastic gradient descent.

In Chapter 8, *Kris De Brabanter, Paola Gloria Ferrario, and László Györfi* investigate the hypothesis that some components of the covariate (feature) vector are ineffective. Consequences of the results for binary pattern recognition are discussed.

In Chapter 9, *Henrik Ohlsson, Allen Yang, Roy Dong, Michel Verhaegen, and Shankar Sastry* extend classical compressive sensing, which assumes a linear relation between samples and the unknowns, to nonlinear models. The case of quadratic relations or second order Taylor expansions is studied. The extension is based on lifting and convex relaxations, where the final formulation takes the form of a semidefinite program. The proposed method of quadratic basis pursuit inherits properties of basis pursuit and classical compressive sensing. Moreover, conditions for perfect recovery are derived. An application to subwavelength diffractive imaging is presented.

In Chapter 10, *Esa Ollila, Hyon-Jung Kim, and Visa Koivunen* provide an overview of recent robust approaches to compressive sensing. Huber iterative hard thresholding avoids the computation of the preliminary estimate of scale which is needed in the generalized iterative hard thresholding method. Good signal reconstruction performance is shown under various noise distributions and signal to noise ratios. Improved performance is observed in comparison with normalized and Lorentzian iterative hard thresholding under non-Gaussian noise.

In Chapter 11, *Theodoros Evgeniou, Massimiliano Pontil, Diomidis Spinellis, and Nick Nassuphis* propose an approach to estimate large autocorrelation portfolios. It employs regularization methods derived from a robust optimization formulation. An iterative optimization learning algorithm that estimates sparse portfolios is discussed. The potential of the method is illustrated on time series of daily S&P 500 stock returns. An extension and algorithm for the more general case of canonical correlation analysis is also discussed.

Motivated by applications of facial feature extraction in image processing, and topic recovery and document classification in text mining and hyperspectral unmixing, in Chapter 12 *Nicolas Gillis* explains about different nonnegative matrix factorization methods. Besides standard NMF algorithms, a recent subclass of NMF problems, referred to as near-separable NMF, which can be solved efficiently in the presence of noise.

In Chapter 13, *Ivan Markovsky* emphasizes the role of structured low-rank approximation problems in data modeling problems. Examples are given in computer vision related to multidimensional scaling, conic section fitting, fundamental matrix estimation, and contour alignment. The contour alignment problem is reduced to the orthogonal Procrustes problem, which is a low-rank approximation problem with an additional orthogonality constraint.

In Chapter 14, *Ryota Tomioka, Taiji Suzuki, Kohei Hayashi, and Hisashi Kashima* review convex optimization based algorithms for tensor decomposition. Notions of rank and multilinear rank are explained and related convex relaxation schemes using trace norm regularization, together with statistical guarantees. ADMM optimization algorithms for overlapped Schatten 1-norm and latent Schatten 1-norm regularization, are discussed. Experimental examples are shown for tensor denoising and tensor completion.

In Chapter 15, *Alessandro Rudi, Guillermo Canas, Ernesto De Vito, and Lorenzo Rosasco* study set learning by analyzing its relations with subspace learning. The latter consists of estimating the smallest linear subspace containing the distribution. They show that the set learning problem can be cast as a subspace learning problem in the associated feature space. In the statistical analysis, novel and sharper sample complexity upper bounds and the consistency of set learning have been established.

In Chapter 16, *Francesco Dinuzzo, Cheng Soon Ong, and Kenji Fukumizu* overview a family of regularization techniques called output kernel learning for learning a multi-task kernel. In these problems the multi-task kernel can be decomposed as the product of a kernel on the inputs and a kernel on the task indices. The output kernel learning methods are illustrated on examples in collaborative filtering, structure discovery in multiclass classification, and pharmacological problems.

In Chapter 17, *Tillmann Falck, Bart De Moor, and Johan Suykens* propose a nuclear norm based regularization scheme for kernel-based identification of systems with multiple outputs using nuclear norm regularization. Based upon least squares support vector machines, the model estimation scheme is formulated as a constrained optimization problem, resulting in a Lagrange dual representation of the model, which employs a positive definite kernel.

In Chapter 18, *Pantelis Bouboulis and Sergios Theodoridis* discuss kernel methods for image denoising, based on the kernel ridge regression scheme and the theory of sparse representations. In the first method a rich basis of functions to model image edges is employed and a regularized ℓ_1 error is taken to account for outliers. In a second method the outliers are explicitly modelled using a sparse vector and the ℓ_0 norm of this vector is minimized. A comparison is made with wavelet methods.

In Chapter 19, *Kun Zhang, Bernhard Schölkopf, Krikamol Muandet, Zhikun Wang, Zhi-Hua Zhou, and Claudio Persello* consider domain adaptation, where both the distribution of the covariate and the conditional distribution of the target given the covariate, change across domains. Target shift, conditional shift, and generalized target shift are studied. Kernel mean embedding of conditional and marginal distributions is proposed to handle the different situations. Applications to remote sensing image classification are given.

In Chapter 20, *Marco Wiering and Lambert Schomaker* present multi-layer support vector machines. Comparisons with deep neural network architecture and multiple-kernel learning algorithms are made. Gradient ascent algorithms for training are discussed. Examples are given on classification, regression, and dimensionality reduction problems.

Motivated by signal processing applications in Chapter 21, *Steven Van Vaerenbergh and Ignacio Santamaría* present on-line regression with kernels. Methods of on-line kernel ridge regression, on-line dictionary learning, kernel recursive least-squares regression, on-line Gaussian processes, and stochastic gradient descent with kernels are discussed.

Acknowledgments

We thank the many chapter authors and co-authors for their enthusiasm and interest in contributing to this edited book.

This edited book has been published within the framework of the European Research Council[2] (ERC) Advanced Grant project[3] 290923 A-DATADRIVE-B. We also gratefully acknowledge support from KU Leuven, FWO, IUAP DYSCO, CoE OPTEC EF/05/006, GOA MANET, iMinds Medical Information Technologies, and the European Union Seventh Framework Programme (FP7 2007-2013) under grant agreement No. 246556.

Johan A.K. Suykens, KU Leuven
Marco Signoretto, KU Leuven
Andreas Argyriou, Ecole Centrale Paris

---

[2]http://erc.europa.eu/
[3]http://www.esat.kuleuven.be/stadius/ADB/

Contributors

Andreas Argyriou
École Centrale Paris
Center for Visual Computing

Annalisa Barla
DIBRIS
Università degli Studi di Genova

Pantelis Bouboulis
Department of Informatics and
　Telecommunications
University of Athens

Guillermo D. Canas
Massachusetts Institute of
　Technology

Kris De Brabanter
Department of Statistics
Department of Computer Science
Iowa State University

Bart De Moor
KU Leuven
ESAT-STADIUS
IMinds Medical IT

Ernesto De Vito
DIMA
Università degli Studi di Genova

Francesco Dinuzzo
IBM Research
Dublin, Ireland

Roy Dong
Department of Electrical Engineering
　and Computer Sciences
University of California, Berkeley

Theodoros Evgeniou
Decision Sciences and Technology
　Management INSEAD
Fontainebleau

Tillmann Falck
KU Leuven
ESAT-STADIUS

Paola Gloria Ferrario
Institut für Medizinische Biometrie
　und Statistik
Universität zu Lübeck

Rémi Flamary
Laboratoire Lagrange
Observatoire de la Côte d'Azur
Université de Nice Sophia-Antipolis

Kenji Fukumizu
The Institute of Statistical
　Mathematics
Tachikawa, Tokyo

Gilles Gasso
LITIS
INSA de Rouen

Nicolas Gillis
Department of Mathematics and
　Operational Research
Faculté Polytechnique
Université de Mons

László Györfi
Department of Computer Science
　and Information Theory
Budapest University of Technology
　and Economics

Kohei Hayashi
National Institute of Informatics
Japan

Martin Jaggi
ETH Zürich

Hisashi Kashima
University of Tokyo
Japan

Hyon-Jung Kim
Aalto University
Finland

Visa Koivunen
Aalto University
Finland

Ivan Markovsky
Department ELEC
Vrije Universiteit Brussel

Giovanni Montana
King's College London

Claire Monteleoni
George Washington University

Krikamol Muandet
Max Planck Institute for Intelligent
  Systems
Tübingen

Nick Nassuphis
31 St. Martin's Lane
London

Henrik Ohlsson
Department of Electrical Engineering
  and Computer Sciences
University of California, Berkeley

Esa Ollila
Aalto University
Finland

Cheng Soon Ong
NICTA
Canberra, Australia

Claudio Persello
Max Planck Institute for Intelligent
  Systems, Tübingen
University of Trento

Massimiliano Pontil
Department of Computer Science
University College London

Alain Rakotomamonjy
LITIS
Université de Rouen

Lorenzo Rosasco
DIBRIS, Università degli Studi di
  Genova
LCSL, Massachusetts Institute of
  Technology
Istituto Italiano di Tecnologia

Alessandro Rudi
DIBRIS, Università degli Studi di
  Genova
LCSL, Massachusetts Institute of
  Technology
Istituto Italiano di Tecnologia

Saverio Salzo
DIMA
Università degli Studi di Genova

Ignacio Santamaría
Department of Communications
  Engineering
University of Cantabria

S. Shankar Sastry
Department of Electrical Engineering
  and Computer Sciences
University of California, Berkeley

Bernhard Schölkopf
Max Planck Institute for Intelligent
 Systems
Tübingen

Lambert R.B. Schomaker
Institute of Artificial Intelligence and
 Cognitive Engineering
University of Groningen

Marco Signoretto
KU Leuven
ESAT-STADIUS

Diomidis Spinellis
Department of Management Science
 and Technology
Athens University of Economics and
 Business

Suvrit Sra
Max Planck Institute Intelligent
 Systems, Tübingen
Carnegie Mellon University,
 Pittsburgh

Johan A.K. Suykens
KU Leuven
ESAT-STADIUS

Taiji Suzuki
Tokyo Institute of Technology
Japan

Cheng Tang
George Washington University

Sergios Theodoridis
Department of Informatics and
 Telecommunications
University of Athens

Ryota Tomioka
Toyota Technological Institute at
 Chicago

Steven Van Vaerenbergh
Department of Communications
 Engineering
University of Cantabria

Michel Verhaegen
Delft Center for Systems and Control
Delft University

Alessandro Verri
DIBRIS
Università degli Studi di Genova

Allen Y. Yang
Department of Electrical Engineering
 and Computer Sciences
University of California, Berkeley

Zhikun Wang
Max Planck Institute for Intelligent
 Systems
Tübingen

Zi Wang
Imperial College London

Marco A. Wiering
Institute of Artificial Intelligence and
 Cognitive Engineering
University of Groningen

Kun Zhang
Max Planck Institute for Intelligent
 Systems
Tübingen

Zhi-Hua Zhou
National Key Laboratory for Novel
 Software Technology
Nanjing University

Chapter 1

An Equivalence between the Lasso and Support Vector Machines

Martin Jaggi[*]

ETH Zürich

Overview

We investigate the relation of two fundamental tools in machine learning and signal processing, which are the support vector machine (SVM) for classification, and the Lasso technique used in regression. We show that the resulting optimization problems are equivalent, in the following sense. Given

---

[*]jaggi@inf.ethz.ch

any instance of an ℓ_2-loss soft-margin (or hard-margin) SVM, we construct a Lasso instance having the same optimal solutions, and vice versa.

As a consequence, many existing optimization algorithms for both SVMs and Lasso can also be applied to the respective other problem instances. Also, the equivalence allows for many known theoretical insights for SVM and Lasso to be translated between the two settings. One such implication gives a simple kernelized version of the Lasso, analogous to the kernels used in the SVM setting. Another consequence is that the sparsity of a Lasso solution is equal to the number of support vectors for the corresponding SVM instance, and that one can use screening rules to prune the set of support vectors. Furthermore, we can relate sublinear time algorithms for the two problems, and give a new such algorithm variant for the Lasso. We also study the regularization paths for both methods.

1.1   Introduction

Linear classifiers and kernel methods, and in particular the support vector machine (SVM) [9], are among the most popular standard tools for classification. On the other hand, ℓ_1-regularized least squares regression, i.e., the Lasso estimator [41], is one of the most widely used tools for robust regression and sparse estimation.

Along with the many successful practical applications of SVMs and the Lasso in various fields, there is a vast amount of existing literature[1] on the two methods themselves, considering both theory and also algorithms for each of the two. However, the two research topics developed largely independently and were not much set into context with each other so far.

In this chapter, we attempt to better relate the two problems, with two main goals in mind. On the algorithmic side, we show that many of the existing algorithms for each of the two problems can be set into comparison, and can be directly applied to the other respective problem. As a particular example of this idea, we can also apply the recent sublinear time SVM algorithm by [8] to any Lasso problem, resulting in a new alternative sublinear time algorithm variant for the Lasso.

On the other hand, we can relate and transfer existing theoretical results between the literature for SVMs and the Lasso. In this spirit, a first example is the idea of the kernel trick. Originally employed for SVMs, this powerful concept has allowed for lifting most insights from the linear classifier setting also to the more general setting of non-linear classifiers, by using an implicit higher dimensional space. Here, by using our equivalence, we propose a simple

---

[1] As of January 2014, Google Scholar returned $300,000$ publications containing the term Support Vector Machine, and over $20,000$ for Lasso regression.

kernelized variant of the Lasso, being equivalent to the well-researched use of kernels in the SVM setting.

As another example, we can also transfer some insights in the other direction, from the Lasso to SVMs. The important datapoints, i.e., those that define the solution to the SVM classifier, are called the support vectors. Having a small set of support vectors is crucial for the practical performance of SVMs. Using our equivalence, we see that the set of support vectors for a given SVM instance is exactly the same as the sparsity pattern of the corresponding Lasso solution.

Screening rules are a way of pre-processing the input data for a Lasso problem, in order to identify inactive variables. We show that screening rules can also be applied to SVMs, in order to eliminate potential support vectors beforehand, and thereby speeding up the training process by reducing the problem size.

Finally, we study the complexity of the solution paths of Lasso and SVMs, as the regularization parameter changes. We discuss path algorithms that apply to both problems, and also translate a result on the Lasso path complexity to show that a single SVM instance, depending on the scaling of the data, can have very different patterns of support vectors.

Support Vector Machines. In this chapter, we focus on large margin linear classifiers, and more precisely on those SVM variants whose dual optimization problem is of the form

$$\min_{x \in \triangle} \|Ax\|^2 . \tag{1.1}$$

Here the matrix $A \in \mathbb{R}^{d \times n}$ contains all n datapoints as its columns, and $\triangle$ is the unit simplex in $\mathbb{R}^n$, i.e., the set of probability vectors, that is the non-negative vectors whose entries sum up to one. The formulation (1.1) includes the commonly used soft-margin SVM with ℓ_2-loss. It includes both the one or two classes variants, both with or without using a kernel, and both using a (regularized) offset term (allowing hyperplanes not passing through the origin) or no offset. We will explain these variants and the large margin interpretation of this optimization problem in more detail in Section 1.2.

Lasso. On the other hand, the Lasso [41] is given by the quadratic program

$$\min_{x \in \blacklozenge} \|Ax - b\|^2 . \tag{1.2}$$

It is also known as the constrained variant of ℓ_1-regularized least squares regression. Here the right hand side b is a fixed vector $b \in \mathbb{R}^d$, and $\blacklozenge$ is the ℓ_1-unit-ball in $\mathbb{R}^n$. Note that the 1-norm $\|.\|_1$ is the sum of the absolute values of the entries of a vector. Sometimes in practice, one would like to have the constraint $\|x\|_1 \leq r$ for some value $r > 0$, instead of the simple unit-norm case $\|x\|_1 \leq 1$. However, in that case it is enough to simply re-scale the input matrix A by a factor of r, in order to obtain our above formulation (1.2) for any general Lasso problem.

In applications of the Lasso, it is important to distinguish two alternative interpretations of the data matrix A, which defines the problem instance (1.2): on one hand, in the setting of *sparse regression*, the matrix A is usually called the dictionary matrix, with its columns $A_{\cdot j}$ being the dictionary elements, and the goal being to approximate the single vector b by a combination of few dictionary vectors. On the other hand, if the Lasso problem is interpreted as *feature-selection*, then each row $A_{i\cdot}$ of the matrix A is interpreted as an input vector, and for each of those, the Lasso is approximating the response b_i to input row $A_{i\cdot}$. The book [3] gives a recent overview of Lasso-type methods and their broad applications. While the penalized variant of the Lasso (meaning that the term $\|x\|_1$ is added to the objective instead of imposed as a constraint) is also popular in applications, here we focus on the original constrained variant.

The Equivalence. We will prove that the two problems (1.1) and (1.2) are indeed equivalent, in the following sense. For any Lasso instance given by (A, b), we construct an equivalent (hard-margin) SVM instance, having the same optimal solution. This will be a simple reduction preserving all objective values. On the other hand, the task of finding an equivalent Lasso instance for a given SVM appears to be a harder problem. Here we show that there always exists such an equivalent Lasso instance, and furthermore, if we are given a weakly separating vector for the SVM (formal definition to follow soon below), then we can explicitly construct the equivalent Lasso instance. This reduction also applies to the ℓ_2-loss soft-margin SVM, where we show that a weakly separating vector is trivial to obtain, making the reduction efficient. The reduction does *not* require that the SVM input data is separable.

Our shown equivalence is on the level of the (SVM or Lasso) *training* formulations; we are not making any claims on the performance of the two different kinds of methods on unseen *test* data.

On the way to this goal, we will also explain the relation to the "non-negative" Lasso variant when the variable vector x is required to lie in the simplex, i.e.,

$$\min_{x \in \triangle} \|Ax - b\|^2 \ . \tag{1.3}$$

It turns out the equivalence of the optimization problems (1.1) and (1.3) is straightforward to see. Our main contribution is to explain the relation of these two optimization problems to the original Lasso problem (1.2), and to study some of the implications of the equivalence.

Related Work. The early work of [18] has already significantly deepened the joint understanding of kernel methods and the sparse coding setting of the Lasso. Despite its title, [18] is not addressing SVM classifiers, but in fact the ε-insensitive loss variant of support vector regression (SVR), which the author proves to be equivalent to a Lasso problem where ε then becomes the ℓ_1-regularization. Unfortunately, this reduction does not apply anymore when $\varepsilon = 0$, which is the case of interest for standard hinge-loss SVR, and also for SVMs in the classification setting, which are the focus of our work here.

Another reduction has been known if the SVR insensitivity parameter ε is chosen close enough to one. In that case, [32] has shown that the SVR problem can be reduced to SVM classification with standard hinge-loss. Unfortunately, this reduction does not apply to Lasso regression.

In a different line of research, [28] have studied the relation of a dual variant of the Lasso to the primal of the so called *potential SVM* originally proposed by [23, 24], which is not a classifier but a specialized method of feature selection.

In the application paper [15] in the area of computational biology, the authors already suggested making use of the "easier" direction of our reduction, reducing the Lasso to a very particular SVM instance. The idea is to employ the standard trick of using two non-negative vectors to represent a point in the ℓ_1-ball [1, 7]. Alternatively, this can also be interpreted as considering an SVM defined by all Lasso dictionary vectors together with their negatives ($2n$ many points). We formalize this interpretation more precisely in Section 1.3.2. The work of [15] does not address the SVM regularization parameter.

Notation. The following three sets of points will be central for our investigations. We denote the unit simplex, and the filled simplex, as well as the ℓ_1-unit-ball in $\mathbb{R}^n$ as follows.

$$
\begin{aligned}
\triangle &:= \left\{ x \in \mathbb{R}^n \mid x \geq \mathbf{0}, \ \textstyle\sum_i x_i = 1 \right\} , \\
\blacktriangle &:= \left\{ x \in \mathbb{R}^n \mid x \geq \mathbf{0}, \ \textstyle\sum_i x_i \leq 1 \right\} , \\
\blacklozenge &:= \left\{ x \in \mathbb{R}^n \mid \|x\|_1 \leq 1 \right\} .
\end{aligned}
$$

The 1-norm $\|.\|_1$ is the sum of the absolute values of the entries of a vector. The standard Euclidean norm is written as $\|.\|$.

For a given matrix $A \in \mathbb{R}^{d \times n}$, we write $A_i \in \mathbb{R}^d$, $i \in [1..n]$ for its columns. We use the notation $AS := \{Ax \mid x \in S\}$ for subsets $S \subseteq \mathbb{R}^d$ and matrices A. The convex hull of a set S is written as $\mathrm{conv}(S)$. By $\mathbf{0}$ and $\mathbf{1}$ we denote the all-zero and all-ones vectors in $\mathbb{R}^n$, and $\mathbf{I}_n$ is the $n \times n$ identity matrix. We write $(A|B)$ for the horizontal concatenation of two matrices A, B.

1.2   Linear Classifiers and Support Vector Machines

Linear classifiers have become the standard workhorse for many machine learning problems. Suppose we are given n datapoints $X_i \in \mathbb{R}^d$, together with their binary labels $y_i \in \{\pm 1\}$, for $i \in [1..n]$.

As we illustrate in Figure 1.1, a linear classifier is a hyperplane that partitions the space $\mathbb{R}^d$ into two parts, such that hopefully each point with a positive label will be on one side of the plane, and the points of negative label will be on the other site. Writing the classifier as the normal vector w of that hyperplane, we can formally write this separation as $X_i^T w > 0$ for those points i with $y_i = +1$, and $X_i^T w < 0$ if $y_i = -1$, assuming we consider hyperplanes that pass through the origin.

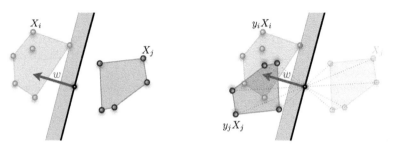

FIGURE 1.1: A linear classifier: illustration of the separation of two point classes, for a normal vector w. Here we highlight two points i, j of different labels, with $y_i = +1$ and $y_j = -1$.

1.2.1   Support Vector Machines

The popular concept of *support vector machines* (SVM) is precisely this linear classifier idea as we have introduced above, with one important addition: instead of being satisfied with any arbitrary one among all separating hyperplanes, we want to find the hyperplane that separates the two point classes by the best possible margin.

The *margin* is defined as the smallest distance to the hyperplane among all the datapoints. Maximizing this margin over all possible linear classifiers w can be simply formalized as the following optimization problem:

$$\max_{w \in \mathbb{R}^d} \min_i \, y_i X_i^T \frac{w}{\|w\|} \, ,$$

i.e., maximizing the smallest projection onto the direction w, among all datapoints.

The SVM optimization problem (1.1) that we defined in the introduction is very closely related to this form of optimization problem, as follows. If we take the Lagrange dual of problem (1.1), we obtain a problem of exactly this margin maximization type, namely

$$\max_{x \in \triangle} \min_i \, A_i^T \frac{Ax}{\|Ax\|} \, . \tag{1.4}$$

Here we think of the columns of A being the SVM datapoints with their signs

$A_i = y_i X_i$. For an overview of Lagrange duality, we refer the reader to [2, Section 5], or see for example [12, Appendix A] for the SVM case here. An alternative and slightly easier way to obtain this dual problem is to compute the simple "linearization" dual function as in [22, 25], which avoids the notion of any dual variables. When starting from the non-squared version of the SVM problem (1.1), this also gives the same formulation (1.4).

No matter if we use problem formulation (1.1) or (1.4), any feasible weight vector x readily gives us a candidate classifier $w = Ax$, represented as a convex combination of the datapoints, because the weight vectors $x \in \triangle \subset \mathbb{R}^n$ lie in the simplex. The datapoints corresponding to non-zero entries in x are called the *support vectors*.

Several other SVM variants of practical interest also have the property that their dual optimization problem is of the form (1.1), as we will discuss in the next Subsection 1.3.2.

Kernelization. A crucial and widely used observation is that both optimization problems (1.1) and (1.4) are formulated purely in terms of the inner products between pairs of datapoints $A_i := y_i X_i$, meaning that they can directly be optimized in the *kernel* case [9], provided that we only have access to the entries of the matrix $A^T A \in \mathbb{R}^{n \times n}$, but not the explicit features $A \in \mathbb{R}^{d \times n}$. The matrix $K = A^T A$ is called the *kernel matrix* in the literature. Using this notation, the SVM dual optimization problem (1.1) becomes

$$\min_{x \in \triangle} x^T K x .$$

Separation, and Approximate Solutions. It is natural to measure the quality of an approximate solution x to the SVM problem as the attained margin, which is precisely the attained value in the above problem (1.4).

Definition 1.1. *A vector $w \in \mathbb{R}^d$ is called σ-weakly separating for the SVM instances (1.1) or (1.4), respectively, for a parameter $\sigma \geq 0$, if it holds that*

$$A_i^T \frac{w}{\|w\|} \geq \sigma \quad \forall i ,$$

meaning that w attains a margin of separation of σ.

The margin value σ in this definition, or also the objective in (1.4), can be interpreted as a useful certificate for the attained optimization quality as follows. If we take some x (the separating vector now being interpreted as $w = Ax$), then the *duality gap* is given by the difference of the margin value from the corresponding objective $\|Ax\|$ in problem (1.1). This gap function is a certificate for the approximation quality (since the unknown optimum must lie within the gap), which makes it a very useful stopping criterion for optimizers; see, e.g., [22, 12, 8, 25].

The simple perceptron algorithm [34] is known to return a σ-weakly separating solution to the SVM after $O(1/\varepsilon^2)$ iterations, for $\varepsilon := \sigma^* - \sigma$ being the additive error, if σ^* is the optimal solution to (1.1) and (1.4).

1.2.2 Soft-Margin SVMs

For the successful practical application of SVMs, the soft-margin concept of tolerating outliers is of central importance. Here we recall that the soft-margin SVM variants using ℓ_2-loss, with regularized offset or no offset, both in the one-class and the two-class case, can also be formulated in the form (1.1). This fact is known in the SVM literature [37, 26, 43], and can be formalized as follows.

The two-class soft-margin SVM with squared loss (ℓ_2-loss), without offset term, is given by the primal optimization problem

$$
\min_{\substack{\bar{w}\in\mathbb{R}^d,\ \rho\in\mathbb{R},\\ \xi\in\mathbb{R}^n}} \quad \tfrac{1}{2}\left\|\bar{w}\right\|^2 - \rho + \tfrac{C}{2}\sum_i \xi_i^2
$$
$$
s.t. \quad y_i \cdot \bar{w}^T X_i \geq \rho - \xi_i \ \ \forall i \in [1..n] \ . \tag{1.5}
$$

For each datapoint, we have introduced a slack-variable ξ_i which is penalized in the objective function in the case that the point violate the margin. Here $C > 0$ is the regularization parameter, steering the tradeoff between large margin and punishing outliers. In the end, $\rho/\|\bar{w}\|$ will be the attained margin of separation. Note that in the classical SVM formulation, the margin parameter ρ is usually fixed to one instead, while ρ is explicitly used in the equivalent ν-SVM formulation known in the literature; see, e.g., [37]. The equivalence of the soft-margin SVM dual problem to the optimization problem (1.1) is stated in the following Lemma:

Lemma 1.1. *The dual of the soft-margin SVM (1.5) is an instance of the classifier formulation (1.1), that is, $\min_{x\in\triangle}\left\|Ax\right\|^2$, with*

$$
A := \begin{pmatrix} Z \\ \tfrac{1}{\sqrt{C}}\mathbf{I}_n \end{pmatrix} \in \mathbb{R}^{(d+n)\times n}
$$

where the data matrix $Z \in \mathbb{R}^{d\times n}$ consists of the n columns $Z_i := y_i X_i$.

Proof. Given in Appendix 1.6 for completeness, using standard Lagrange duality. □

Not all SVM variants have our desired structure of the dual problem. For example, the hinge-loss (or ℓ_1-loss) SVM refers to the setting where the outliers are penalized according to their margin-violation ξ_i, not the squared values ξ_i^2 as we use here. Changing the loss function in the primal optimization problem also affects the dual problem, so that the dual of the ℓ_1-loss SVM is *not* of the form (1.1). However in practice, it is known that both SVM variants with ℓ_1- or ℓ_2-loss do perform similarly well for most applications [27, 6].

Obtaining a Weakly Separating Vector for the ℓ_2-Loss Soft-Margin SVM. By the above lemma, we observe that a weakly separating vector is trivial to obtain for the ℓ_2-loss SVM. This holds without any assumptions on the original input data (X_i, y_i). We set $w := \begin{pmatrix} 0 \\ \tfrac{1}{\sqrt{n}}\mathbf{1} \end{pmatrix} \in \mathbb{R}^{d+n}$ to the all-one vector only on the second block of coordinates, rescaled to unit length. Clearly, this direction w attains a separation margin of $A_i^T \frac{w}{\|w\|} = \left(\frac{y_i X_i}{\sqrt{C} e_i}\right)^T \left(\frac{0}{\frac{1}{\sqrt{n}}\mathbf{1}}\right) = \frac{1}{\sqrt{nC}} > 0$ for all points i in Definition 1.1.

Incorporating an Offset Term. Our above SVM formulation also allows the use of an *offset* (or bias) variable $b \in \mathbb{R}$ to obtain a classifier that does not necessarily pass through the origin. Formally, the separation constraints then become $y_i \cdot (w^T X_i + b) \geq \rho - \xi_i \ \forall i \in [1..n]$. There is a well-known trick to efficiently emulate such an offset parameter while still using our formulation (1.5), by simply increasing the dimensionality of X_i and w by one, and adding a fixed value of one as the last coordinate to each of the datapoints X_i; see, e.g., [26, 43]. As a side-effect, the offset b^2 is then also regularized in the new term $\|w\|^2$. Nevertheless, if desired, the effect of this additional regularization can be made arbitrarily weak by re-scaling the fixed additional feature value from one to a larger value.

One-Class SVMs. All mentioned properties in this section also hold for the case of *one-class SVMs*, by setting all labels y_i to one, resulting in the same form of optimization problems (1.1) and (1.4). One class SVMs are popular, for example, for anomaly or novelty detection applications.

1.3 The Equivalence

1.3.1 Warm-Up: Equivalence between SVM and Non-Negative Lasso

Before we investigate the "real" Lasso problem (1.2) in the next two subsections, we will warm-up by considering the non-negative variant (1.3). It is a simple observation that the non-negative Lasso (1.3) is directly equivalent to the dual SVM problem (1.1) by a translation:

Equivalence by Translation. Given a non-negative Lasso instance (1.3), we can translate each column vector of the matrix A by the vector $-b$. Doing so, we precisely obtain an SVM instance (1.1), with the data matrix being

$$\tilde{A} := A - b\mathbf{1}^T \in \mathbb{R}^{d \times n} .$$

Here we have crucially used the simplex domain, ensuring that $b\mathbf{1}^T x = b$ for any $x \in \triangle$. To summarize, for those two optimization problems, the described translation precisely preserves all the objective values of all feasible points for both problems (1.3) and (1.1), that is, for all $x \in \triangle$. This is why we say that the problems are equivalent.

The reduction in the other direction — i.e., reducing an SVM instance (1.1) to a non-negative Lasso instance (1.3) — is made trivial by choosing $b := \mathbf{0}$.

Now to relate the SVM to the "real" Lasso, the same translation idea is of crucial importance. We explain the two reductions in the following subsections.

1.3.2   (Lasso $\preceq$ SVM): Given a Lasso Instance, Constructing an Equivalent SVM Instance

(This reduction is significantly easier than the other direction.)

Parameterizing the ℓ_1-Ball as a Convex Hull. One of the main properties of polytopes — if not the main one — is that every polytope can be represented as the convex hull of its vertices [45]. When expressing an arbitrary point in the polytope as a convex combination of some vertices, this leads to the standard concept of *barycentric* coordinates.

In order to represent the ℓ_1-ball $\blacklozenge$ by a simplex $\triangle$, this becomes particularly simple. The ℓ_1-ball $\blacklozenge$ is the convex hull of its $2n$ vertices, which are $\{\pm\mathbf{e}_i \mid i \in [1..n]\}$, illustrating why $\blacklozenge$ is also called the cross-polytope.

The barycentric representation of the ℓ_1-ball therefore amounts to the simple trick of using two non-negative variables to represent each real variable, which is standard, for example, when writing linear programs in standard form. For ℓ_1 problems such as the Lasso, this representation was known very early [1, 7]. Formally, any n-vector $x_\diamond \in \blacklozenge$ can be written as

$$x_\diamond = (\mathbf{I}_n \,|-\mathbf{I}_n)x_\triangle \ \text{ for } \ x_\triangle \in \triangle \subset \mathbb{R}^{2n} \ .$$

Here $x_\triangle$ is a $2n$-vector, and we have used the notation $(A|B)$ for the horizontal concatenation of two matrices A, B.

Note that the barycentric representation is *not* a bijection in general, as there can be several $x_\triangle \in \triangle \subset \mathbb{R}^{2n}$ representing the same point $x_\diamond \in \mathbb{R}^n$.

The Equivalent SVM Instance. Given a Lasso instance of the form (1.2), that is, $\min_{x \in \blacklozenge} \|Ax - b\|^2$, we can directly parameterize the ℓ_1-ball by the $2n$-dimensional simplex as described above. By writing $(\mathbf{I}_n \,|-\mathbf{I}_n)x_\triangle$ for any $x \in \blacklozenge$, the objective function becomes $\|(A\,|-A)x_\triangle - b\|^2$. This means we have obtained the equivalent non-negative regression problem of the form (1.3) over the domain $x_\triangle \in \triangle$, which, by our above remark on translations, is equivalent to the SVM formulation (1.1), i.e.,

$$\min_{x_\triangle \in \triangle} \big\| \tilde{A} x_\triangle \big\|^2 \ ,$$

where the data matrix is given by

$$\tilde{A} := (A\,|-A) - b\mathbf{1}^T \in \mathbb{R}^{d \times 2n} \ .$$

The additive rank-one term $b\mathbf{1}^T$ for $\mathbf{1} \in \mathbb{R}^{2n}$ again just means that the vector b is subtracted from each original column of A and $-A$, resulting in a translation of the problem. So we have obtained an equivalent SVM instance consisting of $2n$ points in $\mathbb{R}^d$.

Note that this equivalence not only means that the optimal solutions of the Lasso and the SVM coincide, but indeed gives us the correspondence of all feasible points, preserving the objective values: for any points solution $x \in \mathbb{R}^n$ to the Lasso, we have a feasible SVM point $x_\triangle \in \triangle \subset \mathbb{R}^{2n}$ of the *same* objective value, and vice versa.

1.3.3   (SVM $\preceq$ Lasso): Given an SVM Instance, Constructing an Equivalent Lasso Instance

This reduction is harder to accomplish than the other direction we explained before. Given an instance of an SVM problem (1.1), we suppose that we have a (possibly non-optimal) σ-weakly separating vector $w \in \mathbb{R}^d$ available, for some (small) value $\sigma > 0$. Given w, we will demonstrate in the following how to construct an equivalent Lasso instance (1.2).

Perhaps surprisingly, such a weakly separating vector w is trivial to obtain for the ℓ_2-loss soft-margin SVM, as we have observed in Section 1.2.2 (even if the SVM input data is not separable). Also for hard-margin SVM variants, finding such a weakly separating vector for a small σ is still significantly easier than the final goal of obtaining a near-perfect $(\sigma^* - \varepsilon)$-separation for a small precision ε. It corresponds to running an SVM solver (such as the perceptron algorithm) for only a constant number of iterations. In contrast, obtaining a better ε-accurate solution by the same algorithm would require $O(1/\varepsilon^2)$ iterations, as mentioned in Section 1.2.

The Equivalent Lasso Instance. Formally, we define the Lasso instance $(\tilde{A}, \tilde{b})$ as the translated SVM datapoints

$$\tilde{A} := \left\{ A_i + \tilde{b} \mid i \in [1..n] \right\}$$

together with the right hand side

$$\tilde{b} := -\frac{w}{\|w\|} \cdot \frac{D^2}{\sigma} \ .$$

Here $D > 0$ is a strict upper bound on the length of the original SVM datapoints, i.e., $\|A_i\| < D \ \ \forall i$.

By definition of $\tilde{A}$, the resulting new Lasso objective function is

$$\left\| \tilde{A}x - \tilde{b} \right\| \;=\; \left\| (A + \tilde{b}\mathbf{1}^T)x - \tilde{b} \right\| \;=\; \left\| Ax + (\mathbf{1}^T x - 1)\tilde{b} \right\| \ . \qquad (1.6)$$

Therefore, this objective coincides with the original SVM objective (1.1), for any $x \in \triangle$ (meaning that $\mathbf{1}^T x = 1$). However, this does not necessarily hold for the larger part of the Lasso domain when $x \in \blacklozenge \setminus \triangle$. In the following discussion and the main Theorem 1.1, we will prove that all those candidates $x \in \blacklozenge \setminus \triangle$ can be discarded from the Lasso problem, as they do not contribute to any optimal solutions.

As a side-remark, we note that the quantity $\frac{D}{\sigma}$ that determines the magnitude of our translation is a known parameter in the SVM literature. [4, 37] have shown that the VC-dimension of an SVM, a measure of "difficulty" for the classifier, is always lower than $\frac{D^2}{\sigma^2}$. Note that by the definition of separation, $\sigma \leq D$ always holds.

Geometric Intuition. Geometrically, the Lasso problem (1.2) is to compute the smallest Euclidean distance of the set $A\blacklozenge$ to the point $b \in \mathbb{R}^d$. On the other hand the SVM problem — after translating by b — is to minimize the distance of the smaller set $A\triangle \subset A\blacklozenge$ to the point b. Here we have used the notation $AS := \{Ax \mid x \in S\}$ for subsets $S \subseteq \mathbb{R}^d$ and linear maps A (it is easy to check that linear maps do preserve convexity of sets, so that $\mathrm{conv}(AS) = A\,\mathrm{conv}(S)$).

Intuitively, the main idea of our reduction is to mirror our SVM points A_i at the origin, so that both the points and their mirrored copies — and therefore the entire larger polytope $A\blacklozenge$ — do end up lying "behind" the separating SVM margin. The hope is that the resulting Lasso instance will have all its optimal solutions be non-negative, and lying in the simplex. Surprisingly, this can be done, and we will show that all SVM solutions are preserved (and no new solutions are introduced) when the feasible set $\triangle$ is extended to $\blacklozenge$. In the following we will formalize this precisely, and demonstrate how to translate along our known weakly separating vector w so that the resulting Lasso problem will have the same solution as the original SVM.

Properties of the Constructed Lasso Instance. The following theorem shows that for our constructed Lasso instance, all interesting feasible points are contained in the simplex $\triangle$. By our previous observation (1.6), we already know that all those candidates are feasible for both the Lasso (1.2) and the SVM (1.1), and obtain the same objective values in both problems.

In other words, we have a one-to-one correspondence between all feasible points for the SVM (1.1) on one hand, and the subset $\triangle \subset \blacklozenge$ of feasible points of our constructed Lasso instance (1.2), preserving all objective values. Furthermore, we have that in this Lasso instance, all points in $\blacklozenge \setminus \triangle$ are strictly worse than the ones in $\triangle$. Therefore, we have also shown that all optimal solutions must coincide.

Theorem 1.1. *For any candidate solution $x_\diamond \in \blacklozenge$ to the Lasso problem (1.2) defined by $(\tilde{A}, \tilde{b})$, there is a feasible vector $x_\triangle \in \triangle$ in the simplex, of the same or better Lasso objective value γ.*

Furthermore, this $x_\triangle \in \triangle$ attains the same objective value γ in the original SVM problem (1.1).

On the other hand, every $x_\triangle \in \triangle$ is of course also feasible for the Lasso, and attains the same objective value there, again by (1.6).

Proof. The proof follows directly from the two main facts given in Propositions 1.1 and 1.2 below, which state that "flipping negative signs improves the objective", and that "scaling up improves for non-negative vectors", respectively. We will see below why these two facts hold, which is precisely by the choice of the translation $\tilde{b}$ along a weakly separating vector w, in order to define our Lasso instance.

We assume that the given $x_\diamond$ does not already lie in the simplex. Now by applying Propositions 1.1 and 1.2, we obtain $x_\triangle \in \triangle$, of a strictly better objective value γ for problem (1.3). By the observation (1.6) about the Lasso objective, we know that the original SVM objective value attained by this $x_\triangle$ is equal to γ. □

Proposition 1.1 (Flipping negative signs improves the objective). *Consider the Lasso problem (1.2) defined by $(\tilde{A}, \tilde{b})$, and assume that $x_\diamond \in \blacklozenge$ has some negative entries.*

Then there is a strictly better solution $x_\blacktriangle \in \blacktriangle$ having only non-negative entries.

Proof. We are given $x_\diamond \neq 0$, having at least one negative coordinate. Define $x_\blacktriangle \neq \mathbf{0}$ as the vector you get by flipping all the negative coordinates in $x_\diamond$. We define $\delta \in \blacktriangle$ to be the difference vector corresponding to this flipping, i.e., $\delta_i := -(x_\diamond)_i$ if $(x_\diamond)_i < 0$, and $\delta_i := 0$ otherwise, so that $x_\blacktriangle := x_\diamond + 2\delta$ gives $x_\blacktriangle \in \blacktriangle$. We want to show that with respect to the quadratic objective function, $x_\blacktriangle$ is strictly better than $x_\diamond$. We do this by showing that the following difference in the objective values is strictly negative:

$$
\begin{aligned}
\left\| \tilde{A}x_\blacktriangle - \tilde{b} \right\|^2 - \left\| \tilde{A}x_\diamond - \tilde{b} \right\|^2 &= \|c + d\|^2 - \|c\|^2 \\
&= c^T c + 2c^T d + d^T d - c^T c = (2c + d)^T d \\
&= 4(\tilde{A}x_\diamond - \tilde{b} + \tilde{A}\delta)^T \tilde{A}\delta \\
&= 4(\tilde{A}(x_\diamond + \delta) - \tilde{b})^T \tilde{A}\delta
\end{aligned}
$$

where in the above calculations we have used that $\tilde{A}x_\blacktriangle = \tilde{A}x_\diamond + 2\tilde{A}\delta$, and we substituted $c := \tilde{A}x_\diamond - \tilde{b}$ and $d := 2\tilde{A}\delta$. Interestingly, $x_\diamond + \delta \in \blacktriangle$, since this addition just sets all previously negative coordinates to zero.

The proof then follows from Lemma 1.3 below. □

Proposition 1.2 (Scaling up improves for non-negative vectors). *Consider the Lasso problem (1.2) defined by $(\tilde{A}, \tilde{b})$, and assume that $x_\blacktriangle \in \blacktriangle$ has $\|x_\blacktriangle\|_1 < 1$.*

Then we obtain a strictly better solution $x_\triangle \in \triangle$ by linearly scaling $x_\blacktriangle$.

Proof. The proof follows along similar lines as the above proposition. We are given $x_\blacktriangle \neq 0$ with $\|x_\blacktriangle\|_1 < 1$. Define $x_\triangle$ as the vector we get by scaling up

$x_\triangle := \lambda x_\blacktriangle$ by $\lambda > 1$ such that $\|x_\triangle\|_1 = 1$. We want to show that with respect to the quadratic objective function, $x_\triangle$ is strictly better than $x_\blacktriangle$. As in the previous proof, we again do this by showing that the following difference in the objective values is strictly negative:

$$
\begin{aligned}
\left\|\tilde{A}x_\triangle - \tilde{b}\right\|^2 - \left\|\tilde{A}x_\blacktriangle - \tilde{b}\right\|^2 &= \|c+d\|^2 - \|c\|^2 \\
&= c^T c + 2c^T d + d^T d - c^T c \;=\; (2c+d)^T d \\
&= \lambda'(2\tilde{A}x_\blacktriangle - 2b + \lambda'\tilde{A}x_\blacktriangle)^T \tilde{A}x_\blacktriangle \\
&= 2\lambda'\big(\tilde{A}(1+\tfrac{\lambda'}{2})x_\blacktriangle - \tilde{b}\big)^T \tilde{A}x_\blacktriangle
\end{aligned}
$$

where in the above calculations we have used that $\tilde{A}x_\triangle = \lambda\tilde{A}x_\blacktriangle$ for $\lambda > 1$, and we substituted $c := \tilde{A}x_\blacktriangle - \tilde{b}$ and $d := \tilde{A}x_\triangle - \tilde{A}x_\blacktriangle = (\lambda - 1)\tilde{A}x_\blacktriangle =: \lambda'\tilde{A}x_\blacktriangle$ for $\lambda' := \lambda - 1 > 0$. Note that $x_\triangle := (1+\lambda')x_\blacktriangle \in \triangle$ so $(1+\tfrac{\lambda'}{2})x_\blacktriangle \in \blacktriangle$.

The proof then follows from Lemma 1.3 below. $\square$

Definition 1.2. *For a given axis vector $w \in \mathbb{R}^d$, the* cone *with axis w, angle $\alpha \in (0, \frac{\pi}{2})$ with tip at the origin is defined as* $\mathrm{cone}(w,\alpha) := \{x \in \mathbb{R}^d \mid \measuredangle(x,w) \le \alpha\}$, *or equivalently* $\frac{x^T w}{\|x\|\|w\|} \ge \cos\alpha$. *By* $\overset{\circ}{\mathrm{cone}}(w,\alpha)$ *we denote the interior of the convex set* $\mathrm{cone}(w,\alpha)$, *including the tip* $\mathbf{0}$.

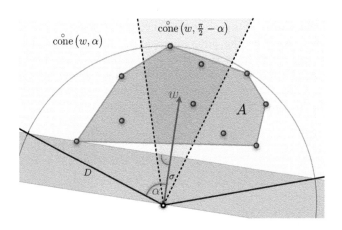

FIGURE 1.2: Illustration of the separation idea from Lemma 1.2, showing the cone of vectors that are still weakly separating for the set of points A. Here we used the angle $\alpha := \arccos(\frac{\sigma}{D})$.

Lemma 1.2 (Separation). *Let w be some σ-weakly separating vector for the SVM (1.1) for $\sigma > 0$. Then*

 i) *$A\blacktriangle \subseteq \overset{\circ}{\mathrm{cone}}(w, \arccos(\frac{\sigma}{D}))$*

 ii) *Any vector in $\mathrm{cone}(w, \arcsin(\frac{\sigma}{D}))$ is still σ'-weakly separating for A for some $\sigma' > 0$.*

Proof. i) Definition 1.1 of weakly separating, and using that $\|A_i\| < D$.

ii) For any unit length vector $v \in \text{cone}(w, \arcsin(\frac{\sigma}{D}))$, every other vector having a zero or negative inner product with this v must have angle at least $\frac{\pi}{2} - \arcsin(\frac{\sigma}{D}) = \arccos(\frac{\sigma}{D})$ with the cone axis w. However, by using i), we have $A\triangle \subseteq \overset{\circ}{\text{cone}}(w, \arccos(\frac{\sigma}{D}))$, so every column vector of A must have strictly positive inner product with v, or in other words v is σ'-weakly separating for A (for some $\sigma' > 0$). See also the illustration in Figure 1.2. $\qquad\square$

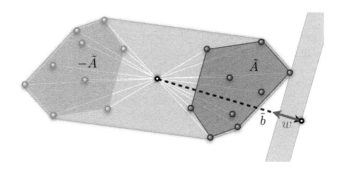

FIGURE 1.3: Illustration of Lemma 1.3. Recall that the translated points are defined by $\tilde{A} := \{A_i + \tilde{b} \mid i \in [1..n]\}$, where the translation is $\tilde{b} := -\frac{w}{\|w\|} \cdot \frac{D^2}{\sigma}$.

Lemma 1.3. *Let w be some σ-weakly separating vector for the SVM for $\sigma > 0$. Then we claim that the translation by the vector $\tilde{b} := -\frac{w}{\|w\|} \cdot \frac{D^2}{\sigma}$ has the following properties. For any pair of vectors $x, \delta \in \blacktriangle, \delta \neq \mathbf{0}$, we have that $(\tilde{A}x - \tilde{b})^T(-\tilde{A}\delta) > 0$.*

Proof. (See also Figure 1.3). By definition of the translation $\tilde{b}$, we have that the entire Euclidean ball of radius D around the point $-\tilde{b}$ — and therefore also the point set $-\tilde{A}\blacktriangle$ and in particular $v := -\tilde{A}\delta$ — is contained in $\text{cone}(w, \arcsin(\frac{\sigma}{D}))$. Therefore by Lemma 1.2 ii), v is separating for A, and by translation v also separates $\tilde{A}$ from $\tilde{b}$. This establishes the result $(\tilde{A}x - \tilde{b})^T v > 0$ for any $x \in \triangle$.

To extend this to the case $x \in \blacktriangle$, we observe that by definition of $\tilde{b}$, the point $\mathbf{0} - \tilde{b}$ also has strictly positive inner product with v. Therefore the entire convex hull of $\tilde{A}\triangle \cup \mathbf{0}$ and thus the set $\tilde{A}\blacktriangle$ has the desired property. $\qquad\square$

1.4   Implications and Remarks

In the following, we will explain a handful of implications of the shown equivalence, by relating both algorithms as well as known theoretical results for the Lasso or the SVM to the respective other method.

1.4.1   Sublinear Time Algorithms for Lasso and SVMs

The recent breakthrough SVM algorithm of [8, 21] in time $O(\varepsilon^{-2}(n + d)\log n)$ returns an ε-accurate solution to problem (1.1). Here ε-accurate means $(\sigma^* - \varepsilon)$-weakly separating. The running time of the algorithm is remarkable since for large data, it is significantly smaller than even the size of the input matrix, being $d \cdot n$. Therefore, the algorithm does not read the full input matrix $\tilde{A}$. More precisely, [8, Corollary III.2] proves that the algorithm provides (with high probability) a solution $p^* \in \triangle$ of additive error at most ε to

$$\min_{p \in \triangle} \max_{\substack{w \in \mathbb{R}^d, \\ \|w\| \leq 1}} w^T \tilde{A} p \ .$$

This is a reformulation of $\min_{p \in \triangle} p^T \tilde{A}^T \tilde{A} p$, which is exactly our SVM problem (1.1), since for given p, the inner maximum is attained when $w = \tilde{A}p$. Therefore, using our simple trick from Section 1.3.2 of reducing any Lasso instance (1.2) to an SVM (1.1) (with its matrix $\tilde{A}$ having twice the number of columns as A), we directly obtain a sublinear time algorithm for the Lasso. Note that since the algorithm of [8, 21] only accesses the matrix $\tilde{A}$ by simple entry-wise queries, it is not necessary to explicitly compute and store $\tilde{A}$ (which is a preprocessing that would need linear time and storage). Instead, every entry $\tilde{A}_{ij}$ that is queried by the algorithm can be provided on the fly, by returning the corresponding (signed) entry of the Lasso matrix A, minus b_i.

It will be interesting to compare this alternative algorithm to the recent more specialized sublinear time Lasso solvers in the line of work of [5, 20], which are only allowed to access a constant fraction of the entries (or features) of each row of A. If we use our proposed reduction here instead, the resulting algorithm from [8] has more freedom: it can (randomly) pick arbitrary entries of A, without necessarily accessing an equal number of entries from each row.

On the other hand, it is an open research question if a sublinear SVM algorithm exists that only accesses a constant small fraction of each datapoint, or of each feature of the input data.

1.4.2   A Kernelized Lasso

Traditional kernel regression techniques [38, 36, 35] try to learn a real-valued function f from the space $\mathbb{R}^d$ of the datapoints, such that the resulting real value for each datapoint approximates some observed value. The regression model is chosen as a linear combination of the (kernel) inner products with few existing landmark datapoints (the support vectors).

Here, as we discuss a kernelization of the Lasso that is in complete analogy to the classical kernel trick for SVMs, our goal is different. We are not trying to approximate n many individual real values (one for each datapoint, or row of A), but instead we are searching for a linear combination of our points in the kernel space, such that the resulting combination is *close* to the lifted point b, measured in the kernel space norm. Formally, suppose our kernel space $\mathcal{H}$ is given by an inner product $\kappa(y, z) = \langle \Phi(y), \Phi(z) \rangle$ for some implicit mapping $\Phi : \mathbb{R}^d \to \mathcal{H}$. Then we define our kernelized variant of the Lasso as

$$\min_{x \in \blacklozenge} \left\| \sum_i \Phi(A_i) x_i - \Phi(b) \right\|_{\mathcal{H}}^2 . \tag{1.7}$$

Nicely, analogous to the SVM case, here this objective function also is determined purely in terms of the pairwise (kernel) inner products $\kappa(\cdot, \cdot)$.

An alternative way to see this is to observe that our simple "mirror-and-translate" trick from Section 1.3.2 also works the very same way in any kernel space $\mathcal{H}$. Here, the equivalent SVM instance is given by the $2n$ new points $\{\pm \Phi(A_i) - \Phi(b) \mid i \in [1..n]\} \subset \mathcal{H}$. The crucial observation is that the (kernel) inner product of any two such points is

$$\langle s_i \Phi(A_i) - \Phi(b), s_j \Phi(A_j) - \Phi(b) \rangle$$
$$= s_i s_j \kappa(A_i, A_j) - s_i \kappa(A_i, b) - s_j \kappa(A_j, b) + \kappa(b, b) .$$

Here $s_i, s_j \in \pm 1$ are the signs corresponding to each point. Therefore we have completely determined the resulting $2n \times 2n$ kernel matrix K that defines the kernelized SVM (1.1), namely $\min_{x \in \triangle} x^T K x$, which solves our equivalent Lasso problem (1.7) in the kernel space $\mathcal{H}$.

Discussion. While traditional kernel regression corresponds to a lifting of the *rows* of the Lasso matrix A into the kernel space, our approach (1.7) by contrast is lifting the *columns* of A (and the r.h.s. b). We note that it seems indeed counter-intuitive to make the regression "more difficult" by artificially increasing the dimension of b. Using, e.g., a polynomial kernel, this means that we also want the higher moments of b to be well approximated by our estimated x. On the other hand, increasing the dimension of b naturally corresponds to adding more data rows (or measurements) to a classical Lasso instance (1.2).

In light of the success of the kernel idea for the classification case with its existing well-developed theory, we think it will be interesting to relate these results to the above proposed kernelized version of the Lasso, and to study how different kernels will affect the solution x for applications of the Lasso. Using a different connection to SVMs, the early work of [18] has studied a similar kernelization of the penalized version of the Lasso; see also [11]. For applications in image retrieval, [40] has recently applied a similar Lasso kernelization idea.

1.4.3 The Pattern of Support Vectors, in the View of Lasso Sparsity

Using our construction of the equivalent Lasso instance for a given SVM, we can translate sparsity results for the Lasso to understand the pattern of support vectors of SVMs.

The motivation here is that a small number of support vectors is crucial for the efficient application of SVMs in practice, in particular in the kernelized case, because the cost to evaluate the resulting classifier is directly proportional to the number of support vectors. Furthermore, the support vectors are the most informative points for the classification task, while the non-support vectors could be safely discarded from the problem.

Using Sparse Recovery Results. There has been a vast amount of literature studying the sparsity of solutions to the Lasso and related ℓ_1-regularized methods, in particular the study of the sparsity of x when A and b are from distributions with certain properties. For example, in the setting known as *sparse recovery*, the goal is to approximately recover a sparse solution x using a Lasso instance A, b (consisting of only a small number of rows). Here b is interpreted as a noisy or corrupted linear measurement $A\hat{x}$, and the unknown original $\hat{x}$ is sparse. Classical recovery results then show that under weak assumptions on A, b and the sparsity of $\hat{x}$, the optimal Lasso solution x must be identical to the (unknown) sparse $\hat{x}$; see, e.g., [7, 33].

Now our construction of the equivalent Lasso instance for a given SVM allows us to translate such sparsity results to the pattern of SVM support vectors. More precisely, any result that characterizes the Lasso sparsity for some distribution of matrices A and suitable b, will also characterize the patterns of support vectors for the equivalent SVM instance (and in particular the number of support vectors). This assumes that a Lasso sparsity result is applicable for the type of translation b that we have used to construct the equivalent Lasso instance. However, this is not hopeless. For example, existence of a weakly separating vector that is a sparse convex combination of the SVM datapoints is sufficient, since this results in a translation b that satisfies $b \propto A\hat{x}$ for a sparse weight vector $\hat{x}$. It remains to investigate which distributions and corresponding sparsity results are of most practical interest from the SVM perspective, in order to guarantee a small number of support vectors.

Lasso Sparsity in the View of SVMs. In another way, sparsity has also been studied for SVMs in the literature, for example in the work of [39], which analyzes the asymptotic regime $n \to \infty$. Using the simpler one of our reductions, the same results also hold for the Lasso, when the number of variables n grows.

1.4.4   Screening Rules for Support Vector Machines

For the Lasso, *screening rules* have been developed recently. This approach consists of a pre-processing of the data A, in order to immediately discard those predictors A_i that can be guaranteed to be inactive for the optimal solution. Provable guarantees for such rules were first obtained by [14], and also studied in the later work [44], or the heuristic paper [42].

Translated to the SVM setting by our reduction, any such existing Lasso screening rule can be used to permanently discard input points before the SVM optimization is started. The screening rule then guarantees that any discarded point will not be a support vector, so the resulting optimal classifier remains unchanged. We are not aware of screening rules in the SVM literature so far, with the exception of the more recent paper of [30].

While the two previous subsections have mainly made use of the more complicated direction of our reduction (SVM $\preceq$ Lasso) from Section 1.3.3, we can also gain some insights into the pattern of support vectors of SVMs by using the other (simpler) direction of reduction, as we will do next.

1.4.5   Regularization Paths and Homotopy Methods

For most machine learning methods — including SVMs and Lasso — one of the main hurdles in practice is the selection of the right free parameters. For SVMs and Lasso, the main question boils down to how to select the best value for the regularization parameter, which determines the trade-off between the best fit of the model, and the model complexity. For the SVM, this is the soft-margin parameter C, while in the Lasso, the regularization parameter is the value r for the required bound $\|x\|_1 \le r$.

Since naive grid-search for the best parameter is error-prone and comes without guarantees between the grid-values, algorithms that follow the solution path — as the regularization parameter changes — have been developed for both SVMs [19] and Lasso [31, 10], and have become popular in particular on the Lasso side.

In light of our joint investigation of Lasso and SVMs here, we can gain the following insights on path methods for the two problems:

General Solution Path Algorithms. We have observed that both the primal Lasso (1.2) and the dual ℓ_2-SVM (1.1) are in fact convex optimization problems over the simplex. This enables us to apply the same solution path methods to both problems. More precisely, for problems of the form $\min_{x \in \triangle} f(x, t)$, general path following methods are available that can maintain an approximation guarantee along the entire path in the parameter t, as shown in [16] and more recently strengthened by [17]. These methods do apply for objective functions f that are convex in x and continuous in the parameter t (which in our case is r for the Lasso and C for the SVM).

Path Complexity. The exact solution path for the Lasso is known to be piecewise linear. However, the number of pieces, i.e., the complexity of the path, is not easy to determine. The recent work of [29] has constructed a Lasso instance $A \in \mathbb{R}^{d \times n}$, $b \in \mathbb{R}^n$, such that the complexity of the solution path (as the parameter r increases) is exponential in n. This is inspired by a similar result by [13] which holds for the ℓ_1-SVM.

Making use of the easier one of the reductions we have shown above, we ask if similar complexity worst-case results could also be obtained for the ℓ_2-SVM. For every constraint value $r > 0$ for the Lasso problem, we have seen that the corresponding equivalent SVM instance (1.1) as constructed in Section 1.3.2 is $\min_{x \in \Delta} \left\| \tilde{A}_{(r)} x \right\|^2$, with

$$\tilde{A}_{(r)} := r(A\,|{-}A) - b\mathbf{1}^T \in \mathbb{R}^{d \times 2n}. \tag{1.8}$$

Therefore, we have obtained a hard-margin SVM, with the datapoints moving in the space $\mathbb{R}^d$ as the Lasso regularization parameter r changes. The movement is a simple linear rescaling by r, relative to the reference point $b \in \mathbb{R}^d$. The result of [29] shows that essentially all sparsity patterns do occur in the Lasso solution as r changes[2], i.e., that the number of patterns is exponential in n. For each pattern, we know that the SVM solution $x_\diamond = (\mathbf{I}_n\,|{-}\mathbf{I}_n)x_\Delta$ is identical to the Lasso solution, and in particular also has the same sparsity pattern. Therefore, we also have the same (exponential) number of different sparsity patterns in the simplex parameterization x_Δ for the SVM (one could even choose more in those cases where the mapping is not unique).

To summarize, we have shown that a simple rescaling of the SVM data can have a very drastic effect, in that *every* pattern of support vectors can potentially occur as this scaling changes. While our construction is still a worst-case result, note that the operation of rescaling is not unrealistic in practice, as it

---

[2]The result of [29] applies to the penalized formulation of the Lasso, that is the solution path of $\min_{x \in \mathbb{R}^n} \|Ax - b\|^2 + \lambda \|x\|_1$ as the parameter $\lambda \in \mathbb{R}_+$ varies. However, here we are interested in the constrained Lasso formulation, that is $\min_{x,\,\|x\|_1 \le r} \|Ax - b\|^2$ where the regularization parameter is $r \in \mathbb{R}_+$.

Luckily, the penalized and the constrained problem formulations are known to be equivalent by standard convex optimization theory, and have the same regularization paths, in the following sense. For every choice of $\lambda \in \mathbb{R}_+$, there is a value for the constraint parameter r such that the solutions to the two problems are identical (choose for example $r_{(\lambda)} := \left\| x^*_{(\lambda)} \right\|_1$ for some optimal $x^*_{(\lambda)}$, then the same vector is obviously also optimal for the constrained problem). This mapping from λ to $r_{(\lambda)}$ is monotone. As the regularization parameter λ is weakened (i.e., decreased), the corresponding constraint value $r_{(\lambda)}$ only grows larger.

The mapping in the other direction is similar: for every r, we know there is a value $\lambda_{(r)}$ (in fact this is the Lagrange multiplier of the constraint $\|x\|_1 \le r$), such that the penalized formulation has the same solution, and the mapping from r to $\lambda_{(r)}$ is monotone as well.

Having both connections, it is clear that the two kinds of regularization paths must be the same, and that the support patterns that occur in the solutions — as we go along the path — are also appearing in the very same order as we track the respective other path.

is similar to popular data preprocessing by re-normalizing the data, e.g., for zero mean and variance one.

However, note that the constructed instance (1.8) is a hard-margin SVM, with the datapoints moving as the parameter r changes. It does not directly correspond to a soft-margin SVM, because the movement of points is different from changing the regularization parameter C in an ℓ_2-SVM. As we have seen in Lemma 1.1, changing C in the SVM formulation (1.5) has the effect that the datapoints $\tilde{A}_{(C)}$ in the dual problem $\min_{x \in \triangle} \left\| \tilde{A}_{(C)} x \right\|^2$ move as follows (after re-scaling the entire problem by the constant factor C):

$$\tilde{A}_{(C)} := \begin{pmatrix} \sqrt{C} X \\ \mathbf{I}_n \end{pmatrix} \in \mathbb{R}^{(d+n) \times n}.$$

In conclusion, the reduction technique here does unfortunately not yet directly translate the regularization path of the Lasso to a regularization path for an ℓ_2-SVM. Still, we have gained some more insight as to how "badly" the set of SVM support vectors can change when the SVM data is simply re-scaled. We hope that the correspondence will be a first step to better relate the two kinds of regularization paths in the future. Similar methods could also extend to the case when other types of parameters are varied, such as, e.g., a kernel parameter.

1.5 Conclusions

We have investigated the relation between the Lasso and SVMs, and constructed equivalent instances of the respective other problem. While obtaining an equivalent SVM instance for a given Lasso is straightforward, the other direction is slightly more involved in terms of proof, but still simple to implement, in particular, e.g., for ℓ_2-loss SVMs.

The two reductions allow us to better relate and compare many existing algorithms for both problems. Also, it can be used to translate a lot of the known theory for each method to the respective other method. In the future, we hope that the understanding of both types of methods can be further deepened by using this correspondence.

1.6   Appendix: Some Soft-Margin SVM Variants That Are Equivalent to (1.1)

We include the derivation of the dual formulation to the ℓ_2-loss soft-margin SVM (1.5) for n datapoints $X_i \in \mathbb{R}^d$, together with their binary class labels $y_i \in \{\pm 1\}$, for $i \in [1..n]$, as defined above in Section 1.2.2.

The equivalence to (1.1) directly extends to the one- and two-class case, without or with (regularized) offset term, and as well for the hard-margin SVM. These equivalent formulations have been known in the SVM literature; see, e.g., [37, 26, 43, 12], and the references therein.

Lemma 1.4. *The dual of the soft-margin SVM (1.5) is an instance of the classifier formulation (1.1), that is* $\min_{x \in \triangle} \|Ax\|^2$ *, with*

$$A := \begin{pmatrix} Z \\ \frac{1}{\sqrt{C}} \mathbf{I}_n \end{pmatrix} \in \mathbb{R}^{(d+n) \times n}$$

where the data matrix $Z \in \mathbb{R}^{d \times n}$ *consists of the* n *columns* $Z_i := y_i X_i$.

Proof. The Lagrangian [2, Section 5] of the soft-margin SVM formulation (1.5) with its n constraints can be written as

$$L(w, \rho, \xi, \alpha) := \tfrac{1}{2} \|w\|^2 - \rho + \tfrac{C}{2} \sum_i \xi_i^2 + \sum_i \alpha_i \left(-w^T Z_i + \rho - \xi_i \right).$$

Here we introduced a non-negative Lagrange multiplier $\alpha_i \geq 0$ for each of the n constraints. Differentiating L with respect to the primal variables, we obtain the KKT optimality conditions

$$\begin{aligned} \mathbf{0} &\overset{!}{=} \tfrac{\partial}{\partial w} &&= w - \sum_i \alpha_i Z_i \\ \mathbf{0} &\overset{!}{=} \tfrac{\partial}{\partial \rho} &&= 1 - \sum_i \alpha_i \\ \mathbf{0} &\overset{!}{=} \tfrac{\partial}{\partial \xi} &&= C\xi - \alpha \,. \end{aligned}$$

When plugged into the Lagrange dual problem $\max_\alpha \min_{w, \rho, \xi} L(w, \rho, \xi, \alpha)$, these give us the equivalent formulation (sometimes called the *Wolfe-Dual*)

$$\begin{aligned} \max_\alpha \quad & \tfrac{1}{2} \alpha^T Z^T Z \alpha - \rho + \tfrac{C}{2} \tfrac{1}{C^2} \alpha^T \alpha \\ & - \alpha^T Z^T Z \alpha + \rho - \tfrac{1}{C} \alpha^T \alpha \\ = & -\tfrac{1}{2} \alpha^T Z^T Z \alpha - \tfrac{1}{2C} \alpha^T \alpha \,. \end{aligned}$$

In other words, the dual is

$$\begin{aligned} \min_\alpha \quad & \alpha^T \left(Z^T Z + \tfrac{1}{C} \mathbf{I}_n \right) \alpha \\ s.t. \quad & \alpha \geq 0 \\ & \alpha^T \mathbf{1} = 1 \,. \end{aligned}$$

This is directly an instance of our first SVM formulation (1.1) used in the introduction, if we use the extended matrix

$$A := \begin{pmatrix} Z \\ \frac{1}{\sqrt{C}} \mathbf{I}_n \end{pmatrix} \in \mathbb{R}^{(d+n) \times n} \ .$$

$\square$

Note that the optimal primal solution w can directly be obtained from any dual optimal α by using the optimality condition $w = A\alpha$.

Bibliography

[1] P Bloomfield and W L Steiger. *Least absolute deviations: theory, applications, and algorithms.* Progress in probability and statistics. Birkhäuser, 1983.

[2] Stephen P Boyd and Lieven Vandenberghe. *Convex optimization.* Cambridge University Press, 2004.

[3] Peter Bühlmann and Sara van de Geer. *Statistics for High-Dimensional Data - Methods, Theory and Applications.* Springer Series in Statistics 0172-7397. Springer, Berlin, Heidelberg, 2011.

[4] Christopher Burges. A Tutorial on Support Vector Machines for Pattern Recognition. *Data Mining and Knowledge Discovery*, 2(2):121–167, 1998.

[5] Nicolò Cesa-Bianchi, Shai Shalev-Shwartz, and Ohad Shamir. Efficient Learning with Partially Observed Attributes. *The Journal of Machine Learning Research*, 12:2857–2878, 2011.

[6] Kai-Wei Chang, Cho-Jui Hsieh, and Chih-Jen Lin. Coordinate Descent Method for Large-scale L2-loss Linear Support Vector Machines. *JMLR*, 9:1369–1398, 2008.

[7] Scott Shaobing Chen, David L Donoho, and Michael A Saunders. Atomic Decomposition by Basis Pursuit. *SIAM Journal on Scientific Computing*, 20(1):33, 1998.

[8] Kenneth L Clarkson, Elad Hazan, and David P Woodruff. Sublinear Optimization for Machine Learning. *FOCS 2010 - 51st Annual IEEE Symposium on Foundations of Computer Science*, 2010.

[9] Corinna Cortes and Vladimir Vapnik. Support-Vector Networks. *Machine Learning*, 20(3):273–297, 1995.

[10] Bradley Efron, Trevor Hastie, Iain Johnstone, and Robert Tibshirani. Least angle regression. *Annals of Statistics*, 32(2):407–499, 2004.

[11] Theodoros Evgeniou, Massimiliano Pontil, and Tomaso Poggio. Regularization Networks and Support Vector Machines. *Advances in Computational Mathematics*, 13(1):1–50, 2000.

[12] Bernd Gärtner and Martin Jaggi. Coresets for polytope distance. *SCG '09: Proceedings of the 25th annual symposium on computational geometry*, 2009.

[13] Bernd Gärtner, Martin Jaggi, and Clément Maria. An Exponential Lower Bound on the Complexity of Regularization Paths. *Journal of Computational Geometry*, 3(1):168–195, 2012.

[14] Laurent El Ghaoui, Vivian Viallon, and Tarek Rabbani. Safe Feature Elimination for the LASSO and Sparse Supervised Learning Problems. *arXiv.org*, 2010.

[15] Debashis Ghosh and Arul M Chinnaiyan. Classification and Selection of Biomarkers in Genomic Data Using LASSO. *Journal of Biomedicine and Biotechnology*, 2005(2):147–154, 2005.

[16] Joachim Giesen, Martin Jaggi, and Sören Laue. Approximating parameterized convex optimization problems. *ACM Transactions on Algorithms*, 9(10):1–17, 2012.

[17] Joachim Giesen, Jens Müller, Soeren Laue, and Sascha Swiercy. Approximating Concavely Parameterized Optimization Problems. In *NIPS*, 2012.

[18] Federico Girosi. An Equivalence Between Sparse Approximation and Support Vector Machines. *Neural Computation*, 10(6):1455–1480, 1998.

[19] Trevor Hastie, Saharon Rosset, Robert Tibshirani, and Ji Zhu. The Entire Regularization Path for the Support Vector Machine. *The Journal of Machine Learning Research*, 5:1391–1415, 2004.

[20] Elad Hazan and Tomer Koren. Linear Regression with Limited Observation. In *ICML*, 2012.

[21] Elad Hazan, Tomer Koren, and Nathan Srebro. Beating SGD: Learning SVMs in Sublinear Time. In *NIPS*, 2011.

[22] Donald W Hearn. The gap function of a convex program. *Operations Research Letters*, 1(2):67–71, 1982.

[23] Sepp Hochreiter and Klaus Obermayer. Gene Selection for Microarray Data. In Bernhard Schölkopf, Jean-Philippe Vert, and Koji Tsuda, editors, *Kernel Methods in Computational Biology*, page 319. MIT Press, 2004.

[24] Sepp Hochreiter and Klaus Obermayer. Support Vector Machines for Dyadic Data. *Neural Computation*, 18(6):1472–1510, 2006.

[25] Martin Jaggi. Revisiting Frank-Wolfe: Projection-Free Sparse Convex Optimization. In *ICML*, 2013.

[26] S Sathiya Keerthi, Shirish K Shevade, Chiranjib Bhattacharyya, and K R K Murthy. A fast iterative nearest point algorithm for support vector machine classifier design. *IEEE Transactions on Neural Networks*, 11(1):124–136, 2000.

[27] Yuh-Jye Lee and Olvi L Mangasarian. RSVM: Reduced Support Vector Machines. In *SDM 2001 - Proceedings of the first SIAM international conference on data mining*, Philadelphia, 2001.

[28] Fan Li, Yiming Yang, and Eric P Xing. From Lasso regression to Feature vector machine. In *NIPS*, 2005.

[29] Julien Mairal and Bin Yu. Complexity Analysis of the Lasso Regularization Path. In *ICML*, 2012.

[30] Kohei Ogawa, Yoshiki Suzuki, and Ichiro Takeuchi. Safe Screening of Non-Support Vectors in Pathwise SVM Computation. In *ICML*, pages 1382–1390, 2013.

[31] Michael R Osborne, Brett Presnell, and Berwin A Turlach. A new approach to variable selection in least squares problems. *IMA Journal of Numerical Analysis*, 20(3):389–403, 2000.

[32] Massimiliano Pontil, Ryan Rifkin, and Theodoros Evgeniou. From Regression to Classification in Support Vector Machines. Technical Report A.I. Memo No. 1649, MIT, 1998.

[33] Ely Porat and Martin J Strauss. Sublinear time, measurement-optimal, sparse recovery for all. In *SODA '12: Proceedings of the Twenty-Third Annual ACM-SIAM Symposium on Discrete Algorithms*. SIAM, 2012.

[34] Frank Rosenblatt. The Perceptron: A Probabilistic Model for Information Storage and Organization in the Brain. *Psychological Review*, 65(6):386–408, 1958.

[35] Volker Roth. The Generalized LASSO. *IEEE Transactions on Neural Networks*, 15(1):16–28, 2004.

[36] Craig Saunders, Alexander Gammerman, and Volodya Vovk. Ridge Regression Learning Algorithm in Dual Variables. In *ICML*. Morgan Kaufmann Publishers Inc, 1998.

[37] Bernhard Schölkopf and Alex J Smola. *Learning with kernels.* support vector machines, regularization, optimization, and beyond. The MIT Press, 2002.

[38] Alex J Smola and Bernhard Schölkopf. A tutorial on support vector regression. *Statistics and Computing*, 14:199–222, 2004.

[39] Ingo Steinwart. Sparseness of Support Vector Machines—Some Asymptotically Sharp Bounds. In *NIPS*, 2003.

[40] Jayaraman J Thiagarajan, Karthikeyan Natesan Ramamurthy, and Andreas Spanias. Local Sparse Coding for Image Classification and Retrieval. Technical report, ASU, 2012.

[41] Robert Tibshirani. Regression Shrinkage and Selection via the Lasso. *Journal of the Royal Statistical Society. Series B (Methodological)*, pages 267–288, 1996.

[42] Robert Tibshirani, Jacob Bien, Jerome Friedman, Trevor Hastie, Noah Simon, Jonathan Taylor, and Ryan J. Tibshirani. Strong rules for discarding predictors in lasso-type problems. *Journal of the Royal Statistical Society: Series B (Statistical Methodology)*, 74(2):245–266, 2011.

[43] Ivor W Tsang, James T Kwok, and Pak-Ming Cheung. Core Vector Machines: Fast SVM Training on Very Large Data Sets. *Journal of Machine Learning Research*, 6:363–392, 2005.

[44] Jie Wang, Binbin Lin, Pinghua Gong, Peter Wonka, and Jieping Ye. Lasso Screening Rules via Dual Polytope Projection. In *NIPS*, 2013.

[45] Günter M Ziegler. *Lectures on Polytopes*, volume 152 of *Graduate Texts in Mathematics*. Springer Verlag, 1995.

Chapter 2

Regularized Dictionary Learning

Annalisa Barla

DIBRIS, Università degli Studi di Genova

Saverio Salzo

DIMA, Università degli Studi di Genova

Alessandro Verri

DIBRIS, Università degli Studi di Genova

2.1 Introduction

In dictionary learning, given a set of signals belonging to a certain class, one wishes to extract relevant information by identifying the generating causes, that is, recovering the elementary signals (atoms) that efficiently represent the data. Generally this goal is achieved by imposing some kind of sparseness constraint on the coefficients of the representation. Moreover, one typically puts priors on the atoms themselves. Ultimately, this gives rise to an optimization problem whose objective function is composed by a data fit term, which accounts for the goodness of the representation, and several penalization terms and constraints on the coefficients and atoms that explain the prior knowledge at hand. Since the pioneering work [25], different instances of dictionary learning problems have been proposed. This encompasses sparse coding [25, 35], ℓ^p-sparse coding [16], hierarchical sparse coding [13], elastic-net based dictionary learning [19], and fused-lasso based dictionary learning [34, 24], among others.

Theoretical studies on dictionary learning that fit the problem into a statistical learning framework and justify the above-mentioned optimization problem are given in [23, 36, 12, 10]. In fact, in the previous framework, we minimize an empirical average over the training data, whereas the idealized task would be to optimize an expected cost function over the underlying (and unknown) distribution that generated the data. We point out that in [10] this study is pursed with general coefficient penalties and dictionary constraints, covering all the types of dictionary learning problems mentioned above.

Concerning the algorithmic aspects, several studies have appeared in literature dealing with different dictionary learning problems [25, 1, 18, 13, 24]. In all these works the optimization task is solved by a procedure that alternatively minimizes over the coefficients and atoms separately. The different contributions, in fact, rely on the way the partial minimization subproblems are tackled.

Dictionary learning has been found effective on a variety of applied problems, such as image denoising [20], audio processing [11, 9], and image classification [29], to name a few. We refer to the aforementioned papers and the references therein for a complete overview of the state-of-the-art on applications of dictionary learning, which is beyond the scope of our work.

In this chapter we focus on the algorithmic and applied aspects of dictionary learning. Based on the work in [2], we present an alternating proximal algorithm suitable for a general dictionary learning framework with composite convex penalization terms. Such an algorithm is attractive for it has stronger convergence properties with respect to the simpler alternating minimization scheme, as established in [2] and recalled in Section 2.4.1. Here, we couple that algorithm with an efficient dual algorithm for the computation of the related proximity operators, which keeps the overall procedure still effective. An analysis of the problem of computing the proximity operators for general sums of composite functions is provided. In this respect, we extend related results given in [38] which also allows us to consider hard constraints. Finally, we note that, to the best of our knowledge, in the context of dictionary learning, the alternating proximal algorithm represents a novelty.

Next, we give an application of the proposed algorithm in the context of genome-wide data understanding, in the setting of [24]. This application was already presented in [22, 21]. The problem consists of identifying latent features (atoms) in array-based comparative genomic hybridization (aCGH) data, which may reveal a genotype-phenotype relationship. Thanks to the generality of the proposed algorithm, we revise the model proposed by [24], modifying and adding penalties in order to treat the whole genomic signal and to select more representative atoms. More precisely: (*a*) we constrain the coefficients to be positive, so that the complexity of the matrix of coefficients is reduced and the resulting atoms become more informative; (*b*) we employ a *weighted* total variation penalty for the atoms, in order to treat the signal of the genome as a whole, still guaranteeing independency among chromosomes; (*c*) we force a structured sparsity on the matrix of coefficients and atoms by using ℓ^1–ℓ^2 mixed norms, which better suits the actual purposes of dictionary learning — this contrasts with the majority of similar models that instead use a simple ℓ^1 norm inducing just a global sparsity on the matrices.

We set up two experiments with properly designed synthetic and simulated data mimicking the properties of aCGH data. The purpose is to assess the effectiveness of the proposed model in identifying the relevant latent features. Then, we consider one further experiment based on real breast cancer aCGH data, with the aim of evaluating the improvements in a clustering scenario. We always compare the results with the ones obtained by the model and algorithm proposed in [24].

The remainder of the chapter has the following structure. In the next section we begin by giving some basic mathematical definitions. Section 2.3 formalizes the problem of dictionary learning. In Section 2.4 we present the algorithm. In Section 2.5 we discuss the application to a relevant problem of computational biology, providing three experiments on synthetic, simulated and real data.

2.2 Basic Notation and Definitions

Let us first introduce some basic notation (for a detailed account of convex analysis, see [4]). Hereafter, $\mathcal{H}$ is a real Hilbert space and we denote with $\langle \cdot, \cdot \rangle_{\mathcal{H}}$ and $\|\cdot\|_{\mathcal{H}}$ its scalar product and associated norm. The extended real line is denoted by $\overline{\mathbb{R}}$. Let $C \subseteq \mathcal{H}$, then the *indicator function* of C is

$$\delta_C \colon \mathcal{H} \to \overline{\mathbb{R}}, \quad \delta_C(x) = \begin{cases} 0 & \text{if } x \in C, \\ +\infty & \text{if } x \notin C, \end{cases} \tag{2.1}$$

and its *support function* is

$$\sigma_C \colon \mathcal{H} \to \overline{\mathbb{R}}, \quad \sigma_C(u) = \sup_{x \in X} \langle x, u \rangle_{\mathcal{H}}.$$

Let $f \colon \mathcal{H} \to \overline{\mathbb{R}}$ be an extended real valued function. The *domain* of f is the set $\operatorname{dom} f = \{x \in \mathcal{H} \mid f(x) < +\infty\}$. The function f is *proper* if for every $x \in \mathcal{H}$ $f(x) > -\infty$ and $\operatorname{dom} f \neq \varnothing$ and is *coercive* if $f(x) \to +\infty$ as $\|x\|_{\mathcal{H}} \to +\infty$. The function f is said to be *positively homogeneous* if for every $\lambda \geq 0$ and $x \in \mathcal{H}$, $f(\lambda x) = \lambda f(x)$. The *(Fenchel) conjugate* of f is the function $f^* \colon \mathcal{H} \to \overline{\mathbb{R}}$, such that for every $u \in \mathcal{H}$, $f^*(u) = \sup_{x \in \mathcal{H}}(\langle x, u \rangle_{\mathcal{H}} - f(x))$. If f is proper and lower semicontinuous, the *proximity operator* of f is the (nonlinear) operator $\mathrm{prox}_f \colon \mathcal{H} \to \mathcal{H}$ which maps every $x \in \mathcal{H}$ to the unique minimizer of the function $f + \|\cdot - x\|_{\mathcal{H}}^2 / 2$. Clearly $\mathrm{prox}_{\delta_C} = P_C$, the projection operator onto C. We recall the Moreau decomposition formula

$$x = \mathrm{prox}_{\gamma f}(x) + \gamma \, \mathrm{prox}_{f^*/\gamma}(x/\gamma), \qquad (2.2)$$

valid for every $x \in \mathcal{H}$ and $\gamma \in \mathbb{R}$. Finally one can show that if g is proper convex and lower semicontinuous, then $f^{**} = f$, hence the Moreau decomposition formula (2.2) also writes $x = \mathrm{prox}_{\gamma f^*}(x) + \gamma \, \mathrm{prox}_{f/\gamma}(x/\gamma)$.

Throughout the chapter, vectors will be denoted by bold small letters, whereas matrices will be denoted by bold capital letters, and their entries by the corresponding small plain letters. The space of real matrices of dimensions N by M is denoted $\mathbb{R}^{N \times M}$. The Frobenius norm is $\|\mathbf{A}\|_F = \sqrt{\sum_{i,j} |a_{i,j}|^2}$. Moreover, using a MATLAB-like notation, if $\mathbf{A} \in \mathbb{R}^{N \times M}$, for every $(n, m) \in \{1, \ldots, N\} \times \{1, \ldots, M\}$, we will set $\mathbf{A}(:, m) = (a_{n',m})_{1 \leq n' \leq N}$ and $\mathbf{A}(n, :) = (a_{n,m'})_{1 \leq m' \leq M}$.

2.3   The Problem of Dictionary Learning

We are given $S \in \mathbb{N}$ samples $(\mathbf{y}_s)_{1 \leq s \leq S}$, with $\mathbf{y}_s \in \mathbb{R}^L$. Then, one seeks J atoms $(\mathbf{b}_j)_{1 \leq j \leq J}$, $\mathbf{b}_j \in \mathbb{R}^L$, which possibly give complete representation of all samples, in the sense that

$$\mathbf{y}_s \cong \sum_{j=1}^J \theta_{js} \mathbf{b}_j \quad \forall s = 1, \ldots, S \qquad (2.3)$$

for some vectors of coefficients $\boldsymbol{\theta}_s = (\theta_{js})_{i \leq j \leq J} \in \mathbb{R}^J$.

Usually, some kind of sparseness on the coefficients is enforced as well as constraints on the atoms, and the quality of the representation is assessed by some ℓ^q norm with $1 \leq q \leq +\infty$. Hereafter, for the sake of brevity, we define the matrices $\mathbf{Y} = \begin{bmatrix} \mathbf{y}_1 & \mathbf{y}_2 & \cdots & \mathbf{y}_S \end{bmatrix} \in \mathbb{R}^{L \times S}$, $\mathbf{B} = \begin{bmatrix} \mathbf{b}_1 & \mathbf{b}_2 & \cdots & \mathbf{b}_J \end{bmatrix} \in \mathbb{R}^{L \times J}$, and $\boldsymbol{\Theta} = \begin{bmatrix} \boldsymbol{\theta}_1 & \boldsymbol{\theta}_2 & \cdots & \boldsymbol{\theta}_S \end{bmatrix} \in \mathbb{R}^{J \times S}$, which gather the data, atoms and coefficients, respectively.

Then, dictionary learning leads to the following minimization problem

$$\min_{\mathbf{B}, \boldsymbol{\Theta}} \sum_{s=1}^S \left(\|\mathbf{y}_s - \mathbf{B}\boldsymbol{\theta}_s\|_q^q + h(\boldsymbol{\theta}_s) \right) \quad s.t. \ \mathbf{B} \in \mathcal{B}, \ \boldsymbol{\Theta} \in \mathbb{R}^{J \times S}, \qquad (2.4)$$

where $h : \mathbb{R}^J \to \overline{\mathbb{R}}$ is a penalization function promoting sparsity in the coefficients and $\mathcal{B} \subseteq \mathbb{R}^{L \times J}$ is a constraint set for the matrix of atoms.

In the literature, different instances of h and $\mathcal{B}$ have been considered. We list some important examples:

Sparse coding [25, 18]. $h(\boldsymbol{\theta}) = \tau \|\boldsymbol{\theta}\|_1$, and $\mathcal{B} = \{\mathbf{B} \mid (\forall j) \ \|\mathbf{b}_j\|_2 = 1\}$ in [25] and $\mathcal{B} = \{\mathbf{B} \mid (\forall j) \ \|\mathbf{b}_j\|_2 \leq c\}$ in [18].

ℓ^p sparsity [16]. $h(\boldsymbol{\theta}) = \|\boldsymbol{\theta}\|_p$, with $0 < p \leq 1$, and $\mathcal{B} = \{\mathbf{B} \mid \|\mathbf{B}\|_F = 1\}$ or $\mathcal{B} = \{\mathbf{B} \mid (\forall j) \ \|\mathbf{b}_j\|_2^2 = 1/J\}$.

Hierarchical Sparse Coding [13]. $h = \sum_{C \in \mathcal{C}} \|\boldsymbol{\theta}_{|C}\|_2$, where $\boldsymbol{\theta}_{|C} = (\theta_s)_{s \in C}$ and $\mathcal{C}$ is a tree-structured set of indices in $\{1, \ldots, S\}$, and the dictionary constraint is $\mathcal{B} = \{\mathbf{B} \mid (\forall j) \ \mu \|\mathbf{b}_j\|_1 + (1-\mu) \|\mathbf{b}_j\|_2^2 \leq 1\}$, with $\mu \in [0, 1]$.

K-SVD [1]. $h = \delta_{\{\boldsymbol{\theta} \mid \|\boldsymbol{\theta}\|_0 \leq \tau\}}$, where $\|\cdot\|_0$ is the semi-norm that counts the non-zero entries of a vector, and $\mathcal{B} = \{\mathbf{B} \mid (\forall j) \ \|\mathbf{b}_j\|_2 = 1\}$.

Elastic-net [19]. $h(\boldsymbol{\theta}) = \tau \|\boldsymbol{\theta}\|_1 + \eta \|\boldsymbol{\theta}\|_2^2$ and $\mathcal{B} = \{\mathbf{B} \mid (\forall j) \ \|\mathbf{b}_j\|_2 \leq 1\}$.

We note that in the sparse coding example above one can take for h the hard constraint variant, i.e., $h = \delta_{\{\boldsymbol{\theta} \mid \|\boldsymbol{\theta}\|_1 \leq \tau\}}$. Similarly, in the K-SVD example, one can consider the soft constraint variant $h(\boldsymbol{\theta}) = \tau \|\boldsymbol{\theta}\|_0$.

2.4 The Algorithm

We consider a general dictionary learning framework whose objective function is as follows

$$\varphi(\boldsymbol{\Theta}, \mathbf{B}) = \frac{1}{2} \|\mathbf{Y} - \mathbf{B}\boldsymbol{\Theta}\|_F^2 + h(\boldsymbol{\Theta}) + g(\mathbf{B}), \qquad (2.5)$$

where $g: \mathbb{R}^{J \times S} \to \overline{\mathbb{R}}$ and $h: \mathbb{R}^{J \times S} \to \overline{\mathbb{R}}$ are extended real valued *convex* functions. This model assumes convex regularization terms, hence it cannot cover some dictionary learning problems described in Section 2.3. On the other hand, we also add a penalization term g for the atoms, allowing more flexibility in setting prior knowledge in the dictionary. Clearly, we recover model (2.4) by choosing g as the indicator function of the set $\mathcal{B}$. In this work we assume the penalties g and h have the following form:

$$h(\boldsymbol{\Theta}) = \sum_{m=1}^{M_1} \phi_m(Q_m(\boldsymbol{\Theta})), \quad g(\mathbf{B}) = \sum_{m=1}^{M_2} \psi_m(T_m(\mathbf{B})), \qquad (2.6)$$

where ϕ_m and ψ_m are proper extended real valued lower semicontinuous and convex functions, and Q_m and T_m are linear operators acting on spaces of matrices of appropriate dimensions. We emphasize that problem (2.5), even with h and g convex, is nonconvex and in general nonsmooth, too. Moreover, we remark that in the data fit term of (2.5), for simplicity, we considered the Frobenius (Euclidean) norm, but the proposed algorithm can be easily adapted to handle the ℓ^q norms with $1 \leq q \leq +\infty$.

2.4.1   An Alternating Proximal Algorithm

Very often, in the context of dictionary learning, the minimization problem (2.5) has been solved by employing an *alternating minimization algorithm* (AMA) [25, 1, 18, 13, 35, 24], which is nothing but a two-block Gauss-Seidel method. In fact, within this algorithmic scheme, depending on the application at hand, the contributions of the papers mainly focus on the different strategies to solve the partial minimization subproblems.

Here we propose an *alternating proximal algorithm* of the following form

$$\left|\begin{array}{l} \eta_n, \zeta_n \in [\rho_1, \rho_2] \\[4pt] \boldsymbol{\Theta}_{n+1} = \text{prox}_{\eta_n \varphi(\cdot, \mathbf{B}_n)}(\boldsymbol{\Theta}_n) \\[4pt] \mathbf{B}_{n+1} = \text{prox}_{\zeta_n \varphi(\boldsymbol{\Theta}_{n+1}, \cdot)}(\mathbf{B}_n) \, . \end{array}\right. \tag{2.7}$$

where $0 < \rho_1 \leq \rho_2$. The formulas for $\boldsymbol{\Theta}_{n+1}$ and $\mathbf{B}_{n+1}$ are written down explicitly as follows

$$\begin{aligned} \boldsymbol{\Theta}_{n+1} &= \underset{\boldsymbol{\Theta}}{\text{argmin}} \left\{ \varphi(\boldsymbol{\Theta}, \mathbf{B}_n) + \frac{1}{2\eta_n} \|\boldsymbol{\Theta} - \boldsymbol{\Theta}_n\|_F^2 \right\} \\[4pt] \mathbf{B}_{n+1} &= \underset{\mathbf{B}}{\text{argmin}} \left\{ \varphi(\boldsymbol{\Theta}_{n+1}, \mathbf{B}) + \frac{1}{2\zeta_n} \|\mathbf{B} - \mathbf{B}_n\|_F^2 \right\}. \end{aligned} \tag{2.8}$$

Thus, algorithm (2.7) actually consists of an alternating regularized minimization procedure.

In [2] a deep analysis of algorithm (2.7) is presented and the following result can be worked out (see Lemma 3.1 and Theorem 3.2 in [2]).

Theorem 2.1. *Suppose that both the functions g and h, in (2.5), are coercive and satisfy the Kurdyka-Łojasiewicz property. If $(\boldsymbol{\Theta}_0, \mathbf{B}_0) \in \mathbb{R}^{J \times S} \times \mathbb{R}^{L \times J}$, and η_n, ζ_n and $(\boldsymbol{\Theta}_n, \mathbf{B}_n)_{n \in \mathbb{N}}$ are defined according to (2.7), then $(\varphi(\boldsymbol{\Theta}_n, \mathbf{B}_n))_{n \in \mathbb{N}}$ is decreasing and $(\boldsymbol{\Theta}_n, \mathbf{B}_n)_{n \in \mathbb{N}}$ converges to a critical point of φ.*

In Theorem 2.1, the coercivity property serves to ensure the boundedness of the sequence $(\boldsymbol{\Theta}_n, \mathbf{B}_n)_{n \in \mathbb{N}}$ and, in this context, it is standard and always satisfied. The Kurdyka-Łojasiewicz property is a kind of metric regularity property for nonsmooth functions and has been deeply studied in [7] and applied to minimization problems in [2, 3]. We briefly recall here the definition. Let $f : \mathbb{R}^N \to \,]-\infty, +\infty]$ be a proper and lower semicontinuous function. Then,

f satisfies the Kurdyka-Łojasiewicz property, if for every critical point $\bar{x}$ of f, it holds

$$\forall u \in \partial(\varphi \circ (f - f(\bar{x})))(x) \quad \|u\| \geq 1$$

for a suitable continuously differentiable concave and strictly increasing function $\varphi : [0, \eta[\to \mathbb{R}_+$, and for every $x \in f^{-1}(]0, \eta[)$ sufficiently close to $\bar{x}$. This requires the function f to be sharp up to a reparameterization of its values.

The significance of Theorem 2.1 stands on the fact that the class of functions satisfying the Kurdyka-Łojasiewicz property is large enough to encompass semi-algebraic and tame functions and ultimately most of the penalization functions used in statistical learning and inverse problems, as for instance power of norms, and indicator functions of norm balls.

This result shows that the alternating proximal algorithm (2.7) has stronger convergence properties with respect to AMA. Indeed, in general, AMA provides convergence to stationary points only for a *subsequence* of $(\mathbf{\Theta}_n, \mathbf{B}_n)_{n \in \mathbb{N}}$ [6].

Algorithm (2.7) requires the computation of the proximity operator of the following partial functions

$$\varphi(\cdot, \mathbf{B}_n)(\mathbf{\Theta}) = \frac{1}{2}\|\mathbf{Y} - \mathbf{B}\mathbf{\Theta}\|_F^2 + \sum_{m=1}^{M_1} \phi_m(Q_m(\mathbf{\Theta})) \tag{2.9}$$

$$\varphi(\mathbf{\Theta}_{n+1}, \cdot)(\mathbf{B}) = \frac{1}{2}\|\mathbf{Y} - \mathbf{B}\mathbf{\Theta}\|_F^2 + \sum_{m=1}^{M_2} \psi_m(T_m(\mathbf{B})). \tag{2.10}$$

The partial functions (2.9)-(2.10) have the form of a sum of composite penalties and for that reason, in general, there is no closed form expression available for their proximity operators. Therefore, a further algorithm is needed that takes into account the particular structure of the functions. Moreover, hopefully the algorithm should be fast enough to keep the alternating proximal algorithm still effective.

2.4.2 The Proximity Operator of Composite Penalties

In view of the discussion presented in the previous section, it is desirable to *efficiently* compute proximity operators for penalties that are sums of composite functions. In this section we answer that issue in a general setting, presenting an algorithm based on a dual approach.

Let $f : \mathcal{H} \to \overline{\mathbb{R}}$ be a function of the following form

$$f(x) = \sum_{m=1}^{M} \omega_m(A_m x), \quad (\forall x \in \mathcal{H}) \tag{2.11}$$

where, for every $m = 1, 2, \ldots, M$, $A_m : \mathcal{H} \to \mathcal{G}_m$ are bounded linear operators between Hilbert spaces and $\omega_m : \mathcal{G}_m \to \overline{\mathbb{R}}$ are proper convex and lower semi-continuous functions. Our purpose is to show how to compute the proximity operator $\mathrm{prox}_{\lambda f} : \mathcal{H} \to \mathcal{H}$ for $\lambda > 0$, in terms of the mappings A_m and the proximity operators of ω_m.

First of all, we note that f is actually of the form

$$f(x) = \omega(Ax), \tag{2.12}$$

for a suitable operator $A : \mathcal{H} \to \mathcal{G}$ and $\omega : \mathcal{G} \to \overline{\mathbb{R}}$. Indeed, it is sufficient to consider the direct sum of the Hilbert spaces $(\mathcal{G}_i)_{1 \le m \le M}$

$$\mathcal{G} := \bigoplus_{m=1}^{M} \mathcal{G}_m \quad \langle u, v \rangle_{\mathcal{G}} := \sum_{m=1}^{M} \langle u_m, v_m \rangle_{\mathcal{G}_m},$$

and define the linear operator $A : \mathcal{H} \to \mathcal{G}$, $Ax = (A_m x)_{1 \le m \le M}$ and the function

$$\omega : \mathcal{G} \to \overline{\mathbb{R}}, \quad \omega(v) = \sum_{m=1}^{M} \omega_m(v_m).$$

Therefore, computing $\mathrm{prox}_{\lambda f}(y)$ aims to solve the following minimization problem

$$\min_{x \in \mathcal{H}} \omega(Ax) + \frac{1}{2\lambda} \|x - y\|_{\mathcal{H}}^2 := \Phi_\lambda(x). \tag{2.13}$$

Its dual problem (in the sense of Fenchel-Rockafellar duality [4]) is

$$\min_{v \in \mathcal{G}} \frac{1}{2\lambda} \|y - \lambda A^* v\|_{\mathcal{H}}^2 + \omega^*(v) - \frac{1}{2\lambda} \|y\|_{\mathcal{H}}^2 := \Psi_\lambda(v). \tag{2.14}$$

We note that in (2.14) the adjoint operator $A^* : \mathcal{G} \to \mathcal{H}$ and the Fenchel conjugate $\omega^* : \mathcal{G} \to \overline{\mathbb{R}}$ are both decomposable, meaning that for every $v = (v_m)_{1 \le m \le M} \in \mathcal{G}$, it holds that

$$A^* v = \sum_{i=m}^{M} A_m^* v_m, \quad \omega^*(v) = \sum_{i=m}^{M} \omega_m^*(v_m). \tag{2.15}$$

From the separability of ω^*, stated in (2.15), it follows that for $\gamma > 0$

$$\mathrm{prox}_{\gamma \omega^*}(v) = \underset{u \in \mathcal{G}}{\operatorname{argmin}} \left\{ \omega^*(u) + \frac{1}{2\gamma} \|u - v\|_{\mathcal{G}}^2 \right\}$$
$$= \left(\mathrm{prox}_{\gamma \omega_m^*}(v_m) \right)_{1 \le m \le M}, \tag{2.16}$$

and the proximity operator of $\gamma \omega^*$ can be decomposed component-wise, too.

Due to the aforementioned decomposability property, the dual problem (2.14) is possibly easier to solve. Thus, instead of solving the primal problem (2.13), we tackle the dual problem (2.14).

The primal and dual problem are connected by the so-called *duality gap* function $G : \mathcal{H} \times \mathcal{G} \to [0, +\infty]$. It is defined for every $(x, v) \in \mathcal{H} \times \mathcal{G}$ as

$$G(x, v) = \Phi_\lambda(x) + \Psi_\lambda(v) \tag{2.17}$$

$$= \sum_{m=1}^{M} \omega_m(A_m x) + \omega_m^*(v_m) + \frac{1}{\lambda}\langle x - y, x \rangle_\mathcal{H} + \frac{1}{2\lambda}\left(\|x_v\|_\mathcal{H}^2 - \|x\|_\mathcal{H}^2 \right),$$

where (the *primal variable*) $x_v = y - \lambda A^* v = y - \lambda \sum_{m=1}^{M} A_m^* v_m \in \mathcal{H}$. The following proposition links minimizing sequences for the dual problem with those of the primal problem in a general setting and it is at the basis of the proposed dual method. This result extends Theorem 5.1 in [38], in the sense that here it is not required that ω has full domain, so hard constraints are allowed.

Proposition 2.1. *Let $\omega \colon \mathcal{G} \to \overline{\mathbb{R}}$ be proper, convex, and lower semicontinuous. Assume that the domain of ω is closed and that the restriction of ω to its domain is continuous[1]. Moreover, suppose ω is continuous in Ax_0 for some x_0[2]. Let $(v_k)_{k \in \mathbb{N}}$ be such that $\Psi_\lambda(v_k) \to \inf \Psi_\lambda$. If we define the primal variables as*

$$x_k = P_{A^{-1}(\mathrm{dom}\,\omega)}(y - \lambda A^* v_k), \tag{2.18}$$

where $P_{A^{-1}(\mathrm{dom}\,\omega)}$ stands for the projection operator onto the closed convex set $A^{-1}(\mathrm{dom}\,\omega)$, then

$$\Phi_\lambda(x_k) - \inf \Phi_\lambda \leq G(x_k, v_k) \to 0. \tag{2.19}$$

Proof. Set $\hat{x} = \mathrm{prox}_{\lambda f}(y)$, which is the solution of the primal problem (2.13), let $\hat{v}$ be a solution of the dual problem (2.14), and set $z_k = y - \lambda A^* v_k$. One can prove (see Theorem 6.1 in [38]) that

$$(\forall\, k \in \mathbb{N}) \quad \frac{1}{2\lambda}\|z_k - \hat{x}\|_\mathcal{H}^2 \leq \Psi_\lambda(v_k) - \Psi_\lambda(\hat{v}).$$

Clearly $\mathrm{dom}\,\Phi_\lambda = A^{-1}(\mathrm{dom}\,\omega)$ and since $P_{\mathrm{dom}\,\Phi_\lambda}$ is continuous, and $\Psi_\lambda(v_k) \to \inf \Psi_\lambda$ by hypothesis, we have

$$x_k = P_{\mathrm{dom}\,\Phi_\lambda}(z_k) \to P_{\mathrm{dom}\,\Phi_\lambda}(\hat{x}) = \hat{x} \quad (\text{as } k \to +\infty)$$

and hence $Ax_k \to A\hat{x}$, $Ax_k, A\hat{x} \in \mathrm{dom}\,\omega$. Now from the continuity of $\omega_{|\mathrm{dom}\,\omega}$ it follows that $\Phi_\lambda(x_k) \to \Phi_\lambda(\hat{x}) = \inf \Phi_\lambda$. Since $\Phi(\hat{x}) = -\Psi(\hat{v})$, we have

$$G(x_k, v_k) = \Phi_\lambda(x_k) + \Psi_\lambda(v_k)$$

$$= \underbrace{\Phi_\lambda(x_k) - \Phi_\lambda(\hat{x})}_{\geq 0} + \underbrace{\Psi_\lambda(v_k) - \Psi_\lambda(\hat{v})}_{\geq 0} \to 0. \tag{2.20}$$

and the statement follows. $\square$

---

[1]The continuity is for the relative topology of $\mathrm{dom}\,\omega$. This is equivalent to require that for every sequence $(v_k)_{k \in \mathbb{N}}, v_k \in \mathrm{dom}\,\omega$ and $v \in \mathrm{dom}\,\omega$, $v_k \to v \implies \omega(v_k) \to \omega(v)$.

[2]This hypothesis is needed just to ensure $\inf \Phi_\lambda + \inf \Psi_\lambda = 0$. However, weaker conditions can be employed, as $0 \in \mathrm{sri}(R(A) - \mathrm{dom}\,\omega)$, where sri stands for strong relative interior [4].

We remark that Proposition 2.1 also gives a stopping criterion for the algorithm. Indeed from (2.19) it follows that one can control the objective function values of the primal problem by controlling the values of the duality gap — which are explicitly computable by (2.17).

Thus, we are justified in solving the dual problem (2.14) by whatever algorithm just provides a minimizing sequence. We underline that, for the dual problem, no convergence on the minimizers is required, but convergence in value is sufficient. Taking advantage of the component-wise decomposition (2.15)-(2.16), we present a generalization to the sum of composite functions of the algorithm given in [38, Section 5.1], which corresponds to applying FISTA [5] to the dual problem (2.14). Set $\boldsymbol{u}_0 = \boldsymbol{v}_0 = 0, t_0 = 1$ and for every $k \in \mathbb{N}$ define

$$
\left|
\begin{aligned}
&x_{tmp} = y - \lambda \sum_{m=1}^{M} A_m^* u_{k,m} \\
&0 < \gamma_k \le (\lambda \|A\|^2)^{-1} \\
&\text{for } m = 1, \ldots, M \\
&\quad v_{k+1,m} = \text{prox}_{\gamma_k \omega_m^*} \left(u_{k,m} + \gamma_k A_m x_{tmp} \right) \\
&x_{k+1} = P_{A^{-1}(\text{dom}\,\omega)} \left(y - \lambda \sum_{m=1}^{M} A_m^* v_{k+1,m} \right) \\
&t_{k+1} = \left(1 + \sqrt{1 + 4t_k^2} \right)/2 \\
&\text{for } m = 1, \ldots, M \\
&\quad u_{k+1,m} = v_{k+1,m} + \frac{t_k - 1}{t_{k+1}} (v_{k+1,m} - v_{k,m}).
\end{aligned}
\right.
\tag{2.21}
$$

Remark 2.1. *In the case that ω_m is positively homogeneous, it holds that $\omega_m^* = \delta_{S_m}$ the indicator function of $S_m = \partial \omega_m(0)$ and $\text{dom}\,\omega_m = \mathcal{G}_m$. Hence $\text{prox}_{\gamma_k \omega_m^*} = P_{S_m}$ and $v_{k+1,m}$ is computed by*

$$
v_{n+1,m} = P_{S_m} \left(u_{k,m} + \gamma_k A_m z_k \right).
\tag{2.22}
$$

If $\omega_m = \delta_{S_m}$ for a closed convex set $S_m \subseteq \mathcal{G}_m$, then $\omega_m^* = \sigma_{S_m}$ the support function of S_m and using the Moreau decomposition formula $\text{prox}_{\gamma_k \omega_m^*}(y) = y - P_{\gamma_k S_m} y$, the vectors $v_{k+1,m}$ in (2.21) can be computed by the formula

$$
v_{k+1,m} = (I - P_{\gamma_k S_m})(u_{k,m} + \gamma_k A_m z_k).
\tag{2.23}
$$

Note that in this case the restriction of ω_m to $\text{dom}\,\omega_m$ is continuous.

2.5   Application to Computational Biology

In this section we provide an application of the proposed algorithm to a significant problem of computational biology, the identification of latent features in array-based comparative genomic hybridization data (aCGH). We present an enhancement of the model proposed in [24] as well as three different experiments to test our approach. The first two consider synthetic and simulated aCGH data. The third experiment analyzes real breast cancer aCGH data in the context of clustering.

2.5.1   A Model for aCGH Data Latent Feature Detection

Array-based comparative genomic hybridization is a modern whole-genome measuring technique that evaluates the occurrence of *copy number variations* (CNVs) across the genome and extends the original CGH technology [14]. CNVs are alterations of the DNA that result in the cell having an abnormal number of copies of one or more sections of the DNA and ultimately may indicate an oncogene or a tumor suppressor gene. A signal measured with an aCGH technology is made of a piecewise constant component plus some composite noise.

The typical analysis on such data is *segmentation*, which is the automatic detection of *loci* where copy number alterations (amplifications or deletions) occur. Beyond that, it is crucial to understand how these alterations co-occur. This turns to *identifying shared patterns* (latent features) in the data, which may reveal a genotype-phenotype relationship.

Many methods have been proposed for the extraction of CNVs based on different principles like filtering (or smoothing), breakpoint detection, and calling, taking into account one sample at a time [17]. Some interesting recent results exploit the possibility of adopting regularization methods for a joint segmentation of many aCGH profiles. The works proposed by [24, 33, 37] follow this stream, and are based on total variation (TV) or fused lasso signal approximation.

Thanks to the generality of the framework proposed in Section 2.4, we present E-FLLat[3] (enhanced fused Lasso latent feature model), a refinement of FLLat (fused Lasso latent feature model)[24], a dictionary learning model that was originally proposed for aCGH data segmentation and latent features identification. We improve several modelling aspects that, in the end, provide a straightforward way to reveal the elementary patterns in the signal, increasing the interpretability of the results. In particular, we choose more complex penalty terms that better suit the prior knowledge on the problem.

We choose the same application context of FLLat, that is aCGH data segmentation and extraction of latent features (atoms). The E-FLLat model is based on the minimization of a functional combining several penalties, properly designed to treat the whole genomic signal and to select more representative atoms. We first recall FLLat and then describe E-FLLat.

---

[3]In [22], we call it CGHDL.

The FLLat model is written as follows:

$$\min_{\boldsymbol{\theta}_s, \mathbf{b}_j} \sum_{s=1}^{S} \left\| \mathbf{y}_s - \sum_{j=1}^{J} \theta_{js} \mathbf{b}_j \right\|^2 + \lambda \sum_{j=1}^{J} \|\mathbf{b}_j\|_1 + \mu \sum_{j=1}^{J} TV(\mathbf{b}_j)$$

$$\text{s.t.} \sum_{s=1}^{S} \theta_{j,s}^2 \leq 1 \quad \forall j = 1, \ldots, J$$

(FLLat)

where $\lambda, \mu > 0$ are regularization parameters. The problem can be put in matrix form as in (2.5), where

$$g(\boldsymbol{B}) = \lambda \sum_{j=1}^{J} \|\mathbf{B}(:,j)\|_1 + \mu \sum_{j=1}^{J} TV(\mathbf{B}(:,j)),$$

$$h(\boldsymbol{\Theta}) = \sum_{j=1}^{J} \delta_{B_1}(\boldsymbol{\Theta}(j,:)),$$

(2.24)

and δ_{B_1} is the indicator function of the euclidean unit ball B_1 of $\mathbb{R}^S$.

These penalties are chosen in order to model the prior knowledge on the atoms, which are expected to be *simple*, in the sense that they resemble *step functions*. This is achieved by employing the sum of the *fused lasso* penalties $\sum_{j=1}^{J}(\lambda \|\mathbf{b}_j\|_1 + \mu TV(\mathbf{b}_j))$. The ℓ^1 penalization term forces each atom $\mathbf{b}_j$ to be sparse and the total variation term $TV(\mathbf{b}_j) = \sum_{l=1}^{L-1} |b_{l+1,j} - b_{l,j}|$ induces small variations in the atoms. The hard constraints on the coefficients θ_j. are imposed for consistency and identifiability of the model. Indeed, multiplying a particular feature $\mathbf{b}_j$ by a constant, and dividing the corresponding coefficients by the same constant, leaves the fit unchanged, but reduces the penalty.

E-FLLat is an enhancement of FLLat, driven by the following optimization problem depending on the three regularization parameters $\lambda, \mu, \tau > 0$:

$$\min_{\boldsymbol{\theta}_s, \mathbf{b}_j} \sum_{s=1}^{S} \left\| \mathbf{y}_s - \sum_{i=1}^{J} \theta_{js} \mathbf{b}_j \right\|^2 + \lambda \sum_{j=1}^{J} \|\mathbf{b}_j\|_1^2 + \mu \sum_{j=1}^{J} TV_{\mathbf{w}}(\mathbf{b}_j) + \tau \sum_{s=1}^{S} \|\boldsymbol{\theta}_s\|_1^2$$

$$\text{s.t.} \; 0 \leq \theta_{js} \leq \theta_{\max}, \quad \forall j = 1, \ldots, J \quad \forall s = 1, \ldots, S$$

(E-FLLat)

where

$$TV_{\mathbf{w}}(\mathbf{b}_j) = \sum_{l=1}^{L-1} w_l |b_{l+1,j} - b_{l,j}|, \quad \mathbf{w} = (w_l)_{1 \leq l \leq L-1} \in \mathbb{R}^{L-1}$$

is the *weighted total variation*. E-FLLat may be turned into problem (2.5) by choosing g and h as follows:

$$g(\boldsymbol{B}) = \lambda \sum_{j=1}^{J} \|\mathbf{B}(:,j)\|_1^2 + \mu \sum_{j=1}^{J} TV_{\mathbf{w}}(\mathbf{B}(:,j)),$$

$$h(\boldsymbol{\Theta}) = \delta_{\Delta^{J \times S}}(\boldsymbol{\Theta}) + \tau \sum_{s=1}^{S} \|\boldsymbol{\Theta}(:,s)\|_1^2 ,$$

(2.25)

and $\delta_{\Delta^{J \times S}}$ is the indicator function of the box set $\Delta^{J \times S} = [0, \theta_{max}]^{J \times S}$. In the experiments we will always set $\theta_{max} = 1$. This choice forces the atoms to be of the same amplitude as the original data.

This model improves FLLat, in several aspects. First it employs the penalization terms

$$\lambda \sum_{i=1}^{J} \|\mathbf{B}(:,j)\|_1^2, \quad \sum_{s=1}^{S} \|\boldsymbol{\Theta}(:,s)\|_1^2,$$

which are sum of squares of ℓ^1 norms (i.e., mixed norms). This possibly forces a structured sparsity only along the columns of the matrix of atoms $\mathbf{B}$ and of coefficients $\boldsymbol{\Theta}$. This choice is more faithful to the actual purposes of dictionary learning and contrasts with the majority of similar models [24] that instead use a global ℓ^1 norm inducing just a scattered sparsity on the matrices.

Secondly in E-FLLat, the total variation term is indeed a generalized total variation due to the presence of the weights $\mathbf{w}$. This modification is introduced in order to relax at some points the constraint of *small jumps* on the atoms. When dealing with aCGH data, this allows us to treat the signal of the whole genome, giving the capability of identifying concomitant alterations on different chromosomes, still guaranteeing their independency. This is achieved by setting the weights $\mathbf{w}_l$ equal to zero at the chromosomes' borders, and one elsewhere. We remark that taking into account the geometry of the platform and the actual distance between probes on the chromosomes may lead to a better strategy to set the weights. Similarly, the weights corresponding to the probes lying in the centromere region may as well be set to zero. This is in line with the work [8]. Conversely, in FLLat, the analysis is supposed to be performed on each chromosome separately (see [24] Section 2.1). This leads to selecting a different set of atoms for each chromosome preventing identifying concomitant alterations on different chromosomes.

Finally, we constrain the coefficients to be positive. This reduces the complexity of the matrix of coefficients $\boldsymbol{\Theta}$ and forces the matrix of atoms $\boldsymbol{B}$ to be more informative: *e.g.*, for deletions and amplifications occurring in different samples but on the same *locus* on the chromosome, different atoms may be selected. Ultimately the interpretability of the results is improved.

We solve the optimization problem connected to the E-FLLat model by means of the algorithm studied in Section 2.4. This amounts to performing the proximal alternating algorithm (2.7) as outer loop and computing the proximity operators $\mathrm{prox}_{\eta_n \varphi(\cdot, \mathbf{B}_n)}$ and $\mathrm{prox}_{\zeta_n \varphi(\boldsymbol{\Theta}_{n+1}, \cdot)}$ with (2.21) as inner loop. This resulting nested algorithm will be named, from now on, E-FLLatPA.

We finally remark that the penalization functions g and h in (2.25) satisfy the Kurdyka-Łojasiewicz property, for they are sums of functions satisfying the Kurdyka-Łojasiewicz property. See Section 4.3 in [2] and Example 5.3 in [3].

2.5.2   Experiment on Synthetic Data

The purpose of this experiment was to evaluate the performance of E-FLLat both in estimating the true signals, and in identifying latent features in the data. We considered here synthetic data, in order to have full control of the relevant alterations. We generated three datasets that were used to test the denoising performance, following analogous experiments in [24]. Next, we built a fourth dataset with a properly designed pattern of alterations and we checked whether E-FLLat made a correct identification.

Data generation The model of the signal follows [24] and the additive noise model follows [26]. The signal is defined as:

$$y_{ls} = \mu_{ls} + \epsilon_{ls}, \quad \mu_{ls} = \sum_{m=1}^{M_s} c_{ms} I_{\{l_{ms} \le l \le l_{ms} + k_{ms}\}}, \quad \epsilon_{ls} \sim N(0, \sigma^2), \quad (2.26)$$

where $l = 1, \ldots, L$, $s = 1, \ldots, S$, μ_{ls} is the mean, and σ is the standard deviation of the noise ϵ_{ls}. The mean signal $\boldsymbol{\mu}_s = (\mu_{ls})_{1 \le l \le L}$ is a step function where M_s is the number of segments (regions of CNVs) generated for sample s and c_{ms}, l_{ms} and k_{ms} are the height, starting position, and length, respectively, for each segment. We chose $M_s \in \{1, 2, 3, 4, 5\}$, $c_{ms} \in \{\pm1, \pm2, \pm3, \pm4, \pm5\}$, $l_{ms} \in \{1, \ldots, L - 100\}$, and $k_{ms} \in \{5, 10, 20, 50, 100\}$, $L = 1000$, $S = 20$. According to this general schema, we generated four types of datasets:

Dataset 1: The samples are generated in order to reduce the chance of sharing CNV regions. Therefore, the choice of the values M_s, c_{ms}, l_{ms}, and k_{ms} is done separately for each sample. This schema follows [24, Section 4.1, Dataset 1].

Dataset 2: Following [24, Section 4.1, Dataset 2], the samples were designed to have common segments of CNVs. Each shared segment appears in the samples according to a fixed proportion randomly picked between $(0.25, 0.75)$. Starting points and lengths were shared among the selected samples, whereas the amplitudes c_{ms} still might vary within samples. The unshared segments were built as in *Dataset 1* for a maximum of 5 segments per sample.

Dataset 3: The atoms $\boldsymbol{\beta}_j$ were generated as the μ_{ls}'s in (2.26), with c_{ms}, l_{ms} and k_{ms} chosen as above. The coefficients θ_{js} were randomly sampled in $[0, 1]$ and the signal was built as $\mathbf{Y} = \mathbf{B}\boldsymbol{\Theta} + \boldsymbol{\varepsilon}$, where $\boldsymbol{\varepsilon}$ is an additive Gaussian noise.

Dataset 4: This dataset was explicitly designed to mimic a real signal composed of different chromosomes. We built three classes of samples. One third of the samples had mean signal as in the upper panel of Figure 2.1, one third had mean signal as in the lower panel of Figure 2.1, and the remaining third was built as in *Dataset 1*.

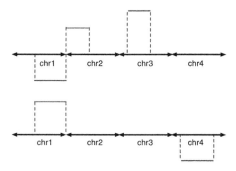

FIGURE 2.1: Mean signals of the two patterns used for *Dataset 4* generation.

Parameter selection The choice of the parameters (J, λ, μ, τ) was done by minimizing the Bayesian information criterion (BIC) [30]. The BIC mitigates the problem of overfitting by introducing a penalty term for the complexity of the model. In our case the BIC is written as:

$$(SL) \cdot \log\left(\frac{\|\mathbf{Y} - \mathbf{B}\mathbf{\Theta}\|_F^2}{SL}\right) + k(\mathbf{B}) \log(SL) \qquad (2.27)$$

and $k(\mathbf{B})$ was computed as the number of jumps in $\mathbf{B}$ and ultimately depends on the parameters (J, λ, μ, τ). Note that, differently from [24], we also used BIC to select the number of atoms J from $\{5, 10, 15, 20\}$.

Results Figure 2.2 shows the performances of the two approaches. Following [24], ROC curves were built by evaluating the correct detection of alterations based on the denoised signal: the results are comparable. Performances on the raw noisy signal are also plotted for reference. Figure 2.3 shows a plot of the solutions obtained by the two approaches on *Dataset 4*. The algorithm implementing (FLLat) achieves good results in denoising, selecting $J = 10$ atoms, but fails in detecting the underlying patterns of Figure 2.1. The selected atoms represent single alterations. Conversely, our approach (right panel in Figure 2.3) selects 5 atoms that clearly comprise the two patterns. Summarizing, these results show that the proposed approach is very effective in identifying the underlying data generating features.

2.5.3 Experiment on Simulated Data

This experiment is in the same vein of that described in Section 2.5.2 on *Dataset 4*, but performed on simulated data, mimicking a real aCGH signal.

Data generation The dataset mimics a real aCGH signal composed of different chromosomes. We simulated a signal affected by spatial bias [15] using chip geometry information and by a wave effect modeled through a sinusoidal function [27]. As chip design, we used the Agilent 44k aCGH platform. To deal with a simpler model, we restricted the dataset to 4 out of 23 chromosomes, choosing those with a relatively small number of probes (i.e., chromosomes 13,

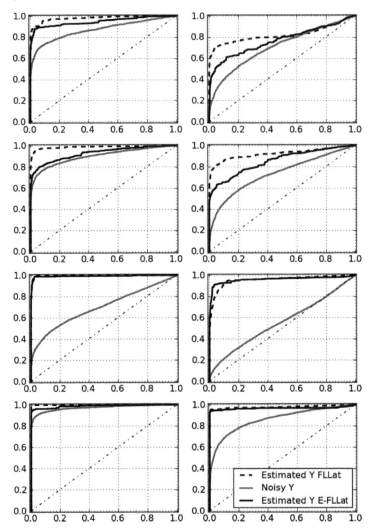

FIGURE 2.2: ROC curves for varying levels of noise (columns) and different dataset type (rows). The left column shows $\sigma = 1.0$, while the right column shows $\sigma = 2.0$. Red lines refer to the performances on the noisy **Y**, dashed and solid lines refer to FLLat and E-FLLat performances, respectively.

15, 18, and 21). Overall, we considered 80 samples composed of 3750 probes. We built three classes of samples. One third of the samples (G1) followed a pattern with a loss on chromosome 13, a gain on chromosome 15, and a gain on the longer arm of chromosome 18. The alterations occur randomly with a probability of 80% each. Similarly, one third of the samples (G2) followed a pattern with a gain on chromosome 13 and a narrow loss on the shorter arm

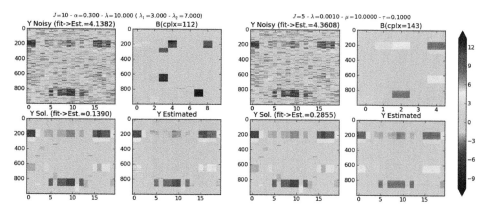

FIGURE 2.3: *Dataset 4* analyzed by FLLat (left panel) and E-FLLat (right panel). Each panel shows four subplots: top left plot represents the noisy data matrix, top right plot shows the atom matrix with atoms as columns, bottom left subplot is the *true* data matrix, and bottom right is the estimated signal.

of chromosome 21. Moreover, in groups G1 and G2, alterations (either gain or loss) on the chromosomes not involved in the patterns can occur randomly with 20% probability. Finally, group G3 had random alterations (either gain or loss) on all chromosomes with low probability (10%).

Parameter selection Keeping in mind that we had three main groups of data, we chose $J = 5$. The parameters (λ, μ, τ) were chosen according to BIC (2.27).

Results Figure 2.4 reports the atoms and coefficients computed by E-FLLatPA on the simulated dataset. One can recognize that the first two atoms capture the pattern of group G1, while the third corresponds to group G2. The fourth atom represents one of the deletions that have been introduced randomly as noise. These results confirm the capability of E-FLLatPA in finding shared patterns in the data.

2.5.4 Clustering of Breast Cancer Data

Here we test the performance of E-FLLat on a clustering problem where a benchmark is available. The clustering procedure takes as input the coefficient matrix or the reconstructed signal matrix computed by E-FLLat and FLLat. We show that E-FLLat allows for a better clustering in terms of intra-group coherence. We performed two different experiments: a single chromosome analysis for chromosomes 17 and 8, and a whole genome analysis.

Data We considered the aCGH dataset from [28], already used by [24] to test FLLat on real data. The dataset consisted of 44 samples of advanced primary breast cancer. Each signal measured the CNV of 6691 human genes.

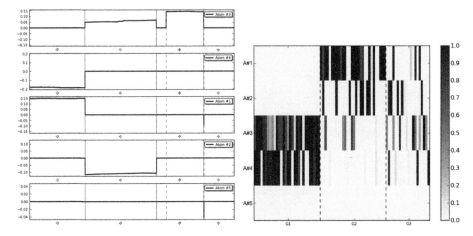

FIGURE 2.4: Profiles of the atoms and coefficient matrix identified by E-FLLatPA for the simulated dataset. The atoms are sorted according to their frequency of occurrence.

The samples were assigned to 3 classes according to tumor grading: 5 samples were assigned to grade 1(G1), 21 to grade 2(G2), 17 to grade 3(G3), and 1 unassigned.

Parameter Selection The number of atoms J was set differently for the considered scenarios. Concerning the analysis based on a single chromosome, for comparison purposes we follow [24] and set $J = 5$ for chromosome 17 and $J = 6$ for chromosome 8. Regarding the whole genome analysis, we consider three different $J \in \{10, 18, 24\}$, which correspond to the number of principal components of $\mathbf{Y}$ able to explain respectively the 50%, 70%, and 80% of the variance. For E-FLLat, the rest of the parameters (μ, λ, τ) were set by searching the best triple in $\{0.01, 0.1, 1.0, 10, 100\} \times \{0.01, 0.1, 1.0, 10, 100\} \times \{0.1, 1.0, 10\}$ according to BIC. Regarding FLLat, the grid of parameters was defined by some heuristics implemented into the given R package.

Experiment Design and Protocol In the first experiment we compared E-FLLat and FLLat focusing on chromosomes 8 (241 mapped genes) and 17 (382 mapped genes), identified by [28] as chromosomes with biologically relevant CNVs. Clustering was performed on $\mathbf{Y}^c$, the original raw noisy data matrix restricted to the chromosome $c \in \{8, 17\}$, on coefficients matrices $\mathbf{\Theta}^c_{EF}$ and $\mathbf{\Theta}^c_F$, and on the denoised samples matrices $\hat{\mathbf{Y}}^c_{EF}$ and $\hat{\mathbf{Y}}^c_F$.

In a second experiment, we compared FLLat and E-FLLat considering the analysis of aCGH data of the whole genome (6691 probes). Clustering was performed on the original raw noisy data matrix $\mathbf{Y}$, on the coefficients matrix $\mathbf{\Theta}_{EF}$, and the denoised samples matrix $\hat{\mathbf{Y}}_{EF}$. We remark that this context is not completely appropriate for FLLat, for it was designed to treat chromosomes one at a time — indeed the unweighted total variation included into its model would unfairly penalize variations on borders of the chromosomes.

For clustering, we adopted a hierarchical agglomerative algorithm, using the *city block* or *Manhattan* distance between points $d(a,b) = \sum_i |a_i - b_i|$ and the *single linkage* criterion $d(\mathcal{A}, \mathcal{B}) = min\{d(a,b) : a \in \mathcal{A}, b \in \mathcal{B}\}$ [31]. The cluster $\mathcal{A}$ is linked with the cluster $\mathcal{B}$ if the distance $d(\mathcal{A}, \mathcal{B})$ is the minimum with respect to all the other clusters $\mathcal{B}'$.

Moreover, to evaluate the coherence of the obtained dendrogram with respect to the groups G1, G2, and G3, we measured the *cophenetic distance* among the samples within each group [32]. For each pair of observations (a, b), the cophenetic distance is the distance between the two clusters that were merged to assign the two points in a single new cluster. The average of the cophenetic distances within each clinical group provides an objective measure of how the resulting dendrogram *describes* the differences between observations, using the clinical grades as ground truth.

Note that, by design, the values contained in the coefficients matrix produced by FLLat and E-FLLat could have a different range of values (in E-FLLat the values are positive and bounded). In order to calculate comparable distance metrics, before clustering and cophenetic distances evaluation, each estimated coefficients matrix $\boldsymbol{\Theta}$ was normalized by is maximum absolute value. The same preprocessing was also applied on the original aCGH signals and on the estimated ones $\hat{\mathbf{Y}} = \mathbf{B}\boldsymbol{\Theta}$.

Results *Analysis restricted to chromosome 17.* In Figure 2.6 (left) we show the means of the cophenetic distances calculated for each group of samples (the unannotated one was not considered) restricted to the chromosome 17. It is clear that the clustering on the coefficients matrix produced by E-FLLat places the samples belonging to homogeneous clinical groups (G1, G2, and G3) closer in the dendrogram. Moreover, also the denoised data matrix $\hat{\mathbf{Y}}^{17}_{EF}$ shows better discriminative performances with respect to $\hat{\mathbf{Y}}^{17}_F$. This may be due to the capability of our model to better detect the main altered patterns in the signals, despite a possibly higher reconstruction error [22]. Such property ultimately induces a more effective clustering. In Table 2.1 (top) we report a summary of the averaged cophenetic distances, also including the clustering on raw signals.

Analysis restricted to chromosome 8. The analysis on chromosome 8 gives similar results. Figure 2.6 (right) shows the means of the cophenetic distances calculated for each group of samples, and Table 2.1 (bottom) shows the corresponding averaged cophenetic distances.

Whole genome analysis. We present the results obtained with $J = 10$. The resulting atoms (see Figure 2.5) describe co-occurrent alterations along different chromosomes but are still fairly simple for a visual interpretation by the domain experts. For different Js we did not note relevant differences in terms of fit and clustering. It is important to note that the four most used atoms of the dictionary extracted by E-FLLat detect the main genomic alterations on chromosomes 8 and 17 as well as a co-occurrence of deletions on chromosome 3 and 5. In [28] all these alterations were already indicated as

TABLE 2.1: Average cophenetic distances after clustering for the analysis restricted to chromosome 17.

	G1	G2	G3
Θ_{EF}^{17}	0.008 ± 0.004	0.079 ± 0.112	0.111 ± 0.124
$\hat{\mathbf{Y}}_{EF}^{17}$	0.022 ± 0.019	0.476 ± 0.720	0.687 ± 0.795
Θ_{F}^{17}	0.178 ± 0.044	0.265 ± 0.173	0.517 ± 0.446
$\hat{\mathbf{Y}}_{F}^{17}$	1.737 ± 0.484	2.945 ± 2.074	5.212 ± 3.851
$\mathbf{Y}^{17}$	19.284 ± 2.374	19.589 ± 3.961	23.941 ± 5.870

	G1	G2	G3
Θ_{EF}^{8}	0.016 ± 0.007	0.054 ± 0.024	0.147 ± 0.142
$\hat{\mathbf{Y}}_{EF}^{8}$	0.222 ± 0.095	0.842 ± 0.410	1.720 ± 1.135
Θ_{F}^{8}	0.301 ± 0.095	0.469 ± 0.236	0.951 ± 0.657
$\hat{\mathbf{Y}}_{F}^{8}$	3.135 ± 1.090	4.962 ± 2.605	9.547 ± 6.638
$\mathbf{Y}^{8}$	12.363 ± 1.165	15.484 ± 4.124	20.150 ± 6.200

being very common but the relation between chromosomes 3 and 5 was not indicated as co-occurrence and needs further biological validation.

TABLE 2.2: Average cophenetic distances after clustering for the analysis extended to all chromosomes with $J = 10$.

	G1	G2	G3
Θ_{EF}	0.738 ± 0.541	0.290 ± 0.213	0.463 ± 0.406
$\hat{\mathbf{Y}}_{EF}$	7.988 ± 4.663	4.191 ± 2.795	5.512 ± 3.632
$\mathbf{Y}$	305.76 ± 39.85	290.26 ± 38.04	302.86 ± 34.76

2.6   Conclusion

We presented an algorithm for dictionary learning, which is based on the alternating proximal algorithm studied by [2] coupled with a fast dual algorithm for the computation of the related proximity operators. This algorithm is suitable for a general dictionary learning model composed of a data fit term that accounts for the goodness of the representation, and several convex penalization terms on the coefficients and atoms, explaining the prior knowledge at hand. As recently proved by [2], an alternating proximal scheme ensures better convergence properties than the simpler alternating minimization. We also show the performance of the proposed algorithm on synthetic, simulated, and real case scenarios.

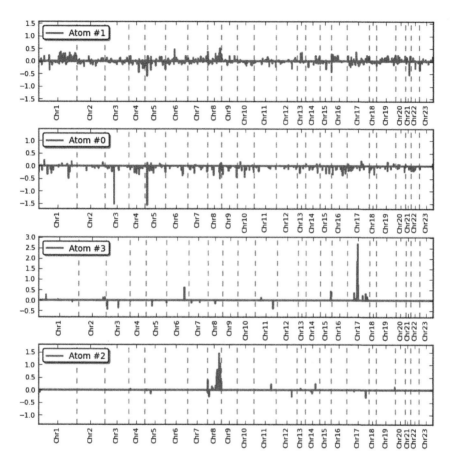

FIGURE 2.5: Profiles of the first four most used atoms for sample reconstruction (sum of the row of Θ) extracted by E-FLLat on all chromosomes. The atom #1 maps a general pattern of alterations, and it is responsible for a high proportion of signal reconstruction. E-FLLat found the alterations on chromosomes 8 and 17, and also detected co-occurring alterations on chromosomes 3 and 5. Vertical lines indicate chromosomes' boundaries.

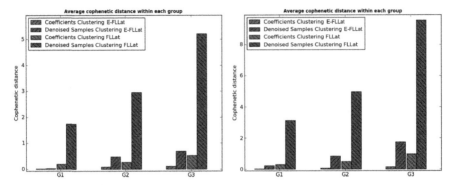

FIGURE 2.6: Average cophenetic distances for the groups G1, G2, and G3 on chromosome 17 (left) and chromosome 8 (right). E-FLLat always has better clustering results (also see Table 2.1). Moreover, it is also interesting to note that clustering the denoised samples by E-FLLat and FLLat, the former has better results, suggesting also a higher quality of the dictionary atoms used to reconstruct the samples.

Bibliography

[1] M. Aharon, M. Elad, and A. Bruckstein. K-SVD: An Algorithm for Designing Overcomplete Dictionaries for Sparse Representation. *IEEE Transactions on Signal Processing*, 54(11):4311–4322, 2006.

[2] H. Attouch, J. Bolte, P. Redont, and A. Soubeyran. Proximal Alternating Minimization and Projection Methods for Nonconvex Problems: An Approach Based on the Kurdyka-Lojasiewicz Inequality. *Mathematics of Operational Research*, 35(2):438–457, 2010.

[3] H. Attouch, J. Bolte, and B. F. Svaiter. Convergence of descent methods for semi-algebraic and tame problems: proximal algorithms, forward-backward splitting, and regularized Gauss-Seidel methods. *Math. Program.*, 137(1-2, Ser. A):91–129, 2013.

[4] H. H. Bauschke and P. L. Combettes. *Convex Analysis and Monotone Operator Theory in Hilbert Spaces*. CMS Books in Mathematics/Ouvrages de Mathématiques de la SMC. Springer, New York, 2011.

[5] A. Beck and M. Teboulle. A Fast Iterative Shrinkage-Thresholding Algorithm for Linear Inverse Problems. *SIAM Journal on Imaging Sciences*, 2(1):183–202, 2009.

[6] D. P. Bertsekas. *Nonlinear Programming*. Athena Scientific, 2nd edition, 1999.

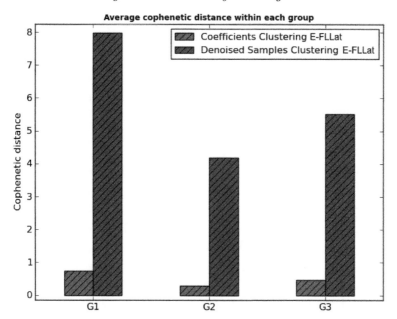

FIGURE 2.7: Average cophenetic distances for the groups G1, G2, and G3 on all chromosomes and $J = 10$ (also see Table 2.2).

[7] J. Bolte, A. Daniilidis, and A. Lewis. The Lojasiewicz inequality for non-smooth subanalytic functions with applications to subgradient dynamical systems. *SIAM J. Optim.*, 17(4):1205–1223 (electronic), 2006.

[8] S. J. Diskin, T. Eck, J.Greshock, Y. P. Mosse, T. Naylor, C. J. Stoeckert, B. L. Weber, J. M. Maris, and G. R. Grant. Stac: A method for testing the significance of DNA copy number aberrations across multiple array-CGH experiments. *Genome Res*, 16(9):1149–58, Sep 2006.

[9] C. Févotte, N. Bertin, and J-L. Durrieu. Nonnegative Matrix Factorization with the Itakura-Saito Divergence: With Application to Music Analysis. *Neural Computation*, 21(3):793–830, 2008.

[10] R. Gribonval, R. Jenatton, F. Bach, M. Kleinsteuber, and M. Seibert. Sample Complexity of Dictionary Learning and Other Matrix Factorizations. *ArXiv e-prints*, 2013.

[11] R. Grosse, R. Raina, H. Kwong, and A. Ng. Shift-invariance sparse coding for audio classification. *Proceedings of the Twenty-Third Annual Conference on Uncertainty in Artificial Intelligence (UAI-07)*, pages 149–158, 2007.

[12] R. Jenatton, R. Gribonval, and F. Bach. Local stability and robustness of sparse dictionary learning in the presence of noise. *ArXiv e-prints*, 2012.

[13] R. Jenatton, J. Mairal, G. Obozinski, and F. Bach. Proximal Methods for Hierarchical Sparse Coding. *Journal of Machine Learning Research*, 12:2297–2334, 2011.

[14] A. Kallioniemi, O.-P. Kallioniemi, D. Sudar, D. Rutovitz, J.W. Gray, F. Waldman, and D. Pinkel. Comparative genomic hybridization for molecular cytogenetic analysis of solid tumors. *Science (New York, N.Y.)*, 258(5083):818–21, 1992.

[15] M. Khojasteh, W.L. Lam, R.K. Ward, and C. MacAulay. A stepwise framework for the normalization of array CGH data. *BMC bioinformatics*, 6:274, 2005.

[16] K. Kreutz-Delgado, J. F. Murray, B. D. Rao, K. Engan, T.-W. Lee, and T. J. Sejnowski. Dictionary Learning Algorithms for Sparse Representation. *Neural Comput.*, 15(2):349–396, 2003.

[17] W. R Lai, M. D Johnson, R. Kucherlapati, and P. J. Park. Comparative analysis of algorithms for identifying amplifications and deletions in array CGH data. *Bioinformatics (Oxford, England)*, 21(19):3763–70, 2005.

[18] H. Lee, A. Battle, R. Raina, and A.Y. Ng. Efficient sparse coding algorithms. *Advances in Neural Information Processing Systems (NIPS)*, 2007.

[19] J. Mairal, F. Bach, and J. Ponce. Task-Driven Dictionary Learning. *Pattern Analysis and Machine Intelligence, IEEE Transactions on*, 34(4):791–804, 2012.

[20] J. Mairal, F. Bach, J. Ponce, G. Sapiro, and A. Zisserman. Non-local sparse models for image restoration. In *Computer Vision, 2009 IEEE 12th International Conference on*, pages 2272–2279, 2009.

[21] S. Masecchia, A. Barla, S. Salzo, and A. Verri. Dictionary learning improves subtyping of breast cancer aCGH data. In *Engineering in Medicine and Biology Society (EMBC), 2013 35th Annual International Conference of the IEEE*, pages 604–607, 2013.

[22] S. Masecchia, S. Salzo, A. Barla, and A. Verri. A dictionary learning based method for aCGH segmentation. *Proc. of ESANN*, 2013.

[23] A. Maurer and M. Pontil. *K*-dimensional coding schemes in Hilbert spaces. *IEEE Trans. Inform. Theory*, 56(11):5839–5846, 2010.

[24] G. Nowak, T. Hastie, J.R. Pollack, and R. Tibshirani. A fused lasso latent feature model for analyzing multi-sample aCGH data. *Biostatistics*, 12(4):776–791, 2011.

[25] B. A. Olshausen and D. J. Field. Sparse coding with an overcomplete basis set: A strategy employed by V1? *Vision Research*, 37(23):3311–3325, 1997.

[26] A. B. Olshen, E. S. Venkatraman, R. Lucito, and M. Wigler. Circular binary segmentation for the analysis of array-based DNA copy number data. *Biostatistics*, 5(4):557–572, 2004.

[27] R. Pique-Regi, A. Ortega, and S. Asgharzadeh. Joint estimation of copy number variation and reference intensities on multiple DNA arrays using GADA. *Bioinformatics*, 25(10):1223–1230, 2009.

[28] J. R. Pollack, T. Sørlie, C. M. Perou, C.A. Rees, S. S. Jeffrey, P. E. Lonning, R. Tibshirani, D. Botstein, A. L. Børresen-Dale, and P. O. Brown. Microarray analysis reveals a major direct role of DNA copy number alteration in the transcriptional program of human breast tumors. *Proceedings of the National Academy of Sciences*, 99(20):12963–12968, 2002.

[29] R. Raina, A. Battle, H. Lee, B. Packer, and A. Y. Ng. Self-taught Learning: Transfer Learning from Unlabeled Data. In *Proceedings of the 24th International Conference on Machine Learning*, ICML '07, pages 759–766, 2007.

[30] G. Schwarz. Estimating the Dimension of a Model. *The Annals of Statistics*, 6(2):461–464, 1978.

[31] R. Sibson. SLINK: An optimally efficient algorithm for the single-link cluster method. *The Computer Journal*, 16(1):30–34, 1973.

[32] R.R. Sokal and F.J. Rohlf. The comparison of dendrograms by objective methods. *Taxon*, 11(2):33–40, 1962.

[33] Z. Tian, H. Zhang, and R. Kuang. Sparse Group Selection on Fused Lasso Components for Identifying Group-specific DNA Copy Number Variations. In *Proc. of IEEE International Conference on Data Mining (ICDM)*, pages 665–674, 2012.

[34] R. Tibshirani, M. Saunders, S. Rosset, J. Zhu, and K. Knight. Sparsity and smoothness via the fused lasso. *Journal of the Royal Statistical Society - Series B: Statistical Methodology*, 67(1):91–108, 2005.

[35] I. Tosic and P. Frossard. Dictionary learning. *Signal Processing Magazine, IEEE*, 28(2):27–38, 2011.

[36] D. Vainsencher, S. Mannor, and A. M. Bruckstein. The sample complexity of dictionary learning. *J. Mach. Learn. Res.*, 12:3259–3281, 2011.

[37] J.-P. Vert and K. Bleakley. Fast detection of multiple change-points shared by many signals using group LARS. *Advances in Neural Information Processing Systems 23*, 1:1–9, 2010.

[38] S. Villa, S. Salzo, L. Baldassarre, and A. Verri. Accelerated and inexact forward-backward algorithms. *SIAM Journal on Optimization*, 23(3):1607–1633, 2013.

Chapter 3

Hybrid Conditional Gradient-Smoothing Algorithms with Applications to Sparse and Low Rank Regularization

Andreas Argyriou[*]

École Centrale Paris, Center for Visual Computing

Marco Signoretto[*]

KU Leuven, ESAT-STADIUS

Johan A.K. Suykens[*]

KU Leuven, ESAT-STADIUS

[*]andreas.argyriou@ecp.fr, marco.signoretto@esat.kuleuven.be, johan.suykens@esat.ku-leuven.be

3.1   Introduction

Conditional gradient methods are old and well-studied optimization algorithms. Their origin dates at least to the 1950s and the Frank-Wolfe algorithm for quadratic programming [18], but they apply to much more general optimization problems. General formulations of conditional gradient algorithms have been studied in the past and various convergence properties of these algorithms have been proven. Moreover, such algorithms have found application in many fields, such as optimal control, statistics, signal processing, computational geometry and machine learning. Currently, interest in conditional gradient methods is undergoing a revival because of their computational advantages when applied to certain large scale optimization problems. Such examples are *regularization* problems involving sparsity or low rank constraints, which appear in many widely used methods in machine learning.

Inspired by such algorithms, in this chapter we study a first-order method for solving certain convex optimization problems. We focus on problems of the form

$$\min \left\{ f(x) + g(Ax) + \omega(x) : x \in \mathcal{H} \right\} \tag{3.1}$$

over a real Hilbert space $\mathcal{H}$. We assume that f is a convex function with *Hölder continuous gradient*, g a *Lipschitz continuous* convex function, A a bounded linear operator, and ω a convex function defined over a *bounded domain*. We also assume that the computational operations available are the *gradient* of f, the *proximity operator* of g, and a *subgradient of the convex conjugate ω^**.[1] A particularly common type of problems covered by (3.1) is

$$\min \left\{ f(x) + g(Ax) : x \in \mathcal{C} \right\}, \tag{3.2}$$

where $\mathcal{C}$ is a bounded, closed, convex subset of $\mathcal{H}$. Common among such examples are regularization problems with one or more penalties in the objective (as the term $g \circ A$) and one penalty as a constraint described by $\mathcal{C}$.

Before presenting the algorithm, we review in Section 3.3 a generic conditional gradient algorithm that has been well studied in the past. This standard algorithm can be used for solving problems of the form (3.1) whenever $g = 0$. However, the conditional gradient algorithm cannot handle problems with a nonzero term g, because it would require computation of a subgradient of a composite convex conjugate function, namely a subgradient of $(g \circ A + \omega)^*$. In many cases of interest, there is no simple rule for such subgradients and the computation itself requires an iterative algorithm.

Thus, in Section 3.4 we discuss an alternative approach that combines ideas from both conditional gradient algorithms and smoothing proximal algorithms, such as Nesterov smoothing. We call the resulting algorithm a *hybrid*

---

[1]For the precise assumptions required, see Assumptions 3.2, 3.3.

conditional gradient-smoothing algorithm, in short, HCGS. This approach involves smoothing the g term, that is, approximating g with a function whose gradient is Lipschitz continuous. Besides this modification, HCGS is similar to the conditional gradient algorithm. We show that, for suitable choices of the smoothing parameter, the estimates of the objective in HCGS converge to the minimum of (3.1). Moreover, the convergence rate is of the order of $\mathcal{O}\left(\frac{1}{\varepsilon^2}\right)$ iterations for attaining an accuracy of ε in terms of the objective. We do not claim originality, however, since similar theoretical results have appeared in recent work by Lan [34].

Our main focus is on highlighting applications of the hybrid approach on certain applications of interest. To demonstrate the applicability of HCGS to regularization problems from machine learning, we present simulations of matrix problems with *simultaneous sparsity and low rank* penalizations. Examples of such applications are *graph denoising*, *link prediction* in social networks, *covariance estimation*, and *sparse PCA*. Each of these problems involves two penalties, an elementwise ℓ_1 norm to promote sparsity, and a *trace norm* to promote low rank. Standard algorithms may not be practical in high dimensional problems of this type. As mentioned above, standard conditional gradient methods require a subgradient computation of a complicated function, whereas proximal algorithms or subgradient-based algorithms require an expensive singular value decomposition per iteration. In contrast, HCGS requires only computation of dominant singular vectors, which is more practical by means of the power iteration. Thus, even though HCGS exhibits a slower asymptotic rate of convergence than conditional gradient algorithms, Nesterov's method or the forward-backward algorithm, it scales much better to large matrices than these methods.

3.2 Preliminaries from Convex Analysis

Throughout the chapter, $\mathcal{H}$ is a real Hilbert space endowed with norm $\|\cdot\|$ and inner product $\langle\cdot,\cdot\rangle$. As is standard in convex analysis, we consider extended value functions $f : \mathcal{H} \to (-\infty, +\infty]$, which can take the value $+\infty$. With this notation, constraints can be written as indicator functions that take the zero value inside the feasible set and $+\infty$ outside.

Definition 3.1. *The domain of a function $f : \mathcal{H} \to (-\infty, +\infty]$ is defined as the set* $\operatorname{dom} f = \{x \in \mathcal{H} : f(x) < +\infty\}$.

Definition 3.2. *The function $f : \mathcal{H} \to (-\infty, +\infty]$ is called* proper *if* $\operatorname{dom} f \neq \varnothing$.

Definition 3.3. *The set of proper lower semicontinuous convex functions from $\mathcal{H}$ to $(-\infty, +\infty]$ is denoted by* $\Gamma_0(\mathcal{H})$.

Definition 3.4. *Let* $f : \mathcal{H} \to [-\infty, +\infty]$. *The* convex conjugate *of* f *is the function* $f^* : \mathcal{H} \to [-\infty, +\infty]$ *defined as*

$$f^*(x) = \sup\{\langle u, x \rangle - f(u) : u \in \mathcal{H}\} \tag{3.3}$$

for every $x \in \mathcal{H}$.

Theorem 3.1. *(Fenchel-Moreau) [4, Thm. 13.32]. Every function* $f \in \Gamma_0(\mathcal{H})$ *is biconjugate,*

$$f^{**} = f. \tag{3.4}$$

Moreover, $f^* \in \Gamma_0(\mathcal{H})$.

Definition 3.5. *Let* $f : \mathcal{H} \to (-\infty, +\infty]$ *be proper. A* subgradient *of* f *at* $x \in \mathcal{H}$ *is a vector* $u \in \mathcal{H}$ *satisfying*

$$\langle u, y - x \rangle + f(x) \leq f(y) \tag{3.5}$$

for every $y \in \mathcal{H}$. *The set of subgradients of* f *at* x *is called the* subdifferential *of* f *at* x *and is denoted as* $\partial f(x)$.

Proposition 3.1. *Let* $f : \mathcal{H} \to (-\infty, +\infty]$ *be proper and* $x \in \mathcal{H}$. *Then* $\partial f(x) \neq \varnothing$ *implies* $x \in \operatorname{dom} f$.

Theorem 3.2. *[4, Thm. 16.23]. Let* $f \in \Gamma_0(\mathcal{H}), x \in \mathcal{H}, u \in \mathcal{H}$. *Then*

$$u \in \partial f(x) \iff x \in \partial f^*(u) \iff f(x) + f^*(u) = \langle x, u \rangle, \tag{3.6}$$

$$(\partial f)^{-1} = \partial f^*. \tag{3.7}$$

Theorem 3.3. *[4, Thm. 16.37]. Let* $f \in \Gamma_0(\mathcal{H})$, $\mathcal{K}$ *a Hilbert space,* $g \in \Gamma_0(\mathcal{K})$ *and* $A : \mathcal{H} \to \mathcal{K}$ *a bounded linear operator. If* $\operatorname{dom} g = \mathcal{K}$ *then*

$$\partial(f + g \circ A) = \partial f + A^* \circ \partial g \circ A.$$

Theorem 3.4. *[4, Thm. 16.2]. Let* $f : \mathcal{H} \to (-\infty, +\infty]$ *be proper. Then*

$$\operatorname{argmin} f = \{x \in \mathcal{H} : 0 \in \partial f(x)\}.$$

Definition 3.6. *The function* $f \in \Gamma_0(\mathcal{H})$ *is called* (p, L)-smooth, *where* $L > 0, p \in (0, 1]$, *if* f *is Fréchet differentiable on* $\mathcal{H}$ *with a Hölder continuous gradient,*

$$\|\nabla f(x) - \nabla f(y)\| \leq L \|x - y\|^p \qquad \forall x, y \in \mathcal{H}.$$

The case $p = 1$ corresponds to functions with Lipschitz continuous gradient, which appear frequently in optimization. The following lemma is sometimes called the *descent lemma* [4, Cor. 18.14].

Lemma 3.1. *If the function $f \in \Gamma_0(\mathcal{H})$ is (p, L)-smooth then*

$$f(x) \leq f(y) + \langle x - y, \nabla f(y) \rangle + \frac{L}{p + 1} \|x - y\|^{p+1} \quad \forall x, y \in \mathcal{H}. \quad (3.8)$$

Theorem 3.5. *Let $\omega \in \Gamma_0(\mathcal{H})$. Then $\operatorname{dom} \omega$ is contained in the ball of radius $\rho \in \mathbb{R}_+$, if and only if the convex conjugate ω^* is ρ-Lipschitz continuous on $\mathcal{H}$.*

Proof. See [47], or [48, Cor. 13.3.3] for a finite-dimensional version. □

Corollary 3.1. *If the function $\omega \in \Gamma_0(\mathcal{H})$ has bounded domain then $\partial \omega^*$ is nonempty everywhere on $\mathcal{H}$.*

Proof. Follows from Theorem 3.5 and [4, Prop. 16.17]. □

3.3 Generalized Conditional Gradient Algorithm

In this section, we briefly review the conditional gradient algorithm in one of its many formulations. We focus on convex optimization problems of a general type and discuss how a generalized conditional gradient algorithm applies to such problems. We should note that this algorithm is not the most generic formulation that has been studied; see, for example [34, 13], but it covers a broad variety of optimization problems in machine learning.

Specifically, we consider the optimization problem

$$\min \{ f(x) + \omega(x) : x \in \mathcal{H} \} \quad (3.9)$$

where we make the following assumptions.

Assumption 3.1.

- $f, \omega \in \Gamma_0(\mathcal{H})$

- f is $(1, L)$-smooth

- $\operatorname{dom} \omega$ is bounded, that is, there exists $\rho \in \mathbb{R}_{++}$ such that $\|x\| \leq \rho, \forall x \in \operatorname{dom} \omega$.

Algorithm 1 Generalized conditional gradient algorithm.

Input $x_1 \in \operatorname{dom} \omega$
for $k = 1, 2, \ldots$ do
$\quad y_k \leftarrow$ an element of $\partial \omega^* (-\nabla f(x_k))$ (I)
$\quad x_{k+1} \leftarrow (1 - \alpha_k) x_k + \alpha_k y_k$ (II)
end for

Remark 3.1. *Under Assumption 3.1, the problem* (3.9) *admits a minimizer. The reason is that, since* $\operatorname{dom} \omega$ *is bounded,* $\lim\limits_{\|x\| \to +\infty} \frac{f(x) + \omega(x)}{\|x\|} = +\infty$ *(supercoercivity).*

The Fenchel dual problem associated with (3.9) is

$$\max \{ -f^*(-z) - \omega^*(z) : z \in \mathcal{H} \}. \tag{3.10}$$

Due to Fenchel's duality theorem and the fact that $\operatorname{dom} f = \mathcal{H}$, the duality gap equals zero and the maximum in (3.10) is attained [4, Thm. 15.23].

Algorithm 1 has been used frequently in the past for solving problems of the type (3.9). It is a generalization of algorithms such as the Frank-Wolfe algorithm for quadratic programming [18, 21] and conditional gradient algorithms [32, 14, 15]. Algorithm 1 applies to the general setting of convex optimization problems of the form (3.9) which satisfy Assumption 3.1. In such general forms, the algorithm has been known and studied for a long time in control theory and several of its convergence properties have been obtained [31, 32, 15, 14, 16]. More recently, interest in the family of conditional gradient algorithms has been revived, especially in theoretical computer science, machine learning, computational geometry and elsewhere [24, 23, 29, 3, 20, 54, 56, 27, 10, 22, 33, 19]. Some of these algorithms have appeared independently in various fields, such as statistics and signal processing, under different names and various guises. For example, it has been observed that conditional gradient methods are related to boosting, greedy methods for sparse problems [10, 51], and to orthogonal matching pursuit [28, 27]. Some very recent papers [3, 53] show an equivalence to the optimization method of mirror descent, which we discuss briefly in Section 3.3.2.

One reason for the popularity and the revival of interest in conditional gradient methods has been their applicability to large scale problems. This advantage is evident, for example, in comparison to proximal methods, see [11] and references therein, and especially in optimization problems involving matrices. Conditional gradient methods generally trade off a slower convergence rate (number of iterations) for lower complexity of each iteration step. The accelerated proximal gradient methods [43] benefit from the "optimal" $\mathcal{O}\left(\sqrt{\frac{1}{\varepsilon}}\right)$ rate (where ε is the accuracy with respect to the optimization objective), whereas conditional gradient methods exhibit a slower $\mathcal{O}\left(\frac{1}{\varepsilon}\right)$ rate. On the other side, each step in the proximal methods requires computation of the *proximity operator* [37, 4] (see Section 3.4), which in some cases can be particularly costly. For example, the proximity operator of the *trace norm* of a matrix $X \in \mathbb{R}^{d \times n}$,

$$\|X\|_{tr} = \sum_{i=1}^{\min\{d,n\}} \sigma_i(X) \,, \tag{3.11}$$

where $\sigma_i(X)$ denote the singular values of X, requires computation of a complete singular value decomposition. In contrast, a conditional gradient method need only compute a dominant pair of left-right singular vectors, and such a computation scales better to large matrices [29].

In general, as Algorithm 1 indicates, conditional gradient methods require computation of *dual subgradients*. Often, this is a much less expensive operation than projection or the proximity operation. In other cases, proximity operations may not be feasible in a finite number of steps, whereas dual subgradients are easy to compute. An obvious such case is ℓ_p or Schatten-ℓ_p regularization; see [28]. Other cases of interest occur when ω is the conjugate of a *max-function*. Then the dual subgradient could be fast to compute while the proximity operation may be complex.

Finally, another advantage of conditional gradient methods is that they build their estimate of the solution incrementally. This implies that, in earlier iterations, time and space costs are low and the algorithm may be stopped once an estimate of the desired parsimony is obtained (this could be, for example, a vector of certain sparsity or a matrix of certain rank). Proximal methods, in contrast, do not necessarily obtain the desired parsimony until later iterations (and even then it is not "exact").

Formulation (3.9) covers many optimization problems studied so far in the conditional gradients literature and provides a concise description of variational problems amenable to the standard conditional gradient algorithm. In Section 3.4 we extend the applicability to problems with multiple penalties, by combining conditional gradients and smoothing techniques.

We remark that formulation (3.9) is valid in a generalized Hilbert space setting, so that it can be applied to infinite dimensional problems. This is particularly useful for *kernel methods* in machine learning, for example, kernelized support vector machines or structured SVMs [52] and nuclear or Schatten-ℓ_p regularization of operators [1].

To motivate Algorithm 1, consider the convex optimization problem (3.9). By Theorems 3.3 and 3.4, $\hat{x} \in \mathcal{H}$ is a minimizer of (3.9) if and only if

$$0 \in \nabla f(\hat{x}) + \partial \omega(\hat{x}) \tag{3.12}$$

or, equivalently,

$$-\nabla f(\hat{x}) \in \partial \omega(\hat{x}) \iff \hat{x} \in \partial \omega^*(-\nabla f(\hat{x})) \,, \tag{3.13}$$

where we have used Theorem 3.2. Thus, step (I) in Algorithm 1 reflects the fixed point equation (3.13). However, $\partial \omega^*(-\nabla f(x_k))$ is not a singleton in general and some elements of this set may be far from the minimizers of the problem. Hence step (II), which weighs the new estimate with past ones, is necessary. With any affine weighting like that of step (II), the fixed point equation (3.13) still holds.

Remark 3.2. *Algorithm 1 is well defined, since the subdifferential at step (I) is always nonempty, due to Corollary 3.1 and Assumption 3.1.*

Finally, let us note that several variants of Algorithm 1 are possible, in the spirit of the extensive literature on conditional gradient methods. For example, there are various techniques (like line search) for the choice of coefficients α_k, more of the past iterates may be used in (II) and so on.

3.3.1   Convergence Rate

Theorem 3.6. *If, for every $k \in \mathbb{N}$, $\alpha_k \in [0, 1]$, then $x_k \in \operatorname{dom} w$ and*

$$f(x_{k+1}) + w(x_{k+1}) - f(x) - w(x) \leq (1 - \alpha_k)\big(f(x_k) + w(x_k) - f(x) - w(x)\big) + 2\alpha_k^2 L\rho^2$$

for every $x \in \operatorname{dom} w$, $k \in \mathbb{N}$.

Theorem 3.6 implies an $\mathcal{O}\left(\frac{1}{k}\right)$ convergence rate with respect to the objective values $f(x_k) + w(x_k) - f(\hat{x}) - w(\hat{x})$, where $\hat{x}$ is a minimizer of (3.9). This rate can be attained, for example, with the choice $\alpha_k = \frac{2}{k+1}$.

Corollary 3.2. *If $\alpha_k = \frac{2}{k+1}$, for every $k \in \mathbb{N}$, then*

$$f(x_{k+1}) + w(x_{k+1}) - f(x) - w(x) \leq \frac{8L\rho^2}{k+1} \qquad (3.14)$$

for every $x \in \operatorname{dom} w$, $k \in \mathbb{N}$.

See [14, 15, 29, 10, 3] and references therein for these and related results, as well as for bounds involving the duality gap estimates. It is also known that the lower bound for conditional gradient and similar algorithms is of the same order [9, 28, 34].

3.3.2   Connections to Mirror Descent and Gradient Descent

It has been observed recently [3, 53] that the conditional gradient algorithm is equivalent to a *mirror descent* algorithm in the dual. The basic mirror descent algorithm [5, 38, 30] may be written as the iteration

$$x_{k+1} \leftarrow \text{ an element of } x_k - t_k \partial \varphi(\nabla \psi^*(x_k)), \qquad (3.15)$$

where $t_k > 0$ are step sizes, ψ is strongly convex on a closed convex set C and φ is convex and Lipschitz continuous on C. Setting $w = (\varphi \circ (-I))^*$, where

I denotes the identity operator, and $f = \psi^*$, algorithm (3.15) rewrites as a variant of Algorithm 1 (in which the update is not a convex combination). The set C can be viewed as the domain of ω.

Consequently, when (3.9) is a proximity computation (that is, when $f = \frac{1}{2\beta}\| \cdot \|^2$, $\beta > 0$) the conditional gradient algorithm 1 is equivalent to a *subgradient descent* in the dual. In such cases $\nabla f = \frac{1}{\beta}I$ and Algorithm 1 becomes

$$x_{k+1} \in (1 - \alpha_k)x_k + \alpha_k \partial \omega^* \left(-\tfrac{1}{\beta}x_k\right) . \tag{3.16}$$

Letting $h = \left(\omega^* \circ \left(-\tfrac{1}{\beta}I\right)\right)^*$, by the chain rule (Theorem 3.3) this iteration is equivalent to

$$x_{k+1} \in x_k - \alpha_k(I + \beta \, \partial h^*)(x_k) . \tag{3.17}$$

In particular, when ω is μ-strongly convex (and hence ω^* is $(1, \frac{1}{\mu})$-smooth) and $\alpha_k \leq \frac{\mu\beta}{1+\mu\beta}$ for every $k \in \mathbb{N}$, the above iteration is equivalent to a *proximal point* algorithm [11, 17, 49] because of [4, Thm. 18.15]. Note that not all cases of subgradient descent are covered, since ω should have bounded domain, implying that the dual objective function should be a quadratic perturbation of a Lipschitz continuous function.

3.4 Hybrid Conditional Gradient-Smoothing Algorithm

We now introduce Algorithm 2, an extension of conditional gradient methods to optimization problems on bounded domains that contain smooth and Lipschitz continuous terms.

3.4.1 Description of the Hybrid Algorithm

Formally, we consider the class of optimization problems of the form

$$\min \{f(x) + g(Ax) + \omega(x) : x \in \mathcal{H}\} \tag{3.18}$$

where we make the following assumptions:

Assumption 3.2.

- $f, \omega \in \Gamma_0(\mathcal{H})$

- $g \in \Gamma_0(\mathcal{K})$, $\mathcal{K}$ *is a Hilbert space*

- $A : \mathcal{H} \to \mathcal{K}$ *is a bounded linear operator*

- f *is (p, L_f)-smooth*

- *g is L_g-Lipschitz continuous on $\mathcal{K}$*

- *$\mathrm{dom}\,\omega$ is bounded, that is, there exists $\rho \in \mathbb{R}_{++}$ such that $\|x\| \leq \rho, \forall x \in$ $\mathrm{dom}\,\omega$*

Remark 3.3. *Under Assumption 3.2, problem (3.18) admits a minimizer. As in Remark 3.1, the reason is growth of the objective function at infinity (the objective equals $+\infty$ outside the feasible set, which is bounded).*

In order for the algorithm to be practical, we require that

Assumption 3.3.

- *the gradient of f is simple to compute at every $x \in \mathcal{H}$,*

- *a subgradient of ω^* is simple to compute at every $x \in \mathcal{H}$,*

- *the proximity operator of βg is simple to compute for every $\beta > 0, x \in \mathcal{H}$.*

The *proximity operator* was introduced by Moreau [37] as the (unique) minimizer

$$\mathrm{prox}_g(x) = \mathrm{argmin}\left\{\frac{1}{2}\|x - u\|^2 + g(u) : u \in \mathcal{H}\right\}. \qquad (3.19)$$

For a review of the numerous applications of proximity operators to optimization, see, for example, [12, 11] and references therein.

The following are some examples of optimization problems that belong to the general class (3.18).

Example 3.1. *Regularization with two norm penalties:*

$$\min\left\{f(x) + \lambda \|x\|_a : \|x\|_b \leq B, x \in \mathbb{R}^d\right\} \qquad (3.20)$$

where f is (p, L_f)-smooth, $\lambda > 0$ and $\|\cdot\|_a, \|\cdot\|_b$ can be any norms on $\mathbb{R}^d$.

Example 3.2. *Regularization with a linear composite penalty and a norm:*

$$\min\left\{f(x) + \lambda \|Ax\|_a : \|x\|_b \leq B, x \in \mathbb{R}^d\right\} \qquad (3.21)$$

where f is (p, L_f)-smooth, $\lambda > 0$, $\|\cdot\|_a, \|\cdot\|_b$ are norms on $\mathbb{R}^\delta$, $\mathbb{R}^d$, respectively, and $A \in \mathbb{R}^{\delta \times d}$.

Example 3.3. *Regularization with multiple linear composite penalties and a norm:*

$$\min\left\{f(x) + \sum_{i=1}^n \lambda_i \|A_i x\|_{a_i} : \|x\|_b \leq B, x \in \mathbb{R}^d\right\} \qquad (3.22)$$

where f is (p, L_f)-smooth and, for all $i \in \{1, \ldots, n\}$, $\lambda_i > 0$, $\|\cdot\|_{a_i}, \|\cdot\|_b$ are norms on $\mathbb{R}^{\delta_i}$, $\mathbb{R}^d$, respectively, and $A_i \in \mathbb{R}^{\delta_i \times d}$. Such problems can be seen as special cases of Example 3.2 by applying the classical direct sum technique,

$$\delta = \textstyle\sum_{i=1}^n \delta_i, \ A = \begin{pmatrix} A_1 \\ \vdots \\ A_n \end{pmatrix}, \ \|(v_i)_{i=1}^n\|_a = \textstyle\sum_{i=1}^n \lambda_i \|v_i\|_{a_i}.$$

Algorithm 2 Hybrid conditional gradient-smoothing algorithm.

Input $x_1 \in \text{dom}\,\omega$
for $k = 1, 2, \ldots$ **do**

$$z_k \leftarrow -\nabla f(x_k) - \frac{1}{\beta_k} A^* A x_k + \frac{1}{\beta_k} A^* \text{prox}_{\beta_k g}(A x_k) \qquad \text{(I)}$$

$$y_k \leftarrow \text{an element of } \partial \omega^*(z_k) \qquad \text{(II)}$$

$$x_{k+1} \leftarrow (1 - \alpha_k) x_k + \alpha_k y_k \qquad \text{(III)}$$

end for

We propose to solve problems like the above with Algorithm 2.[2] We call it a *hybrid conditional gradient-smoothing* algorithm (HCGS in short), because it involves a smoothing of function g with parameter β_k. For any $\beta > 0$, the β-smoothing of g is the Moreau envelope of g, that is, the function g_β defined as

$$g_\beta(x) := \min \left\{ \frac{1}{2\beta} \|x - u\|^2 + g(u) : u \in \mathcal{H} \right\} \qquad \forall x \in \mathcal{H} \,. \qquad (3.23)$$

The function g_β is a smooth approximation to g (in fact, the best possible approximation of $\frac{1}{\beta}$ smoothness), as summarized in the following lemmas from the literature.

Lemma 3.2 (Proposition 12.29 in [4]). *Let* $g \in \Gamma_0(\mathcal{H}), \beta > 0$. *Then* g_β *is* $(1, \frac{1}{\beta})$-*smooth and its gradient can be obtained from the proximity operator of* g *as:* $\nabla g_\beta(x) = \frac{1}{\beta} \left(x - \text{prox}_{\beta g}(x) \right)$.

Lemma 3.3. *Let* $g \in \Gamma_0(\mathcal{H})$ *be* L_g-*Lipschitz continuous and* $\beta > 0$. *Then*

- $g_\beta \leq g \leq g_\beta + \frac{1}{2}\beta L_g^2$

- *if* $\beta \geq \beta' > 0$, *then* $g_\beta \leq g_{\beta'} \leq g_\beta + \frac{1}{2}(\beta - \beta')L_g^2$.

Proof. Define $\Psi_x(u) = \frac{1}{2\beta} \|x - u\|^2 + g(u)$. We have $g_\beta(x) = \min_u \Psi_x(u) \leq \Psi_x(x) = g(x)$, and this proves the left hand side of the first property. For the other side of the inequality, we have

$$\Psi_x(u) = \frac{1}{2\beta} \|x - u\|^2 + g(u) - g(x) + g(x) \geq \frac{1}{2\beta} \|x - u\|^2 + g(x) - L_g \|x - u\|$$

where we have used the Lipschitz property of g. This implies that

$$g_\beta(x) = \min_u \Psi_x(u) \geq g(x) + \inf_u \left(\frac{1}{2\beta} \|x - u\|^2 - L_g \|x - u\| \right) = g(x) - \frac{1}{2}\beta L_g^2 \,.$$

The second property follows from the first one and $(g_{\beta'})_{\beta - \beta'} = g_\beta$ (Proposition 12.22 in [4]). $\qquad \square$

---

[2] Algorithm 2 is well-defined (see Remark 3.2).

Thus, the smoothing parameter β controls the tradeoff between the smoothness and the quality of approximation.

At each iteration, the hybrid Algorithm 2 computes the gradient of the *smoothed* part $f + g_{\beta_k} \circ A$, where β_k is the adaptive smoothing parameter. By Lemma 3.2 and the chain rule, its gradient equals

$$\nabla(f + g_{\beta_k} \circ A)(x) = \nabla f(x) + \frac{1}{\beta_k} A^* \left(Ax - \text{prox}_{\beta_k g}(Ax)\right) .$$

The function f is (p, L_f)-smooth and the function $g_{\beta_k} \circ A$ is $(1, \frac{1}{\beta_k}\|A\|^2)$-smooth. By selecting β_k that approaches 0 as k increases, we ensure that g_{β_k} approaches g.

Algorithm 2 can be viewed as an extension of conditional gradient algorithms and of Algorithm 1. But besides conditional gradient methods, the algorithm also exploits ideas from proximal algorithms obtained by smoothing Lipschitz terms in the objective. These methods are primarily due to Nesterov and have been successfully applied to many problems [42, 41, 44, 25, 7]. The smoothing we apply here is a type of Moreau envelope as in the variational problem (3.23), which is connected to Nesterov smoothing; see, for example, [44, 7].

However, unlike Nesterov's smoothing and other proximal methods, in our method we choose not to smooth function ω, or apply any other proximity-like operation to it. We do this because computation of the proximity operator of ω is not available in the settings that we consider here.[3] For example, if ω expresses a trace norm constraint, the proximity computation requires a *full* singular value decomposition, which does not scale well with the size of the matrix. In contrast, the dual subgradient requires only computation of a single pair of dominant singular vectors and this is feasible even for very large matrices using the power method or Lanczos algorithms.

3.4.2   Convergence Rate

A bound on the convergence rate of the objective function can be obtained by first bounding the convergence of the smoothed objective in a recursive way. The required number of iterations is a function of p and ε, where p is the smoothing exponent of f and ε is the accuracy in terms of the objective function. Regarding the proof technique, we should note that the HCGS Algorithm 2 and the proof of its convergence properties are mostly related to conditional gradient methods. On the other side, the proof technique does not share similarities with proximal methods such as ISTA or FISTA [6, 39, 40].

Theorem 3.7. *Suppose that, for every $k \in \mathbb{N}$, $\alpha_k \in [0, 1]$ and $\beta_k \geq \beta_{k+1} > 0$.*

---

[3] Note that Nesterov's smoothing would require ω to be Lipschitz continuous, and hence it does not apply directly to the case of bounded dom ω but to a similar regularization problem with ω as a penalty in the objective.

Let $F = f + g \circ A$, $F_k = f + g_{\beta_k} \circ A$. Then $x_k \in \operatorname{dom} \omega$, for every $k \in \mathbb{N}$, and

$$F_{k+1}(x_{k+1}) + \omega(x_{k+1}) - F(x) - \omega(x) \le (1 - \alpha_k)\big(F_k(x_k) + \omega(x_k) - F(x) - \omega(x)\big)$$
$$+ \frac{(2\rho)^{p+1} L_f}{p+1} \alpha_k^{p+1} + 2\|A\|^2 \rho^2 \frac{\alpha_k^2}{\beta_k} + \frac{1}{2}(\beta_k - \beta_{k+1}) L_g^2 \quad (3.24)$$

for every $x \in \operatorname{dom} \omega$, $k \in \mathbb{N}$.

Proof. For every $k \in \mathbb{N}$, we apply the descent Lemma 3.1 twice to obtain that

$$f(x_{k+1}) \le f(x_k) + \langle \nabla f(x_k), x_{k+1} - x_k \rangle + \frac{L_f}{p+1} \|x_{k+1} - x_k\|^{p+1}$$
$$= f(x_k) + \alpha_k \langle \nabla f(x_k), y_k - x_k \rangle + \frac{\alpha_k^{p+1} L_f}{p+1} \|y_k - x_k\|^{p+1} \quad (3.25)$$

and

$$g_{\beta_k}(Ax_{k+1}) \le g_{\beta_k}(Ax_k) + \langle \nabla(g_{\beta_k} \circ A)(x_k), x_{k+1} - x_k \rangle + \frac{1}{2\beta_k}\|A\|^2 \|x_{k+1} - x_k\|^2$$
$$= g_{\beta_k}(Ax_k) + \alpha_k \langle \nabla(g_{\beta_k} \circ A)(x_k), y_k - x_k \rangle + \frac{\alpha_k^2}{2\beta_k}\|A\|^2 \|y_k - x_k\|^2 .$$
$$(3.26)$$

Applying Lemma 3.3 to $g_{\beta_k}(Ax_{k+1})$ yields

$$g_{\beta_{k+1}}(Ax_{k+1}) \le g_{\beta_k}(Ax_k) + \alpha_k \langle \nabla(g_{\beta_k} \circ A)(x_k), y_k - x_k \rangle + \frac{\alpha_k^2}{2\beta_k}\|A\|^2 \|y_k - x_k\|^2$$
$$+ \frac{1}{2}(\beta_k - \beta_{k+1}) L_g^2 .$$

Adding (3.25) and (3.26) we obtain that

$$F_{k+1}(x_{k+1}) \le F_k(x_k) + \alpha_k \langle \nabla F_k(x_k), y_k - x_k \rangle + \frac{\alpha_k^{p+1} L_f}{p+1} \|y_k - x_k\|^{p+1}$$
$$+ \frac{\alpha_k^2}{2\beta_k}\|A\|^2 \|y_k - x_k\|^2 + \frac{1}{2}(\beta_k - \beta_{k+1}) L_g^2 . \quad (3.27)$$

By Theorem 3.2 and Proposition 3.1, $y_k \in \operatorname{dom} \omega$ for every $k \in \mathbb{N}$. Since $\alpha_k \in [0, 1]$, applying an induction argument yields that $x_k \in \operatorname{dom} \omega$, for every $k \in \mathbb{N}$. Thus, the values of the objective generated by the algorithm are finite. From the construction of y_k in steps (I), (II) and Theorem 3.2, we obtain that, for every $x \in \operatorname{dom} \omega$,

$$\langle y_k, -\nabla F_k(x_k) \rangle - \omega(y_k) \ge \langle x, -\nabla F_k(x_k) \rangle - \omega(x)$$

and hence that

$$\langle y_k - x_k, -\nabla F_k(x_k) \rangle - \omega(y_k) \ge \langle x - x_k, -\nabla F_k(x_k) \rangle - \omega(x)$$
$$\ge F_k(x_k) - F_k(x) - \omega(x) .$$

Applying Lemma 3.3 to $g_{\beta_k}(Ax)$ yields

$$\langle y_k - x_k, -\nabla F_k(x_k)\rangle - \omega(y_k) \geq F_k(x_k) - F(x) - \omega(x)$$

and, therefore,

$$\alpha_k \langle y_k - x_k, -\nabla F_k(x_k)\rangle - \alpha_k \omega(y_k) \geq \alpha_k F_k(x_k) - \alpha_k(F(x) + \omega(x)) . \quad (3.28)$$

Adding (3.27) and (3.28), we obtain that

$$F_{k+1}(x_{k+1}) + \alpha_k F_k(x_k) - \alpha_k(F(x) + \omega(x)) \leq F_k(x_k) - \alpha_k \omega(y_k)$$
$$+ \frac{\alpha_k^{p+1} L_f}{p+1}\|y_k - x_k\|^{p+1} + \frac{\alpha_k^2}{2\beta_k}\|A\|^2\|y_k - x_k\|^2 + \frac{1}{2}(\beta_k - \beta_{k+1})L_g^2$$

or that

$$F_{k+1}(x_{k+1}) + \omega(x_{k+1}) - F(x) - \omega(x) \leq (1-\alpha_k)\big(F_k(x_k) - F(x) - \omega(x)\big) + \omega(x_{k+1})$$
$$- \alpha_k \omega(y_k) + \frac{\alpha_k^{p+1} L_f}{p+1}\|y_k - x_k\|^{p+1} + \frac{\alpha_k^2}{2\beta_k}\|A\|^2\|y_k - x_k\|^2 + \frac{1}{2}(\beta_k - \beta_{k+1})L_g^2$$
$$\leq (1 - \alpha_k)\big(F_k(x_k) + \omega(x_k) - F(x) - \omega(x)\big) + \frac{\alpha_k^{p+1} L_f}{p+1}\|y_k - x_k\|^{p+1}$$
$$+ \frac{\alpha_k^2}{2\beta_k}\|A\|^2\|y_k - x_k\|^2 + \frac{1}{2}(\beta_k - \beta_{k+1})L_g^2 ,$$

where the last step uses the convexity of ω and (III). Since $x_k, y_k \in \operatorname{dom}\omega$, for every $k \in \mathbb{N}$, it follows that $\|y_k\|, \|x_k\| \leq \rho$ and hence that (3.24) holds. $\square$

Corollary 3.3. *Suppose that, $\alpha_1 = 1$, $\alpha_k \in [0,1]$ and $\beta_k \geq \beta_{k+1} > 0$, for every $k \in \mathbb{N}$. Let $P_j = \prod_{i=j+1}^{k}(1 - \alpha_i)$, for every $j \in \{1,\ldots,k\}$. Then*

$$f(x_{k+1}) + g(Ax_{k+1}) + \omega(x_{k+1}) - f(x) - g(Ax) - \omega(x) \leq \frac{(2\rho)^{p+1} L_f}{p+1}\sum_{j=1}^{k} P_j \alpha_j^{p+1}$$
$$+ 2\|A\|^2\rho^2 \sum_{j=1}^{k} P_j \frac{\alpha_j^2}{\beta_j} + \frac{1}{2}L_g^2 \sum_{j=1}^{k} P_j(\beta_j - \beta_{j+1}) + \frac{1}{2}\beta_{k+1}L_g^2$$

for every $x \in \operatorname{dom}\omega$, $k \in \mathbb{N}$.

Proof. Let $D_k = f(x_k) + g_{\beta_k}(Ax_k) + \omega(x_k) - f(x) - g(Ax) - \omega(x)$. Applying Theorem 3.7, we obtain

$$D_{j+1} \leq (1 - \alpha_j)D_j + \frac{(2\rho)^{p+1} L_f}{p+1}\alpha_j^{p+1} + 2\|A\|^2\rho^2\frac{\alpha_j^2}{\beta_j} + \frac{1}{2}(\beta_j - \beta_{j+1})L_g^2 \quad (3.29)$$

for every $j \in \{1, \ldots, k\}$. Multiplying by P_j and adding up, we obtain

$$D_{k+1} \leq (1 - \alpha_1) P_1 D_1 + \frac{(2\rho)^{p+1} L_f}{p+1} \sum_{j=1}^{k} P_j \alpha_j^{p+1} + 2\|A\|^2 \rho^2 \sum_{j=1}^{k} P_j \frac{\alpha_j^2}{\beta_j}$$

$$+ \frac{1}{2} L_g^2 \sum_{j=1}^{k} P_j (\beta_j - \beta_{j+1})$$

$$= \frac{(2\rho)^{p+1} L_f}{p+1} \sum_{j=1}^{k} P_j \alpha_j^{p+1} + 2\|A\|^2 \rho^2 \sum_{j=1}^{k} P_j \frac{\alpha_j^2}{\beta_j} + \frac{1}{2} L_g^2 \sum_{j=1}^{k} P_j (\beta_j - \beta_{j+1})$$

Applying Lemma 3.3 to $g_{\beta_{k+1}}(Ax_{k+1})$ the assertion follows. $\square$

Corollary 3.4. *If* $\alpha_k = \frac{2}{k+1}$, $\beta > 0$ *and* $\beta_k = \frac{\beta}{\sqrt{k}}$, *for every* $k \in \mathbb{N}$, *then*

$$f(x_{k+1}) + g(Ax_{k+1}) + \omega(x_{k+1}) - f(x) - g(Ax) - \omega(x) \leq$$

$$\frac{(4\rho)^{p+1} L_f}{(p+1)(k+1)^p} + \frac{8\rho^2 \|A\|^2}{\beta\sqrt{k+1}} + \frac{1}{2} L_g^2 \beta \frac{\sqrt{k+2}}{k} + \frac{L_g^2 \beta}{2\sqrt{k+1}}$$

for every $x \in \operatorname{dom} \omega$, $k \in \mathbb{N}$.

Proof. It follows easily from Corollary 3.3 and the computation $P_j = \frac{j(j+1)}{k(k+1)}$. $\square$

We notice that when $p \geq \frac{1}{2}$ the asymptotic rate does not depend on p and translates to $\mathcal{O}\left(\frac{1}{\varepsilon^2}\right)$ iterations, if ε is the precision in terms of the objective function. This rate of convergence is an order of magnitude slower than the rate for the standard conditional gradient algorithm (Corollary 3.2). Thus, the extended flexibility of handling multiple additional penalties (function g) and the Moreau smoothing incur a cost in terms of iterations. In other words, the class of optimization problems to which the hybrid algorithm applies is significantly larger than that of the standard algorithm 1 and a deterioration in the rate of convergence is inevitable. When $0 < p < \frac{1}{2}$, the bound is dominated by the term involving p and the number of iterations required grows as $\mathcal{O}\left(\varepsilon^{-\frac{1}{p}}\right)$.

If there is no $g \circ A$ term ($A = 0, g = 0$) then the algorithm becomes the standard conditional gradient and the corollary reduces to known bounds for standard conditional gradient methods. The number of iterations grows as $\mathcal{O}\left(\varepsilon^{-\frac{1}{p}}\right)$, which ranges from $\mathcal{O}\left(\frac{1}{\varepsilon}\right)$ (for $p = 1$) to impractical when f is too "close" to a Lipschitz continuous function ($p \simeq 0$).

The rate in Corollary 3.4 is also slower than the $\mathcal{O}\left(\frac{1}{\varepsilon}\right)$ rates obtained with smoothing methods, such as [44, 42, 25, 7]. However, smoothing methods require a more powerful computational oracle (the proximity operator of ω instead of the dual subgradient) and hence may be inapplicable in problems like those involving very large matrices, because computation of prox_ω may not scale well. Another $\mathcal{O}\left(\frac{1}{\varepsilon^2}\right)$ alternative is subgradient methods, but these may be inapplicable too for similar reasons. For example, the subgradient of the trace norm as either a penalty term or a constraint requires a full singular value decomposition.

In addition, like other conditional gradient methods or greedy methods and matching pursuits, the HCGS algorithm 2 builds a parsimonious solution in additive fashion rather than starting from a complex solution and then simplifying it. This feature may be desirable in itself whenever a parsimonious solution is sought. For example, in many cases it is more important to obtain a sparse or low rank estimate of the solution rather than a more accurate one with many small nonzero components or singular values. In machine learning problems, especially, this is frequently the case since the optimization objective is just an approximation of the ideal measure of expected risk [51]. Another advantage of such algorithmic schemes is computational. In sparse estimation problems regularized with an ℓ_1 constraint, the data matrix or the dictionary may be huge and hence computation of $\nabla f(x)$ may be feasible only for sparse vectors x (when f is a quadratic function). Moreover, such a computation can be done efficiently since the gradient from the previous iteration can be reused, due to update (III).

3.4.3   Minimization of Lipschitz Continuous Functions

A special case of particular interest occurs when $f = 0$, that is, when there is no smooth part. Then HCGS solves the optimization problem

$$\min \left\{ g(Ax) + \omega(x) : x \in \mathcal{H} \right\} \tag{3.30}$$

under Assumption 3.2 as before. Namely, the objective function consists of a Lipschitz term g and a generic term ω defined on a bounded domain. For example, such a problem is the minimization of a Lipschitz continuous function over a bounded domain. More generally, $g \circ A$ may incorporate a sum of multiple Lipschitz continuous penalties.

The HCGS algorithm specified to problem (3.30) is the same as Algorithm 2 with $\nabla f(x_k)$ removed. In this way, the computational model of conditional gradient methods extends from minimization of smooth functions to minimization of Lipschitz continuous functions. Moreover, the convergence rate deteriorates from $\mathcal{O}\left(\frac{1}{\varepsilon}\right)$ for smooth functions to $\mathcal{O}\left(\frac{1}{\varepsilon^2}\right)$ for Lipschitz functions, which is not surprising, since the latter are in general more difficult to optimize than the former. This fact has been shown recently by Lan for several conditional gradient algorithms [34]. Lan has also shown that these rates coincide with the lower complexity bounds for a family of algorithms involving $\partial\omega^*$ oracles. The above fact is also intriguing in view of the analogy to the results known about Nesterov's proximal methods [40]. Those methods, under a more powerful computational oracle for ω, exhibit an $\mathcal{O}\left(\frac{1}{\sqrt{\varepsilon}}\right)$ rate when g is smooth versus an $\mathcal{O}\left(\frac{1}{\varepsilon}\right)$ rate when g is Lipschitz continuous.

3.4.4 Implementation Details

It is worth noting that the HCGS algorithm does not require knowledge of the Lipschitz constants L_f, L_g and can be implemented with an arbitrary choice of β. An alternative is to optimize the bound in Corollary 3.4 with respect to β, which gives an optimal choice of $\frac{2\sqrt{2}\rho\|A\|}{L_g}$, asymptotically. If the desired accuracy ε can be specified in advance, then the optimal β will also depend on ε. Computing such a β value is possible only if the Lipschitz constant and bound of the optimization problem are available, but for regularization problems these constants can be computed from the regularization parameters.

For $p \geq \frac{1}{2}$, these two constants, ρ and L_g, have the largest influence in the convergence rate, since they appear in the $\mathcal{O}\left(\frac{1}{\sqrt{k}}\right)$ terms that dominate the bound. The constant ρ cannot be changed, since it is a property of the feasibility domain. However, L_g can be reduced by rescaling the objective function and hence it can become independent of the dimensionality of the problem.

Some care may be needed to tackle numerical issues arising from very small values of β_k as k becomes large. These issues affect only step (I), whereas the computation of y_k in step (II) always remains inside the ρ-ball, since ω^* is ρ-Lipschitz continuous. Moreover, for large k, the past estimates dominate the update (III) and hence the effect of any numerical issues diminishes as k grows.

3.5 Applications

We now instantiate the HCGS Algorithm 2 to some special cases that appear in applications and we present the corresponding algorithms. These examples are only a sample and do not cover the whole range of possible applications. First, consider the problem of learning a *sparse and low rank matrix* by regularization with the ℓ_1 norm and a trace norm constraint [46],

$$\min\{f(X) + \lambda\|X\|_1 : \|X\|_{tr} \leq B, X \in \mathbb{R}^{n \times n}\}, \tag{3.31}$$

where $\|\cdot\|_1$ denotes the elementwise ℓ_1 norm of a matrix and $\|\cdot\|_{tr}$ the trace norm (or nuclear norm). The strongly smooth function f expresses an error term (where the dependence on the data is absorbed in f) and may arise by using, for example, the square loss or the logistic loss. This setting

---

Algorithm 3 Hybrid algorithm for sparse-low rank problems.

Input $X_1 \in \mathbb{R}^{n \times n}$ such that $\|X_1\|_{tr} \leq B$
for $k = 1, 2, \ldots$ **do**
    $Z_k \leftarrow -\nabla f(X_k) - \frac{1}{\beta_k} X_k + \frac{1}{\beta_k} \mathcal{S}(X_k; \beta_k \lambda)$
    $(u_k, v_k) \leftarrow$ a left and right pair of singular vectors of Z_k corresponding to the largest singular value
    $Y_k \leftarrow B\, u_k v_k^\top$
    $X_{k+1} \leftarrow (1 - \alpha_k)X_k + \alpha_k Y_k$
end for

---

has been proposed for applications such as graph denoising or prediction of links on a social network. The resulting algorithm (Algorithm 3) depends on the proximity operator of the ℓ_1 norm, also known as the soft thresholding operator,

$$\mathcal{S}(X; \gamma) = \text{sgn}(X) \odot (|X| - \gamma)_+ , \qquad (3.32)$$

where $\text{sgn}, \odot, |\cdot|$ denote elementwise sign, multiplication and absolute value on matrices and $(\cdot)_+$ the positive part elementwise.

Note that the same algorithm can be used for solving a variation of (3.31) that restricts the optimization to the space of symmetric matrices. This may occur, for example, when learning the adjacency matrix of an undirected graph. One should ensure, however, that the initial matrix X_1 is symmetric.

A problem that shares some similarities with the previous one is the convex relaxation of *sparse PCA* proposed in [2],

$$\max\{\langle C, X\rangle - \lambda\|X\|_1 : \text{tr}(X) = 1, X \succeq 0, X \in \mathbb{R}^{n \times n}\} . \qquad (3.33)$$

Solving this optimization problem can be used for finding a dominant sparse eigenvector of C, which is a prescribed $n \times n$ symmetric matrix. The problem falls under the framework (3.18) with f being a linear function and ω the indicator function of the (bounded) spectrahedron $\{X \in \mathbb{R}^{n \times n} : \text{tr}(X) = 1, X \succeq 0\}$. Computation of a dual subgradient amounts to computing a solution of the problem

$$\max\{\langle Y, Z\rangle : \text{tr}(Y) = 1, Y \succeq 0, Y \in \mathbb{R}^{n \times n}\} \qquad (3.34)$$

for a given symmetric matrix $Z \in \mathbb{R}^{n \times n}$. It is easy to see that this computation requires a dominant eigenvector of Z. This results in Algorithm 4.

A related problem is to restrict the sparse-low rank optimization (3.31) to the cone of positive semidefinite matrices. This problem has been proposed for estimating a covariance matrix in [46]. Since the trace norm is equal to the trace on the positive semidefinite cone, the algorithm is similar to Algorithm 4. The only differences are the initialization, a general smooth function f, and a factor of B in the update of Y_k.

Algorithm 4 Hybrid algorithm for sparse PCA relaxation.

Input $X_1 \in \mathbb{R}^{n \times n}$ such that $\text{tr}(X_1) = 1, X_1 \succeq 0$
  for $k = 1, 2, \ldots$ **do**
    $Z_k \leftarrow C - \frac{1}{\beta_k} X_k + \frac{1}{\beta_k} \mathcal{S}(X_k; \beta_k \lambda)$
    $u_k \leftarrow$ a dominant eigenvector of Z_k
    $Y_k \leftarrow u_k u_k^\top$
    $X_{k+1} \leftarrow (1 - \alpha_k) X_k + \alpha_k Y_k$
  end for

A third example of an optimization problem that falls under our framework is a regularization problem with ℓ_1 and additional penalties,

$$\min \left\{ \frac{1}{2} \langle x, Qx \rangle + \langle c, x \rangle + g(Ax) : \|x\|_1 \leq B, x \in \mathbb{R}^d \right\}. \tag{3.35}$$

Here $Q \in \mathbb{R}^{d \times d}$ is a prescribed positive semidefinite matrix, $c \in \mathbb{R}^d$ a prescribed vector, and g, A satisfy Assumption 3.2. For example, (3.35) could arise from an estimation or learning problem, the quadratic part corresponding to the data fit term. The ℓ_1 constraint is used to favor sparse solutions. The penalty terms $g \circ A$ may involve multiple norms whose proximity operator is simple to compute, such as the group Lasso norm [57], total variation norms [50], etc.

The hybrid method, specialized to such problems, is shown in Algorithm 5. In general, several other algorithms may be used for solving problems like (3.35) (smoothing, Douglas-Rachford, subgradient methods, etc.), but here we are interested in cases with *very large dimensionality d*. In such cases, computation of the gradient at an arbitrary vector is $\mathcal{O}(d^2)$ and very costly. On the other side, in the HCGS algorithm, x_{k+1} is $(k + 1)$-sparse and computing the new gradient Qx_{k+1} can be done efficiently by keeping Qx_k in memory and computing Qy_k, which is proportional to the j-th column of Q. The latter requires only $\mathcal{O}(d)$ operations, or $\mathcal{O}(dm)$ if Q is the square of an $m \times d$ data matrix. Thus, HCGS can be applied to such problems at a smaller cost, by starting with an initial cardinality-one vector and stopping before k becomes too large.

There is also an interesting interpretation of Algorithm 5 as an extension of *matching pursuits* [36, 55] to problems with multiple penalties. Assuming that Q is the square of a matrix of dictionary elements (or more generally a Gram matrix of elements from a Hilbert space), then the algorithm shares similarities with orthogonal matching pursuit (OMP). Indeed, such a connection has already been observed for the standard conjugate gradient (which corresponds to absence of the $g \circ A$ term) [28, 27], the main difference from OMP being in the update of x_{k+1}. Similarly, the HCGS algorithm 5 could be phrased as an extension of OMP that imposes additional penalties $g \circ A$, besides sparsity, on the coefficients of the atoms. For example, $g \circ A$ could involve *structured*

Algorithm 5 Hybrid algorithm for sparse multicomposite problems (3.35).

Input $x_1 = Be_i$ for some $i \in \{1, \dots, d\}$
for $k = 1, 2, \dots$ **do**
$\quad z_k \leftarrow -Qx_k - c - \frac{1}{\beta_k} A^* A x_k + \frac{1}{\beta_k} A^* \operatorname{prox}_{\beta_k g}(Ax_k)$
$\quad y_k \leftarrow B \operatorname{sgn}((z_k)_j)e_j,$ where $j \in \operatorname{argmax}_{i=1}^d |(z_k)_j|$
$\quad x_{k+1} \leftarrow (1 - \alpha_k)x_k + \alpha_k y_k$
end for

---

sparsity penalties (such as penalties for group, hierarchical or graph sparsity) and then HCGS would yield a scalable alternative to structured variants of OMP [26] or proximal methods for structured sparsity [35].

3.6   Simulations

3.6.1   Simultaneous Sparse and Low Rank Regularization

In this section we focus on testing Algorithm 3 (HCGS) on the estimation of simultaneously sparse and low rank matrices[4]. Our aim is to compare the procedure with the proximal algorithms proposed, for the same task, in [46]. The experiments illustrate the fact that HCGS scales better than the SVD-based alternatives.

We considered the task of recovering a matrix from a subset of its entries. To this end in each simulation we generated two $N \times 5$ random matrices with entries drawn from the uniform distribution. 90% of the entries corresponding to a subset of uniformly distributed indices were then set to zero. The resulting matrices, denoted by U and V, were then used to obtain a sparse and low rank matrix $UV^\top$. This matrix was corrupted by zero-mean Gaussian noise with variance $\sigma^2 = 10^{-4}$ to obtain the observation matrix Y. A fraction $f \in \{0.05, 0.4\}$ of entries of Y were used for recovery; see Figure 3.1 for an illustration.

We compared HCGS with the two algorithms proposed in [46], namely *generalized forward-backward* (GFB) [45] and *incremental proximal descent* (IPD) [8]. Both these algorithms solve a convex matrix recovery problem that aims at finding a matrix that is simultaneously low rank and sparse. This problem is:

$$\min \left\{ J(X) := \frac{1}{2p} \|\Omega(Y - X)\|_F^2 + \lambda_1 \|X\|_1 + \lambda_2 \|X\|_{tr} \ : \ X \in \mathbb{R}^{N \times N} \right\}$$
$$(3.36)$$

---

[4]Code is available at http://cvn.ecp.fr/personnel/andreas/code/index.html

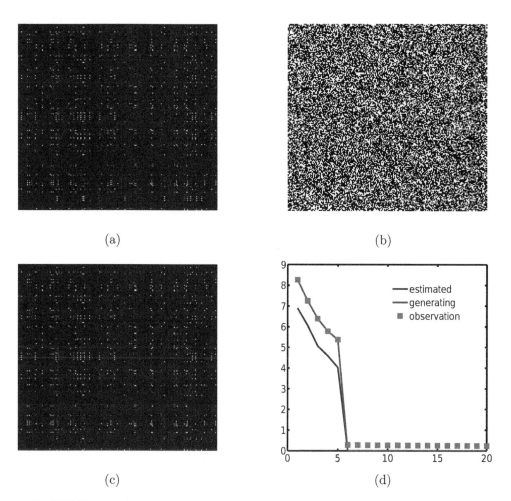

(a)　　　　　　　　　　　　　(b)

(c)　　　　　　　　　　　　　(d)

FIGURE 3.1: An illustration of the synthetic problem for $N = 200$. The generating matrix, simultaneously sparse and low-rank (a), a mask with the observed entries, in white; 40% of the total number of entries are observed (b), the matrix estimated by HCGS (c), the leading singular values for the generating/observation/estimated matrix (d).

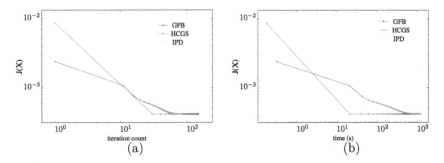

FIGURE 3.2: Comparison of objective values for $N = 400$, (a) as a function of the iteration count, and (b) as a function of time.

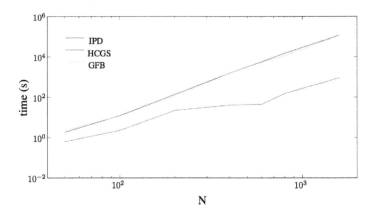

FIGURE 3.3: Required time for convergence as a function of N (40% observed entries), for the sparse-low rank experiment.

where $\|\cdot\|_F$ is the Frobenius norm, p is the number of observed entries, and $\Omega : \mathbb{R}^{N \times N} \rightarrow \mathbb{R}^{N \times N}$ is the sampling operator defined entry-wise by $\Omega(X)_{ij} = X_{ij}$ if the entry indexed by (i, j) is observed, $\Omega(X)_{ij} = 0$ otherwise. In our experiments we set $\lambda_1 = \frac{1}{N^2}$, $\lambda_2 = \frac{10^{-3}}{N^2}$ and used GFB and IPD to obtain optimal estimates $\hat{X}_{\text{GFB}}$ and $\hat{X}_{\text{IPD}}$, respectively. We then set $\tau := \|\hat{X}_{\text{GFB}}\|_{tr}$ and used HCGS to solve the constrained formulation equivalent to (3.36), namely:

$$\min \left\{ \frac{1}{2p} \|\Omega(Y - X)\|_F^2 + \lambda_1 \|X\|_1 \; : \; \|X\|_{tr} \leq \tau, \; X \in \mathbb{R}^{N \times N} \right\} . \tag{3.37}$$

The comparisons were performed on an Intel Xeon with 8 cores and 24GB of memory.

Figure 3.2 shows the evolution of objective values for $N = 400$. Note that for the sake of comparison we have reported the objective of the optimization problem (3.36) even though HCGS actually solves the equivalent problem in (3.37). The same applies to the attained objective function value J_{k^*} in Table 3.1. In this table we have also compared the different algorithms in terms of CPU time[5], relative change in the objective function, and number of iterations upon termination. In all the cases we terminated the algorithms at iteration k^* when the relative change r_{k^*} in the objective value:

$$r_{k^*} = \left| \frac{f_{k^*} - f_{k^*-1}}{f_{k^*-1}} \right| \tag{3.38}$$

was less that 10^{-7}. Note that f in (3.38) refers to the objective function actually minimized by each algorithm; this is not necessarily the objective function J in (3.36). Figure 3.3 shows the time complexity as a function of N. Finally, in Table 3.2 we have reported the average time per iteration as a function of N.

From these figures and tables, we see that the running time of HCGS scales as $\mathcal{O}(N^2)$ with the matrix size, whereas both GFB and IPD scale as $\mathcal{O}(N^3)$.

3.6.2 Sparse PCA

The second set of simulations assesses the computational efficiency of HCGS on the convex relaxation of sparse PCA (3.33). Similar to [2], we generated random matrices C as follows. For each size n, we drew an $n \times n$ matrix U with uniformly distributed elements in $[0, 1]$. Then we generated a vector $v \in \mathbb{R}^n$ from the uniform distribution and set a random 90% of its components to zero. We then set

$$C = UU^\top + 10vv^\top.$$

---

[5]In Figure 3.2b the CPU times also include the evaluation of the objective value in (3.36), which requires computing the singular values of the estimate. In the case of HCGS, this is required only for the sake of comparison with GFB and IPD. In contrast, for Table 3.1 the objective value of (3.36) is computed only upon termination.

TABLE 3.1: Comparison of different algorithms for convex matrix recovery.

5% observed entries

N		$J_{k*} (\times 10^{-4})$	$r_{k*} (\times 10^{-8})$	time (s)	k^*
50	HCGS	6.77	8.35	2.38	1605
	GFB	6.87	5.15	0.09	45
	IPD	6.76	9.07	0.09	50
100	HCGS	4.41	7.52	3.60	1462
	GFB	4.45	9.67	2.57	385
	IPD	4.40	9.80	2.5	388
200	HCGS	4.16	2.42	66	3308
	GFB	4.17	9.95	113	4136
	IPD	4.16	9.83	123	4645
400	HCGS	2.98	3.35	176	4555
	GFB	2.98	9.99	2333	15241
	IPD	2.98	9.99	2389	15665
600	HCGS	2.46	7.15	158	2797
	GFB	2.45	9.99	13157	36049
	IPD	2.45	9.99	13080	36408
800	HCGS	2.20	5.47	478	5197
	GFB	2.20	9.99	40779	61299
	IPD	2.20	9.99	41263	61529

40% observed entries

N		$J_{k*} (\times 10^{-4})$	$r_{k*} (\times 10^{-8})$	time (s)	k^*
50	HCGS	18.51	8.85	0.62	427
	GFB	18.66	8.75	1.82	751
	IPD	18.52	9.96	2.17	1004
100	HCGS	12.09	4.96	2.25	835
	GFB	12.15	9.84	12.1	1681
	IPD	12.10	6.50	11.8	1697
200	HCGS	7.65	9.64	21	1410
	GFB	7.66	9.98	134	3521
	IPD	7.65	7.66	112	3033
400	HCGS	4.39	2.54	40	1281
	GFB	4.39	9.96	1559	7379
	IPD	4.39	9.46	1580	7515
600	HCGS	3.41	7.65	43	760
	GFB	3.41	9.98	5664	10793
	IPD	3.41	9.99	5446	10841
800	HCGS	2.68	1.37	148	1378
	GFB	2.68	9.98	14897	15615
	IPD	2.68	9.99	11242	15647
1600	HCGS	1.62	9.91	929	2931
	GFB	1.62	9.99	119724	36573
	IPD	1.62	9.99	118223	36583

TABLE 3.2: Average time (in seconds) per iteration as a function of N (40% observed entries), for the sparse-low rank experiment.

	N				
	200	400	600	800	1600
HCGS	0.0155	0.032	0.058	0.107	0.317
GFB	0.038	0.211	0.524	0.954	3.273
IPD	0.037	0.210	0.502	0.718	3.231

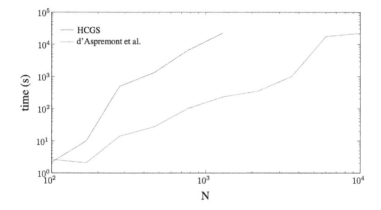

FIGURE 3.4: Computational time (in seconds) versus matrix size for the sparse PCA experiment.

We solved (3.33) for $\lambda = 1$ with HCGS (Algorithm 4) and the Nesterov smoothing method of [2], which optimizes the dual problem of (3.33). We implemented both algorithms in Matlab and used a cluster with 24 cores and sufficient memory. For HCGS we used the power method with a tolerance of 10^{-6}, for computing dominant eigenvectors. We also rescaled the objective function by n in order to keep L_g small enough (see Section 3.4.4). For Nesterov smoothing we used $\mu = \frac{10^{-6}}{2\log n}$.

In Figure 3.4, we plot the computational times required to attain relative change of 10^{-5} in the objective. We note that the objective functions are different, since HCGS optimizes (3.33) whereas the method of [2] optimizes the dual problem. In fact, we have verified that the duality gap estimates are consistently larger for the latter and hence the running times for Nesterov smoothing are optimistic. We observe that the running time scales roughly as $\mathcal{O}(n^2)$ for HCGS whereas Nesterov smoothing scales worse than $\mathcal{O}(n^3)$.

3.7   Conclusion

We have studied the hybrid conditional gradient-smoothing algorithm (HCGS) for solving composite convex optimization problems that contain several terms over a bounded set. Examples of these include regularization problems with several norms as penalties and a norm constraint. HCGS extends conditional gradient methods to cases with multiple nonsmooth terms, in which standard conditional gradient methods may be difficult to apply. The HCGS algorithm borrows techniques from smoothing proximal methods and requires first-order computations (subgradients and proximity operations). Moreover, it exhibits convergence in terms of the objective values at an $\mathcal{O}\left(\frac{1}{\varepsilon^2}\right)$ rate of iterations. Unlike proximal methods, HCGS benefits from the advantages of conditional gradient methods, which render it more efficient on certain large scale optimization problems. We have demonstrated these advantages with simulations on two matrix optimization problems: regularization of matrices with combined ℓ_1 and trace norm penalties; and a convex relaxation of sparse PCA.

Acknowledgments

The research leading to the above results has received funding from the European Union Seventh Framework Programme (FP7 2007-2013) under grant agreement No. 246556. Research was also supported by: Research Council KUL: GOA/10/09 MaNet, PFV/10/002 (OPTEC), several PhD/postdoc and fellow grants; Flemish Government:IOF: IOF/KP/SCORES4CHEM, FWO: PhD/postdoc grants, projects: G.0377.12 (Structured systems), G.083014N (Block term decompositions), G.088114N (Tensor based data similarity), IWT: PhD Grants, projects: SBO POM, EUROSTARS SMART, iMinds 2013, Belgian Federal Science Policy Office: IUAP P7/19 (DYSCO, Dynamical systems, control and optimization, 2012-2017), EU: FP7-SADCO (MC ITN-264735), ERC AdG A-DATADRIVE-B (290923), COST: Action ICO806: IntelliCIS.

Bibliography

[1] J. Abernethy, F. Bach, T. Evgeniou, and J-P. Vert. A new approach to collaborative filtering: Operator estimation with spectral regularization. *Journal of Machine Learning Research*, 10:803–826, 2009.

[2] A. d' Aspremont, L. El Ghaoui, M. I. Jordan, and G. R. G. Lanckriet. A direct formulation for sparse PCA using semidefinite programming. *SIAM Review*, pages 434–448, 2007.

[3] F. Bach. Duality between subgradient and conditional gradient methods. *arXiv preprint arXiv:1211.6302*, 2012.

[4] H. H. Bauschke and P. L. Combettes. *Convex Analysis and Monotone Operator Theory in Hilbert Spaces*. CMS Books in Mathematics. Springer, 2011.

[5] A. Beck and M. Teboulle. Mirror descent and nonlinear projected subgradient methods for convex optimization. *Operations Research Letters*, 31(3):167–175, 2003.

[6] A. Beck and M. Teboulle. A fast iterative shrinkage-thresholding algorithm for linear inverse problems. *SIAM Journal of Imaging Sciences*, 2(1):183–202, 2009.

[7] A. Beck and M. Teboulle. Smoothing and first order methods: a unified framework. *SIAM Journal on Optimization*, 22(2):557–580, 2012.

[8] D. P. Bertsekas. Incremental gradient, subgradient, and proximal methods for convex optimization: A survey. In S. Sra, S. Nowozin, and S. J. Wright, editors, *Optimization for Machine Learning*, pages 85–119. MIT Press, 2011.

[9] M. D. Canon and C. D. Cullum. A tight upper bound on the rate of convergence of Frank-Wolfe algorithm. *SIAM Journal on Control*, 6(4):509–516, 1968.

[10] K. L. Clarkson. Coresets, sparse greedy approximation, and the Frank-Wolfe algorithm. *ACM Transactions on Algorithms*, 6(4):63:1–63:30, 2010.

[11] P. L. Combettes and J. C. Pesquet. Proximal splitting methods in signal processing. In *Fixed-Point Algorithms for Inverse Problems in Science and Engineering*, pages 185–212. Springer, 2011.

[12] P. L. Combettes and V. R. Wajs. Signal recovery by proximal forward-backward splitting. *Multiscale Modeling and Simulation*, 4(4):1168–1200, 2006.

[13] V. F. Dem'yanov and A. B. Pevnyi. Some estimates in minimax problems. *Cybernetics and Systems Analysis*, 8(1):116–123, 1972.

[14] V. F. Dem'yanov and A. M. Rubinov. *Approximate methods in optimization problems*, volume 32. Elsevier, 1970.

[15] J. C. Dunn. Rates of convergence for conditional gradient algorithms near singular and nonsingular extremals. *SIAM Journal on Control and Optimization*, 17(2):187–211, 1979.

[16] J. C. Dunn. Convergence rates for conditional gradient sequences generated by implicit step length rules. *SIAM Journal on Control and Optimization*, 18(5):473–487, 1980.

[17] J. Eckstein and D. P. Bertsekas. On the Douglas-Rachford splitting method and the proximal point algorithm for maximal monotone operators. *Mathematical Programming*, 55(1-3):293–318, 1992.

[18] M. Frank and P. Wolfe. An algorithm for quadratic programming. *Naval research logistics quarterly*, 3(1-2):95–110, 1956.

[19] R. M. Freund and P. Grigas. New analysis and results for the conditional gradient method. Technical report, Massachusetts Institute of Technology, Operations Research Center, 2013.

[20] D. Garber and E. Hazan. Approximating semidefinite programs in sublinear time. In *Advances in Neural Information Processing Systems 24*, pages 1080–1088, 2011.

[21] E. G. Gilbert. An iterative procedure for computing the minimum of a quadratic form on a convex set. *SIAM Journal on Control*, 4(1):61–80, 1966.

[22] Z. Harchaoui, A. Juditsky, and A. Nemirovski. Conditional gradient algorithms for machine learning. In *NIPS Workshop on Optimization for Machine Learning*, 2012.

[23] E. Hazan. Sparse approximate solutions to semidefinite programs. In *Proceedings of the 8th Latin American conference on Theoretical informatics*, LATIN'08, pages 306–316, 2008.

[24] E. Hazan and S. Kale. Projection-free online learning. In *Proceedings of the 29th International Conference on Machine Learning*, 2012.

[25] N. He, A. Juditsky, and A. Nemirovski. Mirror prox algorithm for multiterm composite minimization and alternating directions. arXiv preprint arXiv:1311.1098, 2013.

[26] J. Huang, T. Zhang, and D. Metaxas. Learning with structured sparsity. In *Proceedings of the 26th Annual International Conference on Machine Learning*, pages 417–424. ACM, 2009.

[27] M. Jaggi. Convex optimization without projection steps. Arxiv preprint arXiv:1108.1170, 2011.

[28] M. Jaggi. Revisiting Frank-Wolfe: Projection-free sparse convex optimization. In *Proceedings of the 30th International Conference on Machine Learning*, pages 427–435, 2013.

[29] M. Jaggi and M. Sulovskỳ. A simple algorithm for nuclear norm regularized problems. In *Proceedings of the 27th International Conference on Machine Learning*, pages 471–478, 2010.

[30] A. Juditsky and A. Nemirovski. First order methods for nonsmooth convex large-scale optimization, I: general purpose methods. In *Optimization for Machine Learning*, pages 121–148. MIT Press, 2010.

[31] H. J. Kelley. Method of gradients. *Academic Press*, 2:1578–1580, 1962.

[32] V. Kumar. A control averaging technique for solving a class of singular optimal control problems. *International Journal of Control*, 23(3):361–380, 1976.

[33] S. Lacoste-Julien, M. Jaggi, M. Schmidt, and P. Pletscher. Block-coordinate Frank-Wolfe optimization for structural SVMs. arXiv preprint arXiv:1207.4747, 2012.

[34] G. Lan. The complexity of large-scale convex programming under a linear optimization oracle. arXiv preprint arXiv:1309.5550, 2013.

[35] J. Mairal, R. Jenatton, F. R. Bach, and G. R. Obozinski. Network flow algorithms for structured sparsity. In *Advances in Neural Information Processing Systems*, pages 1558–1566, 2010.

[36] S. G. Mallat and Z. Zhang. Matching pursuits with time-frequency dictionaries. *IEEE Transactions on Signal Processing*, 41(12):3397–3415, 1993.

[37] J.J. Moreau. Proximité et dualité dans un espace hilbertien. *Bulletin de la Société Mathématique de France*, 93(2):273–299, 1965.

[38] A. S. Nemirovsky and D. B. Yudin. *Problem complexity and method efficiency in optimization*. Wiley, 1983.

[39] Y. Nesterov. A method of solving a convex programming problem with convergence rate $O(1/k^2)$. *Soviet Mathematics Doklady*, 27(2):372–376, 1983.

[40] Y. Nesterov. *Introductory lectures on convex optimization: A basic course*. Springer, 2004.

[41] Y. Nesterov. Excessive gap technique in nonsmooth convex minimization. *SIAM J. on Optimization*, 16(1):235–249, May 2005.

[42] Y. Nesterov. Smooth minimization of non-smooth functions. *Mathematical Programming*, 103(1):127–152, 2005.

[43] Y. Nesterov. Gradient methods for minimizing composite objective function. *CORE Discussion Papers*, 2007.

[44] F. Orabona, A. Argyriou, and N. Srebro. PRISMA: PRoximal Iterative SMoothing Algorithm. arXiv preprint arXiv:1206.2372, 2012.

[45] H. Raguet, J. Fadili, and G. Peyré. A generalized forward-backward splitting. *SIAM Journal on Imaging Sciences*, 6(3):1199–1226, 2013.

[46] E. Richard, P.-A. Savalle, and N. Vayatis. Estimation of simultaneously sparse and low rank matrices. In *Proceedings of the 29th International Conference on Machine Learning*, pages 1351–1358, 2012.

[47] R. T. Rockafellar. Level sets and continuity of conjugate convex functions. *Transactions of the American Mathematical Society*, 123:46–63, 1966.

[48] R. T. Rockafellar. *Convex Analysis*. Princeton University Press, 1970.

[49] R. T. Rockafellar. Monotone operators and the proximal point algorithm. *SIAM Journal on Control and Optimization*, 14(5):877–898, 1976.

[50] L. I. Rudin, S. Osher, and E. Fatemi. Nonlinear total variation based noise removal algorithms. *Physica D: Nonlinear Phenomena*, 60(1):259–268, 1992.

[51] S. Shalev-Shwartz, N. Srebro, and T. Zhang. Trading accuracy for sparsity in optimization problems with sparsity constraints. *SIAM Journal on Optimization*, 20(6):2807–2832, 2010.

[52] J. Shawe-Taylor and N. Cristianini. *Kernel Methods for Pattern Analysis*. Cambridge University Press, 2004.

[53] J. Steinhardt and J. Huggins. A greedy framework for first-order optimization. Preprint.

[54] A. Tewari, P. K. Ravikumar, and I. S. Dhillon. Greedy algorithms for structurally constrained high dimensional problems. In *Advances in Neural Information Processing Systems 24*, pages 882–890, 2011.

[55] J. A. Tropp and A. C. Gilbert. Signal recovery from random measurements via orthogonal matching pursuit. *IEEE Transactions on Information Theory*, 53(12):4655–4666, 2007.

[56] Y. Ying and P. Li. Distance metric learning with eigenvalue optimization. *Journal of Machine Learning Research*, 13:1–26, 2012.

[57] M. Yuan and Y. Lin. Model selection and estimation in regression with grouped variables. *Journal of the Royal Statistical Society, Series B (Statistical Methodology)*, 68(1):49–67, 2006.

Chapter 4

Nonconvex Proximal Splitting with Computational Errors

Suvrit Sra

Max Planck Institute for Intelligent Systems and Carnegie Mellon University

4.1 Introduction

In this chapter we study large-scale nonconvex optimization problems with *composite objective functions* that are composed of a differentiable possibly nonconvex cost and a nonsmooth but convex regularizer. More precisely, we consider optimization problems of the form

$$\text{minimize} \quad \Phi(x) := f(x) + r(x), \quad \text{s.t.} \quad x \in \mathcal{X}, \tag{4.1}$$

where $\mathcal{X} \subset \mathbb{R}^n$ is a compact convex set, $f : \mathbb{R}^n \to \mathbb{R}$ is a differentiable cost function, and $r : \mathbb{R}^n \to \mathbb{R}$ is a closed convex function. Further, we assume that the gradient ∇f is *Lipschitz continuous* on $\mathcal{X}$ (denoted $f \in C_L^1(\mathcal{X})$), i.e.,

$$\exists L > 0 \quad \text{s.t.} \quad \|\nabla f(x) - \nabla f(y)\| \le L \|x - y\| \qquad \text{for all} \quad x, y \in \mathcal{X}. \tag{4.2}$$

Throughout this chapter, $\|\cdot\|$ denotes the standard Euclidean norm.

Problem (4.1) generalizes the more thoroughly studied class of *composite convex optimization problems* [30], a class that has witnessed huge interest in machine learning, signal processing, statistics, and other related areas. We refer the interested reader to [21, 2, 3, 37] for several convex examples and recent references. A thread common to existing algorithms for solving composite problems is the remarkably fruitful idea of *proximal-splitting* [9]. Here, nonsmoothness is handled via proximity operators [29], which allows one to treat the nonsmooth objective $f + r$ essentially as a smooth one.

But leveraging proximal-splitting methods is considerably harder for nonconvex problems, especially without compromising scalability. Numerous important problems have appealing nonconvex formulations: matrix factorization [25, 27], blind deconvolution [24], dictionary learning and sparse reconstruction [23, 27], and neural networks [4, 28, 19], to name a few. Regularized optimization within these problems requires handling nonconvex composite objectives, which motivates the material of this chapter.

The focus of this chapter is on a new proximal splitting framework called **N**onconvex **I**nexact **P**roximal **S**plitting, hereafter NIPS. The NIPS framework is *inexact* because it allows for computational errors, a feature that helps it scale to large-data problems. In contrast to typical incremental methods [5] and to most stochastic gradient methods [16, 18] that assume vanishing errors, NIPS allows the computational errors to be *nonvanishing*.

NIPS inherits this capability from the remarkable framework of Solodov [33]. But NIPS not only builds on [33], it strictly generalizes it: Unlike [33], NIPS allows $r \neq 0$ in (4.1). To our knowledge, NIPS is the first nonconvex proximal splitting method that has *both* batch and incremental incarnations; this claim remains true, even if we were to exclude the nonvanishing error capability.[1] We mention some more related work below.

Among batch nonconvex splitting methods an early paper is [14]. Another batch method can be found in the pioneering paper on composite minimization by Nesterov [30], who solves (4.1) via a splitting-like algorithm. Both [14] and [30] rely on monotonic descent (using line-search or otherwise) to ensure convergence. Very recently, [1] introduced a powerful class of "descent-methods" based on Kurdyka-Łojasiewicz theory. In general, the insistence on descent, while theoretically convenient, makes it hard to extend these methods to incremental, stochastic, or online variants. The general proximal framework of [40] avoids strict monotonic descent at each step by using a non-monotonic line-search.

There are some incremental and stochastic methods that apply to (4.1), namely the generalized gradient algorithm of [35] and the stochastic generalized gradient methods of [12, 13], (and the very recent work of [17, 18]). All these approaches are analogous to subgradient methods, and thus face similar practical difficulties (except [18]). For example, it is well-recognized that subgradient methods fail to exploit composite objectives [30, 11]. Moreover, they exhibit the effect of the regularizer only in the limit, which conflicts with early termination heuristics frequently used in practice. If, say, the nonsmooth part of the objective is $\|x\|_1$, then with subgradient-style methods sparse

---

[1]Though very recently, in [18] scalable nonconvex stochastic methods were proposed for smooth nonconvex problems; and even more recently those ideas were extended to cover nonsmooth and accelerated methods [17], though still in the vanishing error framework.

solutions are obtained only in the limit and intermediate iterates may be dense. Thus, like the convex case it may be of substantial practical advantage to use proximal splitting even for (4.1).

4.2   The NIPS Framework

We rewrite (4.1) as an unconstrained problem by introducing the function

$$g(x) := r(x) + \delta(x|\mathcal{X}),$$

where $\delta(x|\mathcal{X})$ is the *indicator function* for set $\mathcal{X}$. Our problem then becomes:

$$\text{minimize} \quad \Phi(x) := f(x) + g(x) \qquad x \in \mathbb{R}^n. \tag{4.3}$$

Since we solve (4.3) via a proximal method, we begin with the definition below.

Definition 4.1 (Proximity operator). *Let $g : \mathbb{R}^n \to \mathbb{R}$ be lower semicontinuous (lsc) and convex. The* proximity operator *for g, indexed by $\eta > 0$, is the nonlinear map [31, Def. 1.22]:*

$$\text{prox}_{g,\eta} : \quad y \mapsto \underset{x \in \mathbb{R}^n}{\text{argmin}} \left(g(x) + \frac{1}{2\eta} \|x - y\|^2 \right). \tag{4.4}$$

Using the operator (4.4), the classic forward-backward splitting (FBS) [8] iteration (for suitable η_k and convex f) is written as

$$x^{k+1} = \text{prox}_{g,\eta_k}(x^k - \eta_k \nabla f(x^k)), \quad k = 0, 1, \ldots. \tag{4.5}$$

The NIPS framework described in this chapter is motivated by the simple form of iteration (4.5). In particular, for this iteration NIPS introduces two powerful generalizations: (i) it permits a nonconvex f; and (ii) it allows computational errors. More precisely, NIPS performs the iteration

$$x^{k+1} = \text{prox}_{g,\eta_k}(x^k - \eta_k \nabla f(x^k) + \eta_k e(x^k)). \tag{4.6}$$

The error vector $e(x^k)$ in (4.6) is the interesting part. It denotes potential error made at step k in the computation of the gradient $\nabla f(x^k)$. It is important to observe that the net error is $\eta_k e(x^k)$, so that the error is scaled by the stepsize. This scaling is made to suggest that the limiting value of the stepsize is what ultimately determines the effective error, and thereby governs convergence.

Remark 4.1. *We warn the reader against a potential pitfall of the notation for error in (4.6). That iteration does* not *mean that NIPS adds an error vector $e(x^k)$ when iterating, but rather that it iterates*

$$x^{k+1} = \text{prox}_{g,\eta_k}(x^k - \eta_k g^k),$$

where g^k is an erroneous computation of the gradient, which is explicitly depicted in (4.6) as $g^k = \nabla f(x^k) - e(x^k)$.

But notice that we *do not* impose the following condition

$$\lim_{k \to \infty} \|e(x^k)\| \to 0 \tag{4.7}$$

on the error vectors, which is typically imposed by stochastic-gradient methods [5]. Since we do not require the errors to vanish in the limit, to make NIPS well-defined we must nevertheless somehow control them. Thus, we impose a mild restriction on the errors: we assume that there is a fixed value $\bar{\eta}$, so that for all stepsizes η smaller than $\bar{\eta}$ the gradient errors satisfy

$$\eta \|e(x)\| \leq \bar{\epsilon}, \quad \text{for some fixed } \bar{\epsilon} \geq 0, \quad \text{and } \forall x \in \mathcal{X}. \tag{4.8}$$

Clearly, condition (4.8) is weaker than the usual requirement (4.7).

Remark 4.2. *We can consider errors in the proximity operator too, i.e., the* $\text{prox}_{g,\eta}$ *computation may also be inexact (for convex optimization inexact proximity operators have been studied for a long time; two recent references are [39, 32]). With inexact proximity operations iteration (4.6) becomes*

$$x^{k+1} = \text{prox}_{g,\eta_k}(x^k - \eta_k \nabla f(x^k) + \eta_k e(x^k)) + \eta_k p(x^k),$$

where $\eta_k p(x^k)$ *is the error in proximity operator. The dependency on* η_k *highlights that the error should eventually shrink in a manner similar to (4.8). This error can be easily incorporated into our analysis below, though at the expense of heavier notation. To avoid clutter we omit details and leave them as an exercise for the interested reader.*

Since errors in the gradient computation need not disappear, we cannot ensure exact stationary points; but we can nevertheless hope to ensure *inexact stationary points*. Let us make this more precise. A point $x^* \in \mathbb{R}^n$ is stationary for (4.3), if and only if it satisfies the inclusion

$$0 \in \partial_C \Phi(x^*) = \nabla f(x^*) + \partial g(x^*), \tag{4.9}$$

where $\partial_C \Phi(x^*)$ is the Clarke subdifferential [7] at x^*. The optimality condition (4.9) may be recast as the fixed-point equation

$$x^* = \text{prox}_{g,\eta}(x^* - \eta \nabla f(x^*)), \quad \text{for } \eta > 0, \tag{4.10}$$

which helps characterize approximate stationarity. Define the *prox-residual*

$$\rho(x) := x - \text{prox}_{g,1}(x - \nabla f(x)); \tag{4.11}$$

then, for stationary x^* the residual norm $\|\rho(x^*)\|$ vanishes. At a point x, let the total perturbation be given by $\epsilon(x) \geq 0$. We define a point $\bar{x}$ to be *ϵ-stationary* if the residual norm satisfies the condition

$$\|\rho(\bar{x})\| \leq \epsilon(\bar{x}). \tag{4.12}$$

Since we cannot measure convergence to an accuracy better than the amount of prevalent noise, we require $\epsilon(x) \geq \eta \|e(x)\|$. By letting η become small enough, we may hope to come arbitrarily close to a stationary point.

4.2.1   Convergence Analysis

In this section, we outline a simple convergence analysis for the NIPS iteration (4.6). Our analysis is structured upon the powerful framework of [33]. But our problem class of composite objectives is more general than the differentiable problems considered in [33] since we allow nonsmooth objective functions. Our analysis leads to the first nonconvex proximal splitting algorithm that allows noisy gradients; also we obtain the first nonconvex incremental proximal splitting algorithm regardless of whether the noise vanishes or not.

For our analysis we make the following standing assumption.

Assumption. The stepsizes η_k satisfy the bounds

$$c \leq \liminf_k \eta_k, \quad \limsup_k \eta_k \leq \min\{1, 2/L - c\}, \quad 0 < c < 1/L. \qquad (4.13)$$

We start our analysis by recalling two well-known facts.

Lemma 4.1 (Descent). *Let f be such that ∇f satisfies (4.2). Then*

$$|f(x) - f(y) - \langle \nabla f(y), x - y \rangle| \leq \tfrac{L}{2} \|x - y\|^2, \quad \forall\, x, y \in \mathcal{X}. \qquad (4.14)$$

Proof. Since $f \in C_L^1$, by Taylor's theorem for $z_t = y + t(x - y)$ we have

$$|f(x) - f(y) - \langle \nabla f(y), x - y \rangle| = |\int_0^1 \langle \nabla f(z_t) - \nabla f(y), x - y \rangle dt|,$$
$$\leq \int_0^1 \|\nabla f(z_t) - \nabla f(y)\| \cdot \|x - y\|\, dt \leq L \int_0^1 t\|x - y\|_2^2 dt = \tfrac{L}{2}\|x - y\|_2^2.$$

We used the triangle-inequality, Cauchy-Schwarz, and that $f \in C_L^1$ above.   □

Lemma 4.2. *The operator* $\mathrm{prox}_{g,\eta}$ *is nonexpansive, that is,*

$$\left\|\mathrm{prox}_{g,\eta} x - \mathrm{prox}_{g,\eta} y\right\| \leq \|x - y\|, \quad \forall\, x, y \in \mathbb{R}^n. \qquad (4.15)$$

Proof. For brevity we drop the subscripted η. After renaming variables, from optimality conditions for the problem (4.4), it follows that $x - \mathrm{prox}_g x \in \eta\partial g(\mathrm{prox}_g x)$. A similar characterization holds for y. Thus, $x - \mathrm{prox}_g x$ and $y - \mathrm{prox}_g y$ are subgradients of g at $\mathrm{prox}_g x$ and $\mathrm{prox}_g y$, respectively. Thus,

$$g(\mathrm{prox}_g x) \geq g(\mathrm{prox}_g y) + \langle y - \mathrm{prox}_g y, \mathrm{prox}_g x - \mathrm{prox}_g y \rangle$$
$$g(\mathrm{prox}_g y) \geq g(\mathrm{prox}_g x) + \langle x - \mathrm{prox}_g x, \mathrm{prox}_g y - \mathrm{prox}_g x \rangle.$$

Adding the two inequalities we obtain *firm nonexpansivity*

$$\left\|\mathrm{prox}_g x - \mathrm{prox}_g y\right\|^2 \leq \langle \mathrm{prox}_g x - \mathrm{prox}_g y, x - y \rangle,$$

from which via Cauchy-Schwarz, we easily obtain (4.15).   □

Next, we prove a crucial monotonicity property of proximity operators.

Lemma 4.3. *Define $P_\eta \equiv \mathrm{prox}_{g,\eta}$; let $y, z \in \mathbb{R}^n$, and $\eta > 0$; define the functions*

$$p(\eta) := \eta^{-1} \| P_\eta(y - \eta z) - y \|, \tag{4.16}$$

$$q(\eta) := \| P_\eta(y - \eta z) - y \|. \tag{4.17}$$

Then, $p(\eta)$ is a decreasing function and $q(\eta)$ is an increasing function of η.

Proof. Our proof relies on well-known results about Moreau envelopes [31, 8]. Consider thus the "deflected" proximal objective

$$m_g(x, \eta; y, z) := \langle z, \, x - y \rangle + \tfrac{1}{2}\eta^{-1} \| x - y \|^2 + g(x), \tag{4.18}$$

to which we associate its *Moreau-envelope*

$$\mathscr{E}_g(\eta) := \inf_{x \in \mathcal{X}} \, m_g(x, \eta; y, z). \tag{4.19}$$

Since m_g is strongly convex in x, and $\mathcal{X}$ is compact, the infimum in (4.19) is attained at a unique point, which is precisely $P_\eta^g(y - \eta z)$. Thus, $\mathscr{E}_g(\eta)$ is a differentiable function of η, and in particular

$$\frac{\partial \mathscr{E}_g(\eta)}{\partial \eta} = -\tfrac{1}{2}\eta^{-2} \| P_\eta(y - \eta z) - y \|^2 = -\tfrac{1}{2}p(\eta)^2.$$

Observe that m_g is jointly convex in (x, η); it follows that $\mathscr{E}_g$ is convex too. Thus, its derivative $\partial \mathscr{E}_g / \partial \eta$ is increasing, whereby $p(\eta)$ is decreasing. Similarly, $\hat{\mathscr{E}}_g(\gamma) := \mathscr{E}_g(1/\gamma)$ is concave in γ (it is a pointwise infimum of linear functions). Thus, its derivative

$$\frac{\partial \hat{\mathscr{E}}_g(\gamma)}{\partial \gamma} = \tfrac{1}{2} \| P_{1/\gamma}(x - \gamma^{-1}y) - x \|^2 = q(1/\gamma),$$

is a decreasing function of γ. Writing $\eta = 1/\gamma$ completes our claim.   □

Remark 4.3. *The monotonicity results (4.16) and (4.17) subsume the monotonicity results for projection operators derived in [15, Lemma 1].*

We now proceed to analyze how the objective function value changes after one step of the NIPS iteration (4.6). Specifically, we seek to derive an inequality of the form (4.20) (where Φ is as in (4.3)):

$$\Phi(x^k) - \Phi(x^{k+1}) \geq h(x^k). \tag{4.20}$$

Our strategy is to bound the potential function $h(x)$ in terms of prox-residual $\|\rho(x)\|$ and the error level $\epsilon(x)$. It is important to note that the potential $h(x)$ may be negative, because we do not insist on monotonic descent.

To reduce clutter, let us introduce brief notation: $u \equiv x^{k+1}$, $x \equiv x^k$, and $\eta \equiv \eta_k$; therewith the main NIPS update (4.6) may be rewritten as

$$u = \mathrm{prox}_{g,\eta}(x - \eta \nabla f(x) + \eta e(x)). \tag{4.21}$$

We are now ready to state the following "descent" theorem.

Theorem 4.1. *Let u, x, η be as in (4.21); assume $\epsilon(x) \geq \eta \|e(x)\|$. Then,*

$$\Phi(x) - \Phi(u) \quad \geq \quad \frac{2 - L\eta}{2\eta} \|u - x\|^2 - \frac{1}{\eta}\epsilon(x)\|u - x\|. \qquad (4.22)$$

Proof. Let m_g be as in (4.18); consider its directional derivative dm_g with respect to x in direction w; at $x = u$ it satisfies the optimality condition

$$dm_g(u, \eta; y, z)(w) = \langle z + \eta^{-1}(u - y) + s,\, w \rangle \geq 0, \quad s \in \partial g(u). \qquad (4.23)$$

In (4.23), substitute $z = \nabla f(x) - e(x)$, $y = x$, and $w = x - u$ to obtain

$$\langle \nabla f(x) - e(x),\, u - x \rangle \leq \langle \eta^{-1}(u - x) + s,\, x - u \rangle. \qquad (4.24)$$

From Lemma 4.1 we know that $\Phi(u) \leq f(x) + \langle \nabla f(x),\, u - x \rangle + \frac{L}{2}\|u - x\|^2 + g(u)$; now add and subtract $e(x)$ to this and combine with (4.24) to obtain

$$
\begin{aligned}
\Phi(u) &\leq f(x) + \langle \nabla f(x) - e(x),\, u - x \rangle + \tfrac{L}{2}\|u - x\|^2 + g(u) + \langle e(x),\, u - x \rangle \\
&\leq f(x) + \langle \eta^{-1}(u - x) + s,\, x - u \rangle + \tfrac{L}{2}\|u - x\|^2 + g(u) + \langle e(x),\, u - x \rangle \\
&= f(x) + g(u) + \langle s,\, x - u \rangle + \left(\tfrac{L}{2} - \tfrac{1}{\eta}\right)\|u - x\|^2 + \langle e(x),\, u - x \rangle \\
&\leq f(x) + g(x) - \tfrac{2 - L\eta}{2\eta}\|u - x\|^2 + \langle e(x),\, u - x \rangle \\
&\leq \Phi(x) - \tfrac{2 - L\eta}{2\eta}\|u - x\|^2 + \|e(x)\|\,\|u - x\|, \\
&\leq \Phi(x) - \tfrac{2 - L\eta}{2\eta}\|u - x\|^2 + \tfrac{1}{\eta}\epsilon(x)\|u - x\|.
\end{aligned}
$$

The third inequality follows from convexity of g, the fourth one from Cauchy-Schwarz, and the last one from the definition of $\epsilon(x)$.

$\square$

To further analyze (4.22), we derive two-sided bounds on $\|x - u\|$ below.

Lemma 4.4. *Let x, u, and η be as in Theorem 4.1, and c as in (4.13). Then,*

$$c\|\rho(x)\| - \epsilon(x) \leq \|x - u\| \leq \|\rho(x)\| + \epsilon(x). \qquad (4.25)$$

Proof. Lemma 4.3 implies that for $\eta > 0$ we have the crucial bounds

$$1 \leq \eta \implies q(1) \leq q(\eta), \quad \text{and} \quad 1 \geq \eta \implies p(1) \leq p(\eta). \qquad (4.26)$$

Let $y \leftarrow x, z \leftarrow \nabla f(x)$. Note that $q(1) = \|\rho(x)\| \leq \|P_\eta(x - \eta\nabla f(x)) - x\|$ if $\eta \geq 1$, while if $\eta \leq 1$, we get $\eta p(1) \leq \|P_\eta(x - \eta\nabla f(x)) - x\|$. Compactly, we may therefore write

$$\min\{1, \eta\}\|\rho(x)\| \leq \|P_\eta(x - \eta\nabla f(x)) - x\|.$$

Using the triangle inequality and nonexpansivity of prox we see that

$$
\begin{aligned}
\min\{1, \eta\}\|\rho(x)\| &\leq \|P_\eta(x - \eta\nabla f(x)) - x\| \\
&\leq \|x - u\| + \|u - P_\eta(x - \eta\nabla f(x))\| \\
&\leq \|x - u\| + \eta\|e(x)\| \quad \leq \quad \|x - u\| + \epsilon(x).
\end{aligned}
$$

As $c \leq \liminf_k \eta_k$, for large enough k it holds that $\|x - u\| \geq c \|\rho(x)\| - \epsilon(x)$. An upper-bound on $\|x - u\|$ may be obtained as follows:

$$\|x - u\| \leq \|x - P_\eta(x - \eta \nabla f(x))\| + \|P_\eta(x - \eta \nabla f(x)) - u\|$$
$$\leq \max\{1, \eta\} \|\rho(x)\| + \eta \|e(x)\| \quad \leq \quad \|\rho(x)\| + \epsilon(x),$$

where we again used Lemma 4.3 and nonexpansivity.   □

Theorem 4.1 and Lemma 4.4 have done the hard work; they imply the following corollary, which is a key component of the convergence framework of [33] that we ultimately will also invoke.

Corollary 4.1. *Let x, u, η, and c be as above and k sufficiently large so that c and $\eta \equiv \eta_k$ satisfy (4.13). Then, $\Phi(x) - \Phi(u) \geq h(x)$ holds for*

$$h(x) := \tfrac{L^2 c^3}{2(2-2Lc)} \|\rho(x)\|^2 - \left(\tfrac{L^2 c^2}{2-cL} + \tfrac{1}{c}\right) \|\rho(x)\| \epsilon(x) - \left(\tfrac{1}{c} - \tfrac{L^2 c}{2(2-cL)}\right) \epsilon(x)^2. \quad (4.27)$$

Proof. We prove (4.27) by showing that $h(x)$ can be chosen as

$$h(x) := a_1 \|\rho(x)\|^2 - a_2 \|\rho(x)\| \epsilon(x) - a_3 \epsilon(x)^2, \quad (4.28)$$

where the constants a_1, a_2, and a_3 satisfy

$$a_1 = \tfrac{L^2 c^3}{2(2-2Lc)}, \quad a_2 = \tfrac{L^2 c^2}{2-cL} + \tfrac{1}{c}, \quad a_3 = \tfrac{1}{c} - \tfrac{L^2 c}{2(2-cL)}. \quad (4.29)$$

Note that by construction the scalars $a_1, a_2, a_3 > 0$. For sufficiently large k, condition (4.13) implies that

$$c < \eta < \tfrac{2}{L} - c \quad \Longrightarrow \quad \tfrac{1}{\eta} > \tfrac{L}{2-Lc}, \quad \tfrac{1}{\eta} < \tfrac{1}{c}, \quad \text{and} \quad 2 - L\eta > Lc, \quad (4.30)$$

which in turn shows that

$$\tfrac{2-L\eta}{2\eta} > \tfrac{(2-L\eta)L}{2(2-Lc)} > \tfrac{L^2 c}{2(2-Lc)} \quad \text{and} \quad -\tfrac{1}{\eta} > -\tfrac{1}{c}.$$

We can plug this into (4.22) to obtain

$$\Phi(x) - \Phi(u) \geq \tfrac{L^2 c}{2(2-Lc)} \|x - u\|^2 - \tfrac{1}{c} \epsilon(x) \|x - u\|.$$

Apply to this the two-sided bounds (4.25), so that we get

$$\Phi(x) - \Phi(u) \geq \tfrac{L^2 c}{2(2-Lc)} \left(c \|\rho(x)\| - \epsilon(x)\right)^2 - \tfrac{1}{c} \epsilon(x) \left(\|\rho(x)\| + \epsilon(x)\right)$$
$$= \tfrac{L^2 c^3}{2(2-Lc)} \|\rho(x)\|^2 - \left(\tfrac{L^2 c^2}{2-Lc} + \tfrac{1}{c}\right) \|\rho(x)\| \epsilon(x) - \left(\tfrac{1}{c} - \tfrac{L^2 c}{2(2-Lc)}\right) \epsilon(x)^2 \quad =: h(x).$$

All that remains to show is that the said coefficients of $h(x)$ are positive. Since $2 - Lc > 0$ and $c > 0$, positivity of a_1 and a_2 is immediate. Inequality $a_3 = \tfrac{1}{c} - \tfrac{L^2 c}{2(2-Lc)} > 0$, holds as long as $0 < c < \tfrac{\sqrt{5}-1}{L}$, which is obviously true since $c < 1/L$ by assumption (4.13). Thus, $a_1, a_2, a_3 > 0$.   □

Theorem 4.2 (Convergence). *Let $f \in C_L^1(\mathcal{X})$ such that $\inf_{\mathcal{X}} f > -\infty$ and g be lsc, convex on $\mathcal{X}$. Let $\{x^k\} \subset \mathcal{X}$ be a sequence generated by (4.6), and let condition (4.8) hold. Then, there exists a limit point x^* of the sequence $\{x^k\}$, and a constant $K > 0$, such that $\|\rho(x^*)\| \leq K\epsilon(x^*)$. Moreover, if the sequence $\{f(x^k)\}$ converges, then for every limit point x^* of $\{x^k\}$ it holds that $\|\rho(x^*)\| \leq K\epsilon(x^*)$.*

Proof. Theorem 4.1, Lemma 4.4, and Corollary 4.1 have shown that the net change in objective from one step to the next is lower bounded by a quadratic function with suitable positive coefficients, which makes the analysis technique of the differentiable case treated by [33, Thm. 2.1] applicable to the setting (the exact nature of the quadratic bound derived above is crucial to the proof, which essentially shows that for a large enough iteration count, this bound must be positive, which ensures progress); we omit the details for brevity. $\square$

Theorem 4.2 says that we can obtain an approximate stationary point for which the norm of the residual is bounded by a linear function of the error level. The statement of the theorem is written in a conditional form, because nonvanishing errors $e(x)$ prevent us from making a stronger statement. In particular, once the iterates enter a region where the residual norm falls below the error threshold, the behavior of $\{x^k\}$ may be arbitrary. This, however, is a small price to pay for having the added flexibility of nonvanishing errors. Under the stronger assumption of vanishing errors (and suitable stepsizes), we can also ensure exact stationarity.

4.3 Scaling Up: Incremental Proximal Splitting

We now apply NIPS to a large-scale setting. Here, the objective function $f(x)$ is assumed to be decomposable, that is

$$f(x) := \sum_{t=1}^{T} f_t(x), \tag{4.31}$$

where $f_t : \mathbb{R}^n \to \mathbb{R}$ is in $C_{L_t}^1(\mathcal{X})$ (set $L \geq L_t$ for all t), and we solve

$$\min \quad f(x) + g(x), \quad x \in \mathcal{X}, \tag{4.32}$$

where g and $\mathcal{X}$ are as before (4.3).

It has long been known that for decomposable objectives it can be advantageous to replace the full gradient $\nabla f(x)$ by an *incremental gradient* $\nabla f_{r(t)}(x)$, where $r(t)$ is some suitably chosen index. Nonconvex incremental methods have been extensively analyzed in the setting of backpropagation algorithms [5, 33], which correspond to $g(x) \equiv 0$ in (4.32). For $g(x) \neq 0$, the stochastic generalized gradient methods of [13] or the perturbed generalized methods of [35] apply. As previously mentioned, these approaches fail to exploit the composite structure of the objective function, which can be a disadvantage already in the convex case [11].

In contrast, we exploit the composite structure of (4.31), and propose the following incremental nonconvex proximal-splitting method:

$$x^{k+1} = \mathcal{M}\Big(x^k - \eta_k \sum_{t=1}^{T} \nabla f_t(z^t)\Big)$$
$$z^1 = x^k, \quad z^{t+1} = \mathcal{O}(z^t - \eta \nabla f_t(z^t)), \quad t = 1, \ldots, T-1.$$
(4.33)

Here, $\mathcal{O}$ and $\mathcal{M}$ are appropriate nonexpansive maps, choosing which we get different algorithms. For example, when $\mathcal{X} = \mathbb{R}^n$, $g(x) \equiv 0$, and $\mathcal{M} = \mathcal{O} = \mathrm{Id}$, then (4.33) reduces to the problem class considered in [34]. If $\mathcal{X}$ is a closed convex set, $g(x) \equiv 0$, $\mathcal{M} = \Pi_{\mathcal{X}}$, and $\mathcal{O} = \mathrm{Id}$, then (4.33) reduces to a method that is essentially implicit in [34]. Note, however, that in this case, the constraints are enforced *only once* every major iteration; the intermediate iterates (z^t) may be infeasible.

Depending on the application, we may implement either of the four variants of (4.33) in Table 4.1. Which of these one prefers depends on the complexity of the constraint set $\mathcal{X}$ and on the cost of applying P_η^g. In the first two examples $\mathcal{X}$ is not bounded, which complicates the convergence analysis; the third variant is also of practical importance, but the fourth variant allows a more instructive analysis, so we only discuss that one.

TABLE 4.1: Different variants of incremental NIPS (4.33). 'P' indicates penalized, 'U' indicates 'unconstrained', while 'C' refers to a constrained problem; 'CCvx' signifies 'Compact convex'.

$\mathcal{X}$	g	$\mathcal{M}$	$\mathcal{O}$	Penalty	Proximity operator calls
$\mathbb{R}^n$	$\not\equiv 0$	prox_g	Id	P,U	once every *major* (k) iteration
$\mathbb{R}^n$	$\not\equiv 0$	prox_g	prox_g	P,U	once every *minor* (k,t) iteration
CCvx	$h(x) + \delta(x|\mathcal{X})$	prox_g	Id	P,C	once every major (k) iteration
CCvx	$h(x) + \delta(x|\mathcal{X})$	prox_g	prox_g	P,C	once every minor (k,t) iteration

4.3.1   Convergence

Our analysis is inspired by [34] with the obvious difference that we are dealing with a nonsmooth problem. First, as is usual with incremental methods, we also rewrite (4.33) in a form that matches the main iteration (4.6)

$$x^{k+1} = \mathcal{M}\Big(x^k - \eta_k \sum_{t=1}^{T} \nabla f_t(z^t)\Big) \quad = \quad \mathcal{M}\big(x^k - \eta_k \nabla F(x^k) + \eta_k e(x^k)\big).$$

The error term at a general x is then given by $e(x) := \sum_{t=1}^{T}\big(f_t(x) - f_t(z^t)\big)$. Since we wish to reduce incremental NIPS to a setting where the analysis of the batch method applies, we must ensure that the norm of the error term is bounded. Lemma 4.7 proves such a bound; but first we need to prove two auxiliary results.

Lemma 4.5 (Bounded increment). *Let z^{t+1} be computed by (4.33). Then,*

$$\text{if} \quad \mathcal{O} = \text{Id}, \quad \text{then} \quad \left\| z^{t+1} - z^t \right\| = \eta \left\| \nabla f_t(z^t) \right\| \tag{4.34}$$

$$\text{if} \quad \mathcal{O} = \Pi_{\mathcal{X}}, \quad \text{then} \quad \left\| z^{t+1} - z^t \right\| \leq \eta \left\| \nabla f_t(z^t) \right\| \tag{4.35}$$

$$\text{if} \quad \mathcal{O} = \text{prox}_g^\eta, \quad s^t \in \partial g(z^t), \quad \text{then} \quad \left\| z^{t+1} - z^t \right\| \leq 2\eta \left\| \nabla f_t(z^t) + s^t \right\|. \tag{4.36}$$

Proof. Relation (4.34) is obvious, and (4.35) follows immediately from nonexpansivity of projections. To prove (4.36), notice that definition (4.4) implies the inequality

$$\tfrac{1}{2} \left\| z^{t+1} - z^t + \eta \nabla f_t(z^t) \right\|^2 + \eta g(z^{t+1}) \leq \tfrac{1}{2} \left\| \eta \nabla f_t(z^t) \right\|^2 + \eta g(z^t),$$

$$\tfrac{1}{2} \left\| z^{t+1} - z^t \right\|^2 \leq \eta \langle \nabla f_t(z^t), z^t - z^{t+1} \rangle + \eta(g(z^t) - g(z^{t+1})).$$

Since g is convex, $g(z^{t+1}) \geq g(z^t) + \langle s_t, z^{t+1} - z^t \rangle$ for $s_t \in \partial g(z^t)$. Moreover,

$$\tfrac{1}{2} \left\| z^{t+1} - z^t \right\|^2 \leq \eta \langle s^t, z^t - z^{t+1} \rangle + \eta \langle \nabla f_t(z^t), z^t - z^{t+1} \rangle$$

$$\leq \eta \left\| s_t + \nabla f_t(z^t) \right\| \left\| z^t - z^{t+1} \right\|$$

$$\implies \quad \left\| z^{t+1} - z^t \right\| \leq 2\eta \left\| \nabla f_t(z^t) + s^t \right\|. \qquad \square$$

Lemma 4.6 (Incrementality error). *Let $x \equiv x^k$, and define*

$$\epsilon_t := \left\| \nabla f_t(z^t) - \nabla f_t(x) \right\|, \quad t = 1, \ldots, T. \tag{4.37}$$

Then, for each $t \geq 2$, the following bound on the error holds:

$$\epsilon_t \leq 2\eta L \sum\nolimits_{j=1}^{t-1} (1 + 2\eta L)^{t-1-j} \left\| \nabla f_j(x) + s^j \right\|, \quad t = 2, \ldots, T. \tag{4.38}$$

Proof. The proof extends the differentiable case treated in [34]. We proceed by induction. The base case is $t = 2$, for which we have

$$\epsilon_2 = \left\| \nabla f_2(z^2) - \nabla f_2(x) \right\| \leq L \left\| z^2 - x \right\| = L \left\| z^2 - z^1 \right\| \stackrel{(4.36)}{\leq} 2\eta L \left\| \nabla f_1(x) + s^1 \right\|.$$

Assume inductively that (4.38) holds for $t \leq r < T$, and consider $t = r + 1$. Then,

$$\epsilon_{r+1} = \left\| \nabla f_{r+1}(z^{r+1}) - \nabla f_{r+1}(x) \right\| \leq L \left\| z^{r+1} - x \right\|$$

$$= L \left\| \sum\nolimits_{j=1}^{r} (z^{j+1} - z^j) \right\| \leq L \sum\nolimits_{j=1}^{r} \left\| z^{j+1} - z^j \right\|$$

$$\stackrel{\text{Lemma 4.5}}{\leq} 2\eta L \sum\nolimits_{j=1}^{r} \left\| \nabla f_j(z^j) + s^j \right\|. \tag{4.39}$$

To complete the induction, first observe that $\|\nabla f_t(z^t) + s^t\| \leq \|\nabla f_t(x) + s^t\| + \epsilon_t$, so that on invoking the induction hypothesis we obtain for $t = 2, \ldots, r$,

$$\|\nabla f_t(z^t)\| \leq \|\nabla f_t(x)\| + 2\eta L \sum_{j=1}^{t-1} (1 + 2\eta L)^{t-1-j} \|\nabla f_j(x) + s^j\|. \quad (4.40)$$

Combining inequality (4.40) with (4.39) we further obtain

$$\epsilon_{r+1} \leq 2\eta L \sum_{j=1}^{r} \left(\|\nabla f_j(x) + s^j\| + 2\eta L \sum_{l=1}^{j-1} (1 + L\eta)^{j-1-l} \|\nabla f_l(x) + s^l\| \right).$$

Writing $\beta_j \equiv \|\nabla f_j(x) + s^j\|$ a simple manipulation of the above inequality yields

$$
\begin{aligned}
\epsilon_{r+1} &\leq\ 2\eta L \beta_r + \sum_{l=1}^{r-1} \left(2\eta L + 4\eta^2 L^2 \sum_{j=l+1}^{r} (1 + 2\eta L)^{j-l-1} \right) \beta_l \\
&=\ 2\eta L \beta_r + \sum_{l=1}^{r-1} \left(2\eta L + 4\eta^2 L^2 \sum_{j=0}^{r-l-1} (1 + 2\eta L)^{j} \right) \beta_l \\
&=\ 2\eta L \beta_r + \sum_{l=1}^{r-1} 2\eta L (1 + 2\eta L)^{r-l} \beta_l \quad =\quad 2\eta L \sum_{l=1}^{r} (1 + 2\eta L)^{r-l} \beta_l.
\end{aligned}
$$

$\square$

Now we are ready to bound the error, which is done by Lemma 4.7 below.

Lemma 4.7 (Bounded error). *If for all $x \in \mathcal{X}$, $\|\nabla f_t(x)\| \leq M$ and $\|\partial g(x)\| \leq G$, then $\|e(x)\| \leq K$ for some constant $K > 0$.*

Proof. If z^{t+1} is computed by (4.33), $\mathcal{O} = \mathrm{prox}_g$, and $s^t \in \partial g(z^t)$, then

$$\|z^{t+1} - z^t\| \ \leq\ 2\eta \|\nabla f_t(z^t) + s^t\|. \quad (4.41)$$

Using (4.41) we can bound the error incurred upon using z^t instead of x^k. Specifically, if $x \equiv x^k$, and

$$\epsilon_t := \|\nabla f_t(z^t) - \nabla f_t(x)\|, \quad t = 1, \ldots, T, \quad (4.42)$$

then Lemma 4.6 shows the following bound

$$\epsilon_t \leq 2\eta L \sum_{j=1}^{t-1} (1 + 2\eta L)^{t-1-j} \|\nabla f_j(x) + s^j\|, \quad t = 2, \ldots, T. \quad (4.43)$$

Since $\epsilon_1 = 0$, we have

$$
\begin{aligned}
\|e(x)\| \leq \sum_{t=2}^{T} \epsilon_t &\overset{(4.43)}{\leq} 2\eta L \sum_{t=2}^{T} \sum_{j=1}^{t-1} (1 + 2\eta L)^{t-1-j} \beta_j \\
&= 2\eta L \sum_{t=1}^{T-1} \beta_t \left(\sum_{j=0}^{T-t-1} (1 + 2\eta L)^{j} \right) \\
&= \sum_{t=1}^{T-1} \beta_t \left((1 + 2\eta L)^{T-t} - 1 \right) \\
&\leq \sum_{t=1}^{T-1} (1 + 2\eta L)^{T-t} \beta_t \\
&\leq (1 + 2\eta L)^{T-1} \sum_{t=1}^{T-1} \|\nabla f_t(x) + s^t\| \\
&\leq C_1 (T-1)(M + G) =: K.
\end{aligned}
$$

$\square$

Thanks to the error bounds established above, convergence of incremental NIPS follows immediately from Theorem 4.2; we omit details for brevity.

4.4 Application to Matrix Factorization

The main contribution of our paper is the new NIPS framework, and a specific application is not one of the prime aims of this paper. We do, however, provide an illustrative application of NIPS to a challenging nonconvex problem: *sparsity regularized low-rank matrix factorization*

$$\min_{X,A\geq 0} \quad \tfrac{1}{2}\|Y - XA\|_{\mathrm{F}}^2 + \psi_0(X) + \sum_{t=1}^{T} \psi_t(a_t), \tag{4.44}$$

where $Y \in \mathbb{R}^{m\times T}$, $X \in \mathbb{R}^{m\times K}$ and $A \in \mathbb{R}^{K\times T}$, with $a_1,\ldots,a_T$ as its columns. Problem (4.44) generalizes the well-known nonnegative matrix factorization (NMF) problem of [25] by permitting arbitrary Y (not necessarily nonnegative), and adding regularizers on X and A. A related class of problems was studied in [27], but with a crucial difference: the formulation in [27] *does not* allow nonsmooth regularizers on X. The class of problems studied in [27] is in fact a subset of those covered by NIPS. On a more theoretical note, [27] considered stochastic-gradient like methods whose analysis requires computational errors and stepsizes to vanish, whereas our method is deterministic and allows nonvanishing stepsizes and errors.

Following [27] we also rewrite (4.44) in a form more amenable to NIPS. We eliminate A and consider the nonnegatively constrained optimization problem

$$\min_X \quad \Phi(X) := \sum_{t=1}^{T} f_t(X) + g(X), \quad \text{where} \quad g(X) := \psi_0(X) + \delta(X|\geq 0), \tag{4.45}$$

and where each $f_t(X)$ for $1 \leq t \leq T$ is defined as

$$f_t(X) := \min_a \quad \tfrac{1}{2}\|y_t - Xa\|^2 + g_t(a), \tag{4.46}$$

where $g_t(a) := \psi_t(a) + \delta(a|\geq 0)$. For simplicity, assume that (4.46) attains its unique[2] minimum, say a^*, then $f_t(X)$ is differentiable and we have $\nabla_X f_t(X) = (Xa^* - y_t)(a^*)^T$. Thus, we can instantiate (4.33), and all we need is a subroutine for solving (4.46).[3]

We present empirical results on the following two variants of (4.45): (i) pure unpenalized NMF ($\psi_t \equiv 0$ for $0 \leq t \leq T$) as a baseline; and (ii) sparsity penalized NMF where $\psi_0(X) \equiv \lambda\|X\|_1$ and $\psi_t(a_t) \equiv \gamma\|a_t\|_1$. Note that without the nonnegativity constraints, (4.45) is similar to sparse-PCA.

---

[2]If not, then at the expense of more notation, we can add a strictly convex perturbation to ensure uniqueness; this error can be absorbed into the overall computational error.

[3]In practice, we use *mini-batches* for all the algorithms.

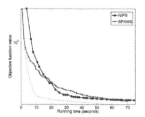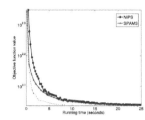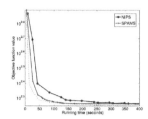

FIGURE 4.1: Running times of NIPS (Matlab) versus SPAMS (C++) for NMF on RAND, CBCL, and YALE datasets. Initial objective values and tiny runtimes have been suppressed for clarity.

We use the following datasets and parameters:

(i) RAND: 4000×4000 dense random (uniform $[0, 1]$); rank-32 factorization; $(\lambda, \gamma) = (10^{-5}, 10)$;

(ii) CBCL: CBCL database [38]; 361×2429; rank-49 factorization;

(iii) YALE: Yale B Database [26]; 32256×2414 matrix; rank-32 factorization;

(iv) WEB: Web graph from Google; sparse 714545×739454 (empty rows and columns removed) matrix; ID: 2301 in the sparse matrix collection [10]); rank-4 factorization; $(\lambda = \gamma = 10^{-6})$.

On the NMF baseline (Fig. 4.1), we compare NIPS against the well optimized state-of-the-art C++ toolbox SPAMS (version 2.3) [27]. We compare against SPAMS only on dense matrices, as its NMF code seems to be optimized for this case. Obviously, the comparison is not fair: unlike SPAMS, NIPS and its subroutines are all implemented in MATLAB, and they run equally easily on large sparse matrices. Nevertheless, NIPS proves to be quite competitive: Fig. 4.1 shows that our MATLAB implementation runs only slightly slower than SPAMS. We expect a well-tuned C++ implementation of NIPS to run at least 4–10 times faster than the MATLAB version—the dashed line in the plots visualizes what such a mere 3X-speedup to NIPS might mean.

Figure 4.2 shows numerical results comparing the stochastic generalized gradient (SGGD) algorithm of [13] against NIPS, when started at the same point. As is well-known, SGGD requires careful stepsize tuning; so we searched over a range of stepsizes, and have reported the best results. NIPS too requires some stepsize tuning, but to a much lesser extent than SGGD. As predicted, the solutions returned by NIPS have objective function values lower than SGGD, and have greater sparsity.

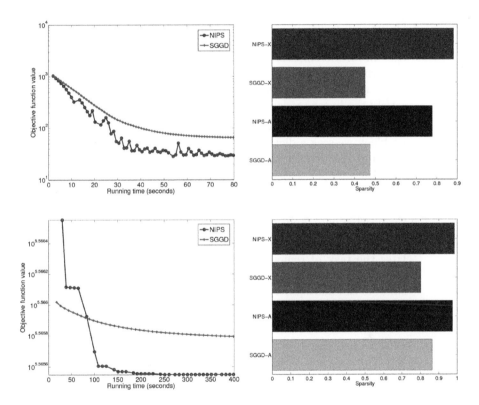

FIGURE 4.2: Sparse NMF: NIPS versus SGGD. The bar plots show the sparsity (higher is better) of the factors X and A. Left plots for RAND dataset; right plots for WEB. SGGD yields slightly worse objective function values and significantly less sparse solutions than NIPS.

4.5   Other Applications

We mention below a few other applications where we have used the NIPS framework successfully. While it lies outside the scope of this chapter to cover the details of these applications, we refer the reader to research articles that include the requisite details.

- Online multiframe blind deconvolution [20]. In this application, a slightly modified version of NIPS is used for processing a stream of blurry images in an incremental fashion. The end goal is to obtain a sharp reconstruction of a single underlying image. The optimization problem is nonconvex because given observations $y_1, \ldots, y_T$, which are assumed to satisfy the linear model $y_i \approx A_i x$, we need to recover both A_i and x.

- Generalized dictionary learning for positive definite tensors [36]. In this problem, we seek a dictionary whose "atoms" can be sparsely combined to reconstruct a set of matrices. The key difference from ordinary dictionary learning [23] is that the observations are positive definite matrices, so the dictionary atoms must be positive definite, too. The problem fits in the NIPS framework (as for NMF, subproblems relied on a nonnegative least-squares solver [22] and a nonsmooth convex solver [21]).

- Denoising signals with spiky (sparse) noise [6]. This application formulates the task of removing spiky noise from signals by formulating it as a nonconvex problem with sparsity regularization, and was hence a suitable candidate for NIPS.

4.6   Discussion

This chapter discussed a general optimization framework called NIPS that can solve a broad class of nonconvex composite objective (regularized) problems. Our analysis is inspired by [33], and we extend the results of [33] to admit problems that are *strictly* more general by handling nonsmooth components via proximity operators. NIPS permits nonvanishing perturbations, which is a useful practical feature. We exploited the perturbation analysis to derive both batch and incremental versions of NIPS. Finally, experiments with medium to large matrices showed that NIPS is competitive with state-of-the-art methods; NIPS was also seen to outperform the stochastic generalized gradient method.

We conclude by mentioning that NIPS includes numerous algorithms and problem settings as special cases. Example are: forward-backward splitting with convex costs, incremental forward-backward splitting (convex), gradient projection (both convex and nonconvex), the proximal-point algorithm, and so on. Thus, it will be valuable to investigate if some of the theoretical results for these methods can be carried over to NIPS.

The most important theoretical question worth pursuing at this point is a less pessimistic convergence analysis for the scalable incremental version of NIPS than implied by Lemma 4.7.

Bibliography

[1] H. Attouch, J. Bolte, and B. Svaiter. Convergence of descent methods for semi-algebraic and tame problems: proximal algorithms, forward-backward splitting, and regularized Gauss-Seidel methods. *Mathematical Programming*, 137(1-2):91–129, 2013.

[2] F. Bach, R. Jenatton, J. Mairal, and G. Obozinski. Convex optimization with sparsity-inducing norms. In S. Sra, S. Nowozin, and S. J. Wright, editors, *Optimization for Machine Learning*. MIT Press, 2011.

[3] A. Beck and M. Teboulle. A Fast Iterative Shrinkage-Thresholding Algorithm for Linear Inverse Problems. *SIAM J. Imgaging Sciences*, 2(1):183–202, 2009.

[4] D. P. Bertsekas. *Nonlinear Programming*. Athena Scientific, second edition, 1999.

[5] D. P. Bertsekas. Incremental Gradient, Subgradient, and Proximal Methods for Convex Optimization: A Survey. In S. Sra, S. Nowozin, and S. J. Wright, editors, *Optimization for Machine Learning*. MIT Press, 2011.

[6] A. Cherian, S. Sra, and N. Papanikolopoulos. Denoising sparse noise via online dictionary learning. In *IEEE International Conference on Acoustics, Speech, and Signal Processing (ICASSP)*, May 2011.

[7] F. H. Clarke. *Optimization and nonsmooth analysis*. John Wiley & Sons, Inc., 1983.

[8] P. L. Combettes and V. R. Wajs. Signal recovery by proximal forward-backward splitting. *Multiscale Modeling and Simulation*, 4(4):1168–1200, 2005.

[9] Patrick L. Combettes and Jean-Christophe Pesquet. Proximal splitting methods in signal processing. In *Fixed-Point Algorithms for Inverse Problems in Science and Engineering*, pages 185–212. Springer, 2011.

[10] Timothy A. Davis and Yifan Hu. The University of Florida sparse matrix collection. *ACM Transactions on Mathematical Software (TOMS)*, 38(1):1, 2011.

[11] J. Duchi and Y. Singer. Online and Batch Learning using Forward-Backward Splitting. *J. Mach. Learning Res. (JMLR)*, 10: 2873–2898, Dec 2009.

[12] Y. M. Ermoliev and V. I. Norkin. Stochastic generalized gradient method with application to insurance risk management. Technical Report IR-97-021, IIASA, Austria, Apr. 1997.

[13] Y. M. Ermoliev and V. I. Norkin. Stochastic generalized gradient method for nonconvex nonsmooth stochastic optimization. *Cybernetics and Systems Analysis*, 34:196–215, 1998.

[14] M. Fukushima and H. Mine. A generalized proximal point algorithm for certain non-convex minimization problems. *Int. J. Systems Science*, 12(8):989–1000, 1981.

[15] E. M. Gafni and D. P. Bertsekas. Two-metric projection methods for constrained optimization. *SIAM Journal on Control and Optimization*, 22(6):936–964, 1984.

[16] A. A. Gaivoronski. Convergence properties of backpropagation for neural nets via theory of stochastic gradient methods. Part 1. *Optimization Methods and Software*, 4(2):117–134, 1994.

[17] Saeed Ghadimi and Guanghui Lan. Accelerated gradient methods for nonconvex nonlinear and stochastic programming. *arXiv:1310.3787*, 2013.

[18] Saeed Ghadimi and Guanghui Lan. Stochastic First- and Zeroth-order Methods for Nonconvex Stochastic Programming. *arXiv:1309.5549*, 2013.

[19] S. Haykin. *Neural Networks: A Comprehensive Foundation*. Prentice Hall PTR, 1st edition, 1994.

[20] M. Hirsch, S. Harmeling, S. Sra, and B. Schölkopf. Online multi-frame blind deconvolution with super-resolution and saturation correction. *Astronomy & Astrophysics (AA)*, Feb. 2011. 11 pages.

[21] D. Kim, S. Sra, and I. S. Dhillon. A scalable trust-region algorithm with application to mixed-norm regression. In *International Conference on Machine Learning (ICML)*, 2010.

[22] D. Kim, S. Sra, and I. S. Dhillon. A non-monotonic method for large-scale non-negative least squares. *Optimization Methods and Software (OMS)*, Dec. 2011. 28 pages.

[23] K. Kreutz-Delgado, J. F. Murray, B. D. Rao, K. Engan, T.-W. Lee, and T. J. Sejnowski. Dictionary learning algorithms for sparse representation. *Neural Computation*, 15:349–396, 2003.

[24] D. Kundur and D. Hatzinakos. Blind image deconvolution. *IEEE Signal Processing Magazine*, 13(3):43–64, May 1996.

[25] D. D. Lee and H. S. Seung. Algorithms for nonnegative matrix factorization. In *Advances in Neural Information Processing Systems (NIPS)*, pages 556–562, 2000.

[26] K.C. Lee, J. Ho, and D. Kriegman. Acquiring linear subspaces for face recognition under variable lighting. *IEEE Transactions on Pattern Analysis and Machine Intelligence*, 27(5):684–698, 2005.

[27] J. Mairal, F. Bach, J. Ponce, and G. Sapiro. Online Learning for Matrix Factorization and Sparse Coding. *Journal of Machine Learning Research (JMLR)*, 11:10–60, 2010.

[28] O. L. Mangasarian. Mathematical Programming in Neural Networks. *Informs J. Computing*, 5(4):349–360, 1993.

[29] J. J. Moreau. Proximité et dualité dans un espace hilbertien. *Bull. Soc. Math. France*, 93:273–299, 1965.

[30] Yu. Nesterov. Gradient methods for minimizing composite objective function. Technical Report 2007/76, Université catholique de Louvain, Center for Operations Research and Econometrics (CORE), September 2007.

[31] R. T. Rockafellar and R. J.-B. Wets. *Variational analysis*. Springer, 1998.

[32] M. Schmidt, N. Le Roux, and F. Bach. Convergence rates of inexact proximal-gradient methods for convex optimization. In *Advances in Neural Information Processing Systems (NIPS)*, 2011.

[33] M. V. Solodov. Convergence analysis of perturbed feasible descent methods. *Journal Optimization Theory and Applications*, 93(2):337–353, 1997.

[34] M. V. Solodov. Incremental gradient algorithms with stepsizes bounded away from zero. *Computational Optimization and Applications*, 11:23–35, 1998.

[35] M. V. Solodov and S. K. Zavriev. Error stability properties of generalized gradient-type algorithms. *Journal Optimization Theory and Applications*, 98(3):663–680, 1998.

[36] S. Sra and A. Cherian. Generalized Dictionary Learning for Symmetric Positive Definite Matrices with Application to Nearest Neighbor Retrieval. In *European Conf. Machine Learning (ECML)*, Sept. 2011.

[37] S. Sra, S. Nowozin, and S. J. Wright, editors. *Optimization for Machine Learning*. MIT Press, 2011.

[38] K.-K. Sung. *Learning and Example Selection for Object and Pattern Recognition*. PhD thesis, MIT, Artificial Intelligence Laboratory and Center for Biological and Computational Learning, Cambridge, MA, 1996.

[39] Silvia Villa, Saverio Salzo, Luca Baldassarre, and Alessandro Verri. Accelerated and inexact forward-backward algorithms. *SIAM Journal on Optimization*, 23(3):1607–1633, 2013.

[40] S. J. Wright, R. D. Nowak, and M. A. T. Figueiredo. Sparse reconstruction by separable approximation. *IEEE Trans. Sig. Proc.*, 57(7):2479–2493, 2009.

Chapter 5

Learning Constrained Task Similarities in Graph-Regularized Multi-Task Learning

Rémi Flamary

Laboratoire Lagrange, Observatoire de la Côte d'Azur, Université de Nice Sophia-Antipolis

Alain Rakotomamonjy

LITIS, Université de Rouen

Gilles Gasso

LITIS, INSA de Rouen

This chapter addresses the problem of learning constrained task related-ness in a graph-regularized multi-task learning framework. In such a context, the weighted adjacency matrix of a graph encodes the knowledge on task similarities and each entry of this matrix can be interpreted as a hyperpa-rameter of the learning problem. This task relation matrix is learned via a bilevel optimization procedure where the outer level optimizes a proxy of the generalization errors over all tasks with respect to the similarity matrix and the inner level estimates the parameters of the tasks knowing this similarity matrix. Constraints on task similarities are also taken into account in this optimization framework and they allow the task similarity matrix to be more interpretable, for instance, by imposing a sparse similarity matrix. Since the global problem is non-convex, we propose a non-convex proximal algorithm that provably converges to a stationary point of the problem. Empirical ev-idence illustrates the approach is competitive compared to existing methods that also learn task relation and exhibits an enhanced interpretability of the learned task similarity matrix.

5.1   Introduction

Multi-task learning (MTL) has gained a lot of attention over the past years. Given a set of different but related tasks, the underlying idea is to jointly learn these tasks by exploiting their relationships. Such a procedure has proved useful (in terms of generalization ability) when only a few samples are available for the tasks [5, 1, 26, 21]. One of the most important problems in multi-task learning is the assessment of task relatedness. Existing approaches seek task relationship either in input space, feature space [1, 27], or output space [11]. They consider relations between tasks through dedicated regularizations or based on Bayesian priors.

To be more concrete, when jointly learning the tasks, regularization ap-proaches assume that the tasks share a common low-dimensional latent sub-space or their parameters are close to the average parameter vector [10]. The links between tasks can also be enforced by imposing a joint sparsity pattern [22, 1] across tasks. However, such a similarity is globally imposed for all tasks and may hinder good generalization performance mainly when unrelated tasks are forced to borrow similar characteristics. Hence, more elaborate approaches such as pairwise similar tasks [10], clustering of tasks [10, 15, 16], or hierarchi-cally structured tasks [28, 33] were proposed in order to cope with this issue. For instance, Widmer et al. [33] supposed the knowledge of a hierarchical tree modeling task dependency while Xing et al. [28] used agglomerative clustering to find the tree, which is further applied to learn task parameters. Contrarily, clustering approaches due to [15], [16], [19], or [35] attempt to discover groups of similar tasks along with the estimation of their parameters. In the same vein, Zhang et al. [34] characterized task relatedness via a covariance matrix. Their formulation encompasses various existing methods including pairwise similarity constraints or cluster constraints. Solving for the covariance matrix turns out to be a problem of learning distance metric between tasks.

Most methods for learning task similarities provide good empirical performance, but few of them aim at enhancing the interpretability of the learned task relations. For instance, Zhang et al. [34] learn a dense task covariance matrix that depicts relations between tasks. Since this matrix is dense, it is thus difficult to interpret which task relations are the most relevant for the learning problem. In this work, we look at learning *interpretable* task similarities in multi-task learning problems. We focus on a popular multi-task framework denoted as a graph-based regularized framework [10]. As formally defined in the sequel, in this framework, task relations are represented as a graph whose nodes are the tasks, and the weighted edges encode some knowledge over similarities between tasks. This framework has been shown to be of practical interest [9, 32] and benefits from very efficient algorithms for solving the related multi-task optimization problems [32].

Our objective and proposal in this chapter is to learn the adjacency matrix of the task relations graph, jointly with the task decision function parameters, while making the graph as interpretable as possible. Hence, we may accept some slight loss in generalization performance if the gain in graph interpretability is important. This interpretability of the adjacency matrix is achieved by incorporating in the global learning problem some specific constraints over the graph parameters. The constraints that we consider in the sequel are usually sparsity-inducing penalties that are enforcing the tasks to become unrelated.

Our main contribution hereafter is to propose a novel procedure for learning similarities between tasks in graph-based multi-task learning. As detailed in the sequel, since in this framework the relation between a pair of tasks can be interpreted as a hyper-parameter of the global multi-task model, we address the problem as a hyper-parameter optimization issue ([2, 6]). Typically, when few hyper-parameters have to be optimized, a cross-validation procedure, aiming at optimizing an estimation of the generalization error, is employed in conjunction with a grid-search over the hyper-parameter values [14, 4]. Since in our framework, the number of hyper-parameters (typically $\mathcal{O}(T^2)$ parameters, for T tasks) to optimize make this approach intractable, the method we advocate consists of a bilevel approach: at the outer level, a generalization criterion over all tasks is employed to measure the goodness of task relatedness parameters and this criterion is thus optimized with respect to these parameters under some sparsity-inducing constraints. The inner level is devoted to the optimization of task parameters for a fixed task relation graph. Due to the non-convexity of the generalization errors with respect to task similarity parameters, the overall problem is non-convex. Fortunately, the inner problem we design is convex and, depending on the loss functions considered, it may admit a closed-form solution. We solve this bilevel problem through a non-convex proximal approach with guaranteed convergence properties. The flexibility of the approach allows the use of a broad range of generalization error proxies that can be adapted to the MTL problem at hand. It also allows easy incorporation of constraints over the task relations such as sparsity, *link*, or *cannot-link* constraints in the matrix similarity learning process. As a consequence, the learned task-similarity matrix is sparse and provides improved interpretability of connections between tasks. The experimental results of synthetic and real-world problems clearly support this evidence.

The rest of the chapter is organized as follows: Section 5.2 describes the graph-based multi-task learning setting we are interested in and states how the task parameters are obtained once the task similarity matrix is fixed. The learning of this matrix is explained in Section 5.3 where we formulate the bilevel optimization problem and the non-convex proximal algorithm used to solve it. Finally, empirical comparisons illustrate the compelling performance of the approach. In particular, these experiments show the ability of our algorithm to unravel the underlying structure (groups or manifold) of the tasks and emphasize the interpretability of the results.

5.2   Similarity Based Multi-Task Learning

Before describing how task relatedness is learned, we first present the general multi-task learning framework we are dealing with, as well as the regularizer we have considered for inducing transfer between tasks.

5.2.1   Multi-Task Learning Framework

Assume we are given T learning tasks to be learned from T different datasets $(\mathbf{x}_{i,1}, y_{i,1})_{i=1}^{n_1}, \cdots, (\mathbf{x}_{i,T}, y_{i,T})_{i=1}^{n_T}$, where any $\mathbf{x}_{i,\cdot} \in \mathbb{R}^d$ and $y_{i,\cdot} \in \mathbb{R}$, and n_t denotes the t-th dataset size. In the sequel, we will represent the training examples $\{\mathbf{x}_{i,t}\}_{i=1}^{n_t}$ in a matrix form as $\mathbf{X}_t \in \mathbb{R}^{n_t \times d}$ and the corresponding labels gathered in vector $\mathbf{y}_t \in \mathbb{R}^{n_t}$. For a given task t, we are looking for a linear prediction function $f_t(\mathbf{x})$ of the form

$$f_t(\mathbf{x}) = \mathbf{w}_t^\top \mathbf{x} + b_t \tag{5.1}$$

with $\mathbf{w}_t \in \mathbb{R}^d$ and $b_t \in \mathbb{R}$ being the linear function parameters. Basically, $f_t(\mathbf{x})$ depicts the presumable dependencies between a given example $\mathbf{x}$ and its associated label y.

Multi-task methods aim at learning all T decision functions in a simultaneous way while imposing some constraints that induce relatedness between tasks. Hence, most multi-task learning problems can be cast as the following optimization setup:

$$\min_{\{\mathbf{w}_t\},\{b_t\}} \sum_{t,i} L(f_t(\mathbf{x}_{i,t}), y_{i,t}) + \Omega(\mathbf{w}_1, \cdots, \mathbf{w}_T) \tag{5.2}$$

where $L(f_t(\mathbf{x}), y)$ is a loss function measuring discrepancy between the actual and predicted output related to an example $\mathbf{x}$, and Ω a regularizer inducing task relatedness thus involving all vectors $\{\mathbf{w}_t\}$.

5.2.2 Similarity-Based Regularization

One typical issue in multi-task learning is the choice of a regularization term Ω that efficiently helps in improving the generalization performances of the prediction functions. Indeed, most of the existing MTL regularization terms are based on a strong *prior knowledge* about the task relatedness. We can for instance mention regularizers that enforce similarity of task parameters $\{\mathbf{w}_t\}$ to the average parameter vector $\frac{1}{T}\sum_t \mathbf{w}_t$ [9, 10], that make classifiers belong to a low dimensional linear subspace [1]. Other regularizers induce the classifiers to be agglomerated into clusters [15], or compel tasks to share a common subset of discriminative kernels [23] or even impose tasks to be similar according to pre-defined task networks [17].

Because it encompasses several forms of the above-mentioned regularizers, we focus on the graph-based regularization term proposed by Evgeniou et al. [10] that induces pairwise similarity between tasks. This regularizer is defined as

$$\Omega(\{\mathbf{w}_t\}_{t=1}^T, \{\lambda_t\}, \mathbf{P}) = \sum_t \lambda_t \|\mathbf{w}_t\|_2^2 + \sum_{t,s} \rho_{t,s} \|\mathbf{w}_t - \mathbf{w}_s\|_2^2 \tag{5.3}$$

where $\lambda_t \in \mathbb{R}^+ \setminus \{0\}$ and $\mathbf{P} \in (\mathbb{R}^+)^{T \times T}$, a matrix of general term $\rho_{t,s}$ (i.e., $\mathbf{P} = [\rho_{t,s}]_{t,s=1}^T$), are the regularization hyper-parameters. The first term of this regularizer corresponds to the classical ℓ_2-norm regularization (ridge) while the second one promotes a pairwise task similarity imposed by the $\rho_{t,s}$ parameters. From the graph point of view, the matrix $\mathbf{P}$ is the weighted graph adjacency matrix and it reflects the relationship between tasks, as obviously, a large $\rho_{t,s}$ value enforces tasks s and t to be similar while if $\rho_{t,s} = 0$ then these tasks will likely be unrelated (in the ℓ_2-norm sense). We have imposed $\mathbf{P}(t,t) = 0 \; \forall t$ as these diagonal terms of the matrix have no impact on the cost function. Furthermore, in order to reduce the number of hyper-parameters in the model and because it intuitively makes sense, we also have considered $\mathbf{P}$ to be a symmetric matrix. Graph-regularized multi-task learning problems are denoted as such because the regularizer given in Equation (5.3) can also be interpreted as the following. Indeed, by defining matrix $\mathbf{W} = [\mathbf{w}_1 \cdots \mathbf{w}_T]$, one can notice that Equation (5.3) equivalently writes

$$\Omega(\mathbf{W}, \{\lambda_t\}, \mathbf{P}) = \text{trace}\left(\mathbf{W}\Gamma\mathbf{W}^\top\right) \tag{5.4}$$

where $\Gamma = \Lambda + \mathbf{L}$ and Λ is a diagonal matrix with entries λ_t and $\mathbf{L}$ is the

Laplacian of the graph where the vertices are corresponding to the tasks. Assuming the edges are parameters $\rho_{t,s}$, the Laplacian matrix writes $\mathbf{L} = \mathbf{D} + \mathbf{P}$ where $\mathbf{D}$ is a diagonal matrix with elements $\mathbf{D}(t,t) = \sum_{s=1}^{T} \rho_{t,s}$.

Our main contribution is to propose a framework for learning this matrix $\mathbf{P}$ of task relatedness and to make this matrix as interpretable as possible by imposing some constraints on its entries. Because the matrix $\mathbf{P}$ can also be considered as a matrix of hyper-parameters, we introduce here a problem where these hyper-parameters are learned with respect to a proxy of the generalization error. This contribution is of importance for obtaining prediction functions with good generalization capabilities as well as a task similarity matrix that is interpretable. Indeed, using our novel formulation of the problem, it becomes easy to impose single or group sparsity-inducing constraints over entries of $\mathbf{P}$.

Before providing the details of how these task relations are learned, we show in the next paragraph how Problem (5.2) can be solved for fixed λ_t and matrix $\mathbf{P}$.

5.2.3   Solving the Graph-Regularized Multi-Task Learning Problem

We focus now on solving problem (5.2) with the regularization term defined as in Equation (5.3). For the sake of clarity, we restrict ourselves to the squared loss function, denoted as $L(f(\mathbf{x}), y) = \frac{1}{2}(f(\mathbf{x}) - y)^2$, although our algorithm can be applied to other loss functions such as the hinge loss. We discuss this point in the sequel.

Using a quadratic loss function and matrix notation, and based on the regularization given in Equation (5.3), Problem (5.2) reads

$$\min_{\{\mathbf{w}_t\},\{b_t\}} \quad J(\{\mathbf{w}_t\}, \{b_t\}, \mathbf{P}, \{\lambda_t\}) \qquad (5.5)$$

where the objective function is

$$J(\cdot) = \frac{1}{2}\sum_t \|\mathbf{y}_t - \mathbf{X}_t\mathbf{w}_t - b_t\,\mathbb{1}_t\|_2^2 + \sum_t \lambda_t \|\mathbf{w}_t\|_2^2 + \sum_{t,s} \rho_{t,s}\|\mathbf{w}_t - \mathbf{w}_s\|_2^2 \quad (5.6)$$

with $\mathbb{1}_t \in \mathbb{R}^{n_t}$ being a vector of ones. Note that for $\lambda_t > 0, \forall t$ and $\rho_{t,s} \geq 0, \forall t, s$, this problem is strictly convex and admits a unique solution. For fixed parameters $\{\lambda_t\}_{t=1}^{T}$ and $\mathbf{P}$, a closed-form solution of this problem can be obtained by solving the linear system related to the normal equations. The gradient of $J(\cdot)$ with respect to the prediction function parameters of task k is given by:

$$\nabla_{\mathbf{w}_k} J = \mathbf{Q}_k\mathbf{w}_k + b_k\mathbf{X}_k^{\top}\mathbb{1}_k - 4\sum_t \rho_{t,k}\mathbf{w}_t - \mathbf{c}_k \qquad (5.7)$$

where $\mathbf{I}$ is the identity matrix, $\mathbf{Q}_k = \mathbf{X}_k^{\top}\mathbf{X}_k + (2\lambda_k + 4p_k)\mathbf{I} \in \mathbb{R}^{d\times d}$,

$\mathbf{c}_k = \mathbf{X}_k^\top \mathbf{y}_k \in \mathbb{R}^d$ and finally $p_k = \sum_t \rho_{t,k}$. Similarly, the gradient of $J(\cdot)$ with respect to the bias term b_k takes the form

$$\nabla_{b_k} J = \mathbb{1}_k^\top \mathbf{X}_k \mathbf{w}_k + n_k b_k - \mathbb{1}_k^\top \mathbf{y}_k. \tag{5.8}$$

From the gradients (5.7) and (5.8) and the resulting optimality conditions, solution of Problem (5.5) is obtained by solving the system:

$$\mathbf{A}\boldsymbol{\beta} = \mathbf{c} \tag{5.9}$$

where $\boldsymbol{\beta} = \begin{bmatrix} \tilde{\mathbf{w}}_1^\top & \cdots & \tilde{\mathbf{w}}_T^\top \end{bmatrix}^\top \in \mathbb{R}^{(d+1)\cdot T}$ is the vector containing all the prediction function parameters with $\tilde{\mathbf{w}}_k = \begin{bmatrix} \mathbf{w}_k^\top & b_k \end{bmatrix}^\top \in \mathbb{R}^{d+1}$, $\mathbf{c} = \begin{bmatrix} \tilde{\mathbf{c}}_1^\top & \cdots & \tilde{\mathbf{c}}_T^\top \end{bmatrix}^\top \in \mathbb{R}^{(d+1)\cdot T}$ with $\tilde{\mathbf{c}}_k = \begin{bmatrix} \mathbf{c}_k^\top & \mathbb{1}_k^\top \mathbf{y}_k \end{bmatrix}^\top$, and $\mathbf{A} \in \mathbb{R}^{(d+1)\cdot T \times (d+1)\cdot T}$ is a matrix of the form:

$$\mathbf{A} = \begin{bmatrix} \tilde{\mathbf{Q}}_1 & -4\rho_{1,2}\tilde{\mathbf{I}} & \cdots & -4\rho_{1,T}\tilde{\mathbf{I}} \\ -4\rho_{2,1}\tilde{\mathbf{I}} & \tilde{\mathbf{Q}}_2 & \cdots & -4\rho_{2,T}\tilde{\mathbf{I}} \\ \vdots & \vdots & \ddots & \vdots \\ -4\rho_{T,1}\tilde{\mathbf{I}} & -4\rho_{T,2}\tilde{\mathbf{I}} & \cdots & \tilde{\mathbf{Q}}_T \end{bmatrix} \tag{5.10}$$

involving the matrices

$$\tilde{\mathbf{Q}}_k = \begin{bmatrix} \mathbf{Q}_k & \mathbf{X}_k^\top \mathbb{1} \\ \mathbb{1}^\top \mathbf{X}_k & n_k \end{bmatrix}, \text{ and } \tilde{\mathbf{I}} = \begin{bmatrix} \mathbf{I} & 0 \\ 0 & 0 \end{bmatrix}.$$

Notice that for $\lambda_t > 0 \, \forall t$, the matrix $\mathbf{A}$ is full-rank regardless of the matrices $\{\mathbf{X}_t\}$ and thus the linear system (5.9) has a unique solution that provides the optimal parameters of all prediction functions $\{f_t\}_{t=1}^T$. Note that depending on the number of tasks and the dimensionality of the training examples this matrix $\mathbf{A}$ can be pretty large. In such a case, it can be beneficial to take advantage of the sparse structure of $\mathbf{A}$ for the linear system (5.9) resolution. For instance, a Gauss-Seidel procedure, which consists of optimizing alternatingly over the parameters of a given task, can be considered.

For optimizing the task similarity matrix $\mathbf{P}$, the linear system $\mathbf{A}\boldsymbol{\beta} = \mathbf{c}$ will be of paramount importance since it defines an implicit function that relates the optimal task parameters $\{\mathbf{w}_t, b_t\}_{t=1}^T$ to the entries of $\mathbf{P}$. Indeed, the bilevel approach we apply to determine $\mathbf{P}$ (see Equations (5.12) and (5.13)) in the next section requires the gradient of the estimated generalization error measure in function of $\mathbf{P}$. Owing to this equation we will be able to compute this gradient via the explicit expression of the gradient of $\boldsymbol{\beta}$ with respect to $\mathbf{P}$. Consequently, while we have stated that other loss functions can be considered in our approach, the solution of the graph-regularizer multi-task learning has to satisfy a linear system of the form $\mathbf{A}\boldsymbol{\beta} = \mathbf{c}$. For instance, the hinge loss function satisfies this property if the learned problem is solved in the dual [18, 31, 9].

5.3   Non-Convex Proximal Algorithm for Learning Similarities

The multi-task approach with fixed hyperparameters described in the above section is interesting in itself. However, it may be limited by the large number of regularization parameters to be chosen, namely all the $\{\lambda_t\}_{t=1}^T$ and the matrix of task similarities $\mathbf{P}$. When there exists a strong *prior knowledge* concerning task similarities, the $\mathbf{P}$ matrix might be pre-defined beforehand. When no prior information is available, $\mathbf{P}$ can be learned from training data as done by Zhang et al. [34]. In addition, when one's objective is also to gain some insights over the structure of the tasks and how they are related, then constraints on task similarities have to be imposed. In what follows, we describe our algorithm for learning the matrix $\mathbf{P}$ as well as the regularization parameters $\{\lambda_t\}$ in the context of graph-regularized multi-task learning.

5.3.1   Bilevel Optimization Framework

We learn the matrix task similarity $\mathbf{P}$ as well as the regularization parameters $\{\lambda_t\}_{t=1}^T$ by considering them as hyper-parameters and by minimizing an estimate of a generalization error denoted as $E(\cdot)$. This estimate E is naturally a function of all decision function parameters β. For addressing this problem, we consider a bilevel optimization problem similar to the one of Bennett et al. [3]: the outer level of the problem consists of minimizing E with respect to $\mathbf{P}$ and all $\{\lambda_t\}_{t=1}^T$ and the inner level aims at learning all decision function parameters β.

While several choices of E can be considered, for the sake of clarity, we have set $E(\cdot)$ to be a validation error of the form :

$$E(\beta^\star) = \sum_{t=1}^T \sum_i L_v(\tilde{y}_{i,t}, \tilde{\mathbf{x}}_{i,t}^\top \mathbf{w}_t^\star + b_t^\star) \qquad (5.11)$$

where $\beta^\star$ is a vector including the optimal decision function parameters $\mathbf{w}_t^\star$ and $b_t^\star$ for all tasks $t = 1, \ldots, T$. The sets $\{\tilde{\mathbf{x}}_{i,t}, \tilde{y}_{i,t}\}_{t=1}^T$ refer to some validation examples and $L_v(\cdot, \cdot)$ is a twice differentiable loss function that measures the discrepancy between the real and predicted output associated to an input example. Two kinds of loss have been considered in this work, one more adapted to regression tasks $L_v(y, \hat{y}) = (y - \hat{y})^2$ and another one more suited to classification tasks, which is a non-convex sigmoid function that smoothly approximates the $0 - 1$ loss function $L_v(y, \hat{y}) = \frac{1}{1+e^{\kappa y \hat{y}}}$ with $\kappa > 0$. Note that, at the expense of introducing some cumbersome notations, it is straightforward to modify Equation (5.11), so that generalization error estimate is a leave-one-out error or a k-fold cross-validation error.

Now that $E(\cdot)$ has been formally defined, we are in a position to state the bilevel optimization we are interested in. Indeed, since the vector β depends on the matrix $\mathbf{P}$ and the hyper-parameters λ_t, the bilevel optimization problem can be expressed as:

$$\min_{\boldsymbol{\theta}} \quad E(\boldsymbol{\beta}^*(\boldsymbol{\theta})) + \Omega_\theta(\boldsymbol{\theta}) \tag{5.12}$$

$$\text{with } \boldsymbol{\beta}^*(\boldsymbol{\theta}) = \operatorname*{argmin}_{\boldsymbol{\beta}} \quad J(\boldsymbol{\theta}, \boldsymbol{\beta}) \tag{5.13}$$

with $\boldsymbol{\theta} = \left[\lambda_1, \ldots, \lambda_T, \{\rho_{i,j}\}_{i=1,j>i}^{T} \right]^{\top}$, a vector of size $D = T + \frac{T(T-1)}{2}$ (considering symmetry of $\mathbf{P}$ and $\mathbf{P}(t,t) = 0, \forall t$), $J(\cdot)$ the objective function defined in Equation (5.6), and Ω_θ being a regularizer over the parameters $\boldsymbol{\theta}$. Typically, Ω_θ is related to the projection onto some convex and closed subset Θ of $\mathbb{R}^D$ that defines some constraints over the vector $\boldsymbol{\theta}$. The bilevel Problem (5.12) has a particular structure in that the inner Problem (5.13), which is actually Problem (5.5), is strictly convex for $\lambda_t > 0$, a condition that is guaranteed by some specific choice of Θ. This strict convexity is of primary importance since it allows us to compute the unique $\boldsymbol{\beta}^*$ for a given $\boldsymbol{\theta}$. In general cases, this Problem (5.12) is non-convex, but its structure suggests that a non-convex proximal splitting can be of interest. Indeed, since $E(\cdot)$ is supposed to be twice differentiable and Ω_θ a non-smooth function, a non-convex forward-backward splitting algorithm, such as the one proposed by [29], can be successfully lifted to our purpose, especially if the proximal operator of Ω_θ can be simply computed. For a proper convex function, this proximal operator is defined as [8]:

$$P_{\Omega_\theta}(\hat{\boldsymbol{\theta}}) = \arg\min_{\boldsymbol{\theta}} \frac{1}{2}\|\boldsymbol{\theta} - \hat{\boldsymbol{\theta}}\|_2^2 + \Omega_\theta(\boldsymbol{\theta}).$$

Hence, if Ω_θ is defined as the indicator over a convex set Θ, the proximal operator boils down to be a projection onto the set Θ.

According to Sra [29], the non-convex proximal algorithm we use for solving (5.12) is based on the following simple iterative scheme:

$$\boldsymbol{\theta}^{k+1} = P_{\Omega_\theta}\left(\boldsymbol{\theta}^k - \eta_k \nabla_{\boldsymbol{\theta}} E(\boldsymbol{\beta}^*(\boldsymbol{\theta}^k)) \right) \tag{5.14}$$

where P_{Ω_θ} is the proximal operator of Ω_θ, η_k a step size that should satisfy $\eta_k \leq \frac{1}{L}$, L is the Lipschitz constant of $\nabla_{\boldsymbol{\theta}} E$, and $\boldsymbol{\beta}^*(\boldsymbol{\theta}^k)$ denotes the vector of all optimal decision function parameters given the fixed matrix $\mathbf{P}$, and $\{\lambda_t\}$ as defined by $\boldsymbol{\theta}^k$.

Convergence of this iterative scheme to a stationary point of Problem (5.12) can be formally stated according to the following proposition:

Proposition 5.1. *For compact sets Θ that guarantee $\lambda_t > 0$, $\forall t$ and for any loss functions L_v that are continuous and twice differentiable on Θ, $E(\beta(\boldsymbol{\theta}))$ is continuous and gradient Lipschitz on Θ. Hence, the sequence of $\{\theta^{(k)}\}$ obtained using the iteration given by Equation (5.14) converges towards a stationary point of Problem (5.12).*

Proof. (Sketch) The proof proceeds by showing that $E(\cdot)$ is continuous and its gradient is Lipschitz and then by directly applying Theorem 2 in Sra [29]. For showing smoothness of $E(\cdot)$ with respect to $\boldsymbol{\theta}$, we compute $\frac{\partial^2 E}{\partial \theta_k \partial \theta_s}$ and show that each of these components of the Hessian is continuous with respect to $\boldsymbol{\theta}$. Once continuity has been shown, we use arguments on compactness of Θ to prove that absolute value of all these components is bounded. This implies that the Frobenius norm of the Hessian is bounded and so is the largest eigenvalue of the Hessian. Hence, we can state that $E(\cdot)$ is indeed gradient Lipschitz and this concludes the proof. Details about continuity and differentiability of $E(\cdot)$ as well as Hessian entries computation are given in the appendix. $\qquad\square$

5.3.2   Gradient Computation

Like all gradient proximal splitting algorithms, our approach needs the gradient of the objective function $E(\cdot)$. The next paragraphs explain how it can be efficiently computed.

At first, we apply the chain rule of differentiation [7] in order to obtain the gradient of $E(\cdot)$ with respect to $\boldsymbol{\theta}$. This leads to the general expression of the partial derivatives of $E(\cdot)$:

$$\frac{\partial E(\boldsymbol{\beta}(\boldsymbol{\theta}))}{\partial \theta_k} = \sum_{s=1}^{(d+1)\times T} \frac{\partial E(\boldsymbol{\beta}(\boldsymbol{\theta}))}{\partial \beta_s(\boldsymbol{\theta})} \frac{\partial \beta_s(\boldsymbol{\theta})}{\partial \theta_k} = \nabla_{\boldsymbol{\beta}} E(\boldsymbol{\beta})^\top \dot{\boldsymbol{\beta}}_k \qquad (5.15)$$

with $\dot{\boldsymbol{\beta}}_k$ a vector containing the partial derivatives $\{\frac{\partial \beta_s}{\partial \theta_k}\}$. These latter partial derivatives can be obtained through the implicit function defined by Equation (5.9) relating the optimal values of $\boldsymbol{\beta}$ to the parameters $\boldsymbol{\theta}$. Differentiating (5.9) with respect to θ_k leads to

$$\mathbf{A}\dot{\boldsymbol{\beta}}_k + \dot{\mathbf{A}}_k \boldsymbol{\beta} = \dot{\mathbf{c}}_k,$$

with $\dot{\mathbf{A}}_k$ and $\dot{\mathbf{c}}_k$ being, respectively, the matrix and vector of component-wise derivative of $\mathbf{A}$ and $\mathbf{c}$ with respect to θ_k (detailed expression of the matrix $\dot{\mathbf{A}}_k$ is given in the appendix). By rearranging the equation and taking into account the fact that $\dot{\mathbf{c}}_k = 0$, we have

$$\dot{\boldsymbol{\beta}}_k = -\mathbf{A}^{-1}(\dot{\mathbf{A}}_k \boldsymbol{\beta}) \qquad (5.16)$$

Note that for small-size problems, computing this gradient can be relatively cheap since the inverse matrix $\mathbf{A}^{-1}$ may be obtained as a by-product of the resolution of Problem (5.9). However, if $\mathbf{A}^{-1}$ has not been pre-computed, obtaining the complete gradient of $E(\cdot)$ requires solving $D = \frac{T^2+T}{2}$ linear systems of size $(d+1)\cdot T$ and this can rapidly become intractable. In order to render the problem tractable, a simple trick proposed in Keerthi et al. [18] can also be used here. Indeed, by plugging Equation (5.16) back into (5.15),

$\frac{\partial E}{\partial \theta_k}$ can be reformulated as:

$$\begin{aligned}
\frac{\partial E(\boldsymbol{\beta}(\boldsymbol{\theta}))}{\partial \theta_k} &= \nabla_{\boldsymbol{\beta}} E(\boldsymbol{\beta})^\top \mathbf{A}^{-1}(-\dot{\mathbf{A}}_k \boldsymbol{\beta}) \\
&= \mathbf{d}^\top (-\dot{\mathbf{A}}_k \boldsymbol{\beta})
\end{aligned} \tag{5.17}$$

with $\mathbf{d}$ being the solution of the linear system:

$$\mathbf{A}^\top \mathbf{d} = \nabla_{\boldsymbol{\beta}}, E(\boldsymbol{\beta}) \tag{5.18}$$

which does not depend on the variable θ_k used for differentiation. Hence, according to this formulation of the partial derivative $\frac{\partial E(\boldsymbol{\beta}(\boldsymbol{\theta}))}{\partial \theta_k}$, only a single linear system has to be solved for computing the full gradient of $E(\cdot)$ with respect to $\boldsymbol{\theta}$.

5.3.3 Constraints on P and λ_t

Let us now discuss the choice of the regularizer Ω_θ. Typically, Ω_θ is defined as the indicator function over a set Θ, formally

$$\Omega_\theta(\theta) = \mathbf{I}_\Theta(\boldsymbol{\theta}) = \begin{cases} 0 & \text{if } \theta \in \Theta \\ \infty & \text{otherwise} \end{cases}$$

where the set Θ defines some constraints we want to impose on the matrix similarity $\mathbf{P}$ and the hyper-parameters $\{\lambda_t\}_{t=1}^T$.

This set Θ can be defined as the intersection of several constraints and it typically translates some prior knowledge we have over the task relatedness. If no knowledge on task similarities are given, the simplest set one may choose is

$$\Theta = \begin{cases} \rho_{t,s} & : 0 \leq \rho_{t,s} \leq M, \forall t, s \\ \lambda_t & : m_\lambda \leq \lambda_t \leq M, \forall t \end{cases} \tag{5.19}$$

with $0 < m_\lambda$ and M being some lower and upper bounds. The small quantity m_λ ensures that a minimal smoothness constraint is enforced on all task parameters $\{\mathbf{w}_t\}_{t=1}^T$. The set defined by (5.19) imposes on the λ_t to be strictly positive as required for convergence of the algorithm and it lets the algorithm fix all task similarity parameters. While rather simple, this choice already proves to provide good multi-task performance as well as excellent interpretability of the task similarity since it induces sparsity of the matrix $\mathbf{P}$ because negative correlations of pairwise tasks are ignored by our regularizer defined in Equation (5.3). Ignoring these negative correlations can surely induce a lack of information transfer between tasks and thus may induce slight loss of generalization performance. However, this is inherently due to the graph-based regularization framework and cannot be alleviated by our learning algorithm.

If some tasks are known to be respectively unrelated, strongly related, and with unknown relatedness, the following set Θ can be considered instead:

$$\Theta = \begin{cases} \rho_{t,s} & : \rho_{t,s} = 0, \text{ for non-similar tasks} \\ \rho_{t,s} & : m_\rho \leq \rho_{t,s} \leq M, \text{for must-be-similar tasks} \\ \rho_{t,s} & : 0 \leq \rho_{t,s} \leq M, \text{for all other pairwise tasks} \\ \lambda_t & : m_\lambda \leq \lambda \leq M, \forall t. \end{cases} \tag{5.20}$$

Note that in this set, we have lower-bounded some task similarities with $m_\rho \gg 0$ for tasks that are known to be related since this will indeed force the parameters of these related tasks to be close.

In order to enhance interpretability, sparsity of matrix $\mathbf{P}$ can be further increased by considering in the regularization term an ℓ_1 regularizer in addition to the projection on the set Θ. In such a case, we may have

$$\Omega_{\Theta - \ell_1}(\boldsymbol{\theta}) = \lambda_\theta \sum_{k=1}^{\frac{T(T-1)}{2}} |\theta_k| + \mathbf{I}_\Theta(\boldsymbol{\theta}) \tag{5.21}$$

where $\mathbf{I}_\Theta(\boldsymbol{\theta})$ stands for the indicator function of the set Θ. From simple algebras, one can show that for the above-given convex sets Θ, the proximal operator of $\Omega_{\Theta - \ell_1}$ consists of a component-wise application of a soft-thresholding operator $S(\theta) = \text{sign}(\theta)(|\theta| - \lambda_\theta)_+$ followed by a projection on the set Θ with the function $(z)_+ = \max(0, z)$.

According to the iterative scheme, it is easy to consider other kinds of constraints on the matrix task similarities as long as their proximal operators are simple to obtain. For instance, we could have dropped the positivity constraints on $\rho_{t,s}$ and instead impose positive definiteness constraint on the task covariance matrix Γ defined in Equation 5.4. In this context, the proximal operator on the set of positive definite matrices would have been in play. We could also have combined sparsity constraints on components of $\mathbf{P}$ in addition to the positive definiteness of Γ, resulting in a more involved but computable proximal operator. However, these constraints would considerably increase the computational burden of the overall optimization scheme as the related proximals involve spectral decomposition of Γ and we have not considered them in this work.

5.3.4   Computational Complexity

The global algorithm for learning task similarities and regularization parameters is presented in Algorithm 6. It is difficult to evaluate the number of iterations needed before convergence. We can note, however, that for each iteration of the algorithm, the main computational bulk resides in solving the MTL problem for fixed task-similarity matrix $\mathbf{P}$. In our case, this consists of solving (5.9) and the linear system needed for obtaining $\mathbf{d}$. A plain implementation of these two linear systems would lead to a global complexity of the order of $\mathcal{O}\left((d+1)^3 T^3\right)$. Nonetheless, we believe that the specific structure of

---

Algorithm 6 Non-convex Proximal Splitting for Learning Task Similarities

---

1: $k \leftarrow 0$
2: initialize $\boldsymbol{\theta}^0 \in \Theta$
3: choose step size $\eta \leq \frac{1}{L}$ (or do backtracking)
4: **repeat**
5:    % steps for computing $\nabla_{\boldsymbol{\theta}} E$
6:    compute $\boldsymbol{\beta}(\boldsymbol{\theta}^k)$ by solving (5.9)
7:    compute $\nabla_{\boldsymbol{\beta}} E$
8:    $\mathbf{d} \leftarrow$ solution of $\mathbf{A}^\top \mathbf{d} = \nabla_{\boldsymbol{\beta}} E$
9:    **for** all k **do**
10:       compute $\dot{\mathbf{A}}_k$
11:       $\frac{\partial E}{\partial \theta_k} \leftarrow -\mathbf{d}^\top \dot{\mathbf{A}}_k \boldsymbol{\beta}$
12:    **end for**
13:    % proximal step
14:    $\boldsymbol{\theta}^{k+1} \leftarrow P_\Theta(\boldsymbol{\theta}^k - \eta \nabla_{\boldsymbol{\theta}} E)$
15: **until** convergence criterion is met

$\mathbf{A}$ can be exploited for achieving better complexity, or that specific efficient algorithms for graph-based regularized multi-task learning can be developed. Such an algorithm already exists for hinge loss function and it can be adapted to the square loss. From another perspective, block iterative methods such as Gauss-Seidel methods can be implemented for solving the linear system (5.9). But this implementation study is left to future works.

5.4 Numerical Experiments

The approach we propose for learning the task similarity matrix $\mathbf{P}$ as well as the model hyper-parameters λ_t has been tested on several numerical problems including toy examples and three real-world problems.

Besides reporting regression and classification performance, we also provide results on the interpretability of the task relations learned by our algorithm. Indeed, the similarity matrix $\mathbf{P}$ can be understood as an adjacency matrix of the task relation graph. Hence, it can be nicely plotted and its sparsity pattern analyzed.

Note that for all experiments we have considered, the loss function of the inner level is the quadratic loss function, thus $J(\cdot)$ is exactly the one given in Equation (5.6). While one may argue that such a loss function is inadequate for classification problems, Rifkin et al. [25] and Suykens and co-authors [30, 13, 12], however, stated that it is still competitive in many of these problems.

5.4.1   Toy Problems

The toy problems we consider here aim at only proving that our algorithm can learn the intrinsic structure of the tasks and how they are related. We show that even for the simple constraints Θ (5.19) we have imposed on $\boldsymbol{\theta}$, our approach is able to learn different task-relation structures such as clusters of tasks or tasks living in a non-linear manifold. The problem is built as follows: given a vector $\bar{\mathbf{w}}^\top = [1, 2]$, a rotation of angle γ_t is applied to $\bar{\mathbf{w}}$ so as to obtain the actual linear model parameters $\bar{\mathbf{w}}_t$ for task t. Examples $\{\mathbf{x}_{i,t}\}$ are drawn from a two-dimensional zero-mean and unit variance normal distribution and the corresponding $y_{i,t}$ are obtained according to the equation

$$y_{i,t} = \mathbf{x}_{i,t}^\top \bar{\mathbf{w}}_t + \epsilon_{i,t}$$

where $\epsilon_{i,t} \sim \mathcal{N}(0, 0.5)$ is some additive noise added to the output.

Two specific synthetic problems illustrate our points. In the first one, tasks are structured in two clusters by randomly applying a rotation of $\gamma_t = 0$ or $\gamma_t = \pi/2$. Hence, our algorithm should be able to recover this clustered structure. For the other example, we apply a rotation whose angle γ_t is uniformly drawn from $[0, \pi]$. Hence, tasks are supposed to be similar only to few neighbors and the adjacency matrix should reflect these local similarities of tasks. For each of these problems, 40 tasks were built and 20 examples for the learning set and 20 examples for the validation set are randomly drawn. We have reported an example of the qualitative results obtained using Ω_Θ as the indicator of the set given in Equation (5.19) as well as $\Omega_{\Theta-\ell_1}$ with $m_\lambda = 0.1$, $M = 1000$, and $\lambda_\theta = 0.05$. We also report the obtained results while applying the approach of Zhang et al. [34], named hereafter metric-MTL and based on the estimation of a dense task covariance matrix.

Figure 5.1 provides an example of results that can be obtained by our approach as well as the competitor's on the *clustered tasks* toy problem. We note that our algorithm using Ω_Θ is able to nicely infer the tasks relation since only 23% of the adjacency matrix $\mathbf{P}$ entries are non-zero and among those coefficients, 98% corresponds to links between tasks from the same cluster. When a sparsity-inducing regularizer is further added to the constraints, the adjacency matrix is more sparse and links between tasks from different clusters disappear.

For the *manifold-based tasks* problem, results are depicted in Figure 5.2. Again, we can clearly see that our approach using both types of constraints is able to learn the underlying task structure. Indeed, we can note that, when using only Ω_Θ, the ratio of non-zero coefficients in the adjacency matrix is about 14% which shows that pairwise relationships were found. In addition, as desired, links between tasks mainly exist between neighbor tasks, thus providing evidence that the manifold structure of the task has been recovered. Using $\Omega_{\Theta-\ell_1}$ as a regularizer gives a similar result although with a more aggressive sparsity pattern.

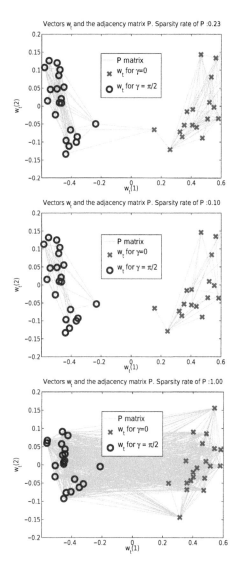

FIGURE 5.1: Learned weight vectors $\{\mathbf{w}_t\}$ and adjacency graph $\mathbf{P}$ for the *clustered tasks* toy problem. The graphs plot these 2D vectors learned for each task and the adjacency graph inferred by our approach is materialized as lines between the vectors. The task parameters are theoretically split in two clusters with either $\gamma_t = 0$ or $\gamma_t = \pi/2$. Results obtained by our algorithm using (top) Ω_Θ, (middle) $\Omega_{\Theta-\ell_1}$, and (bottom) metric-MTL (method due to Zhang et al. [34]). For the latter method, the links depict non-zero entries in the inverse covariance task matrix. Sparsity rate refers to the proportion of non-zero entries of the learned similarity matrix.

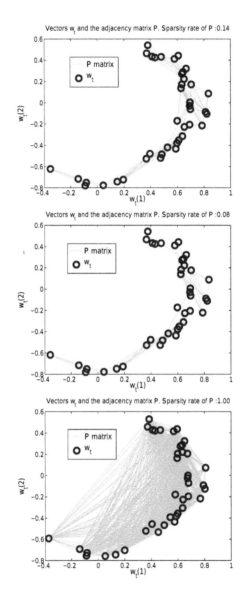

FIGURE 5.2: Learned weight vectors $\{\mathbf{w}_t\}$ and adjacency graph $\mathbf{P}$ for the *manifold-based tasks* toy problem. The graphs plot these 2D vectors learned for each task and the adjacency graph inferred by our approach is materialized as lines between the vectors. The task parameters theoretically lie on a manifold. Results obtained by our algorithm using (top) Ω_Θ, (middle) $\Omega_{\Theta-\ell_1}$, and (bottom) metric-MTL. For the latter method, the links depict non-zero entries in the inverse covariance task matrix. The sparsity rate is the proportion of non-zero entries of the similarity matrix.

These two examples illustrate that our approach is able to learn complex relationships between tasks such as non-linear manifold and that the proposed constraints induce sparsity in the similarity matrix and thus enhance inter-pretability of the task relations. In comparison, looking at the rightmost plots of Figures 5.1 and 5.2, we can see that metric-MTL is also able to recover the complex geometrical relationship between tasks but these relationships are completely hidden by the dense structure of the learned similarity matrix. Note that for this problem involving 40 tasks and a total number of 1600 training/validation examples, the optimization is of the order of 10 seconds on a recent Intel processor with non-optimized MATLAB source code. Our approach can then estimate an optimal $\mathbf{P}$ matrix in a reasonable amount of time. This could not have been performed using a classical cross-validation procedure on the corresponding 580 $\rho_{t,s}$ parameters.

5.4.2   Real-World Datasets

The approach we proposed has also been experimented on several real-world datasets. The results we have achieved are presented hereafter detailing the experimental set-up.

5.4.2.1   Experimental Set-Up

Several multi-task learning algorithms have been compared in terms of performance as well as in terms of interpretability of the learned task rela-tionships if the latter is applicable.

The baseline approach, denoted as "ridge indep.", is an ensemble of ridge regression problems trained independently on each task. MTL approaches that learn task relations have also been considered. This includes the metric-MTL of Zhang et al. [34] and clustered multi-task learning (cluster-MTL in the remainder), an approach proposed by Jacob et al. [15] where task similarities are also learned through the inference of the underlying metric between tasks.

For a fair comparison between our approach, named "CoGraph-MTL" (for Constrained Graph-regularized MTL) and the other methods, we selected competitor hyper-parameters by maximizing their performance on the val-idation set. Note that our bilevel approach also uses the validation set for selecting hyper-parameters but they are optimized through our non-convex proximal method in the outer level.

Depending on the datasets, a squared function $E_{\ell_2}(\cdot, \cdot)$ or a sigmoid func-tion $E_{sig}(\cdot, \cdot)$ with $\kappa = 1$ is used as the outer loss function $L_v(\cdot, \cdot)$. In addition, for all problems, unless specified, we have used Ω_Θ as defined in Equation (5.19) with $m_\lambda = 1$ and $M = 1000$ as well as $\Omega_{\Theta - \ell_1}$ (see Equation 5.21) with $\lambda_\theta = 0.05$ for constraining our graph-regularized MTL method to be sparser.

For each dataset, 10 random splits have been generated and averaged per-formance measure, mean square error or area under the ROC curve (AUC), on the test set was reported. We also performed a signed rank Wilcoxon test

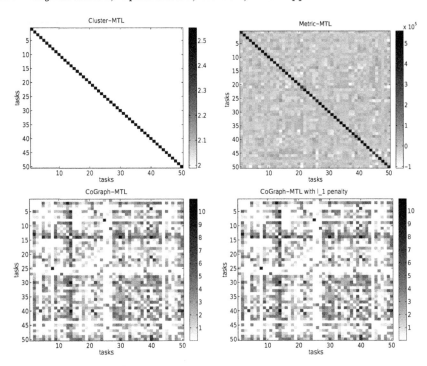

FIGURE 5.3: Example of task similarity matrices on the *school* dataset: (top-left) cluster-MTL; (top-right) metric-MTL inverse covariance task matrix; (bottom-left) our CoGraph-MTL with Ω_Θ; (bottom-right) our CoGraph-MTL with $\Omega_{\Theta-\ell_1}$.

to evaluate the statistical difference in performance between the two variants of our method and the best performing competitor.

5.4.2.2   School Dataset

We have also tested our approach on the well known school dataset that is available online and consists of predicting the examination score of students from different schools in London. This problem can be addressed as a multi-task problem since differences between schools have to be taken into account, for instance, by learning a prediction function per school. We refer the reader to Argyriou et al. [1] for a more complete description of the data and focus instead on the feature extraction we used. In their work, they have shown that the tasks might share a common linear subspace. Hence, we took this knowledge into account and performed a PCA on the whole dataset and kept the 10 principal components out of 27 features. We learned the prediction function of the 50 tasks having the largest number of examples. For each task, we have an average number of ≈ 170 samples and we randomly selected 50

TABLE 5.1: Mean square error on the *school* dataset. p-value of a Wilcoxon signrank test with respect to the performance of the best competitor as well as the sparsity of the resulting task relation matrix are also reported.

Method	MSE	p-value	Sparsity (%)
Ridge Indep	118.27±2.97	-	-
Cluster-MTL	110.78±2.62	-	100.0
Metric-MTL	108.27±2.51	-	100.0
CoGraph-MTL	107.31±2.24	0.002	56.6
CoGraph-MTL $\Omega_{\Theta-l_1}$	107.32±2.24	0.002	55.8

of them for the training set, 50 for the validation set, and the remaining ones for the test set.

Mean square error (MSE) obtained for all the described methods are reported in Table 5.1, as well as the sparsity of resulting task similarity matrices. We can note that the two variants of our approach achieve the lowest prediction error and they are statistically better than the competitor according to a Wilcoxon sign-rank test.

Learned similarity matrices are depicted in Figure 5.3. We can remark that the matrices retrieved by the cluster-MTL and metric-MTL are rather dense and present a very similar structure: large diagonal terms and nearly constant off diagonal components. According to these matrices, we may conclude that examination scores of one given school are related to those of all other schools. The matrices learned by our method also have a similar structure but their entries are more sparse. Hence, we can understand that a given school is related only to few other ones. A deeper understanding of these similarities may then be carried out if some more information, like geographical or social aspects, was available about all the schools.

5.4.2.3   Brain Computer Interface Dataset

In this brain computer interface (BCI) problem, our objective is to recognize the presence of an event-related potential (ERP) in a recorded signal during the use of a virtual keyboard. The dataset has been recorded by the Neuroimaging Laboratory of Universidad Autonoma Metropolitana (UAM, Mexico) [20] on 30 subjects performing P300 spelling tasks on a 6×6 virtual keyboard. We consider each subject as a task and learn all classifiers simultaneously. For each subject, we have approximately 4000 single trial samples that have been pre-processed according to the following steps: first a low-pass filtering is applied to the 10 channel signals followed by a decimation; we kept a 1s time window (6 temporal samples) following the stimulus as features leading to trials containing 60 features [24]. Finally for each data split, we randomly selected for each task 100 trials for the training set, 100 trials for the validation set, and the remaining samples for the test set. Note that

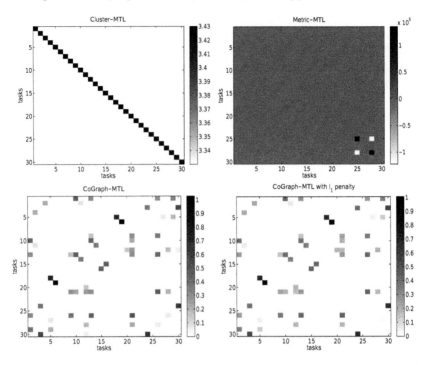

FIGURE 5.4: Example of task similarity matrices on the *BCI* dataset: (top-left) cluster-MTL; (top-right) metric-MTL inverse covariance task matrix; (bottom-left) our CoGraph-MTL with Ω_Θ; (bottom-right) our CoGraph-MTL with $\Omega_{\Theta-\ell_1}$.

for P300 classification, since the datasets are highly imbalanced, we decided to use the area under the ROC curve as a measure of performance. For this classification task, we have used a sigmoid function as the outer loss function.

We note in Table 5.2 that all multi-task learning approaches provided statistically similar performance measures and they all perform far better than a method that learns each task independently from all others. Interestingly, the two variants of our method output similarity matrices that are considerably sparse. Hence, for each subject, instead of considering that all other subjects were similar to that one, they were able to retrieve a few others that provide sufficient information for transfer learning. Figure 5.4 provides some examples of learned task relation matrix for the different algorithms. We can see there how our method is able to extract relevant and interpretable information from the data. Indeed, the task relation matrices retrieved by our two variants are very sparse. It is thus possible to find which BCI users are related.

TABLE 5.2: Area under the ROC curve (AUC) on the *BCI* dataset. p-value of a Wilcoxon signrank test with respect to the performance of the best competitor as well as the sparsity of the resulting task relation matrix are also reported.

Method	AUC	p-value	Sparsity (%)
Ridge Indep	0.65±0.01	-	-
Cluster-MTL	0.78±0.00	-	100.0
Metric-MTL	0.78±0.01	-	100.0
CoGraph-MTL	0.78±0.00	0.625	6.9
CoGraph-MTL $\Omega_{\Theta-l_1}$	0.77±0.00	0.232	6.4

TABLE 5.3: Area under the ROC curve (AUC) on the *OCR* dataset. p-value of a Wilcoxon signrank test with respect to the performance of the best competitor as well as the sparsity of the resulting task relation matrix are also reported.

Method	AUC	p-value	Sparsity
Ridge Indep	0.94±0.01	-	-
Cluster-MTL	0.96±0.01	-	100.0
Metric-MTL	0.98±0.01	-	100.0
CoGraph-MTL	0.96±0.01	0.002	11.1
CoGraph-MTL $\Omega_{\Theta-l_1}$	0.95±0.01	0.002	9.8
CoGraph-MTL Ω_{Θ_G}	0.97±0.01	0.375	51.7

5.4.2.4   OCR Dataset

Finally we evaluate our approach on an OCR classification problem. We used the same OCR dataset as in the works of [22]. Here, the aim is to learn a binary classifier for each writer so as to take into account writer variability. The main difficulty in this dataset is that we have only a few examples but still want to learn robust classifiers.

We focus here on two binary classification problems : "e" vs. "c" and "a" vs. "g" for 20 different writers. We want to learn simultaneously these 40 classification tasks. The data is a raw bitmap of size 8×16 that was pre-processed as follows. We performed PCA on the raw data and kept 20 principal components. We randomly select 8 samples for the training set, 4 samples for the validation set, and the remaining 8 for the test set. Moreover, in order to prove how easy it is to add prior information in our approach, we integrated group knowledge in the learning problem by imposing specific constraints on $\boldsymbol{\theta}$, denoted as Ω_{Θ_G}. Indeed, we force a link between tasks from the same binary classification problem by imposing $0.01 \leq \rho_{t,k} \leq 1000$ (which is rather a weak constraint).

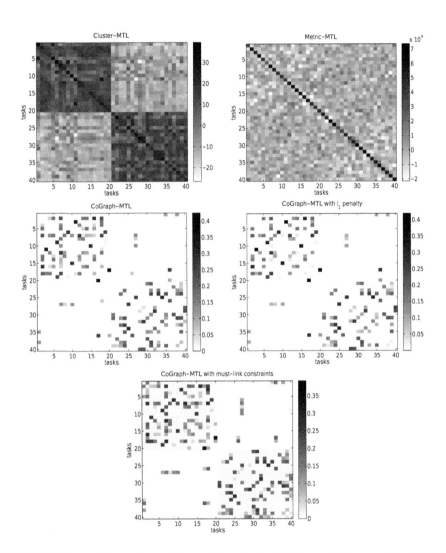

FIGURE 5.5: Examples of learned task similarity matrices on the *OCR* dataset: (top-left) cluster-MTL; (top-right) metric-MTL inverse covariance task matrix; (middle-left) our CoGraph-MTL with Ω_Θ; (middle-right) our CoGraph-MTL with $\Omega_{\Theta-\ell_1}$; (bottom) our CoGraph-MTL with Ω_{Θ_G}.

Performance of the different methods can be seen in Table 5.3. We remark that our method using Ω_{Θ_G} achieves equivalent performance with metric-MTL whereas the two other variants of our method are slightly but significantly worse than the best performing competitor. Indeed, adding prior knowledge helps in learning robust classifiers.

We should however emphasize that the metric-MTL approach is not able to retrieve the correct relation between tasks as shown in Figure 5.5. Indeed, the learned similarity matrix tells us that for metric-MTL all the tasks are related. Cluster-MTL is able to infer that they form two specific clusters of tasks in the problem. Our methods are also capable of detecting these clusters and in addition yield sparse similarity matrices as very few relations between tasks from two different clusters were uncovered. Note that even though some *must-link* constraints have been imposed when using Ω_{Θ_G} as a regularizer, some hyper-parameters have still been optimized by our algorithm.

5.5   Conclusion

We have proposed a novel framework for learning task similarities in multi-task learning problems. Unlike other previous works on this topic, we learn these similarities so as to optimize a proxy on the generalization errors of all tasks. For this purpose, we introduce a bilevel optimization problem that involves both the minimization of the generalization error and the optimization of the task parameters. The global optimization is solved by means of a non-convex proximal splitting algorithm, which is simple to implement and can easily be extended to situations where different constraints on task similarities have to be imposed.

Experimental results on toy problems clearly show that the method we propose can help in learning complex structures of task similarities. On real-world problems, our method clearly achieves performances similar to other multi-task learning algorithms that learn tasks similarities while providing task similarity matrices that are far more interpretable.

Acknowledgments

This work has been partly supported by the French ANR Agency Grant ANR-11-JS02-0010 Lemon.

Appendix

Continuity and differentiability of E with respect to θ_k

First, note that β^* is implicitly defined by the linear system $\mathbf{A}\beta = \mathbf{c}$. Hence, we have

$$\beta = \mathbf{A}^{-1}\mathbf{c}$$

Existence and unicity of β is guaranteed by invertibility of $\mathbf{A}$ since we have imposed that $\lambda_t > 0, \forall t$. Hence, since $\mathbf{A}$ and $\mathbf{c}$ are both continuous and differentiable with respect to θ_k, so are $\mathbf{A}^{-1}$ and each component of β. Using similar arguments, we can show that each component of $\dot{\beta}_k = -\mathbf{A}^{-1}(\dot{\mathbf{A}}_k \beta)$ is continuous.

Continuity and differentiability of $E(\cdot)$ is easily obtained since E is built from differentiability-preserving operations over differentiable functions.

Hessian Computation. In order to prove the gradient Lipschitz property of E, we show that the Hessian of the function $E(\beta(\theta))$ is bounded. The Hessian is a matrix of general term:

$$\frac{\partial E(\beta(\theta))}{\partial \theta_k \partial \theta_s} = \frac{\partial}{\partial \theta_s} \left(\mathbf{d}^\top (\dot{\mathbf{c}}_k - \dot{\mathbf{A}}_k \beta) \right) \qquad (5.22)$$

$$= -\left[\frac{\partial \mathbf{d}^\top}{\partial \theta_s} (\dot{\mathbf{A}}_k \beta) + \mathbf{d}^\top \dot{\mathbf{A}}_k \frac{\partial \beta}{\partial \theta_s} \right] \qquad (5.23)$$

where we used the fact that $\mathbf{c}$ does not depend on θ so $\dot{\mathbf{c}}_k = \mathbf{0}$ and $\dot{\mathbf{A}}_k$ is a constant matrix whose components are equal to 0 when differentiated. By using the definition of $\mathbf{d}$ in Equation 5.18, it is easy to see that

$$\frac{\partial \mathbf{d}}{\partial \theta_s} = (\mathbf{A}^\top)^{-1} \left[\frac{\partial}{\partial \theta_s} \nabla_\beta E(\beta) - \left(\frac{\partial \mathbf{A}}{\partial \theta_s} \right)^\top \mathbf{d} \right] \qquad (5.24)$$

Now, we can note that all components of the Hessian general term (5.23) are continuous with respect to θ. For instance, $\mathbf{d}$ is continuous since it is the product of a continuous matrix $\mathbf{A}^{-1}$ and continuous vector $\nabla_\beta E$ (by hypothesis on the loss function). Similar arguments can be employed for showing continuity of all other terms.

Expression of $\dot{\mathbf{A}}_k$. The expression of this gradient takes different forms according to the type of hyper-parameter. Using Definition (5.10), we get the expression

$$\dot{\mathbf{A}}_k = \begin{bmatrix} \mathbf{0} & \cdots & \mathbf{0} & \cdots & \mathbf{0} & \cdots & \mathbf{0} \\ \vdots & \cdots & \vdots & \cdots & \vdots & \cdots & \vdots \\ \mathbf{0} & \cdots & 4\tilde{\mathbf{I}} & & -4\tilde{\mathbf{I}} & & \mathbf{0} \\ \vdots & \cdots & \mathbf{0} & \cdots & \mathbf{0} & \cdots & \mathbf{0} \\ \mathbf{0} & \cdots & -4\tilde{\mathbf{I}} & & 4\tilde{\mathbf{I}} & \cdots & \mathbf{0} \\ \vdots & \cdots & \mathbf{0} & \cdots & \mathbf{0} & \cdots & \vdots \\ \mathbf{0} & \cdots & \mathbf{0} & \cdots & \mathbf{0} & \cdots & \mathbf{0} \end{bmatrix}, \quad \text{for} \quad \theta_k = \rho_{i,j}, \text{with } j > i$$

and

$$\dot{\mathbf{A}}_k = \begin{bmatrix} \mathbf{0} & \cdots & \mathbf{0} & \cdots & \mathbf{0} \\ \vdots & \cdots & \vdots & \cdots & \vdots \\ \mathbf{0} & \cdots & 2\tilde{\mathbf{I}} & \mathbf{0} & \vdots \\ \vdots & \cdots & \mathbf{0} & \cdots & \vdots \\ \mathbf{0} & \cdots & \mathbf{0} & \cdots & \mathbf{0} \end{bmatrix}, \quad \text{for} \quad \theta_k = \lambda_t, \text{with } t = 1, \ldots, T$$

Bibliography

[1] A. Argyriou, T. Evgeniou, and M. Pontil. Convex multi-task feature learning. *Machine Learning*, 73(3):243–272, 2008.

[2] Y. Bengio. Gradient-based optimization of hyperparameters. *Neural Computation*, 12:1889–1900, 2000.

[3] K.P. Bennett, J. Hu, X. Ji, G. Kunapuli, and J.S. Pang. Model selection via bilevel optimization. In *Neural Networks, International Joint Conference on*, pages 1922–1929, 2006.

[4] J. Bergstra and Y. Bengio. Random search for hyper-parameters optimization. *Journal of Machine Learning Research*, 13:281–305, 2012.

[5] R. Caruana. Multi-task learning. *Machine Learning*, 28:41–75, 1997.

[6] O. Chapelle, V. Vapnik, O. Bousquet, and S. Mukerjhee. Choosing multiple parameters for SVM. *Machine Learning*, 46(1-3):131–159, 2002.

[7] B. Colson, P. Marcotte, and G. Savard. An overview of bilevel optimization. *Annals of operations research*, 153(1):235–256, 2007.

[8] P.L. Combettes and J.C. Pesquet. Proximal splitting methods in signal processing. *Fixed-Point Algorithms for Inverse Problems in Science and Engineering*, pages 185–212, 2011.

[9] T. Evgeniou and M. Pontil. Regularized multi-task learning. In *Proceedings of the Tenth Conference on Knowledge Discovery and Data Mining*, 2004.

[10] Theodoros Evgeniou, Charles A. Micchelli, and Massimiliano Pontil. Learning multiple tasks with kernel methods. *Journal of Machine Learning Research*, 6:615–637, 2005.

[11] S. Feldman, B. A. Frigyk, M. R. Gupta, L. Cazzanti, and P. Sadowski. Multi-task output space regularization. *arXiv*, 2011.

[12] T. Van Gestel, J.A.K. Suykens, B. Baesens, S. Viaene, J. Vanthienen, G. Dedene, B. De Moor, and J. Vandewalle. Benchmarking least squares support vector machine classifiers. *Machine Learning*, 54(1):5–32, 2004.

[13] Tony Van Gestel, Johan A. K. Suykens, Gert R. G. Lanckriet, Annemie Lambrechts, Bart De Moor, and Joos Vandewalle. Bayesian framework for least-squares support vector machine classifiers, Gaussian processes, and kernel Fisher discriminant analysis. *Neural Computation*, 14(5):1115–1147, 2002.

[14] T. Hastie, R. Tibshirani, and J. Friedman. *The Elements of Statistical Learning*. Springer-Verlag, 2001.

[15] L. Jacob, F. Bach, and J.-P. Vert. Clustered multi-task learning: A convex formulation. In *Advances in Neural Information Processing Systems, NIPS*, 2008.

[16] Zhuoliang Kang, Kristen Grauman, and Fei Sha. Learning with whom to share in multi-task feature learning. In *Proceedings of the 28th ICML*, pages 521–528. ACM, June 2011.

[17] T. Kato, H. Kashima, M. Sugiyama, and K. Asai. Multi-task learning via conic programming. In *Advances in Neural Information Processing Systems*, 2008.

[18] S. Sathiya Keerthi, Vikas Sindhwani, and Olivier Chapelle. An efficient method for gradient-based adaptation of hyperparameters in svm models. In *Advances in Neural Information Processing Systems 19*, pages 673–680. MIT Press, 2007.

[19] A. Kumar and H. Daumé III. Learning task grouping and overlap in multi-task learning. In *Proceedings of the International Conference on Machine Learning*, 2012.

[20] Claudia Ledesma-Ramirez, Erik Bojorges Valdez, Oscar Yáñez Suarez, Carolina Saavedra, Laurent Bougrain, and Gerardo Gabriel Gentiletti. An Open-Access P300 Speller Database. In *Fourth International Brain-Computer Interface Meeting*, 2010.

[21] A. Maurer, M. Pontil, and B. Romera-Paredes. Sparse coding for multi-task and transfer learning. In *Proceedings of the International Conference on Machine Learning*, 2013.

[22] Guillaume Obozinski, Ben Taskar, and Michael I. Jordan. Joint covariate selection and joint subspace selection for multiple classification problems. *Statistics and Computing*, 20:231–252, April 2010.

[23] A. Rakotomamonjy, R. Flamary, G. Gasso, and S. Canu. lp-lq penalty for sparse linear and sparse multiple kernel multi-task learning, *IEEE Transactions on Neural Networks*, 22(8):1307–1320, 2011.

[24] A. Rakotomamonjy and V. Guigue. BCI competition III: Dataset II - ensemble of SVMs for BCI P300 speller. *IEEE Trans. Biomedical Engineering*, 55(3):1147–1154, 2008.

[25] R. Rifkin, G. Yeo, and T. Poggio. Regularized least squares classification. In *Advances in Learning Theory : Methods, Model and Applications*, pages 131–153. IOS Press, 2003.

[26] B. Romera-Paredes, A. Argyriou, N. Bianchi-Berthouze, and M. Pontil. Exploiting unrelated tasks in multi-task learning. In *JMLR Proceeding track*, volume 22, pages 951–959, 2012.

[27] B. Romera-Paredes, M. Hane Aung, N. Bianchi-Berthouze, and M. Pontil. Multilinear multitask learning. In *Proceedings of the International Conference on Machine Learning*, 2013.

[28] Kim Seyoung and Eric P. Xing. Tree-guided group lasso for multi-task regression with structured sparsity. In *ICML*, pages 543–550, 2010.

[29] S. Sra. Nonconvex proximal splitting: batch and incremental algorithms. In *Advances in Neural Information Processing Systems (NIPS)*, 2012.

[30] J. A. K. Suykens, T. Van Gestel, J. De Brabanter, B. De Moor, and J. Vandewalle. *Least squares support vector machines*. World Scientific, 2002.

[31] S. V. N. Vishwanathan, A. J. Smola, and M. Murty. SimpleSVM. In *International Conference on Machine Learning*, 2003.

[32] C. Widmer, M. Kloft, N. Goernitz, and G. Raetsch. Efficient training of graph-regularizer multi-task svm. In *Proceedings of the European Conference on Machine Learning (ECML)*, 2012.

[33] Christian Widmer, Nora C. Toussain, Yasemin Altun, and Rätsch Gunnar. Inferring latent task structure for multitask learning by multiple kernel learning. *BMC Bioinformatics*, 11(Suppl 8):S5, 2010.

[34] Y. Zhang and D.Y. Yeung. A convex formulation for learning task relationships in multiple task learning. In *Proceedings of Uncertainty and Artificial Intelligence*, 2010.

[35] L. Zhong and J. Kwok. Convex multitask learning with flexible task clusters. In *Proceedings of the 29th International Conference on Machine Learning (ICML), 2012*, 2012.

Chapter 6

The Graph-Guided Group Lasso for Genome-Wide Association Studies

Zi Wang

Imperial College London

Giovanni Montana

King's College London

6.1 Introduction

Genetic variation in human DNA sequences can cause alterations in an individual's traits, which may be indicators of disease status or quantitative measures of disease risk or physical properties. The associations between genetic variations and human traits have traditionally been investigated through inheritance studies in families [9]. While this has proved useful for single gene disorders, the associations identified with "complex" diseases involving multiple genetic determinants are hard to reproduce [1]. In the last decade, the number of genome-wide association studies (GWAs) has increased immensely [14, 10, 7, 19]. GWAs search common genetic variants across the human genome in unrelated individuals that are associated with a trait. They have been widely applied and have succeeded in identifying reproducible associations with thousands of human diseases and traits [18]. In this work we assume the genetic variants are single-nucleotide polymorphisms (SNPs), and we develop a statistical method to identify SNPs associated to a univariate trait. We refer to the SNPs/genes having a non-random association with the trait as "causal SNPs/genes", which are unknown and to be inferred.

Conventional GWAs involve a case-control study design, in which the subjects are partitioned into case and control groups, and hypothesis tests (e.g., based on χ^2 statistic) are carried out for each SNP to examine if the frequencies of the genotypes (commonly denoted as aa, Aa, and AA) are significantly altered between the case and control groups [5, 6]. Some limitations of this approach include: substantial estimation biases and spurious associations due to violation of the assumptions in case-control design [28]; insufficient sample size in comparison to the number of variables, typically sample sizes vary between a few hundred and a few thousand, whereas the number of SNPs to be investigated is about 1 million and may go beyond 10 million [6, 24]; and multiple testing issues [28].

In this work, we let the "trait of interest" be a continuous random variable, and formulate the variable selection problem in a linear regression context. We define the SNPs as predictors and the trait as a univariate response, thereby the strength of associations can be quantified by the coefficients from the linear regression model. When the number of subjects is smaller than the number of SNPs, the coefficients can be estimated by minimizing the empirical loss function (the squared loss). However, in GWAs, the number of SNPs greatly exceeds the number of subjects, which raises computational issues in coefficients estimation. Many regularized methods have been proposed to tackle this problem, which involve adding a penalty term of the coefficient vector to the objective function. Taking the ℓ_1 norm of the coefficient vector as the penalty term, amongst others, some entries in the estimated coefficients are set to zero and a set of important SNPs can be automatically selected that correspond to the non-zero entries [38, 3, 15, 33, 27]. We refer to such models as penalized linear regression models.

There appears to be two trends in the recent development of the penalized linear regression models for GWAs. The first is developing fast computation algorithms that can be applied to "big data" problems [32, 31], where the sample size is as large as $200,000$ or beyond [11]; the second trend is developing new penalties to impose biologically informed sparsity patterns. The motivation is that the variability in disease related traits often arises from the joint contribution from multiple loci within a gene or the joint action of genes within a pathway [22, 34]. Although some genetic variants may only facilitate a moderate or weak effect on the trait individually, their joint effect could be influential. Unfortunately, the set of SNPs extracted by purely data-driven methods often account for only a small proportion of the genetic variants associated with the trait, which may provide very limited insight into the biological mechanisms underpinning the complex disease [23]. One promising approach involves incorporating prior biological information on the functional relations between the genetic variants to guide the selection of causal predictors. The hope is that the penalty function acting on this information will encourage the selection of co-functioning genetic variants, hence producing a set of important predictors explanatory of the variability in the trait as well as biologically plausible [41, 39, 36, 35].

Structured knowledge on co-functioning SNPs and genes can be organized in many ways. One of these is partitioning the candidate SNPs into non-overlapping groups according to the gene-SNP mapping, where each group corresponds to a set of SNPs that constitute a gene or a pathway. Using the group lasso penalty [42], the model can identify a set of important genes or pathways that explain the variability in the disease related trait [43, 40]. Using the sparse group lasso penalty, the most influential SNPs within the important genes or pathways can be extracted, which may give further insight of the biological process [13]. When groups overlap, for instance when an SNP is mapped to more than one gene, the set of selected SNPs belong to the union of some important genes [17, 36, 34].

Another way to organize the structured knowledge of the co-functioning genes is to present the pairwise relations in a network. Typically, the nodes represent genes (or proteins), and two nodes are connected by an edge if the associated genes belong to the same genetic pathway or share related functions [37]. In previous GWAs, such networks were often used after the set of important genes had been selected and to examine the roles of these genes in the biological process [20, 30]. Nonetheless, the reverse engineering that uses the gene network as prior information to guide the selection of interacting genes has also been exploited. The selected genes often correspond to densely connected regions in the given network and this has been shown to improve the accuracy of variable selection [21, 2, 29].

In this work we consider the case where heterogeneous prior information is available for GWAs: SNPs are grouped into genes, and genes are organized into a weighted gene network encoding the functional relatedness between all pairs of genes. We believe that integrating prior knowledge from the two levels will lead to improved variable selection accuracy of genes and SNPs while facilitating biologically informed models. We propose a penalized regression model, the graph-guided group lasso (GGGL), which selects functionally related genes and SNPs within these genes that are associated with a quantitative trait. To the best of our knowledge, this is the first model in GWAs that uses graph and grouping structure on hierarchical biological variants to drive variable selection at multiple levels.

The rest of this chapter is organized as follows: In Section 6.2 we present two versions of the GGGL models, which differ on the integration of structured information at heterogeneous levels, and we show their theoretical properties; in Section 6.3 we present a serial and a parallel computation algorithm for each version of GGGL; we study the power of variable and group selection of the GGGL models using simulated data in Section 6.4; a brief conclusion summarizing our contribution is given in Section 6.5.

6.2   Method

6.2.1   The Graph-Guided Group Lasso (GGGL)

Let X be the $n \times p$ design matrix containing n independent samples for which p SNPs have been observed, and y be the n-dimensional vector containing the univariate quantitative trait. We normalize the columns of X to have zero sum and unit length and center y, such that the predictors are scaled and the intercept term can be dropped from the linear regression model which seeks coefficient vector β to minimize the squared loss: $\|y - X\beta\|_2^2$. In a penalized linear regression model, the estimated coefficients $\hat{\beta}$ is obtained by minimizing:

$$\frac{1}{2}\|y - X\beta\|_2^2 + P(\beta)$$

where $P(\beta)$ in (6.1) is a function of β, called "the penalty function". For example, $P(\beta) = \|\beta\|_1$ is the "lasso penalty" [38].

Suppose that the SNPs are grouped into mutually exclusive genes $\mathcal{R} = \{R_1, R_2, ...\}$, in which each element R_I is a set of SNP indexes. The size of gene R_I is denoted by $|R_I|$. We let X_I denote the $n \times |R_I|$ sub-matrix of X, where the columns correspond to SNPs in R_I, and let X_i denote the ith column of X. Let $\mathcal{G} = \mathcal{G}(V, E)$ be the gene network with vertex set V corresponding to the $|\mathcal{R}|$ genes in $\mathcal{R}$. We shall use the terms "network" and "graph" interchangeably in the following. For convenience, when referring to the node in $\mathcal{G}$ corresponding to the gene R_I, we shall write I for short. The weight of the edge connecting I and J is denoted by w_{IJ}, which can be either binary or continuous. In the case it is continuous, the magnitude measures the strength of the relatedness between two genes: a larger magnitude indicates the two genes are more likely to involve in the same biological process. For simplicity, we assume that all weights are non-negative. The network $\mathcal{G}$ may consist of several disjoint subgraphs, whose vertex set and edge set are subsets of $V(\mathcal{G})$ and $E(\mathcal{G})$, respectively. Each of these disjoint subgraphs is called a "component". For the GGGL, regression coefficients are obtained by minimizing the least square loss plus a composite penalty term:

$$\frac{1}{2}\|y - X\beta\|_2^2 + P_1(\beta) + P_2(\beta) + P_3(\beta) \tag{6.1}$$

where:

$$P_1(\beta) = \lambda_1 \sum_{K=1}^{|\mathcal{R}|} \sqrt{|R_K|}\|\beta_K\|_2, \quad P_2(\beta) = \lambda_2\|\beta\|_1 \tag{6.2}$$

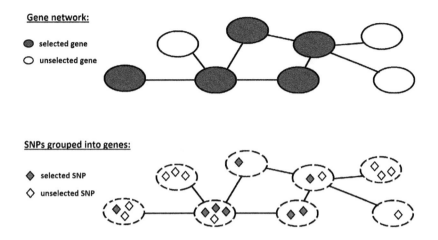

FIGURE 6.1: Illustration of the sparsity patterns allowed by the GGGL model: SNPs are grouped into genes, and the pairwise relations between the genes are represented by a network. The model selects genes that explain the variance observed in the response variable and the most influential SNPs within these genes. The network structure is accounted such that the model encourages the selection of genes that are connected in the network.

$$P_3(\beta) = \frac{1}{2}\mu \sum_{i \in R_I, j \in R_J, I \sim J} w_{IJ}(\beta_i - \beta_j)^2 \qquad (6.3)$$

the constants λ_1, λ_2, and μ are non-negative regularization parameters, and $I \sim J$ if and only if there is an edge between nodes I and J in $\mathcal{G}$. In P_1, β_K denotes the sub-vector extracted from the $R_K{}^{th}$ entries of β, while the ℓ_2 norm measures the magnitude of the effect on y from each gene. The sum of ℓ_2 norms is referred as an ℓ_1/ℓ_2 norm, which induces sparsity at genes' level [42]. The number of selected genes is controlled via λ_1: as it increases, less genes are included in the model. The scaling factor $\sqrt{|R_K|}$ is added to correct the bias towards selecting genes consisting of a large number of SNPs. The ℓ_1 norm penalty P_2 is added to enable the model to identify the most influential SNPs within the selected genes [13]. When we are only interested in selecting SNP-sets rather than SNPs, we let $\lambda_2 = 0$ and no sparsity within the genes is imposed. The sparsity pattern imposed by $P_1 + P_2$ in (6.1) typically looks like:

$$\hat{\beta} = ([0.2, 0, 0], [0, 0, 0, 0], [0, 0.1], ...)$$

where coefficients in "[ ]" indicate SNPs belonging to the same gene. Applying P_1 and P_2 alone, the solution to (6.1) is solely determined by data matrices X and y, ignoring the rich structured information between the interacting genes. This prior knowledge is accounted for by P_3 in (6.3), which is a standard

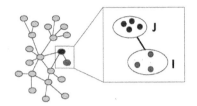
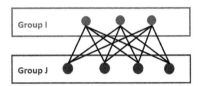

FIGURE 6.2: Insight of the graph penalty in GGGL-1: If $I \sim J$, then penalty P_3 in (6.3) is equivalent to reformulating $\mathcal{G}$ by a composite bipartite graph with vertex sets I and J. All edge weights in this bipartite graph are set to w_{IJ}.

Laplacian penalty penalizing the squared difference between all related SNPs. Two SNPs i and j are "related" if the genes they belong to are connected in the given network $\mathcal{G}$. The effect of the standard Laplacian penalty is to smooth the coefficient estimates $\hat{\beta}_i$ and $\hat{\beta}_j$ towards a value between the two, known as "coefficient smoothing". In particular, if $\hat{\beta}_i > 0$ and $\hat{\beta}_j = 0$ when the composite penalty involves P_1 and P_2 only, introducing P_3 will drive $\hat{\beta}_j$ towards a nonzero value, and consequently encourages the selection of both genes that the two SNPs belong to [21, 4]. The strength of this coefficient smoothing is controlled by μ: When μ is large, we have a stronger belief that the given network truly describes the underlying biological process and that the selected genes should be densely clustered in $\mathcal{G}$. When μ is sufficiently large, coefficients associated to SNPs from interacting genes tend to obtain the same value, in which case the coefficients reflect the averaged effect from the multiple components in $\mathcal{G}$.

Note that through P_3, we make use of structured information at the genes level to guide the selection of genes and SNPs, which involves integrating information at heterogeneous levels. The choice of P_3 defined in (6.3) is equivalent to a re-formulation of $\mathcal{G}$ from the genes level to the SNPs level. This is done by constructing a complete bipartite graph $K(I, J)$ on the vertex set $I \cup J$ whenever there is an edge between genes R_I and R_J in $\mathcal{G}$, as illustrated in Figure 6.2. A side effect of this is that coefficients of SNPs belonging to the same gene are also smoothed, thus the estimated coefficients of causal SNPs and non-causal SNPs tend to have similar values, which may increase the chances of selecting non-causal SNPs within the selected genes.

We also propose a second version of the GGGL. This version, called GGGL-2, addresses this problem arisen when imposing sparsity within the selected genes. In GGGL-2, P_3 is replaced by:

$$P_3(\beta) = \frac{1}{2}\mu \sum_{I \sim J} w_{IJ}(\bar{\beta}_I - \bar{\beta}_J)^2 \tag{6.4}$$

where $\bar{\beta}_I$ denotes the average coefficient of variables in R_I. We further require the coefficients to be non-negative, so that the averaged coefficients can represent the average effect of the grouped variables. Intuitively, if the group R_J is selected and $\hat{\beta}_J \neq 0$, then for its neighbouring group R_I, $\hat{\beta}_I$ will be smoothed towards a nonzero value and in this way the model encourages the selection of the group R_I. This coefficient smoothing on averaged coefficients will be formally presented in the next subsection. In addition, we show the choice of P_3 in (6.4) does not smooth the coefficients of the SNPs belonging to the same gene, which is the primary difference between the two versions of the GGGL model.

6.2.2 Properties

We shall present properties regarding the smoothing effect of the two versions of GGGL, differed by the choice of P_3 in (6.3) and (6.4), respectively. By smoothing effect we refer to the shrinkage of the difference between a pair of coefficient estimates toward the difference between the weighted average of their respective neighbors, as the regularization parameter μ increases [16]. This further indicates the genes interacting with a large common set of genes have similar magnitude of coefficients, and are encouraged to be selected together. Since the group lasso penalty $P_1 + P_2$ induces no smoothing effect, it is only necessary to consider P_3 alone.

6.2.2.1 GGGL-1: Smoothing Effect

In this subsection we show GGGL-1 imposes a smoothing effect on $\hat{\beta}_i$ and $\hat{\beta}_j$ as long as the associated genes of ith and the jth SNPs belong to the same component in $\mathcal{G}$. Thus the coefficients are smoothed both within and between interacting SNP-sets.

Proposition 6.1. *Given data matrices X and y, where X is column-wise normalized and y is centered, let $i \in R_I$ and $j \in R_J$ and assume R_I and R_J belong to the same component in the given network $\mathcal{G}$. For fixed μ, let $\hat{\beta}$ be the vector that minimizes:*

$$L^1(\beta) := \|y - X\beta\|_2^2 + \mu \sum_{k,l:k \in R_K, l \in R_L, K \sim L} w_{KL}(\beta_k - \beta_l)^2$$

Define the following:

$$\rho_{ij} = X_i^T X_j$$

$$C_I = \sum_{K \sim I} w_{IK} |R_K|$$

$$\Gamma_I = \frac{\sum_{k \in R_K, K \sim I} w_{IK} \hat{\beta}_k}{C_I}$$

$$D_\mu(i,j) = |(\hat{\beta}_i - \hat{\beta}_j) - (\Gamma_I - \Gamma_J)| \tag{6.5}$$

Then:

$$D_\mu(i,j) \le \frac{\|y\|_2}{\mu} \left(\frac{\sqrt{2(1-\rho_{ij})}}{C_I} + \left| \frac{1}{C_I} - \frac{1}{C_J} \right| \right). \tag{6.6}$$

To prove (6.6), we need the following lemmas:

Lemma 6.1. *Let a, b, c, $d \in \mathbb{R}$, and b, $d \ne 0$, the following inequality holds:*

$$\left| \frac{a}{b} - \frac{c}{d} \right| \le \left| \frac{a-c}{b} \right| + |c| \cdot \left| \frac{1}{b} - \frac{1}{d} \right|.$$

Proof.

$$\left| \frac{a}{b} - \frac{c}{d} \right| = \left| \frac{ad - bc + cd - cd}{bd} \right|$$

$$= \left| \frac{ad - cd}{bd} + \frac{cd - bc}{bd} \right|$$

$$\le \left| \frac{a-c}{b} \right| + |c| \cdot \left| \frac{1}{b} - \frac{1}{d} \right|.$$

□

Lemma 6.2. *Let X be an $n \times p$ matrix where each column is normalized to have zero mean and unit length. Let y be an n-dimensional vector that is centered. Let $\hat{r} = y - X\hat{\beta}$ and assuming $\|\hat{r}\|_2 \le \|y\|_2$, the following inequalities hold:*

$$|(X_i^T - X_j^T)\hat{r}| \le \sqrt{2(1-\rho_{ij})}\|y\|_2$$

$$|X_i^T \hat{r}| \le \|y\|_2$$

The proofs were given in [44].
Now we give the proof for Proposition 6.1:

Proof. Suppose $i \in R_I$, solving $\frac{\partial L^1}{\partial \beta_i} = 0$ gives:

$$-2X_i^T\hat{r} + 2\mu \cdot \underbrace{\left(\sum_{\substack{k:\ k \in R_K, \\ K \sim I}} w_{IK} \right)}_{C_I} \cdot \hat{\beta}_i - 2\mu \cdot \underbrace{\left(\sum_{\substack{k:\ k \in R_K, \\ K \sim I}} w_{IK}\hat{\beta}_k \right)}_{\Gamma_I \cdot C_I} = 0$$

where $\hat{r} = y - X\hat{\beta}$. Rearranging to give:

$$\hat{\beta}_i = \frac{X_i^T\hat{r}}{\mu C_I} + \Gamma_I$$

Similarly for $j \in R_J$, we have:

$$\hat{\beta}_j = \frac{X_j^T\hat{r}}{\mu C_J} + \Gamma_J$$

Thus:

$$|(\hat{\beta}_i - \hat{\beta}_j) - (\Gamma_I - \Gamma_J)| = \left| \frac{X_i^T \hat{r}}{\mu C_I} - \frac{X_j^T \hat{r}}{\mu C_J} \right|.$$

Applying Lemma 1 and 2, we can obtain an upper bound for the right hand side of (6.6):

$$
\left| \frac{X_i^T \hat{r}}{\mu C_I} - \frac{X_j^T \hat{r}}{\mu C_J} \right| \leq \left| \frac{(X_i^T - X_j^T)\hat{r}}{\mu C_I} \right| + |X_j^T \hat{r}| \cdot \left| \frac{1}{\mu C_I} - \frac{1}{\mu C_J} \right| \quad \text{by Lemma 1}
$$

$$
\leq \frac{\sqrt{2(1 - \rho_{ij})}\|y\|_2}{\mu C_I} + \frac{\|y\|_2}{\mu} \cdot \left| \frac{1}{C_I} - \frac{1}{C_J} \right| \quad \text{by Lemma 2.}
$$

$\square$

Remark 6.1. *Proposition 6.1 does not require $I \sim J$, and it also holds when $R_I = R_J$, namely i and j belong to the same gene.*

Note C_I depends on the topology of $\mathcal{G}$ and gene size only, therefore C_I and C_J are constants. Γ_I is the weighted average estimated coefficients of the SNPs whose associated genes are connected with gene R_I in $\mathcal{G}$. The quantity $D_\mu(i, j)$ measures the deviation between the difference of $\hat{\beta}_i$ and $\hat{\beta}_j$ and the difference of the weighted average estimated coefficients of their corresponding neighbors. In other words, it measures the discrepancy between the centered coefficients at the ith and j^{th} predictor. When we have a strong belief that interacting genes and the corresponding SNPs have similar functions, we would expect smoother coefficients of the interacting genes and smaller $D_\mu(i, j)$ as μ increases. Letting μ tend to infinity in (6.6), we indeed have $D_\infty(i, j) \to 0$, thus proved the smoothing effect from GGGL-1.

In the case where i and j belong to the same gene such that $R_I = R_J$, we can deduce the following:

Corollary 6.1. *Under the same setting of Proposition 6.1 and assuming i and j belong to the same gene R_I, define the partial residual:*

$$\hat{r}_{ij} = y - X\hat{\beta} + X_i \hat{\beta}_i + X_j \hat{\beta}_j \tag{6.7}$$

the estimated coefficients $\hat{\beta}$ satisfy:

$$|\hat{\beta}_i - \hat{\beta}_j| = \frac{(X_i^T - X_j^T)\hat{r}_{ij}}{1 - \rho_{ij} + \mu C_I}. \tag{6.8}$$

Moreover, the left hand side of (6.8) is bounded above by:

$$\frac{\sqrt{2(1 - \rho_{ij})}\|y\|_2}{\mu C_I}. \tag{6.9}$$

From (6.9), it can be observed when μ increases, the coefficient estimates of all SNPs in the same gene are pushed towards the same value, possibly making it more difficult to identify the causal SNPs within the selected gene.

6.2.2.2   GGGL-2: Smoothing Effect

In this subsection we show GGGL-2 imposes a smoothing effect on the average coefficients $\bar{\beta}_I$ and $\bar{\beta}_J$ provided R_I and R_J belong to the same component in $\mathcal{G}$. However, it does not impose any smoothing effects on the coefficients of SNPs within the same gene.

Proposition 6.2. *Given data matrices X and y, where X is column-wise normalized and y is centered, let $i \in R_I$ and $j \in R_J$ and assume R_I and R_J belong to the same component in the given network $\mathcal{G}$. Denote the vertex degree of R_I in $\mathcal{G}$ by d_I, and let $\bar{\beta}_I = \frac{1}{|R_I|}\sum_{i \in R_I} \beta_i$. For fixed μ, let $\hat{\beta}$ be the vector that minimizes:*

$$L^2(\beta) := \|y - X\beta\|_2^2 + \mu \sum_{K \sim L} w_{KL}(\bar{\beta}_K - \bar{\beta}_L)^2 \ . \tag{6.10}$$

Define the following:

$$\Theta_I = \sum_{K \sim I} \frac{w_{IK}}{d_I} \bar{\hat{\beta}}_K \ , \quad and$$

$$D_\mu(I, J) = |(\bar{\hat{\beta}}_I - \bar{\hat{\beta}}_J) - (\Theta_I - \Theta_J)| \tag{6.11}$$

Then:

$$D_\mu(I, J) \leq \frac{\|y\|_2}{\mu}\left(\frac{2|R_I|}{d_I} + \left|\frac{|R_I|}{d_I} - \frac{|R_J|}{d_J}\right|\right) \ . \tag{6.12}$$

Proof. Suppose $i \in R_I$, solving $\frac{\partial L^2}{\partial \beta_i} = 0$ gives:

$$-2X_i^T \hat{r} + 2\mu \sum_{I \sim K} w_{IK}\frac{1}{|R_I|}(\bar{\beta}_I - \bar{\beta}_J) = 0 \ .$$

Rewriting the summation of the differences as the difference of two sums:

$$-X_i^T\hat{r} + \frac{\mu}{|R_I|} \cdot \underbrace{\left(\sum_{K \sim I} w_{IK}\right)}_{d_I} \cdot \bar{\beta}_I - \frac{\mu}{|R_I|} \cdot \underbrace{\left(\sum_{K \sim I} w_{IK}\bar{\beta}_K\right)}_{\Theta_I \cdot d_I} = 0$$

Rearrange the above equation to give:

$$\bar{\beta}_I = \frac{X_i^T\hat{r}}{\frac{\mu d_I}{|R_I|}} + \Theta_I$$

and similarly for $j \in R_J$ we have:

$$\bar{\beta}_J = \frac{X_j^T\hat{r}}{\frac{\mu d_J}{|R_J|}} + \Theta_J \ .$$

Hence, we can derive an upper bound for the left hand side of (6.12):

$$
\begin{aligned}
|(\bar{\beta}_I - \bar{\beta}_J) - (\Theta_I - \Theta_J)| &= \left| \frac{X_i^T \hat{r}}{\frac{\mu d_I}{|R_I|}} - \frac{X_j^T \hat{r}}{\frac{\mu d_J}{|R_J|}} \right| \\
&\leq \frac{|(X_i^T - X_j^T)\hat{r}|}{\frac{\mu d_I}{|R_I|}} + |X_j^T \hat{r}| \cdot \left| \frac{|R_I|}{\mu d_I} - \frac{|R_J|}{\mu d_J} \right| \\
&\qquad \text{by Lemma 1} \\
&\leq \frac{\sqrt{2(1 - \rho_{ij})} \|y\|_2 \cdot |R_I|}{\mu d_I} + \frac{\|y\|_2}{\mu} \cdot \left| \frac{|R_I|}{d_I} - \frac{|R_J|}{d_J} \right| \\
&\qquad \text{by Lemma 2} \\
&= \frac{\|y\|_2}{\mu} \left(\frac{\sqrt{2(1 - \rho_{ij})} \cdot |R_I|}{d_I} + \left| \frac{|R_I|}{d_I} - \frac{|R_J|}{d_J} \right| \right) \\
&\leq \frac{\|y\|_2}{\mu} \left(\frac{2|R_I|}{d_I} + \left| \frac{|R_I|}{d_I} - \frac{|R_J|}{d_J} \right| \right).
\end{aligned}
$$

$\square$

Remark 6.2. *Proposition 6.2 does not require $I \sim J$, and it also holds when $R_I = R_J$, namely i and j belong to the same gene.*

Note Θ_I is a weighted mean of the coefficients of the SNPs that are functionally related to the gene R_I, and $D_\mu(I, J)$ measures the strength of the smoothing effect on the averaged coefficient estimates for the genes. The upper bound obtained in (6.12) is determined by the data, the network topology, and the gene size, for fixed μ. Letting $\mu \to \infty$, we have $D_\infty(I, J) \to 0$, thus having proved the smoothing effect from GGGL-2. In the case where i and j belong to the same gene such that $R_I = R_J$, the following corollary can be deduced following the same proof line as Proposition 6.2.

Corollary 6.2. *Under the same setting of Proposition 6.2 and assuming i and j belong to the same gene, recalling the partial residual defined in (6.7), the estimated coefficients $\hat{\beta}$ satisfy:*

$$
|\hat{\beta}_i - \hat{\beta}_j| = \frac{|(X_i^T - X_j^T)\hat{r}_{ij}|}{1 - \rho_{ij}} . \tag{6.13}
$$

The right hand side of (6.13) does not involve μ, which means $|\hat{\beta}_i - \hat{\beta}_j|$ is data-driven, via a trade-off between the loss function and graph penalty in (6.10). We conclude that the graph penalty does not enforce the coefficient estimates of SNPs within the same gene to take the same value.

6.3   Estimation Algorithms

In this section we present model estimation algorithms for the two versions of the GGGL model, based on block coordinate descent methods. The first is a serial version that updates the coefficients one block (gene) at a time. The second is a parallel algorithm that updates a subset of the blocks (i.e., multiple genes) in parallel in each step. This enables the program to be run independently on multiple processors (e.g., graphics processing units), so that the amount of computation required by each processor is reduced and thus speeds up the computation. The parallel algorithm is suitable for "big data" in which both n and p are large.

6.3.1   Serial Block Coordinate Descent Algorithm

Note both versions of the GGGL models seek to minimise (6.1) in which sparsity penalties are defined in (6.2) and graph penalties are defined in (6.3) and (6.4), respectively. Note when $\mu = 0$ and thus $P_3 = 0$, the objective function reduces to that of the sparse group lasso, which can be efficiently solved using a block coordinate descent algorithm [13]. We thus attempt to re-formulate (6.1) into a sparse group lasso problem and apply the existing efficient algorithm to solve it.

6.3.1.1   Serial Algorithm for GGGL-1

By definition, the summation part in (6.3) can be rewritten as:

$$\sum_{i \in R_I, j \in R_J, I \sim J} w_{IJ}(\beta_i - \beta_j)^2 = \sum_{i \leq j} w_{ij}(\beta_i - \beta_j)^2 \qquad (6.14)$$

where w_{ij} is defined as:

$$w_{ij} = \begin{cases} 0 & \text{if} \quad i \text{ and } j \text{ belongs to the same gene} \\ w_{IJ} & \text{if} \quad i \in R_I, j \in R_J \neq R_I \end{cases} \qquad (6.15)$$

Let L be a $p \times p$ matrix whose (i, j)th entry is:

$$(L)_{ij} = \begin{cases} d_i & \text{if} \quad i = j \in R_I \\ -w_{ij} & \text{if} \quad i \neq j \end{cases} \qquad (6.16)$$

where $d_i = \sum_{j=1}^{p} w_{ij}$. Using L, the right hand side of (6.14) can be re-formulated into: [4]

$$\sum_{i \leq j} w_{ij}(\beta_i - \beta_j)^2 = \beta^T L \beta .$$

Up to this point, we have:

$$\|y - X\beta\|_2^2 + \mu \sum_{i \in R_I, j \in R_J, I \sim J} w_{IJ}(\beta_i - \beta_j)^2 = \|y - X\beta\|_2^2 + \mu \beta^T L \beta . \qquad (6.17)$$

By construction, L is positive semi-definite, therefore we can find $p \times p$ matrix U such that: $L = UU^T$, using for instance, Cholesky decomposition. We then construct the $(n + p) \times 1$ matrix y^* and the $(n + p) \times p$ matrix X^* by:

$$y* = \begin{pmatrix} y_{n \times 1} \\ 0_{p \times 1} \end{pmatrix}, \qquad X^* = \begin{pmatrix} X \\ \sqrt{\mu} U^T \end{pmatrix}. \tag{6.18}$$

Consequently, (6.17) can be re-formulated into: $\|y^* - X^*\beta\|_2^2$, as in [8, 21, 44]. The objective function (6.1) with penalties (6.2) (6.3) can be re-written as:

$$\|y^* - X^*\beta\|_2^2 + 2\lambda_1 \sum_{K=1}^{|\mathcal{R}|} \sqrt{|R_K|} \|\beta_K\|_2 + 2\lambda_2 \|\beta\|_1 \tag{6.19}$$

where β_K is the $|R_K| \times 1$ matrix containing the entries corresponding to variables in R_K only. Note (6.19) is exactly the objective function for the sparse group lasso [13] on data matrices (X^*, y^*), which can be efficiently solved by the block coordinate descent algorithm. The full procedure is given in Algorithm 7. We remark that the one-dimensional optimization problem corresponding to steps 12 to 15 in Algorithm 7 can be dealt by other methods, for example, the successive parabolic interpolation, as done in [13].

Algorithm 7 GGGL Version 1

Input: data X, y; parameter λ_1, λ_2, μ; partition of predictors $\mathcal{R}$, which consists of $|\mathcal{R}|$ groups; weight matrix on groups w_{IJ}; starting value $\tilde{\beta}$; ϵ

Output: column vector $\hat{\beta}$

  1: Define $S_\lambda(z) := \sigma(z) \cdot \max\{|z| - \lambda, 0\}$, where $\sigma(z) = 1$ if $z > 0$,
       -1 if $z < 0$, 0 if $z = 0$.

  2: Compute $p \times p$ weight matrix w_{ij} according to (6.15).

  3: Compute L from (6.16).

  4: Use Cholesky decomposition to compute U such that $L = UU^T$.

  5: Compute X^* and y^* defined by (6.18).

  6: $\beta^0 \leftarrow \tilde{\beta}$

  7: **for** I in 1:$|\mathcal{R}|$ **do**

  8:      $r_I \leftarrow y^* - \sum_{K \neq I} X_K^* \tilde{\beta}_K$

  9:      **if** $\|S_{\lambda_2}(X_I^* r_I)\|_2 \leq \lambda_1 |R_I|$ **then**

10:          $\hat{\beta}_I = 0$

11:      **else**

12:          **for** i in R_I **do**

13:             $r_i \leftarrow r_I - \sum_{j:j \in R_I/\{i\}} X_j^* \tilde{\beta}_j$

14:             $\hat{\beta}_i \leftarrow (1 - \dfrac{\lambda_1 \sqrt{|R_I|}}{\sqrt{\sum_{j:j \in R_I}(S_{\lambda_2}(X_j^{*T} r_j))^2}}) \cdot \dfrac{S_{\lambda_2}(X_i^{*T} r_i)}{X_i^{*T} X_i^*}$

15:          **end for**

16:          **if** $\|\hat{\beta}_I - \tilde{\beta}_I\|_2 \leq \epsilon$ **then**

17:             **Return:** $\hat{\beta}_I$

18:          **else**

19:             $\tilde{\beta}_I \leftarrow \hat{\beta}_I$; go back to 12

20:          **end if**

21:      **end if**

22: **end for**

23: **if** $\|\hat{\beta} - \beta^0\|_2 \leq \epsilon$ **then**

24:      **Return:** $\hat{\beta}$

25: **else**

26:      $\beta^0 \leftarrow \hat{\beta}$; go back to 7

27: **end if**

6.3.1.2 Serial Algorithm for GGGL-2

Considering the graph penalty (6.4) in GGGL-2, we notice:

$$
\begin{aligned}
w_{IJ}(\bar{\beta}_I - \bar{\beta}_J)^2 &= w_{IJ}\left(\frac{1}{|R_I|}\sum_{i:i\in R_I}\beta_i - \frac{1}{|R_J|}\sum_{j:j\in R_J}\beta_j\right)^2 \quad \text{by definition} \\
&= \frac{w_{IJ}}{|R_I|^2}\sum_{i:i\in R_I}\beta_i^2 + \frac{w_{IJ}}{|R_J|^2}\sum_{j:j\in R_J}\beta_j^2 \\
&\quad + \frac{w_{IJ}}{|R_I|^2}\sum_{i:i\in R_I}\sum_{s:s\in R_I\setminus\{i\}}\beta_i\beta_s \\
&\quad + \frac{w_{IJ}}{|R_J|^2}\sum_{j:j\in R_J}\sum_{t:t\in R_J\setminus\{j\}}\beta_j\beta_t \\
&\quad - \frac{2\cdot w_{IJ}}{|R_I|\cdot|R_J|}\sum_{i:i\in R_I}\sum_{j:j\in R_J}\beta_i\beta_j
\end{aligned}
\tag{6.20}
$$

after expanding the brackets.

From (6.20) it can be deduced that for $i \in R_I$, the terms involving β_i in $\sum_{I\sim J}w_{IJ}(\bar{\beta}_I - \bar{\beta}_J)^2$ are:

$$
\left(\sum_{J:J\sim I}\frac{w_{IJ}}{|R_I|^2}\right)\beta_i^2 + 2\left(\sum_{J:J\sim I}\frac{w_{IJ}}{|R_I|^2}\right)\sum_{s:s\in R_I\setminus\{i\}}\beta_i\beta_s
$$
$$
- 2\sum_{J:J\sim I}\left\{\left(\frac{w_{IJ}}{|R_I|\cdot|R_J|}\right)\sum_{j:j\in R_J}\beta_i\beta_j\right\}.
$$

It is then straightforward to deduce that there exists a $p \times p$ matrix $\mathcal{L}$ such that $\sum_{I\sim J}w_{IJ}(\bar{\beta}_{R_I} - \bar{\beta}_{R_J})^2 = \beta^T\mathcal{L}\beta$, where $\mathcal{L}$ is defined as:

$$
(\mathcal{L})_{ij} = \begin{cases} \sum_{\{K:K\sim I\}}\frac{w_{IK}}{|R_I|^2} & \text{if } i\in R_I, j\in R_I \\ -\frac{w_{IJ}}{|R_I|\cdot|R_J|} & \text{if } i\in R_I, j\in R_J \end{cases}
\tag{6.21}
$$

Then solution to GGGL-2 can be computed according to Algorithm 7, replacing steps 2 and 3 by (6.21). The non-negative coefficients constraint in GGGL-2 is tackled by further requiring $\hat{\beta}_i$ in step 14 to be non-negative. Thus if a negative estimate is obtained, the corresponding entry in $\hat{\beta}$ is set to zero.

6.3.2 Parallel Coordinate Descent Algorithm

In this subsection we describe the implementation of a parallel algorithm that can be used to solve (6.19). The algorithm presented below is an application of the general framework proposed in [32], which considers the problem of minimizing:

$$
F(\beta) := f(\beta) + \Omega(\beta)
\tag{6.22}
$$

where $\beta \in \mathbb{R}^p$, $f(x)$ denotes the loss function, and $\Omega(\beta)$ denotes the composite penalty function. If there exists a partition $\bigcup_{i=1}^{m} U_i$ of the variables $\{\beta_1, \beta_2, ..., \beta_p\}$ such that $F(\beta)$ can be written as the sum of $m \geq 2$ functions, each depending on a unique set U_i of the variables, we say $F(\beta)$ is "separable", and the minimization problem (6.22) can be decomposed into m independent optimization problems, which can be solved in parallel. Therefore if there are ρ ($2 \leq \rho \leq m$) processors available for the computation, it is expected to get a ρ times speed-up by parallelizing the computation. However, sometimes, e.g. in (6.19), only the composite penalty term is "separable", making it non-trivial to parallelize.

In addition to the usual requirement of smoothness and convexity, the parallel coordinate descent method (PCDM) presented in [32] relaxes the separability condition on $f(\beta)$ to f being a (block) partially separable function. By coordinate descent method (CDM) we refer to the strategy of cyclically updating the coordinates of β one coordinate (or one block of coordinates) at each iteration [12]. CDMs are fast to compute at each iteration but usually take more iterations to converge than competing methods, e.g., gradient methods [26]; however, their performance on big data optimisation is generally more efficient [32]. An outline of this PCDM is given in Algorithm 8.

Algorithm 8 Parallel Coordinate Descent Method

Input: data X, y; parameter λ_1, λ_2; partition of predictors $\mathcal{R}$, which consists of $|\mathcal{R}|$ blocks; starting value $\tilde{\beta}$; ϵ; number of blocks to be updated in each step ρ (which is not less than the number of processors available).

Output: column vector $\hat{\beta}$

  1:   Compute constants ω_{R_I} for each block R_I and the constant B.

  2:   Choose initial estimate $\hat{\beta}^{(0)}$.

  3:   $k \leftarrow 0$

  4:   Randomly generate a set of blocks from $\mathcal{R}$: $k_1, k_2, ...$

  5:   In parallel do: $\hat{\beta}_{R_{k_I}}^{(k+1)} \leftarrow \phi(\hat{\beta}^{(k)}, k_I)$, for $I = 1, 2, ...$

  6:   Collect estimates from the processors to obtain $\hat{\beta}^{(k+1)}$.

  7:   Set $k \leftarrow k + 1$ and go back to 4 until a stopping criterion is met.

In Algorithm 8 step 4, the blocks to be updated in each iteration are randomly generated from a uniform sampling scheme, so that the probability of any block being updated in step k is constant and independent of previous updates. In step 5, the blockwise update of $\hat{\beta}$ in step $(k+1)$ is obtained from a function ϕ, which corresponds to another optimization problem depending on the full estimate from step k. For notation simplicity, denote $\phi(\hat{\beta}^{(k)}, k_I)$ by ϕ_I, which is a column vector with $|R_{k_I}|$ entries. Without loss of generality we assume a processor only updates one block of variables in each iteration, so that the index I corresponds to R_I in (6.19). Then ϕ_I minimizes the function:

$$< \nabla_I f, \ \Phi > + \frac{B\omega_I}{2} \|\Phi_I\|_2^2 + \Omega(\beta_I^{(k)} + \Phi_I) \qquad (6.23)$$

with respect to Φ_I, where $<,>$ denotes the inner product, ∇ denotes the gradient operation, and B and ω_i are constants to be determined.

The consequence of the conditions on f and Ω, the uniform sampling scheme, and the computation in (6.23) with pre-determined constants B and ω_I is that we obtain a monotonically decreasing sequence in terms of the expected value of $F(\hat{\beta})$ after each iteration. Specifically:

$$\mathbb{E}[F(\hat{\beta}^{(k+1)})|\hat{\beta}^{(k)}] \leq \mathbb{E}[F(\hat{\beta}^{(k)})] . \qquad (6.24)$$

Since this monotonically decreasing sequence is bounded below (certainly by 0), the algorithm is expected to converge.

We now re-state the assumptions necessary to validate this PCDM algorithm, as originally presented in [32], and apply it to (6.19). These include:

Partial separability of f: Let $\bigcup_{i=1}^m U_i$ be a partition of the variables $\{\beta_1, \beta_2, ..., \beta_p\}$, then f is *partially separable of degree* δ if it can be rewritten as:

$$f(\beta) = \sum_{J \in \mathcal{J}} f_J(\beta) \qquad (6.25)$$

where $\mathcal{J}$ is a subset of $\{U_1, U_2, ..., U_m\}$ and each J is non-empty (possibly overlaps with others), $|J| \leq \delta$ for all $J \in \mathcal{J}$. f_J are differentiable convex functions that depend on the blocks of β in J only.

Note the algorithm requires no knowledge on the exact decomposition of (6.25); it is only necessary to compute δ, which can be taken as the maximum number of non-zero elements in the rows of X^*, when $f(\beta) = \|y^* - X^*\beta\|_2^2$ as in (6.19).

Smoothness of f: This requires $\nabla_i f$ to satisfy the Lipschitz condition with Lipschitz constants L_i for $i = 1, 2, ..., m$.

Separability and convexity of Ω: Let $\bigcup_{i=1}^m U_i$ be a partition of the variables $\{\beta_1, \beta_2, ..., \beta_p\}$, $\Omega(\beta)$ can be written as:

$$\Omega(\beta) = \sum_{i=1}^m \Omega_i(\beta_{[U_i]}) \qquad (6.26)$$

where $\beta_{[U_i]}$ depends on the entries of β in U_i only, and all Ω_i are convex and closed.

It then can be easily verified that these conditions are satisfied in (6.19), where the partition of variables can be readily obtained as $\{U_1, U_2, ..., U_m\} =$

$\mathcal{R}$. Now, to apply PCDM, note: $f = \frac{1}{2}\|y - X\beta\|_2^2$ and $\Omega(\beta) = \lambda_1 \sum_I \sqrt{|R_I|}\|\beta_I\|_2 + \lambda_2\|\beta\|_1$. Substituting into (6.23) we obtain the optimization problem on ϕ_I for block R_I: (for notation simplicity we drop * on X and y)

$$-(y - X\beta)^T X_I \phi_I + \frac{B\omega_I}{2}\|\phi_I\|_2^2 + \lambda_1 \sqrt{|R_I|} \cdot \|\beta_I + \phi_I\|_2 + \lambda_2\|\beta_I + \phi_I\|_1 . \quad (6.27)$$

Let $i \in R_I$, differentiating (6.27) with respect to the ith coordinate of ϕ_I, denoted by ϕ_i, and setting to zero, we see ϕ_i must satisfy:

$$-(y - X\beta)^T X_i + B\omega_I \phi_i + \lambda_1 \sqrt{|R_I|} \cdot s_i + \lambda_2 t_i = 0 \quad (6.28)$$

where:

$$s_i \begin{cases} = \frac{\beta_i + \phi_i}{\|\beta_I + \phi_I\|_2} & \text{if} \quad \beta_I + \phi_I \neq 0 \\ \in [-1, 1] & \text{if} \quad \beta_I + \phi_I = 0 \end{cases} \quad (6.29)$$

$$t_i \begin{cases} = \sigma(\beta_i + \phi_i) & \text{if} \quad \beta_i + \phi_i \neq 0 \\ \in [-1, 1] & \text{if} \quad \beta_i + \phi_i = 0 \end{cases} \quad (6.30)$$

where σ is the sign function.

Note (6.28), (6.29), and (6.30) are the sub-gradient equations on $\beta_i + \phi_i$ of a sparse group lasso model as in [13], which can be solved by a block CDM. Specifically, set $\hat{\phi}_I = -\beta_I$ if:

$$\|S_{\lambda_2}[(y - X\beta)^T X_I + B\omega_I\beta_I]\|_2 \leq \lambda_1 \sqrt{|R_I|} \quad (6.31)$$

where $S_\lambda(z) := \sigma(z) \cdot max\{|z| - \lambda, 0\}$. Otherwise, for $i \in R_I$, set $\hat{\phi}_i = -\beta_i$ if:

$$|(y - X\beta)^T X_i + B\omega_I\beta_i| \leq \lambda_2 . \quad (6.32)$$

If neither (6.31) or (6.32) holds, let $\gamma_i = \beta_i + \phi_i$ and Equations (6.28), (6.29), and (6.30) merge to:

$$(y - X\beta)^T X_i + B\omega_I\beta_i = B\omega_I\gamma_i + \lambda_1 \sqrt{|R_I|} \cdot \frac{\gamma_i}{\|\gamma_I\|_2} + \lambda_2\sigma(\gamma_i)$$

which can be solved by:

$$\hat{\gamma}_i = \frac{1}{B\omega_I}(1 - \frac{\lambda_1 \sqrt{|R_I|}}{\|S_{\lambda_2}[(y - X\beta)^T X_I + B\omega_I\beta_I]\|_2})$$
$$\times S_{\lambda_2}[(y - X\beta)^T X_i + B\omega_I\beta_i] . \quad (6.33)$$

To summarize, step 5 in Algorithm 8 can be computed by Algorithm 9.

The constants B and ω_I are determined by the data and the sampling scheme involved in step 4 of Algorithm 8. In [32], several sampling methods are described and studied. In our implementation, we use "nice sampling", which randomly selects ρ groups from $\mathcal{R}$ and each group is to be selected with equal probability. In this case we have: $B = \min\{\delta, \rho\}$ and ω_I is the Lipschitz constant associated with R_I, which is taken as $\|X_I^T X_I\|_\mathcal{F}$.

Algorithm 9 PCDM Step 5

Input: The group to be updated R_{k_I}, denoted by R_I for short; data X, y (superscript * removed for simplicity); parameter λ_1, λ_2; estimated coefficients before step k: $\beta^{(k)}$, denoted by β for convenience, ϵ

Output: column vector $\hat{\beta}^{(k+1)}_{R_{k_I}}$, denoted by $\hat{\beta}_I$ for convenience.

  1:   **if** (6.31) holds **then**
  2:       $\hat{\beta}_I = (0, 0, ..., 0)^T$
  3:   **else**
  4:       **for** $i \in R_i$ **do**
  5:           **if** (6.32) holds **then**
  6:             $\hat{\beta}_i = 0$
  7:           **else**
  8:             $\hat{\beta}_i = \hat{\gamma}_i$ as in (6.33).
  9:           **end if**
10:       **end for**
11:       Repeat 4-10 until $\|\hat{\gamma}_I - \hat{\beta}_I\|_2 \leq \epsilon$
12:   **end if**

6.4   Simulation Studies

In this section we present simulation studies to assess the power of detecting true signal-carrying predictor groups (e.g., genes) and predictors (e.g., SNPs). We compare the performance of the GGGL models with models that do not account for the prior information between the predictor groups, as encoded by the network. We carry out two sets of simulations: in (I) we compare the GGGL-1 with the group lasso [42] on group selection; and in (II) we compare the GGGL-2 with the sparse group lasso [13] on both group and predictor selection. Throughout the simulation studies we fix $n = 200$, $p = 1000$, and $|\mathcal{R}| = 100$, which have equal sizes of 10. In set (I), we consider five cases of simulations, corresponding to signal-to-noise ratios (SNRs) at $0.15, 0.125, 0.10, 0.075, 0.05$; in set (II), due to the extensive computation it involves, we perform one case of simulation in which SNR is set at 0.075. Here the SNR is defined as the ratio of the mean of the response variable to the standard deviation of the noise. For each case of the simulations, we generate 500 data sets each consisting of X, y, $\mathcal{R}$, and $\mathcal{G}$. We generate the $n \times p$ matrix X assuming the predictors follow independent standard normal distributions. The indexes of the signal-carrying variables are fixed and known in each set of the simulations so that the performance of each model can be evaluated by comparing the number of correctly identified predictors or predictor groups while controlling for the number of falsely detected ones. Specifically, we let signal-carrying predictors fall into 20 groups, and define coefficient vector v

in which the non-zero entries, which correspond to the signal-carrying predictors, are generated from a uniform distribution in the interval $(0.1, 1)$. In (I), we assume 70% of the variables in the signal-carrying groups have non-zero entries in v, while in (II) we assume a smaller proportion at 50%. We define the response vector: $y = Xv + \eta \cdot e$, where e is the vector of random errors generated from independent standard normal distributions, and η is a positive real number set to achieve the desired SNRs, computed as the ratio of the mean of Xv to the standard deviation of $(\eta \cdot e)$. Finally, we normalize the columns of X to have zero mean and unit Euclidean norm and center y as required by the models.

As for the network generation, we randomly partition the signal carrying groups into clusters of ten, and likewise partition the non signal-carrying groups, resulting in ten clusters each of size ten. Assuming that the probability for a pair of nodes being directly connected is independent of all other pairs, we generate the network $\mathcal{G}$ using a set of three probability parameters: p_C is the probability of connection between groups belonging to the same cluster; p_{CC} is the probability of connection between signal-carrying or non signal-carrying groups belonging to different clusters; and p_{SN} is the probability of connection between a signal-carrying group and a non signal-carrying group. We generate networks such that signal-carrying groups are relatively densely connected whereas there are very few links between these groups and the non signal-carrying groups. Specifically, we set $p_C = 0.4$, $p_{CC} = 0.15$, and $p_{SN} = 0.03$.

In (I) we evaluate the performance of the GGGL-1 and use the group lasso [42] as a benchmark, which essentially drops P_2 and P_3 in (6.2) and (6.3) from (6.1). We assess the performance of the two models by group selection. When fitting the GGGL-1, we fix $\mu = 100$, $\lambda_2 = 0$, and tune λ_1 such that the model selects 25 groups on average and all variables in these groups have non-zero coefficients. For group lasso, we tune λ_1 such that the same number of groups are selected. Note the sparsity level does not affect the relative performance of the models as long as the same sparsity level is imposed so that the results are comparable. We compute the empirical selection probabilities for all groups within each case of the simulations. By varying the threshold from 0 to 1 and defining the important groups as those with selection probabilities greater than the threshold, we can construct the receiver operating characteristic (ROC) curve on the $|\mathcal{R}|$ groups. In a ROC curve, the proportion of signal-carriers classified as important groups (true positive rate; TPR) is plotted against the proportion of non signal-carriers classified as important groups (false positive rate; FPR). The area under curve (AUC) can be interpreted as the probability that a randomly chosen signal-carrying group has larger selection frequency than a randomly chosen group carrying no signal, which will be used as the evaluation criterion of model comparison. We plot the AUC against SNR in Figure 6.3. We observe when signal is strong, both models perform equally well; yet as the SNR continues to decrease, the loss in power of the group lasso is at a faster rate compared with GGGL-1. When

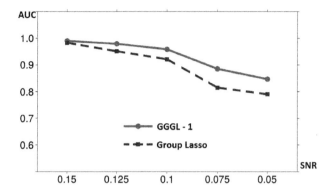

FIGURE 6.3: Plot of area under ROC curve (AUC) against signal-to-noise ratio (SNR), where variable selection is carried out at groups (genes) level. As the SNR continues to fall, using the prior knowledge encoded by the network to guide group selection, GGGL-1 (red line) gains additional power compared to the group lasso (dotted blue).

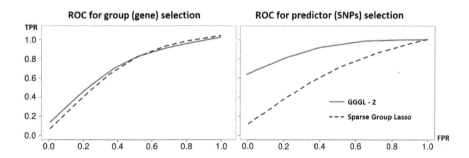

FIGURE 6.4: ROC curves plotting true positive rate (TPR) against false positive rate (FPR). The left column refers to selection of groups (genes) and the right column refers to selection of predictors (SNPs). Using the network, GGGL-2 (red line) retains the power of group selection yet improves on predictor selection, compared with the sparse group lasso (dotted blue).

SNR is low, the plot shows a clear advantage of incorporating the structured prior knowledge on the predictor groups.

In (II) we compare the GGGL-2 with the sparse group lasso [13], whose penalty terms consist of (6.2), where we impose sparsity on and within the groups. We fix $\mu = 10$ for GGGL-2 and tune λ_1, λ_2 in both models such that on average 25 groups are selected and half of the variables in these groups have non-zero coefficients. The ROC curves for group selection and predictor selection are presented in Figure 6.4. We observe the two models are equally competitive in selecting the predictor groups. However, GGGL-2 enhances the power of selecting the signal-carrying predictors via the smoothing of average coefficients it exercises on the predictor groups. The results in (I) and (II) demonstrate that incorporating graphical prior knowledge on the predictor groups enables the GGGL models to gain additional power compared to the (sparse) group lasso models that ignore such information, in particular when the SNR is low.

6.5   Conclusion

The work presented in this chapter is motivated by GWAs and the need to use structured prior information to guide the selection of SNPs and genes associated with a quantitative trait. The prior information is available at two levels: SNPs grouped into genes via gene-SNPs mapping, and a network encoding the functional relatedness between the genes. We have proposed a penalized linear regression model, the graph-guided group lasso in two versions, which can select functionally related genes and influential SNPs within these genes that explain the variability in the trait.

We have studied some theoretical properties regarding the smoothing effect of the GGGL models, and derived an upper bound for $D_\mu(i, j)$ and $D_\mu(I, J)$, defined in (6.5) and (6.11), respectively, which measure the strength of the smoothing effect. Specifically, Proposition 6.1 shows GGGL-1 smooths the coefficients of the SNPs if they belong to the same gene or functionally related genes; and Proposition 6.2 shows GGGL-2 smooths the average coefficients of the SNPs belonging to functionally related genes. The primary difference is, as stated in Corollaries 6.1 and 6.2, that GGGL-2 does not smooth the coefficients of SNPs within the same gene while GGGL-1 does. From a biological perspective, GGGL-1 assumes all or almost all of the SNPs in a causal gene contribute to the variability in the trait and hence encourage the model to select all of them, whereas GGGL-2 assumes only a subset of the SNPs in a causal gene do so and hence tries to avoid selecting a SNP simply because it belongs to a causal gene. In practice we use GGGL-1 when the primary interest is identifying the causal genes, and we use GGGL-2 when we are also interested in identifying the causal SNPs within the selected genes.

We have presented two estimation algorithms for the GGGL models, one serial algorithm and one parallel algorithm. The latter, adopted from the theoretical framework in [32], has been implemented in CUDA and can be run on graphic processing units (GPUs).

Using simulated data, we have demonstrated that by incorporating the prior information encoded by the network, GGGL-1 gains additional power in identifying the signal-carrying predictor groups compared with the group lasso [42]. In addition, when imposing sparsity within the selected predictor groups, GGGL-2 improves the power of selecting the signal-carrying predictors compared with the sparse group lasso [13].

The models presented here were motivated by GWAs, in which the predictors were assumed to be SNPs and the predictor groups were genes. However, application of the GGGL models need not be restricted to GWAs data. One of the possible different applications is in the context of medical imaging data, where predictors are voxels and predictor groups correspond to regions of interest (ROI). A network on the ROIs may encode the functional correlations between different regions, as presented in [25]. Application of the GGGL models to real GWAs and imaging data will constitute the future work.

Symbol Description

AUC	area under curve	ROC	receiver operating characteristic
CDM	coordinate descent method		
FPR	false positive rate	ROI	region of interest
GGGL	graph-guided group lasso	SNPs	single-nucleotide polymorphisms
GWAs	genome-wide association studies		
PCDM	parallel coordinate descent method	SNR	signal-to-noise ratio
		TPR	true positive rate

Bibliography

[1] Janine Altmüller, Lyle Palmer, Guido Fischer, Hagen Scherb, and Matthias Wjst. Genomewide scans of complex human diseases: True linkage is hard to find. *Am J Hum Genet*, 69(5):936–950, 2001.

[2] Chloé-Agathe Azencott, Dominik Grimm, Mahito Sugiyama, Yoshinobu Kawahara, and Karsten Borgwardt. Efficient network-guided multi-locus association mapping with graph cuts. *Bioinformatics*, 29:i171–i179, 2013.

[3] Lin-S. Chen, Carolyn Hutter, John Potter, Yan Liu, Ross Prentice, Ulrike Peters, and Li Hsu. Insights into colon cancer etiology via a regularized approach to gene set analysis of gwas data. *The American Journal of Human Genetics*, 86(6):860–871, 2010.

[4] F.R.K. Chung. Spectral graph theory. *CBMS regional conference series 92. Amer. Math. Soc., Providence, RI. MR1421568.*, 1997.

[5] Geraldine Clarke, Carl Anderson, Fredrik Pettersson, Lon Cardon, Andrew Morris, and Krina Zondervan. Basic statistical analysis in genetic case-control studies. *Nat Protoc.*, 6(2):121–133, 2011.

[6] The Wellcome Trust Case Control Consortium. Genome-wide association study of 14,000 cases of seven common diseases and 3,000 shared controls. *Nature*, 447(7415):661–678, 2009.

[7] Jing Cui, Eli Stahl, Saedis Saevarsdottir, Corinne Miceli, and Dorothee Diogo *et al.* Genome-wide association study and gene expression analysis identifies cd84 as a predictor of response to etanercept therapy in rheumatoid arthritis. *PLOS Genetics*, 9(3):1–11, 2013.

[8] Z.J. Daye and X.J. Jeng. Shrinkage and model selection with correlated variables via weighted fusion. *Computational Statistics and Data Analysis*, 53(4):1284–1298, 2009.

[9] D.F. Easton, D.T. Bishop, D. Ford, and G.P. Crockford. Genetic linkage analysis in familial breast and ovarian cancer: results from 214 families. The Breast Cancer Linkage Consortium. *Am J Hum Genet.*, 52(4):678–710, 1993.

[10] Stephan Ripke *et al.* Genome-wide association study identifies five new schizophrenia loci. *Nature Genetics*, 43:969–976, 2011.

[11] The International Consortium for Blood Pressure Genome-Wide Association Studies. Genetic variants in novel pathways influence blood pressure and cardiovascular disease risk. *Nature*, 478(7367):103–109, 2011.

[12] Jerome Friedman, Trevor Hastie, Holger Höfling, and Robert Tibshirani. Pathwise coordinate optimization. *Ann. Appl. Stat.*, 2(1):302–332, 2007.

[13] Jerome Friedman, Trevor Hastie, and Robert Tibshirani. A note on the group lasso and a sparse group lasso. *arXiv:1001.0736*, 2010.

[14] Jonathan Haines, Michael Hauser, Silke Schmidt, William Scott, Lana Olson, Paul Gallins, Kylee Spencer, ShuYing Kwan, Maher Noureddine, John Gilbert, Nathalie Schnetz-Boutaud, Anita Agarwal, Eric Postel, and Margaret Pericak-Vance. Complement factor h variant increases the risk of age-related macular degeneration. *Science*, 308:419–421, 2005.

[15] Gabriel Hoffman, Benjamin Logsdon, and Jason Mezey. Puma: A unified framework for penalized multiple regression analysis of gwas data. *PLOS Computational Biology*, 9:(1), 2013.

[16] Jian Huang, Shuangge Ma, Hongzhe Li, and CunHui Zhang. The sparse Laplacian shrinkage estimator for high-dimensional regression. *The Annals of Statistics*, 39(4):2021–2046, 2011.

[17] Laurent Jacob, Guillaume Obozinski, and Jean-Phillippe Vert. Group lasso with overlap and graph lasso. *ICML '09 Proceedings of the 26th Annual International Conference on Machine Learning*, pages 433–440, 2009.

[18] Andrew Johnson and Christopher O'Donnell. An open access database of genome-wide association results. *BMC Med Genet.*, 10:(6), 2009.

[19] CheeSeng Ku, EnYun Loy, Yudi Pawitan, and Kee Seng Chia. The pursuit of genome-wide association studies: where are we now? *J. Hum. Genet.*, 55(4):195–206, 2010.

[20] Younghee Lee, Haiquan Li, Jianrong Li, Ellen Rebman, Ikbel Achour, Kelly Regan, Eric Gamazon, James Chen, Xinan Yang, Nancy Cox, and Yves Lussier. Network models of genome-wide association studies uncover the topological centrality of protein interactions in complex diseases. *J Am Med Inform Assoc*, 20:6:619–629, 2013.

[21] Caiyan Li and Hongzhe Li. Network-constrained regularization and variable selection for analysis of genomic data. *Bioinformatics*, 24(9):1175–1182, 2008.

[22] Li Luo, Gang Peng, Yun Zhu, Hua Dong, Christopher Amos, and Momiao Xiong. Genome-wide gene and pathway analysis. *European Journal of Human Genetics*, 18:1045–1053, 2010.

[23] Teri Manolio, Francis Collins, Nancy Cox, David Goldstein, Lucia Hindorff *et al.* Finding the missing heritability of complex diseases. *Nature*, 461:747–753, 2009.

[24] dbSNP. National Center for Biotechnology Information. *http://www.ncbi.nlm.nih.gov/SNP/*, 2014.

[25] Steven Nelson, Alexander Cohen, Jonathan Power, Gagan Wig, Francis Miezin, Mark Wheeler, Katerina Velanova, David Donaldson, Jeffrey Phillips, Bradley Schlaggar, and Steven Petersen. A parcellation scheme for human left lateral parietal cortex. *Neuron*, 67(1):156–170, 2010.

[26] Yurii Nesterov. Efficiency of coordinate descent methods on huge-scale optimization problems. *SIAM Journal on Optimization*, 22(2):341–362, 2012.

[27] Fabian Ojeda, Marco Signoretto, Raf Van de Plas, Etienne Waelkens, Bart De Moor, and Johan A.K. Suykens. Semi-supervised learning of sparse linear models in mass spectral imaging. *Pattern Recognition in Bioinformatics, 5th IAPR International Conference, PRIB 2010 Nijmegen, The Netherlands*, IV:325–334, 2010.

[28] Thomas Pearson and Teri Manolio.   How to interpret a genome-wide association study. *Journal of the American Medical Association*, 299(11):1335–1344, 2008.

[29] Yu Qian, Søren Besenbacher, Thomas Mailund, and Mikkel Schierup. Identifying disease associated genes by network propagation. *BMC Systems Biology*, 8(Suppl 1):S6, 2014.

[30] Towfique Raj, Joshua Shulman, Brendan Keenan, Lori Chibnik, Denis Evans, David Bennett, Barbara Stranger, and Philip De Jager. Alzheimer disease susceptibility loci: Evidence for a protein network under natural selection. *The American Journal of Human Genetics*, 90(4):720–726, 2012.

[31] Peter Richtárik and Martin Takáč.   Distributed coordinate descent method for learning with big data. *arXiv:1310.2059v1*, 2013.

[32] Peter Richtárik and Martin Takáč. Parallel coordinate descent methods for big data optimization. *arXiv:1212.0873v2*, 2013.

[33] Marco Signoretto, Anneleen Daemen, Carlo Savorgnan, and Johan A.K. Suykens. Variable selection and grouping with multiple graph priors. *in Proc. of the 2nd Neural Information Processing Systems (NIPS) Workshop on Optimization for Machine Learning, Whister, Canada*, 2009.

[34] Matt Silver, Peng Chen, Ruoying Li, Ching-Yu Cheng, Tien-Yin Wong, E-Shyong Tai, Yik-Ying Teo, and Giovanni Montana. Pathways-driven sparse regression identifies pathways and genes associated with high-density lipoprotein cholesterol in two Asian cohorts. *PLoS Genetics*, 9(11):1–28, 2013.

[35] Matt Silver, Eva Janousova, Xue Hua, Paul Thompson, and Giovanni Montana. Identification of gene pathways implicated in Alzheimer's disease using longitudinal imaging phenotypes with sparse regression. *NeuroImage*, 63(3):1681–1694, 2012.

[36] Matt Silver and Giovanni Montana. Fast identification of biological pathways associated with a quantitative trait using group lasso with overlaps. *Statistical Applications in Genetics and Molecular Biology*, 11(1):article 7, pp 1–43, 2012.

[37] Damian Szklarczyk, Andrea Franceschini, Michael Kuhn, Milan Simonovic, Alexander Roth, Pablo Minguez, Tobias Doerks, Manuel Stark, Jean Muller, Peer Bork, Lars Jensen, and Christian von Mering. The string database in 2011: functional interaction networks of proteins, globally integrated and scored. *Nucleic Acids Res.*, 39(Database issue):D561–D568, 2011.

[38] Robert Tibshirani. Regression shrinkage and selection via the lasso. *J.R.Statist.*, Soc.B, 58:267–288, 1996.

[39] Robert Tibshirani, Michael Saunders, Saharon Rosset, Ji Zhu, and Keith Knight. Sparsity and smoothness via the fused lasso. *Journal of the Royal Statistical Society: Series B*, 67(1):91–108, 2005.

[40] Hua Wang, Feiping Nie, Heng Huang, Sungeun Kim, Kwangsik Nho, Shannon Risacher, Andrew Saykin, and Li Shen. Identifying quantitative trait loci via group-sparse multitask regression and feature selection: an imaging genetics study of the adni cohort. *Bioinformatics*, 28(2):229–237, 2012.

[41] Kai Wang, Mingyao Li, and Hakon Hakonarson. Analysing biological pathways in genome-wide association studies. *Nature Reviews Genetics*, 11:843–854, 2010.

[42] Ming Yuan and Yi Lin. Model selection and estimation in regression with grouped variables. *J.R.Statist.*, Soc.B, 68(1):49–67, 2006.

[43] Hua Zhou, Mary Sehl, Janet Sinsheimer, and Kenneth Lange. Association screening of common and rare genetic variants by penalized regression. *Bioinformatics*, 26(19):2375–2382, 2010.

[44] Hui Zou and Trevor Hastie. Regularization and variable selection via the elastic net. *J.R.Statist.*, Series B, 67(2):301–320, 2005.

Chapter 7

On the Convergence Rate of Stochastic Gradient Descent for Strongly Convex Functions

Cheng Tang

George Washington University

Claire Monteleoni

George Washington University

7.1   Introduction

In this chapter we survey recent works on the non-asymptotic analysis of stochastic gradient descent (SGD) on strongly convex functions. We discuss how both strong convexity and smoothness of a strongly convex function, denoted by F, may affect the rate of error-convergence in Section 7.2. This leads us to characterize F by these two properties by presenting a subclassification scheme of strongly convex functions in Section 7.4.1.

7.1.1   Background

Consider a standard binary classification problem: We have a sequence of i.i.d. pairs of random variables, $(Y_1, X_1), \ldots, (Y_n, X_n)$, sampled from some unknown distribution p, where $\forall i = 1, \ldots, n$, $Y_i = 0$ or 1 is the label of $X_i \in \mathcal{X}$, where $\mathcal{X}$ is the sample space. Additionally, suppose our hypothesis space is composed of linear classifiers $w \in W \subset \mathbb{R}^d$ and we are given a loss function $l(w, (Y, X))$ to measure the discrepancy between the classifier's prediction and the true label.

Based on this information, how can we determine the optimal classifier in the hypothesis space? According to statistical learning theory, the optimal classifier is the one that minimizes the true risk $E_p[l(w, (Y, X))]$. Empirical Risk Minimization (ERM) is a commonly used principle to estimate the optimal classifier using a finite data sample. It chooses the best classifier as the one that minimizes the empirical risk or regularized empirical risk:

$$Q_n[l(w, (Y, X))] = \frac{1}{n} \sum_{i=1}^{n} l(w, (Y_i, X_i)) + R(w) \tag{7.1}$$

where R is a regularizer, which penalizes the complexity of w.

When $Q_n[l(w, (Y, X))]$ is convex in $w \in W$, a gradient descent algorithm will find the optimum of this function. For simplification, we temporarily assume $Q_n[l(w, (Y, X))]$ is differentiable. Then, to run gradient descent, we need to compute the gradient $\nabla Q_n[l(w, (y, x))]$. Given the functional form of $Q_n[l(w, (y, x))]$, we see that to find the gradient of the empirical risk, we need to evaluate the loss on every sample in the data set. For large data sets, this imposes a lot of computation.

We study the performance of an efficient version of gradient descent, Stochastic Gradient Descent (SGD). Instead of computing the gradient based on the entire data, SGD estimates $\nabla E_p[l(w, (y, x))]$ by only evaluating the loss, $l(w, (Y, X))$, based on one random sample, (Y, X), of the data. With different learning paradigms, there are two ways to generate the random sample.

The first corresponds to our binary classification model: if we have a batch i.i.d. sample from p, then we can sub-sample the pair (Y, X) from the batch uniformly at random. Alternatively, suppose we are in an online setting and we receive samples according to the true distribution p, then (Y, X) is a random sample from p. Under either model, (Y, X) is distributed according to p (see [14]).

According to the gradient sampling scheme defined by SGD, it is easy to see that $\hat{\nabla}_{SGD}[l(w, (Y, X))]$ is an unbiased estimator of the gradient of the empirical risk in Equation (7.1):

$$E[\hat{\nabla}_{SGD}[l(w, (Y, X))]] = \nabla Q_n[l(w, (Y, X))] \tag{7.2}$$

where the expectation on the left-hand side of the equation is with respect to the uniform random sampling of the batch data when estimating the gradient. Then we may ask how does SGD compare with the deterministic gradient descent in optimizing the empirical risk $Q_n[l(w, (Y, X))$, assuming it is convex in w.

In SGD, data are processed in an online fashion and the computational effort of evaluating the approximated gradient is independent of the data size. Despite the significant reduction in computation and memory, SGD has been shown to converge in expectation to the optimum of a convex function. This advantage contributes to the popularity of SGD for large-scale learning (see [2]).

Note that the setting we consider here is subsumed by and easier than the more general online convex programming setting introduced by [15], where the instant loss function $l_t(w, (Y_t, X_t))$ upon receiving (Y_t, X_t) can be adversarial.

7.1.2 The First-Order Oracle Model for Stochastic Optimization

The optimization mechanism with SGD described earlier is a special case of the following oracle model (see [6]). Given a convex hypothesis space W and a convex function F (not necessarily strongly convex), we run a gradient-based algorithm **A** (not necessarily SGD) on W to find an optimizer of F. At the t-th iteration, **A** chooses $w_t \in W$. Then an oracle provides unbiased estimates of the function value and the gradient of F at w_t. In general, the function F need not be differentiable. However, since F is convex, a subdifferential exists at any point of the convex domain. In that case, the oracle will provide an unbiased estimate of a subgradient of F at w_t. That is, $\hat{g}(w_t)$ is provided by the oracle such that $E[\hat{g}(w_t)] = g(w_t) \in \partial f(w_t)$, where $\partial f(w_t)$ is the subdifferential of F at w_t (see [11] for more details). Then we update the hypothesis to get w_{t+1} using the function and the gradient information.

We focus on studying SGD under the first-order oracle model, where the optimization procedure can be formally stated as:

1. At $T = 1$, choose $w_1 \in W$ arbitrarily.

2. At $T = t$, choose learning rate η_t and use the update,

$$w_{t+1} = \Pi_W(w_t - \eta_t \hat{g}(w_t)) \tag{7.3}$$

where Π is the orthogonal projection operator:

$$\Pi_W(y) = \arg \min_{w \in W} \|w - y\|_{L_2}.$$

The reason for projection is that $w_t - \eta_t \hat{g}(w_t)$ might fall outside of the hypothesis space W. Therefore, a projection step is used to guarantee that w_{t+1} stays in W. We are mainly interested in how the expected optimization error $E[F(w_t) - F(w_{opt})]$ depends on T, the number of oracle queries.

7.1.3   When F is Strongly Convex and Non-Smooth

Different assumptions about F can be made in accordance with different problem formulations. Convexity is usually assumed for any global convergence analysis. In addition, strong convexity of F and smoothness around the optimum w_{opt} are two of the most common assumptions. Formally, F is λ-strongly convex, if $\forall w, w' \in W$, and any subgradient g of F at w, [9]

$$F(w') \geq F(w) + \langle g(w), w' - w \rangle + \frac{\lambda}{2} \|w' - w\|^2. \tag{7.4}$$

An important consequence of the above definition of strong convexity is, locally, with w_{opt} as the reference point,

$$F(w) - F(w_{opt}) \geq \frac{\lambda}{2} \|w - w_{opt}\|^2 \tag{7.5}$$

and

$$\|g(w)\| \geq \lambda \|w - w_{opt}\|. \tag{7.6}$$

On the other hand, let w_{opt} be the optimum of F, then F is μ-smooth with respect to w_{opt} if $\forall w \in W$

$$F(w) - F(w_{opt}) \leq \frac{\mu}{2} \|w - w_{opt}\|^2 \tag{7.7}$$

which implies

$$\|g(w)\| \leq \mu \|w - w_{opt}\|. \tag{7.8}$$

In view of Equation (7.6) and Equation (7.8), strong convexity and smoothness conditions are nothing but lower and upper bounds on $\|g(w)\|$ with respect to $\|w - w_{opt}\|$, the distance of the current hypothesis to the optimum.

Recall the binary classification problem described at the beginning. We may find a classifier using ERM, which can be cast as an optimization problem where $F = Q_n[l(w, (Y, X))] = \frac{1}{n} \sum_{i=1}^{n} l(w, (Y_i, X_i)) + R(w)$. Different losses and regularizers can be chosen with different learning methods. Under linear SVM, the problem is instantiated as

$$\min_{w} \frac{1}{n} \sum_{i=1}^{n} l(w, (Y_i, X_i)) + \frac{\lambda}{2} \|w\|^2 \tag{7.9}$$

where $l(w, (Y_i, X_i)) = \max\{0, 1 - Y_i\langle w, X_i\rangle\}$, i.e., the hinge loss. Under this formulation, [12] derived Pegasos, a primal solver of linear SVM geared towards large-scale classification problems. Note that in Equation (7.9), the objective F is strongly convex and non-smooth.

7.1.4 Contributions

Driven by discussions in Section 7.1.2 and Section 7.1.3, we study the general SGD-oracle model, where F is strongly convex. Our assumptions follow the analyses of [4] and [9]. Our contributions can be summarized as follows:

1. We provide a survey on the current analysis of the convergence rate of gradient-based algorithms on strongly convex functions under the first-order oracle model: We summarize the major techniques used in the analysis of the lower bound of the problem, and we compare variance reduction methods used in developing faster algorithms under this framework (Section 7.2).

2. We demonstrate how both the degree of strong convexity and smoothness affect the convergence rate of $E[F(w_t) - F(w_{opt})]$ and derive a subclassification scheme to characterize F (Section 7.4.1).

3. We illustrate how the new scheme could yield a better analysis of convergence, and we proposed an open problem on whether it is easier to prove the optimal convergence rate of SGD on a "nice" subset of strongly convex functions (Section 7.5.3).

Note that the subclassification of F is only for the analysis of SGD. The algorithm itself does not have information about F, other than that given by the oracle.

7.2 Related Work

To motivate our approach, we review existing analyses of stochastic convex optimization under the oracle model introduced above. In particular, we first discuss how strong convexity is tied to information gain in proving lower bounds. Then we review the optimal algorithms and discuss how variance reduction methods are used to achieve fast convergence.

7.2.1   Strong Convexity and the Lower Bound of Stochastic Optimization

Pertaining to lower bounds for stochastic convex optimization, [1] initiated an information-theoretic approach. Their key observation is that convex functions of the same shape can be parametrized by their optima. Thus, function optimization is tied to the problem of statistical parametric estimation. Based on the dimension of the problem, they construct a core subclass of functions, parametrized by their optima. They show that in order to achieve low optimization error on any of the functions in this subclass, an algorithm needs to estimate which function is being used. This amounts to gaining enough information about the true parameter from the oracle during the optimization procedure. Then they construct a Bernoulli oracle, which outputs an unbiased estimator of a subgradient, while not revealing much information about the true parameter at any iteration. Thus, they show that in order to achieve the desired accuracy of optimization, the algorithm needs to query the oracle a sufficient amount of times. Specifically, they show that the error of any algorithm for any convex F is $\Omega(\frac{1}{\sqrt{T}})$, where T is the number of oracle calls. For strongly convex functions, the lower bound is $\Omega(\frac{1}{T})$.

The work of [8] analyzes the problem by constructing a Markov chain feedback system and uses a white-noise Gaussian oracle instead. Utilizing the Markov chain property, they also use information theoretic inequalities to achieve the same lower bounds. In addition, they point out the difference between an algorithm achieving a desired accuracy arbitrarily and an algorithm whose expected accuracy decreases to zero asymptotically. The latter class, which contains SGD, is called "any-time" algorithms. For any-time algorithms, they show that decrease in errors $\|F(w_t) - F(w_{opt})\|$ and $\|g(w_t)\|$ entails decrease in information gain. This law of diminishing return makes it clear why strong convexity is beneficial to optimization from an information-theoretic perspective; it provides a lower bound on information gain through a lower bound on the subgradients in Equation (7.6). In other words, at each query, a strongly convex function is bound to give more information about the true parameter through its large subgradients.

The relation between information gain and strong convexity is further quantified and generalized in the work of [10]. There, they subclassify the degree to which a function is strongly convex and derive specific lower bounds based on the degree: they show that for a Lipschitz continuous function F such that $\forall w \in W, F(w) - F(w_{opt}) \geq \frac{\lambda}{2} \|w - w_{opt}\|^k$, where $k > 1$ (i.e., k is the exponent of the Hölder condition), the error of the T-th iteration of SGD on $F(w_T) - F(w_{opt})$ is $\Omega(\frac{1}{T^{\frac{k}{2k-2}}})$. When $k = 2$, F is strongly convex as defined in Equation (7.5). When $1 < k < 2$, F is highly strongly convex, and should have faster convergence rate. They also developed an algorithm based on [4], which attains the lower bound for any $k > 1$. However, their adapted algorithm uses the knowledge of the strong convexity parameter k, which is usually not assumed to be given.

Does the previous work suggest strong convexity is always beneficial? Perhaps in view of lower bound, since it means an algorithm could gain more information at each query of the oracle by the strong convexity assumption on F. However, if we restrict ourselves to the simple update rule of SGD as in Equation (7.3), then we may find a possible disadvantage of strong convexity: large subgradients might influence the stability of the predictors w_t's produced by SGD. We will motivate this intuition in the next section and further discuss it in Section 7.5.3.

7.2.2   Variance Reduction Methods and Optimal Algorithms

The convergence rate of SGD is inherently limited by the randomness of the estimated subgradients. The oracle model only assumes that the variance of the estimated gradient, $var[\hat{g}(w_t)]$, is bounded. As a result, if one simply updates the algorithm based on the given gradient, w_t will diverge from w_{opt} asymptotically. Therefore, in order to achieve asymptotic convergence of w_t to w_{opt}, the term η_t is added in the SGD update rule in Equation (7.3) to ensure $var[\eta_t \hat{g}(w_t)] \to 0$ as $t \to \infty$ [7]. Since there are two terms in $\eta_t \hat{g}(w_t)$, two natural variance reduction methods are derived in previous work: choosing η_t as a decreasing function of t, or reducing the variance of the oracle estimate $\hat{g}(w_t)$.

Letting η_t decrease reduces variance by forcing it to go to zero asymptotically. However, by choosing a small η_t, the speed of convergence will be limited due to an update step of size $\eta_t \|\hat{g}(w_t)\|$, which is less than what the gradient suggests. Therefore, a careful choice of η_t is needed. Unfortunately, the optimal η_t is very sensitive to other factors, as we shall see soon, making it a difficult task.

On the other hand, reducing $var[\hat{g}(w_t)]$ directly does not have the disadvantage of limiting the convergence speed. [5] develops a subgradient modification scheme that achieves exponentially fast error decay in expectation. The idea is that the variance of their improved estimate goes to zero asymptotically. Thus, they can choose a constant η, meaning the speed of convergence does not have to be tempered by η_t. However, the analysis of these methods requires additional assumptions on the form of F, and the storage and re-computation of gradients impose more computational demands, which deviates from the original goal of using SGD. Moreover, it violates the assumptions made in the first-order oracle model underlying our setting.

Besides the two obvious methods of variance reduction, predictor averaging is also used; after obtaining a sequence of predictors $w_1, \dots, w_t$ from SGD, we take $\tilde{w}_t = \sum_{i=t-N_t}^{t} w_i$ as our final prediction at each iteration. By predictor averaging, $var[w_t - w_{opt}]$ can be decreased to $var[\tilde{w}_t - w_{opt}]$, thus indirectly reducing $var[F(w_t) - F(w_{opt})]$, the variance of the error in function value. However, since the previous iterates $w_1, \dots, w_{t-1}$ are likely to be farther away from w_{opt}, making N_t too large is also suboptimal. Thus, the method of predictor averaging faces a similar problem as that of choosing η_t.

To reach optimal convergence, one might have to interweave the choice of η_t and N_t. The problem was studied in earlier stochastic approximation literature, where the optimization objective and the algorithm, called Robbins-Monro procedure, resembles SGD under the first-order oracle model (see [7] and references therein). There, they show $\eta_t = \frac{1}{t}$ is optimal for the stochastic procedure in terms of estimator efficiency when $N_t = 0, \forall t$, i.e., no averaging is taken. Interestingly, they also show that when $\eta_t = \frac{1}{t^\gamma}$, where $\frac{1}{2} < \gamma < 1$, averaging the predictors is the optimal choice instead.

The two known optimal algorithms for optimizing general strongly convex functions both extend from simple SGD (as described earlier) and incorporate the choice of η_t and N_t. The first optimal algorithm for a general strongly convex function under the first-order oracle model is given by [4]). Their algorithm chooses a sequence of geometrically decreasing η_t, accompanied by a geometrically increasing N_t. Then [9] subsequently develops a simpler optimal algorithm of a similar flavor; with $\eta_t = \frac{1}{t}$, and fixing the total number of iterations, T, it takes $N_T = \alpha T$ number of averages. The optimality of their algorithms suggests that more stabilization is needed around the optimum for the optimization of a general strongly convex function. Not surprisingly, the optimal degree of stabilization is also related to the functional property. The work in [4] shows that if $N_T = T$, then the convergence rate is $\Omega(\frac{\log T}{T})$, suggesting that too much stabilization is suboptimal for strongly convex functions.

The discussion above seems to lead us to conclude that the variance of SGD-based prediction is only determined by η_t, N_t, and $var[\hat{g}(w_t)]$. However, the degree to which local smoothness of F around w_{opt}, or equivalently, an upper bound assumption on $\|g(w_t)\|$ with respect to $\|w_t - w_{opt}\|$ as in Equation (7.7) and Equation (7.8), affects the convergence rate was also brought to attention recently by [9] and [13]. It can be shown that if F is strongly convex and locally smooth around w_{opt}, then SGD with $\eta_t = \frac{1}{t}$ and $N_t = 0, \forall\, t \geq 1$ attains the optimal convergence rate. The analysis of convergence provided in [9] gives a simple reason for why smoothness is beneficial to SGD: running SGD on a strongly convex function F has the guarantee that $\|w_t - w_{opt}\|^2 = O(\frac{1}{t})$. Thus, by the smoothness assumption in 7.7, one directly obtains $F(w_t) - F(w_{opt}) = O(\frac{1}{t})$. Intuitively, under a certain Lipschitz assumption of F, F can be viewed as a stabilizing transformation of random variable w_t. We will explain this effect further in Section 7.5.3.

7.3   Preliminaries

For the rest of the chapter, we assume that

- The unknown objective function F to be optimized is λ-strongly convex whose (unique) optimum is denoted by w_{opt}.

- The estimated subgradient provided by the oracle at time t satisfies $E[\hat{g}_t] = g(w_t) \in \partial F(w_t)$ and that $\exists G > 0$ such that $E[\|\hat{g}_t\|^2] \leq G$. That is, $\hat{g}(\cdot)$ is an unbiased estimator of a subgradient of $F(\cdot)$, and its second moment is bounded.

- In the SGD algorithm, we set the learning rate η_t as a linear function of $\frac{1}{t}$. That is, $\eta_t = \frac{c}{\lambda t}$, where c is a positive constant and λ is the strong convexity parameter.

Based on the oracle model framework introduced earlier and the above assumptions, we are interested in whether the expected error of the last iterate of SGD, that is, $E[\epsilon_t] := E[F(w_t) - F(w_{opt})]$, achieves the optimal convergence rate, $\Omega(\frac{1}{t})$.

7.4 Our Approach

The observation that strong convexity and smoothness of function F might both affect the convergence rate of SGD leads us to characterize F by these two properties. The benefits and motivations of our subclassification scheme include the following:

1. Since both the degree of strong convexity and the degree of local smoothness affect the performance of SGD, we may examine different subclasses of F and gain further insight into how the property of F affects the convergence rate.

2. Our proposed subclassification scheme provides us with a tool for a careful analysis of the choice of learning rate η_t and the use of predictor averaging, by restricting our focus on a subclass of convex functions (this can already be observed in the analysis of [10]), which might lead to faster algorithms.

3. Our proposed subclassification scheme unifies different assumptions on F, including those used in the past work.

7.4.1 Subclassification of Strongly Convex Functions in Two Ways

Comparing the definitions of strong convexity and smoothness in Equation (7.4) and Equation (7.7), we see that since the inequality signs are reversed, it seems that gaining one advantage means losing the other. However, these definitions only capture functions at the two extremes. Can we balance their advantage and disadvantage when the functions vary from one extreme to the other? We utilize finer characterizations to bridge the gap between strong convexity and smoothness.

7.4.2　Subclassify Strongly Convex Functions by (ν, L_ν)-Weak Smoothness

We first relax the notion of smoothness and define weak smoothness using the Hölder condition, which allows us to consider polynomials of continuous degree.

Definition 7.1 (((ν, L_ν)-Weak smoothness). *Let $\nu \in [0, 1]$ and let F be a strictly convex function. Then, we say F is (ν, L_ν)-weakly smooth, if $\exists L_\nu > 0$ such that for all $w \in W$, any subgradient $g(w) \in \partial F(w)$ and any $g(w_{opt}) \in \partial F(w_{opt})$ satisfy the following inequality,*

$$\|g(w) - g(w_{opt})\| \le L_\nu \|w - w_{opt}\|^\nu$$

where the norm is the Euclidean norm.

Since F is a strictly convex function, $\exists\, g(w_{opt}) = 0$, the inequality above implies

$$\|g(w)\| \le L_\nu \|w - w_{opt}\|^\nu$$

This leads to the following inequality (see, for example, [3])

$$F(w) - F(w_{opt}) \le \frac{L_\nu}{1 + \nu} \|w - w_{opt}\|^{1+\nu}$$

for all $w \in W$.

Naturally, we define the set of H^ν functions as below:

Definition 7.2 (the H^ν class). *$F \in H^\nu$, if F is (ν, L_ν)-weakly smooth.*

7.4.3　Subclassify Strongly Convex Functions by (μ, λ_μ)-Strong Convexity

In the following definition, (μ, λ_μ)-strong convexity is defined almost symmetrically, except for the extra restriction on the boundedness of the subgradients, which is automatically guaranteed for the weakly-smooth functions.

Definition 7.3 (((μ, λ_μ)-strong convexity). *Let $\mu \in [0, 1]$ and F be a strictly convex function, and $\exists\, K > 0$ s.t. $\forall w \in W$, and $\forall\, g(w) \in \partial F(w)$, $\|g(w)\| \le K$. Then, we say F is (μ, λ_μ)-strongly convex, if $\exists \lambda_\mu > 0$ such that for all $w \in W$, any subgradient $g(w) \in \partial F(w)$ and any $g(w_{opt}) \in \partial F(w_{opt})$ satisfy the following inequality,*

$$\|g(w) - g(w_{opt})\| \ge \lambda_\mu \|w - w_{opt}\|^\mu.$$

Similar to the weakly-smooth case, this leads us to the following inequalities,

$$\|g(w)\| \geq \lambda_\mu \|w - w_{opt}\|^\mu$$

$$F(w) - F(w_{opt}) \geq \frac{\lambda_\mu}{1 + \mu} \|w - w_{opt}\|^{1+\mu}$$

for all $w \in W$.

We subsequently define the S^μ functions.

Definition 7.4 (the S^μ class). *$F \in S^\mu$, if F is (μ, λ_μ)-strongly convex.*

Remark

1. By definition, **the class of all strongly convex functions with bounded subgradients coincides with S^1**.

2. It is easy to check that if $\nu_1 \geq \nu_2 \geq 0$, then $F \in H^{\nu_1} \implies F \in H^{\nu_2}$. That is, the sets form a nested family: $H^0 \supseteq H^1 \supseteq H^2, \ldots$ Symmetrically, we have $S^0 \subseteq S^1 \subseteq S^2, \ldots$

7.5 Main Results

In this section, we first illustrate the set relations captured by the strong convexity parameter μ and the weak smoothness parameter ν. Then we discuss why we use the two parameters and how their relation affects the convergence rate of SGD.

7.5.1 Set Relations between S^μ and H^ν

Now we prove the set relations between S^μ and H^ν and discuss why we need two different ways to subclassify strongly convex functions. The set relations are illustrated in Figure 7.1.

Lemma 7.1. $S^1 \subseteq H^0 \setminus \{\cup_{\nu>1, \nu \in \mathbb{Q}} H^\nu\}$.

Proof. That is, we need to show if $F \in S^1$, then $F \in H^0$ and $F \notin H^\nu$ for any $\nu > 1, \nu \in \mathbb{Q}$.

To prove $F \in H^0$, suppose otherwise. Then, $\forall L > 0, \exists w \in W$ s.t. $F(w) - F(w_{opt}) > L \|w - w_{opt}\|$. Then we can choose $L^* > K$ and $w \in W$ s.t. $F(w) - F(w_{opt}) > L^* \|w - w_{opt}\|$. Then, by definition, $\exists$ a subgradient $g(w)$ with $\|g(w)\| > L^* > K$. On the other hand, we also have $\|g(w)\| \leq K, \implies$ contradiction. To prove $F \notin H^\nu$, suppose otherwise. Then by definition of (ν, L_ν)-weak-smoothness, $\exists L_\nu > 0$ s.t. $F(w) - F(w_{opt}) \leq \frac{L_\nu}{1+\nu} \|w - w_{opt}\|^{1+\nu}$. On the other hand, by strong convexity, we also have $\exists \lambda > 0$ s.t. $F(w) - F(w_{opt}) \geq \frac{\lambda}{2} \|w - w_{opt}\|^2$. Since $\nu > 1$, this leads to a contradiction. $\square$

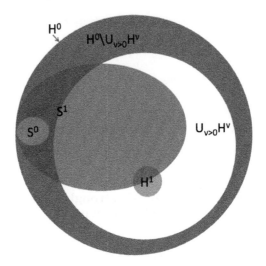

FIGURE 7.1: **Set relations between S^μ and H^ν.** The entire ellipse is the set of strongly convex functions with bounded subgradients; previous results showed optimal convergence on the intersection of the H^1 and S^1.

Remark: We use the restriction $\nu \in \mathbb{Q}$ to avoid the problem of uncountability of real numbers, when taking the infinite unions of sets H^ν.

This lemma shows that the set $H^0 \setminus \{\cup_{\nu>1, \nu \in \mathbb{Q}} H^\nu\}$ contains all strongly convex functions. Note that we use a proof by contradiction to show that $F \in H^0$, instead of simply using the boundedness of subgradients for strongly convex functions in Definition 7.3 because the given bound K need not be the same as the strong-smoothness coefficient L_0 defined in Definition 7.1, which depends on the true property of F. In particular, it is possible that $K > L_0$. That is, K may not exactly characterize the strong-smoothness property of the unknown function F, but only an upper bound.

Lemma 7.2. $S^0 \subseteq \left\{ S^1 \cap \{H^0 \setminus \cup_{\nu>0, \nu \in \mathbb{Q}} H^\nu\} \right\}$

Proof. By the definition of S^μ, $S^0 \subseteq S^1$. Also, since H^0 includes all functions with bounded subgradients, $S^0 \subseteq H^0$. Finally, suppose $S^0 \cap \{\cup_{\nu>0, \nu \in \mathbb{Q}} H^\nu\} \neq \emptyset$. Then $\exists F \in S^0 \cap \{\cup_{\nu>0, \nu \in \mathbb{Q}} H^\nu\}$. That is, $\exists \lambda_0, L_\nu$ such that $L_\nu \|w - w_{opt}\|^\nu \geq \|g(w)\| \geq \lambda_0 \|w - w_{opt}\|^0 = \lambda_0$, where $\nu > 0$, for all $w \in W$. Taking the limit as $w \to w_{opt}$, we have $0 \geq \lambda_0 > 0 \implies$ contradiction. Thus, $S^0 \cap \{\cup_{\nu>0, \nu \in \mathbb{Q}} H^\nu\} = \emptyset$. Therefore, $S^0 \subseteq S^1 \cap \{H^0 \setminus \cup_{\nu>0, \nu \in \mathbb{Q}} H^\nu\}$. $\square$

An example: As a simple example of $F \in S^0$, consider the function $y = |x|$, where $x \in \mathbb{R}$. It satisfies $g(w) = 1, \forall w > 0$ and $g(w) = -1, \forall w < 0$. Thus, $\|g(w)\| = 1, \forall w \neq 0$, which is the optimum. Thus, by Definition 7.3, F is a $(0,1)$-strongly convex function, and $F \in S^0$.

In the subsequent section, we discuss the need for employing both (μ, λ_μ)-strong convexity and (ν, L_ν)-weak smoothness to subclassify the strongly convex functions and demonstrate how the quantitative relation between strong smoothness parameter ν and strong convexity parameter μ characterizes the difficulty of optimization.

7.5.2 The Gap between Strong Convexity and Weak Smoothness

Definitions 7.1 and 7.3 imply that for a function to be strongly convex (or weakly smooth) at w_{opt}, it must be strongly convex (or weakly smooth) at w_{opt} from every direction. Thus, for any $\alpha \in [0,1]$, it is possible that a function F is in S^α and also in $H^\beta \setminus \{\cup_{\nu > \beta, \nu \in \mathbb{Q}} H^\nu\}$ for any $0 \le \beta \le \alpha$; for example, there is a function F that is $(0, \lambda_0)$-strongly convex from one direction at w_{opt}, and thus failing to be in $\{\cup_{\nu > \beta, \nu \in \mathbb{Q}} H^\nu\}$ for any $\beta \in [0,1]$, i.e., $F \in H^0 \setminus \{\cup_{\nu > 0, \nu \in \mathbb{Q}} H^\nu\}$. On the other hand, F can simultaneously be $(1, L_1)$-weakly smooth from the opposite direction at w_{opt}, failing to be in $\{\cup_{\mu < \beta, \mu \in \mathbb{Q}} S^\mu\}$ for any $\beta \in [0,1]$, hence F is also in $S^1 \setminus \{\cup_{\mu < 1, \mu \in \mathbb{Q}} S^\mu\}$. In this case, $F \in \{H^0 \setminus \{\cup_{\nu > 0, \nu \in \mathbb{Q}} H^\nu\}\} \cap \{S^1 \setminus \{\cup_{\mu < 1, \mu \in \mathbb{Q}} S^\mu\}\}$, implying that the tightest upper and lower bound on $\|g(w_t)\|$ derived from the (μ, λ_μ)-strong convexity and (ν, L_ν)-weak smoothness conditions are that $\|g(w_t)\| = O(1)$ and $\|g(w_t)\| = \Omega(\|w - w_{opt}\|)$. These are the functions where there is no α such that $F \in \{H^\alpha \setminus \{\cup_{\nu > \alpha, \nu \in \mathbb{Q}} H^\nu\}\} \cap \{S^\alpha \setminus \{\cup_{\mu < \alpha, \mu \in \mathbb{Q}} S^\mu\}\}$.

This observation justifies our two different ways of subclassifying the strongly convex functions. Combined, they provide a refined way of describing the function around its optimum. This is achieved by locating the function at the intersection of the two sets described by strong convexity and weak smoothness. Specifically, given F, we can find $\beta \le \alpha$ such that $F \in \{H^\beta \setminus \{\cup_{\nu > \beta, \nu \in \mathbb{Q}} H^\nu\}\} \cap \{S^\alpha \setminus \{\cup_{\mu < \alpha, \mu \in \mathbb{Q}} S^\mu\}\}$. It is easy to see that $0 \le \alpha - \beta \le 1$, and the greater this discrepancy is, the larger the gap between the upper and lower bound of $\|g(w_t)\|$ is. Here we present a lemma that describes the subclass of functions that we would not have been able to cover without using parameter ν.

Lemma 7.3. *Let* $X = \{S^1 - S^1 \cap \{\cup_{\nu > 0, \nu \in \mathbb{Q}} H^\nu\}\} \triangle S^0$, *then*

$$X \subseteq \{H^0 \setminus \{\cup_{\nu > 0, \nu \in \mathbb{Q}} H^\nu\} - S^0\}.$$

Proof. Let $\nu \in \mathbb{Q}$ and $\nu \in [0,1]$. $S^1 = \{S^1 \cap \{\cup_{\nu > 0} H^\nu\}\} \cup \{S^1 \cap \{\cup_{\nu > 0} H^\nu\}^c\}$. By Lemma 7.1, $S^1 \subseteq H^0$, so $S^1 \cap \{\cup_{\nu > 0} H^\nu\}^c \subseteq H^0 \cap \{\cup_{\nu > 0} H^\nu\}^c = H^0 \setminus \{\cup_{\nu > 0} H^\nu\}$, since $\cup_{\nu > 0} H^\nu \subseteq H^0$. Therefore, $S^1 \subseteq \{S^1 \cap \{\cup_{\nu > 0} H^\nu\}\} \cup \{H^0 \setminus \{\cup_{\nu > 0} H^\nu\}\}$. Since $S^0 \subset S^1$, and by Lemma 7.2, $S^0 \subseteq H^0 \setminus \{\cup_{\nu > 0} H^\nu\}$, we have $X = \{S^1 - S^1 \cap \{\cup_{\nu > 0} H^\nu\}\} \triangle S^0 = \{S^1 - S^1 \cap \{\cup_{\nu > 0} H^\nu\}\} \setminus S^0 \subseteq \{\{H^0 \setminus \{\cup_{\nu > 0} H^\nu\}\} \setminus S^0\}$. Thus $X \subseteq \{H^0 \setminus \{\cup_{\nu > 0} H^\nu\}\} - S^0$. $\square$

Note that the symbols "$-$" and "$\setminus$" are used interchangeably in the proof above to represent set exclusion.

7.5.3   Convergence Rate and the Quantitative Relation between μ and ν

Here we demonstrate how the relation between the weak smoothness and strong convexity parameters pertain to the analysis of convergence of SGD. Suppose we have a (ν, L_ν)- weakly smooth and $(\mu, , \lambda_\mu)$-strongly convex function F such that $1 \geq \mu \geq \nu \geq 0$. Note that $F \in S^1$, hence it is strongly convex in the usual sense. According to Equation (7.3) and strong convexity, it is not hard to derive that (see Lemma 7.4)

$$E \left\| w_{t+1} - w_{opt} \right\|^2 \leq E \left\| w_t - w_{opt} \right\|^2 - 2\eta_t E[F(w_t) - F(w_{opt})] + \eta_t^2 G$$

where G is an upper bound on $E \left\| \hat{g}(w_t) \right\|^2$, as assumed in Section 7.3. Then, according to our assumptions on F,

$$E[F(w_t) - F(w_{opt})] \geq \frac{\lambda_\mu}{1 + \mu} E \left\| w_t - w_{opt} \right\|^{1+\mu}.$$

Therefore,

$$E \left\| w_{t+1} - w_{opt} \right\|^2 \tag{7.10}$$

$$\leq \quad E \left\| w_t - w_{opt} \right\|^2 - 2\eta_t \frac{\lambda_\mu}{1 + \mu} E \left\| w_t - w_{opt} \right\|^{1+\mu} + \eta_t^2 G. \tag{7.11}$$

A standard analysis uses this recurrence relation to upper bound $E \left\| w_{T+1} - w_{opt} \right\|^2$ by induction for all $t = 1, \ldots, T$. Hence, Inequality (7.10) suggests that $\| w_t - w_{opt} \|$ converges to zero faster if μ is smaller, i.e., if F is more strongly-convex around w_{opt}.

On the other hand, the weak-smoothness assumption implies

$$F(w_t) - F(w_{opt}) \leq \frac{L_\nu}{1 + \nu} \| w_t - w_{opt} \|^{1+\nu}. \tag{7.12}$$

Inequality (7.12) implies that for a fixed quantity $\| w_t - w_{opt} \|$, $F(w_t) - F(w_{opt})$ is smaller when ν is larger, i.e., if F is more smooth around w_{opt}, demonstrating the stabilizing effect of a smooth function F, as introduced at the end of Section 7.2.2.

Combining Inequalities (7.10) and (7.12), we see that when $\mu - \nu$ is small, the convergence of $F(w_t) - F(w_{opt})$ should be faster. This effect is illustrated in the already known optimal convergence analysis of SGD on strongly convex and smooth functions, in which case $\nu = \mu = 1$. We conjecture that the optimal convergence can be achieved for any functions with $\nu = \mu$.

7.5.4   A Technical Lemma

Here, we present a technical lemma that derives some standard inequalities (see [9], for example) by the SGD updating rule and strong convexity.

Lemma 7.4.

1. Suppose the conditions in Section 7.3 hold, then $\forall w \in W$,

$$E\left\|w_{t+1} - w\right\|^2 \le (1 - \lambda\eta_t)E\left\|w_t - w\right\|^2 - 2\eta_t E[F(w_t) - F(w)] + \eta_t^2 E\left\|\hat{g}_t\right\|^2$$

2. Further assume that $\exists d_0 > 0$ s.t. $\forall w_t \ne w_{opt}$, $F(w_t) - F(w_{opt}) \ge d_0 \left\|w_t - w_{opt}\right\|$. Consider SGD with step sizes $\eta_t = \frac{c}{\lambda t}$, with $c \ge 2$. Then for any $T > 1$, it holds that

$$E\left\|w_{t+1} - w_{opt}\right\|^2 \le (1 - \lambda\eta_t)E\left\|w_t - w_{opt}\right\|^2 - 2\eta_t d_0 E\left\|w_t - w_{opt}\right\| + \eta_t^2 G$$

Proof. Since by the SGD update rule and by strong convexity, we have

$$E\left\|w_{t+1} - w\right\|^2 \le E\left\|w_t - \eta_t \hat{g}_t - w\right\|^2$$
$$= E\left\|w_t - w\right\|^2 - 2\eta_t E\langle \hat{g}_t, w_t - w\rangle + \eta_t^2 E\left\|\hat{g}_t\right\|^2$$
$$\le E\left\|w_t - w\right\|^2 - 2\eta_t E[(F(w_t) - F(w) + \frac{\lambda}{2}\left\|w_t - w\right\|^2)] + \eta_t^2 E\left\|\hat{g}_t\right\|^2$$
$$\le (1 - \lambda\eta_t)E\left\|w_t - w\right\|^2 - 2\eta_t E[F(w_t) - F(w)] + \eta_t^2 E\left\|\hat{g}_t\right\|^2 \quad (7.13)$$

Since Equation (7.13) holds for any $w \in W$, we substitute w by w_{opt}, and get

$$E\left\|w_{t+1} - w_{opt}\right\|^2$$
$$\le (1 - \lambda\eta_t)E\left\|w_t - w_{opt}\right\|^2 - 2\eta_t E[F(w_t) - F(w_{opt})] + \eta_t^2 E\left\|\hat{g}_t\right\|^2$$

Since by the our assumptions, we have

$$F(w_t) - F(w_{opt}) \ge d_0 \left\|w_t - w_{opt}\right\|$$

Combining the inequalities, we get

$$E\left\|w_{t+1} - w_{opt}\right\|^2 \le (1 - \lambda\eta_t)E\left\|w_t - w_{opt}\right\|^2 - 2\eta_t d_0 E\left\|w_t - w_{opt}\right\| + \eta_t^2 G.$$

$\square$

7.6 Conclusion and Future Work

In conclusion, we demonstrate how the degree of strong convexity and the degree of smoothness of a function influence the convergence rate of SGD. We introduce a subclassification scheme of strongly convex functions. Then we discuss how this new scheme could help us in the convergence analysis. For future work, we intend to analyze the convergence rate of SGD on different subclasses. As discussed in Section 7.5.2, we believe when $\mu - \nu$ is small, the optimal convergence is more likely to be achieved. Another direction is to use the scheme to unify existing assumptions that capture the properties of commonly used objective functions (for example, Assumption 4.1 in [14]). We also want to further explore how the learning rate η_t, number of averaging terms N_t, oracle induced variance, and functional property together determine the convergence rate of SGD.

Bibliography

[1] Alekh Agarwal, Peter L. Bartlett, Pradeep D. Ravikumar, and Martin J. Wainwright. Information-theoretic lower bounds on the oracle complexity of convex optimization. In *NIPS*, pages 1–9, 2009.

[2] Léon Bottou. Large-scale machine learning with stochastic gradient descent. In Yves Lechevallier and Gilbert Saporta, editors, *Proceedings of the 19th International Conference on Computational Statistics (COMPSTAT'2010)*, pages 177–187, Paris, France, August 2010. Springer.

[3] Olivier Devolder, François Glineur, and Yurii Nesterov. First-order methods of smooth convex optimization with inexact oracle. *CORE Discussion Papers*, 2011.

[4] Elad Hazan and Satyen Kale. Beyond the regret minimization barrier: an optimal algorithm for stochastic strongly-convex optimization. *Journal of Machine Learning Research - Proceedings Track*, 19:421–436, 2011.

[5] Rie Johnson and Tong Zhang. Accelerating stochastic gradient descent using predictive variance reduction. In *NIPS*, pages 315–323, 2013.

[6] A.S. Nemirovsky and D.B. Yudin. *Problem complexity and problem efficiency in optimization*. Wiley-Interscience, 1983.

[7] B.T. Polyak and A.B. Juditsky. Acceleration of stochastic approximation by averaging. *SIAM Journal on Control and Optimization*, 30(4):838–855, 1992.

[8] Maxim Raginsky and Alexander Rakhlin. Information-based complexity, feedback and dynamics in convex programming. *IEEE Transactions on Information Theory*, 57(10):7036–7056, 2011.

[9] Alexander Rakhlin and Ohad Shamir. Making gradient descent optimal for strongly convex stochastic optimization. In *Proceedings of the 29th International Conference on Machine Learning, ICML 2012, Edinburgh, Scotland, UK, June 26–July 1, 2012*. Omnipress, 2012.

[10] Aaditya Ramdas and Aarti Singh. Optimal rates for stochastic convex optimization under Tsybakov noise condition. In *Proceedings of the 30th International Conference on Machine Learning, ICML 2013, Atlanta, GA, USA, 16–21 June 2013*, volume 28 of *JMLR Proceedings*. JMLR.org, 2013.

[11] Tyrrell Rockafellar. *Convex Analysis*. Princeton University Press, Princeton, NJ, USA, 1970.

[12] Shai Shalev-Shwartz, Yoram Singer, and Nathan Srebro. Pegasos: Primal estimated sub-gradient solver for svm. In *Proceedings of the 24th International Conference on Machine Learning*, ICML '07, pages 807–814, New York, NY, USA, 2007. ACM.

[13] Ohad Shamir. Open problem: Is averaging needed for strongly convex stochastic gradient descent? *Open Problems presented at COLT*, 2012.

[14] Tong Zhang. Solving large scale linear prediction problems using stochastic gradient descent algorithms. In *ICML 2004: Proceedings of the Twenty-First International Conference on Machine Learning. Omnipress*, pages 919–926, 2004.

[15] Martin Zinkevich. Online convex programming and generalized infinitesimal gradient ascent. In *ICML*, pages 928–936, 2003.

Chapter 8

Detecting Ineffective Features for Nonparametric Regression

Kris De Brabanter

Department of Statistics and Department of Computer Science, Iowa State University

Paola Gloria Ferrario

Institut für Medizinische Biometrie und Statistik, Universität zu Lübeck

László Györfi

Department of Computer Science and Information Theory, Budapest University of Technology and Economics

In the nonparametric regression setting, we investigate the hypothesis that some components of the covariate (feature) vector are ineffective. Identifying them would enable us to reduce the dimension of the feature vector, without increasing the minimum mean squared error.

Since the proof of the asymptotical normality of the related test statistic is fairly complicated, we present here some attempts, heuristics, and simulations about its behavior. Finally, we show some consequences of the results for binary pattern recognition.

This chapter is organized as follows: After some preliminary definitions and a comparison with available results, we introduce an estimator of the minimum mean squared error. In Section 8.2 we try to test the hypothesis that the minimum mean squared error does not increase by leaving out an ineffective component, say, the k-one, and we make a conjecture about the asymptotic behavior of the related test statistic. In Section 8.3 we are constrained to take a modification of it, by splitting the sample, in order to avoid some difficulties and remain capable of testing. Here we support our conjecture by Monte Carlo experiment. Later, in Section 8.4 we introduce another setting, where the response variable is binary valued and in Section 8.5 we extend the test setting and the results of the previous sections to this classification problem. At the end we conclude with a brief summary of the remaining open questions and we indicate where some additional works remain still to be done.

8.1   Estimate the Minimum Mean Squared Error

Let the label Y be a real valued random variable and let the feature vector $\mathbf{X} = (X_1, \ldots, X_d)$ be a d-dimensional random vector. The regression function m is defined by

$$m(\mathbf{x}) = \mathbf{E}\{Y \mid \mathbf{X} = \mathbf{x}\}.$$

The minimum mean squared error, also called variance of the residual $Y - m(\mathbf{X})$, is denoted by

$$L^* := \mathbf{E}\{(Y - m(\mathbf{X}))^2\} = \min_f \mathbf{E}\{(Y - f(\mathbf{X}))^2\}.$$

The regression function m and the minimum mean squared error L^* cannot be calculated when the distribution of $(\mathbf{X}, Y)$ is unknown. Assume, however, that we observe data

$$\mathcal{D}_n = \{(\mathbf{X}_1, Y_1), \ldots, (\mathbf{X}_n, Y_n)\}$$

consisting of independent and identically distributed copies of $(\mathbf{X}, Y)$. $\mathcal{D}_n$ can be used to produce an estimate of L^*. For nonparametric estimates of the minimum mean squared error $L^* = \mathbf{E}\{(Y - m(\mathbf{X}))^2\}$ see, e.g., Dudoit and van der Laan [8], Kohler [13], Liitiäinen et al. [15], [16], Liitiäinen et al. [17], Müller and Stadtmüller [19], Neumann [21], Pelckmans et al. [23], Stadtmüller and Tsybakov [24], and the literature cited there (see also [9, 11, 22, 25] for related work). Devroye et al. [7] proved that without any tail and smoothness condition, L^* cannot be estimated with guaranteed rate of convergence. Müller, Schick, and Wefelmeyer [20] estimated L^* as the variance of an independent measurement error Z in the model

$$Y = m(\mathbf{X}) + Z \tag{8.1}$$

such that $\mathbf{E}\{Z\} = 0$, and $\mathbf{X}$ and Z are independent. Sometimes this is called the additive noise model or the homoscedastic regression model. Devroye et al. [7] introduced an estimate of the minimum mean squared error L^* by the modified nearest neighbor cross-validation estimate

$$\hat{L}_n = \frac{1}{2n} \sum_{i=1}^{n} (Y_i - Y_{j(i)})^2,$$

where $Y_{j(i)}$ is the label of the modified first nearest neighbor of $\mathbf{X}_i$ among $\mathbf{X}_1, \ldots, \mathbf{X}_{i-1}, \mathbf{X}_{i+1}, \ldots, \mathbf{X}_n$. We adopt the same definition of the first nearest neighbor, but slightly changing the notation in $X_{n,i,1}$, where

$$\{n, i, 1\} := \underset{1 \leq j \leq n, \, j \neq i}{\arg \min} \, \rho(X_i, X_j), \tag{8.2}$$

and ρ is a metric (typically the Euclidean one) in R^d.
The k-th nearest neighbor of X_i among $\mathbf{X}_1, \ldots, \mathbf{X}_{i-1}, \mathbf{X}_{i+1}, \ldots, \mathbf{X}_n$ is defined as $X_{n,i,k}$ via generalization of definition (8.2):

$$\{n, i, k\} := \underset{1 \leq j \leq n, \, j \neq i, \, j \notin \{\{n,i,1\}, \ldots, \{n,i,k-1\}\}}{\arg \min} \, \rho(X_i, X_j), \tag{8.3}$$

by removing the preceding neighbors. If ties (ambiguities in the assignment of the neighbors) occur, a simple possibility to break them is given by taking the minimal index. On the other hand, when $\mathbf{X}$ has a density, the case of ties among nearest neighbor distances occurs with probability zero, and this one will be our setting.

If Y and $\mathbf{X}$ are bounded, and m is Lipschitz continuous

$$|m(\mathbf{x}) - m(\mathbf{z})| \leq C\|\mathbf{x} - \mathbf{z}\|, \tag{8.4}$$

then for $d \geq 3$, Devroye et al. [7] proved that

$$\mathbf{E}\{|\hat{L}_n - L^*|\} \leq c_1 n^{-1/2} + c_2 n^{-2/d}. \tag{8.5}$$

Liitiäinen et al. [15] introduced another estimate of the minimum mean squared error L^* by the first and second nearest neighbor cross-validation

$$L_n = \frac{1}{n} \sum_{i=1}^{n} (Y_i - Y_{n,i,1})(Y_i - Y_{n,i,2}),$$

where $Y_{n,i,1}$ and $Y_{n,i,2}$ are the labels of the first and second nearest neighbors of $\mathbf{X}_i$ among $\mathbf{X}_1, \ldots, \mathbf{X}_{i-1}, \mathbf{X}_{i+1}, \ldots, \mathbf{X}_n$, resp.

If Y and $\mathbf{X}$ are bounded and m is Lipschitz continuous, then for $d \geq 2$ and for L_n, they proved the rate of convergence (8.5). Ferrario and Walk [10] proved that, for bounded Y,

$$L_n \to L^*$$

almost surely (a.s.). Moreover, under the condition $\mathbf{E}\{Y^2\} < \infty$

$$\frac{1}{n} \sum_{i=1}^{n} L_i \to L^*$$

a.s., which is a universal consistency result. Devroye et al. [5] derived a new and simple estimator of L^*, considering the definition

$$L^* = \mathbf{E}\{(Y - m(\mathbf{X}))^2\} = \mathbf{E}\{Y^2\} - \mathbf{E}\{m(\mathbf{X})^2\}.$$

Obviously, $\mathbf{E}\{Y^2\}$ can be estimated by $\frac{1}{n}\sum_{i=1}^{n}Y_i^2$, while we estimate the term $\mathbf{E}\{m(\mathbf{X})^2\}$ by $\frac{1}{n}\sum_{i=1}^{n}Y_iY_{n,i,1}$. Therefore, L^* can be estimated by

$$\tilde{L}_n := \frac{1}{n}\sum_{i=1}^{n}Y_i^2 - \frac{1}{n}\sum_{i=1}^{n}Y_iY_{n,i,1}. \qquad (8.6)$$

They proved the strong universal consistency, and under some regularity condition showed the rate (8.5). In the next section we consider the testing of ineffective features, where the following limit distribution result would be useful:

Conjecture 8.1.

$$Z_n := \frac{1}{\sqrt{n}}\sum_{i=1}^{n}(Y_iY_{n,i,1} - \mathbf{E}\{Y_iY_{n,i,1}\})$$

is asymptotically normal with mean zero and a finite variance σ^2.

The main difficulty here is that Z_n is an average of dependent random variables. However, this dependence has a special property, called exchangeability. A triangular array $V_{n,i}$, $n = 1, 2, \ldots$, $i = 1, \ldots, n$ is called (row-wise) exchangeable, if for any fixed n and for any permutation $\rho(1), \ldots, \rho(n)$ of $1, \ldots, n$, the distributions of

$$(V_{n,1}, \ldots, V_{n,n})$$

and

$$(V_{n,\rho(1)}, \ldots, V_{n,\rho(n)})$$

are equal. There is a classical central limit theorem for exchangeable arrays:

Theorem 8.1. *(Blum et al. [2], Weber [27].) Let $\{V_{n,i}\}$ be a triangular array of exchangeable random variables with zero mean and variance σ^2. Assume that*

(i)

$$\lim_{n\to\infty} n\mathbf{E}\{V_{n,1}V_{n,2}\} = 0,$$

(ii)

$$\lim_{n\to\infty} \mathbf{E}\{V_{n,1}^2V_{n,2}^2\} = \sigma^4,$$

(iii)

$$\mathbf{E}\{|V_{n,1}|^3\} = o(\sqrt{n}).$$

Then

$$\frac{\sum_{i=1}^{n}V_{n,i}}{\sqrt{n}}$$

is asymptotically normal with mean zero and variance σ^2.

In order to prove the asymptotic normality of Z_n we may apply Theorem 8.1 for

$$V_{n,i} = Y_i Y_{n,i,1} - \mathbf{E}\{Y_i Y_{n,i,1}\}.$$

From the conditions of Theorem 8.1, (i) is the most crucial. Equivalently, it requires that

$$\mathbf{Var}(V_{n,1}) = \frac{\mathbf{Var}(\sum_{i=1}^{n} V_{n,i})}{n} + o(1).$$

The verification of the condition (ii) in Theorem 8.1 seems to be easier, while (iii) is satisfied if Y is bounded. According to our simulations (i) is not true such that instead of (i) we have that

$$n\mathbf{E}\{V_{n,1} V_{n,2}\} \to c,$$

where $c \neq 0$ is a constant. This modified condition is equivalent to

$$\frac{\mathbf{Var}(\sum_{i=1}^{n} V_{n,i})}{n} \to \sigma^2 + c.$$

Thus, one has to generalize Theorem 1 as follows:

Conjecture 8.2. *Let $\{V_{n,i}\}$ be a triangular array of exchangeable random variables with zero mean and variance σ^2. Assume that*

(i)

$$\lim_{n \to \infty} n\mathbf{E}\{V_{n,1} V_{n,2}\} = c,$$

(ii)

$$\lim_{n \to \infty} \mathbf{E}\{V_{n,1}^2 V_{n,2}^2\} = \sigma^4,$$

(iii)

$$\mathbf{E}\{|V_{n,1}|^3\} = o(\sqrt{n}).$$

Then

$$\frac{\sum_{i=1}^{n} V_{n,i}}{\sqrt{n}}$$

is asymptotically normal with mean zero and variance $\sigma^2 + c$.

8.2 Tests for the Regression Problem

It is of great importance to be able to estimate the minimum mean squared error L^* accurately, even before one of the regression estimates is applied: in a standard nonparametric regression design process, one considers a finite number of real-valued features $\mathbf{X}^{(I)} = (X_k, k \in I)$, and evaluates whether these suffice to explain Y. In case they suffice for the given explanatory task, an estimation method can be applied on the basis of the features already under consideration; if not, more or different features must be considered. The quality of a collection of features $\mathbf{X}^{(I)}$ is measured by the minimum mean squared error

$$L^*(I) := \mathbf{E}\{(Y - \mathbf{E}\{Y|\mathbf{X}^{(I)}\})^2\}$$

that can be achieved using the features as explanatory variables. $L^*(I)$ depends upon the unknown distribution of $(Y, \mathbf{X}^{(I)})$. The first phase of any regression estimation process therefore heavily relies on estimates of L^* (even before a regression estimate is picked). Another way of dimension reduction would be to detect the ineffective components of the feature vector. Let $\mathbf{X}^{(-k)} = (X_1, \ldots, X_{k-1}, X_{k+1}, \ldots, X_d)$ be the $d-1$ dimensional feature vector such that we leave out the k-th component from $\mathbf{X}$. Then the corresponding minimum error is

$$L^{*(-k)} := \mathbf{E}\{(Y - \mathbf{E}\{Y|\mathbf{X}^{(-k)}\})^2\}.$$

The aim of this section is to test the hypothesis

$$\mathcal{H}_k : L^{*(-k)} = L^*,$$

which means that by leaving out the k-th component the minimum mean squared error does not increase. The hypothesis $\mathcal{H}_k$ means that

$$\mathbf{E}\{\mathbf{E}\{Y \mid \mathbf{X}\}^2\} = \mathbf{E}\{\mathbf{E}\{Y \mid \mathbf{X}^{(-k)}\}^2\}. \tag{8.7}$$

Because of the equality

$$\mathbf{E}\{\mathbf{E}\{Y \mid \mathbf{X}\}^2\} - \mathbf{E}\{\mathbf{E}\{Y \mid \mathbf{X}^{(-k)}\}^2\} = \mathbf{E}\{(\mathbf{E}\{Y \mid \mathbf{X}\} - \mathbf{E}\{Y \mid \mathbf{X}^{(-k)}\})^2\}$$

$\mathcal{H}_k$ happens if and only if

$$m(\mathbf{X}) = \mathbf{E}\{Y \mid \mathbf{X}\} = \mathbf{E}\{Y \mid \mathbf{X}^{(-k)}\} =: m^{(-k)}(\mathbf{X}^{(-k)}) \quad \text{a.s.}$$

Koepke and Bilenko [12] investigated this testing problem such that the hypothesis $\mathcal{H}_k$ is verified from the hypothesis that $Y, \mathbf{X}^{(-k)}$ and X_k are independent. However, under this independence condition we have

$$\mathbf{E}\{Y \mid \mathbf{X}\} = \mathbf{E}\{Y \mid \mathbf{X}^{(-k)}, X_k\} = \mathbf{E}\{Y \mid \mathbf{X}^{(-k)}\}.$$

Unfortunately, this independence condition is too much for our purposes, since a component can be ineffective even in the case when $Y, \mathbf{X}^{(-k)}$ and X_k are dependent.

Next we introduce a new test. Using the data

$$\mathcal{D}_n^{(-k)} = \{(\mathbf{X}_1^{(-k)}, Y_1), \ldots, (\mathbf{X}_n^{(-k)}, Y_n)\},$$

one can estimate $L^{*(-k)}$ by

$$\tilde{L}_n^{(-k)} := \frac{1}{n} \sum_{i=1}^{n} Y_i^2 - \frac{1}{n} \sum_{i=1}^{n} Y_i Y_{n,i,1}^{(-k)},$$

so the corresponding test statistic is

$$\tilde{L}_n^{(-k)} - \tilde{L}_n = \frac{1}{n} \sum_{i=1}^{n} Y_i (Y_{n,i,1} - Y_{n,i,1}^{(-k)}).$$

We can accept the hypothesis $\mathcal{H}_k$ if

$$\tilde{L}_n^{(-k)} - \tilde{L}_n \tag{8.8}$$

is "close" to zero. De Brabanter and Györfi [3] observed that this test statistic is small even in the case when the hypothesis $\mathcal{H}_k$ is not true, since with large probability the first nearest neighbors of $\mathbf{X}_i$ and $\mathbf{X}_i^{(-k)}$ are the same, hence $Y_{n,i,1} - Y_{n,i,1}^{(-k)} = 0$. It implies that according to the asymptotic distribution of $\sqrt{n}(\tilde{L}_n^{(-k)} - \tilde{L}_n)$ the zero is an atom and consequently asymptotic normality is not possible. Their simulations showed that $\mathbf{P}(Y_{n,i,1} = Y_{n,i,1}^{(-k)})$ is decreasing as n increases and increasing as d increases. In the simulations where they considered the function

$$Y = \sum_{i=1}^{d} c_i X_i + Z, \tag{8.9}$$

$\mathbf{X}$ is uniform on $[0,1]^d$ and $Z \sim \mathbb{N}(0,1)$. First, they investigated how this probability depends on the sample size for fixed $d = 5$. Figure 8.1 shows the percentage of $Y_{n,i,1} = Y_{n,i,1}^{(-1)}$ as a function of the sample size n. Figure 8.2 shows the percentage of $Y_{n,i,1} = Y_{n,i,1}^{(-1)}$ as a function of the dimensionality d with sample size $n = 2000$.

Thus, De Brabanter and Györfi [3] modified the test statistic such that

$$(\hat{\mathbf{X}}_{n,i,1}, \hat{\mathbf{X}}_{n,i,1}^{(-k)})$$
$$= \begin{cases} (\mathbf{X}_{n,i,1}, \mathbf{X}_{n,i,1}^{(-k)}) & \text{if } (\mathbf{X}_{n,i,1})^{(-k)} \neq \mathbf{X}_{n,i,1}^{(-k)}, \\ I_i(\mathbf{X}_{n,i,2}, \mathbf{X}_{n,i,1}^{(-k)}) + (1 - I_i)(\mathbf{X}_{n,i,1}, \mathbf{X}_{n,i,2}^{(-k)}) & \text{otherwise} \end{cases}$$

with

$$I_i = \begin{cases} 0 & \text{with probability } 1/2, \\ 1 & \text{with probability } 1/2, \end{cases}$$

and correspondingly

$$(\hat{Y}_{n,i,1}, \hat{Y}_{n,i,1}^{(-k)})$$
$$= \begin{cases} (Y_{n,i,1}, Y_{n,i,1}^{(-k)}) & \text{if } (\mathbf{X}_{n,i,1})^{(-k)} \neq \mathbf{X}_{n,i,1}^{(-k)}, \\ I_i(Y_{n,i,2}, Y_{n,i,1}^{(-k)}) + (1 - I_i)(Y_{n,i,1}, Y_{n,i,2}^{(-k)}) & \text{otherwise} \end{cases}$$

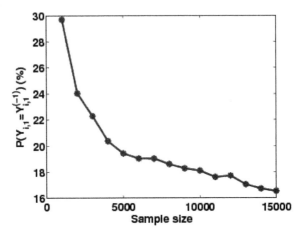

FIGURE 8.1: Percentage of $Y_{n,i,1} = Y_{n,i,1}^{(-k)}$ as a function of the sample size n with $d = 5$.

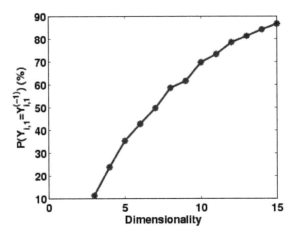

FIGURE 8.2: Percentage of $Y_{n,i,1} = Y_{n,i,1}^{(-k)}$ as a function of the dimensionality d with sample size $n = 2000$.

yielding the test statistic

$$\hat{L}_n^{(-k)} - \hat{L}_n = \frac{1}{n}\sum_{i=1}^{n} Y_i(\hat{Y}_{n,i,1} - \hat{Y}_{n,i,1}^{(-k)}).$$

De Brabanter and Györfi [3] performed some additional simulations, based on which they conjectured that, under $\mathcal{H}_k$,

$$\sqrt{n}(\hat{L}_n^{(-k)} - \hat{L}_n) \tag{8.10}$$

is asymptotically normal with mean zero and finite variance. Such a result would be a surprise, since in this case the smoothness of the regression function m and the dimension d do not count. Under the alternative hypothesis $\mathcal{H}_k^c$, one can prove that

$$\lim_{n\to\infty} \frac{1}{n} \sum_{i=1}^{n} Y_i(\hat{Y}_{n,i,1} - \hat{Y}_{n,i,1}^{(-k)}) = \mathbf{E}\{(m(\mathbf{X}) - m^{(-k)}(\mathbf{X}^{(-k)}))^2\} > 0 \text{ a.s.}$$

8.3 A New Test

The estimate $\hat{L}_n$ is fairly complicated such that the proof of (8.10) seems to be intractable.

Therefore we introduce another test based on the estimate (8.6). In the previous section, we illustrated that the test statistic (8.8) has an atom at zero. This difficulty can be avoided if we apply data splitting. Split therefore the data $\mathcal{D}_{2n} = \{(\mathbf{X}_1, Y_1), ..., (\mathbf{X}_{2n}, Y_{2n})\}$ into two sub-samples $\mathcal{D}_n = \{(\mathbf{X}_1, Y_1), ..., (\mathbf{X}_n, Y_n)\}$ and $\mathcal{D}'_n = \{(\mathbf{X}'_1, Y'_1), ..., (\mathbf{X}'_n, Y'_n)\}$ where

$$(\mathbf{X}'_i, Y'_i) = (\mathbf{X}_{n+i}, Y_{n+i}),$$

$i = 1, ..., n$. One can estimate $L^{*(-k)}$ by

$$\tilde{L}_n'^{(-k)} := \frac{1}{n} \sum_{i=1}^{n} Y_i^2 - \frac{1}{n} \sum_{i=1}^{n} Y_i' Y_{n,i,1}'^{(-k)}. \tag{8.11}$$

The difference between the estimates (8.11) and (8.6)

$$T_n := \frac{1}{n} \sum_{i=1}^{n} Y_i Y_{n,i,1} - \frac{1}{n} \sum_{i=1}^{n} Y_i' Y_{n,i,1}'^{(-k)}. \tag{8.12}$$

can be a test statistic, which avoids the problem of an atom at zero, since the two averages in (8.12) are independent.

If Conjecture 8.1 holds, then under $\mathcal{H}_k$

$$\sqrt{n}(T_n - \mathbf{E}\{T_n\})$$

is asymptotically normal with mean zero and variance $2\sigma^2$. Moreover, using the fact that $m(\mathbf{X}) = m^{(-k)}(\mathbf{X}^{(-k)})$, we guess that under the Lipschitz continuity of m

$$\sqrt{n}\mathbf{E}\{T_n\} = \frac{1}{\sqrt{n}}\sum_{i=1}^{n}\mathbf{E}\{Y_iY_{n,i,1} - Y_i'Y_{n,i,1}'^{(-k)}\}$$
$$= \sqrt{n}\mathbf{E}\{m(\mathbf{X}_1)m(\mathbf{X}_{n,1,1}) - m(\mathbf{X}_1')m^{(-k)}(\mathbf{X}_{n,1,1}'^{(-k)})\}$$
$$= \sqrt{n}\mathbf{E}\{m(\mathbf{X}_1)(m(\mathbf{X}_{n,1,1}) - m^{(-k)}(\mathbf{X}_{n,1,1}^{(-k)}))\}$$
$$\approx 0.$$

Because of independence,

$$\mathbf{Var}(Y_1Y_{n,1,1} - Y_1'Y_{n,1,1}'^{(-k)}) = \mathbf{Var}(Y_1Y_{n,1,1}) + \mathbf{Var}(Y_1'Y_{n,1,1}'^{(-k)}).$$

Introduce the notation

$$M_2(\mathbf{x}) = \mathbf{E}\{Y^2 \mid \mathbf{X} = \mathbf{x}\}.$$

Then

$$\mathbf{Var}(Y_1Y_{n,1,1}) = \mathbf{E}\{Y_1^2Y_{n,1,1}^2\} - \mathbf{E}\{m(\mathbf{X}_1)m(\mathbf{X}_{n,1,1})\}^2$$
$$= \mathbf{E}\{M_2(\mathbf{X}_1)M_2(\mathbf{X}_{n,1,1})\} - \mathbf{E}\{m(\mathbf{X}_1)m(\mathbf{X}_{n,1,1})\}^2$$
$$\to \mathbf{E}\{M_2(\mathbf{X}_1)^2\} - \mathbf{E}\{m(\mathbf{X}_1)^2\}^2,$$

as $n \to \infty$. Similarly,

$$\mathbf{Var}(Y_1'Y_{n,1,1}'^{(-k)}) \to \mathbf{E}\{M_2^{(-k)}(\mathbf{X}_1^{(-k)})^2\} - \mathbf{E}\{m^{(-k)}(\mathbf{X}_1^{(-k)})^2\}^2,$$

where

$$M_2^{(-k)}(\mathbf{x}^{(-k)}) = \mathbf{E}\{Y^2 \mid \mathbf{X}^{(-k)} = \mathbf{x}^{(-k)}\}.$$

If n-times the covariances of the terms in T_n are asymptotically zero, then under $\mathcal{H}_k$

$$\mathbf{Var}(\sqrt{n}T_n) \approx \mathbf{E}\{M_2(\mathbf{X}_1)^2\} + \mathbf{E}\{M_2^{(-k)}(\mathbf{X}_1^{(-k)})^2\} - 2\mathbf{E}\{m(\mathbf{X}_1)^2\}^2$$
$$= 2\mathbf{E}\{M_2(\mathbf{X}_1)^2\} - 2\mathbf{E}\{m(\mathbf{X}_1)^2\}^2$$
$$=: 2\sigma^2.$$

This asymptotic variance can be estimated by

$$\sigma_n^2 := \frac{2}{n}\sum_{i=1}^{n}Y_i^2Y_{n,i,1}^2 - 2\left(\frac{1}{n}\sum_{i=1}^{n}Y_iY_{n,i,1}\right)^2.$$

For an $0 < \alpha < 1$, an asymptotically α-level test accepts the hypothesis $\mathcal{H}_k$ if

$$T_n' := \frac{\sqrt{n}T_n}{\sigma_n} \leq \Phi^{-1}(1 - \alpha),$$

where Φ denotes the standard normal distribution function.

In order to verify the asymptotic normality of the test statistic T'_n we conducted the following Monte Carlo experiment. Consider the regression function (8.9) with $c_2 = 0$, $c_1 = c_3 = c_4 = 1$, $\mathbf{X}$ is uniform on $[0,1]^4$ and $Z \sim \mathbb{N}(0,1)$. The sample size is set to $n = 1000$.

The result of the simulation is represented in Figure 8.3. The density histogram shows the test statistic T'_n based on 10,000 repetitions. We also fitted the best normal density (using maximum likelihood estimation) together with the standard normal density. It shows that the distribution of T'_n is approximately normal with mean zero, but the variance is larger than 1. In fact $\sigma_n^2 = 5.06$ and the asymptotic variance of $\sqrt{n}T_n$ is $\sigma_n'^2 = 6.07$.

Thus, the condition (i) of Theorem 8.1 is not satisfied, and therefore there is a need to prove its extension (Conjecture 8.2).

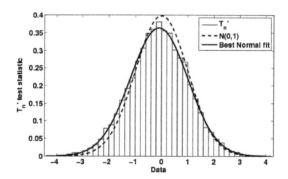

FIGURE 8.3: Test statistic T'_n with best normal fit and standard normal distributed fit.

Under the alternative hypothesis $\mathcal{H}_k^c$, the strong universal consistency result of Devroye et al. [5] implies that

$$
\begin{aligned}
\lim_{n \to \infty} T_n &= \mathbf{E}\{m(\mathbf{X})^2\} - \mathbf{E}\{m^{(-k)}(\mathbf{X}^{(-k)})^2\} \\
&= \mathbf{E}\{(m(\mathbf{X}) - m^{(-k)}(\mathbf{X}^{(-k)}))^2\} > 0 \quad \text{a.s.}
\end{aligned}
$$

The approximate calculations above and the simulation results lead to the following conjecture:

Conjecture 8.3. *If $\mathbf{X}$ and Y are bounded, $\mathbf{X}$ has a density, and m is Lipschitz continuous, then $\sqrt{n}T_n$ is asymptotically normal with mean zero and variance $2(\sigma')^2$ such that $\sigma < \sigma'$.*

We considered the following nonlinear function with 5 uniformly distributed inputs on $[0,1]^5$ with $n = 4000$: $Y = \sin(2\pi X_1)\cos(\pi X_2) + \varepsilon$, with

$\varepsilon \sim N(0, 0.1^2)$. The simulation results (based on 1000 runs) are summarized in Table 8.1. The significance level is set to 0.05.

TABLE 8.1: Appearance frequency of the variables.

X_1	X_2	X_3	X_4	X_5
982	988	42	44	46

We performed some simulations testing whether the k-th component is ineffective ($k = 1, \ldots, 5$). Table 8.1 confirms our conjectures on asymptotic normality. We have to emphasize that the rate of convergence of the distribution of normality depends on d and on the actual regression function.

8.4   Estimation of the Bayes Error Probability

Again, let the feature vector $\mathbf{X} = (X_1, \ldots, X_d)$ be a d-dimensional random vector such that its distribution is denoted by μ, and let the label Y be binary valued. If g is an arbitrary decision function then its error probability is denoted by

$$R(g) = \mathbf{P}\{g(\mathbf{X}) \neq Y\}.$$

Put

$$m(\mathbf{x}) = \mathbf{P}\{Y = 1 \mid \mathbf{X} = \mathbf{x}\} = \mathbf{E}\{Y \mid \mathbf{X} = \mathbf{x}\}.$$

The Bayes decision g^* minimizes the error probability:

$$g^*(\mathbf{x}) = \begin{cases} 1 & \text{if } m(\mathbf{x}) > 1/2 \\ 0 & \text{otherwise,} \end{cases}$$

and

$$R^* = \mathbf{P}\{g^*(\mathbf{X}) \neq Y\} = \mathbf{E}\{\min[m(\mathbf{X}), 1 - m(\mathbf{X})]\}$$

denotes its error probability. Put

$$D(\mathbf{x}) = 2m(\mathbf{x}) - 1$$

then the Bayes decision has the following equivalent form:

$$g^*(\mathbf{x}) = \begin{cases} 1 & \text{if } D(\mathbf{x}) > 0 \\ 0 & \text{otherwise.} \end{cases}$$

In the standard model of pattern recognition, we are given training labelled samples, which are independent and identical copies of $(\mathbf{X}, Y)$: $(\mathbf{X}_1, Y_1), \ldots$, $(\mathbf{X}_n, Y_n)$. Based on these labeled samples, one can estimate the regression function D by D_n, and the corresponding plug-in classification rule is defined by

$$g_n(\mathbf{x}) = \begin{cases} 1 & \text{if } D_n(\mathbf{x}) > 0 \\ 0 & \text{otherwise.} \end{cases}$$

Then

$$\mathbf{E}\{R(g_n)\} - R^* \leq \mathbf{E}\{|D(\mathbf{X}) - D_n(\mathbf{X})|\}$$

(cf. Theorem 2.2 in Devroye, Györfi, Lugosi [6]), therefore we may get an upper bound on the rate of convergence of $\mathbf{E}\{R(g_n)\} - R^*$ via the L_1 rate of convergence of the corresponding regression estimation. However, according to Section 6.7 in Devroye, Györfi, Lugosi [6], the classification is easier than L_1 regression function estimation, since the rate of convergence of the error probability depends on the behavior of the function D in the neighborhood of the decision boundary

$$B = \{\mathbf{x}; D(\mathbf{x}) = 0\}.$$

This phenomenon has been discovered by Mammen and Tsybakov [18], Tsybakov [26], who formulated the so called margin condition:

- *The strong margin condition.* Assume that for all $1 \geq t > 0$,

$$\mathbf{P}\{|D(\mathbf{X})| \leq t\} \leq ct^\alpha, \tag{8.13}$$

  where $\alpha > 0$.

Kohler and Krzyzak [14] introduced the weak margin condition:

- *The weak margin condition.* Assume that for all $1 \geq t > 0$,

$$\mathbf{E}\{I_{\{|D(\mathbf{X})| \leq t\}}|D(\mathbf{X})|\} \leq ct^{1+\alpha}. \tag{8.14}$$

  where $\alpha > 0$ and I denotes the indicator function.

Obviously, the strong margin condition implies the weak margin condition:

$$\mathbf{E}\{I_{\{|D(\mathbf{X})| \leq t\}}|D(\mathbf{X})|\} \leq \mathbf{E}\{I_{\{|D(\mathbf{X})| \leq t\}}t\} = t\mathbf{P}\{|D(\mathbf{X})| \leq t\} \leq ct \cdot t^\alpha.$$

The reverse is not true, for example, let X be uniformly distributed on $[-1, 1]$ and put

$$D(x) = (x - t_0)^+ - (-t_0 - x)^+,$$

where $0 < t_0 < 1$ and x^+ denotes the positive part of x. Then on the one hand, for $0 < t < 1 - t_0$,

$$\mathbf{P}\{|D(X)| \leq t\} = 2\mathbf{P}\{0 \leq X, (X - t_0)^+ \leq t\} = t + t_0,$$

which does not tend to 0 as $t \to 0$, so the strong margin condition is not satisfied. On the other hand,

$$\mathbf{E}\{I_{\{|D(X)| \leq t\}}|D(X)|\} = 2\mathbf{E}\{I_{\{0 \leq X, (X-t_0)^+ \leq t\}}(X - t_0)^+|\} = t^2/2,$$

therefore the weak margin condition is satisfied with $\alpha = 1$ and $c = 1/2$. Audibert and Tsybakov [1] proved that if the plug-in classifier g has been derived from the regression estimate $\tilde{D}$ and the strong margin condition is satisfied, then

$$R(g) - R^* \le \left(\int (\tilde{D}(\mathbf{x}) - D(\mathbf{x}))^2 \mu(d\mathbf{x}) \right)^{\frac{1+\alpha}{2+\alpha}}. \tag{8.15}$$

It is easy to see that (8.15) holds even under weak margin conditions. If D is Lipschitz continuous and $\mathbf{X}$ is bounded then there are regression estimates such that

$$\mathbf{E}\left\{ \int (D_n(\mathbf{x}) - D(\mathbf{x}))^2 \mu(d\mathbf{x}) \right\} \le c_2^2 n^{-\frac{2}{d+2}},$$

therefore for the corresponding plug in rule g_n, (8.15) implies that

$$\mathbf{E}\{R(g_n)\} - R^* \le \left(c_2^2 n^{-\frac{2}{d+2}} \right)^{\frac{1+\alpha}{2+\alpha}} = \left(c_2^2 n^{-\frac{1+\alpha}{d+2}} \right)^{\frac{2}{2+\alpha}}.$$

Kohler and Krzyzak [14] proved that for the standard local majority classification rules (partitioning, kernel, nearest neighbor) we get that the order of the upper bound can be smaller:

$$n^{-\frac{1+\alpha}{d+2}}. \tag{8.16}$$

For the time being, there is a fast estimate of the Bayes error probability R^*. One can estimate R^* using data splitting as follows: For the data $\mathcal{D}_{2n} = \{(\mathbf{X}_1, Y_1), ..., (\mathbf{X}_{2n}, Y_{2n})\}$ let g_n be a pattern recognition rule based on the first n observations $\mathcal{D}_n = \{(\mathbf{X}_1, Y_1), ..., (\mathbf{X}_n, Y_n)\}$ (training data). An estimate of R^* can be obtained by the relative frequency of errors for g_n on the remaining samples (test data). More precisely, we consider the estimate

$$\bar{R}_{2n} = \frac{1}{n} \sum_{i=n+1}^{2n} I_{\{g_n(\mathbf{X}_i) \ne Y_i\}}.$$

Obviously, the rate of convergence of $\bar{R}_{2n}$ to R^* depends on the quality of the classification rule g_n. If D is Lipschitz continuous and $\mathbf{X}$ is bounded then (8.16) implies that

$$\mathbf{E}\{|\bar{R}_{2n} - R^*|\} \le \mathbf{E}\{|\bar{R}_{2n} - R(g_n)|\} + \mathbf{E}\{R(g_n)\} - R^* \le n^{-1/2} + c_3 n^{-\frac{1+\alpha}{d+2}}.$$

How to estimate R^* with faster rate of convergence is an open problem.

8.5   Tests for the Classification Problem

For pattern recognition, the dimension reduction is to detect the ineffective components of the feature vector. As before, let $\mathbf{X}^{(-k)}$ be the $d-1$ dimensional feature vector such that we leave out the k-th component from $\mathbf{X}$. Then the corresponding Bayes error probability is

$$R^{*(-k)} := \mathbf{E}\{\min[\mathbf{E}\{Y \mid \mathbf{X}^{(-k)}\}, 1 - \mathbf{E}\{Y \mid \mathbf{X}^{(-k)}\}]\}.$$

Our aim is to test the hypothesis

$$\mathcal{H}_k : R^{*(-k)} = R^*,$$

which means that leaving out the k-th component the Bayes error probability does not increase. The hypothesis $\mathcal{H}_k$ means that

$$\begin{aligned}
&\mathbf{E}\{\min[\mathbf{E}\{Y \mid \mathbf{X}^{(-k)}\}, 1 - \mathbf{E}\{Y \mid \mathbf{X}^{(-k)}\}]\} \\
&= \mathbf{E}\{\min[\mathbf{E}\{Y \mid \mathbf{X}\}, 1 - \mathbf{E}\{Y \mid \mathbf{X}\}]\}.
\end{aligned}$$

Obviously, if

$$m(\mathbf{X}) = \mathbf{E}\{Y \mid \mathbf{X}\} = \mathbf{E}\{Y \mid \mathbf{X}^{(-k)}\} =: m^{(-k)}(\mathbf{X}^{(-k)}) \qquad (8.17)$$

a.s. then $\mathcal{H}_k$ is satisfied. The function min is not strictly concave, therefore the reverse is not true, i.e., $\mathcal{H}_k$ does not imply (8.17). We have no solution for this testing problem. Instead we modify the hypothesis $\mathcal{H}_k$ such that the Bayes error probability is replaced by the asymptotic error probability of the first nearest neighbor classification rule:

$$R_{NN} = \mathbf{E}\{\mathbf{E}\{Y \mid \mathbf{X}\}(1 - \mathbf{E}\{Y \mid \mathbf{X}\})\}.$$

(cf. Cover, Hart [4]). Because of

$$R_{NN} = \mathbf{E}\{Y\} - \mathbf{E}\{\mathbf{E}\{Y \mid \mathbf{X}\}^2\},$$

the modified hypothesis is defined by

$$\mathbf{E}\{m(\mathbf{X})^2\} = \mathbf{E}\{m^{(-k)}(\mathbf{X}^{(-k)})^2\}. \qquad (8.18)$$

The hypotheses (8.7) and (8.18) are the same, therefore all results of subsections 2, 3, 4 can be applied for the more restrictive hypothesis (8.18) such that here Y is binary valued.

8.6 Conclusions

In this chapter we introduced different statistics, in order to identify if some components of the independent vector $\mathbf{X}$ are ineffective. Since we could only simulate the asymptotic behavior of the test statistic, the complete chain of proof remains an open topic for future works. It would also be interesting to find other strategies to avoid the splitting of the sample.

Acknowledgments

This work was partially supported by the European Union and the European Social Fund through project FuturICT.hu (grant no.: TAMOP-4.2.2.C-11 /1 /KONV-2012-0013).

Bibliography

[1] J-Y. Audibert and A. B. Tsybakov. Fast learning rates for plug-in classifiers. *Annals of Statistics*, 35:608–633, 2007.

[2] J. R. Blum, H. Chernoff, M. Rosenblatt, and H. Teicher. Central limit theorems for interexchangeable processes. *Canadian J. of Mathematics*, 10:222–2229, 1958.

[3] K. De Brabanter and L. Györfi. Feature Selection via Detecting Ineffective Features. In *International Workshop on Advances in Regularization, Optimization, Kernel Methods and Support Vector Machines: Theory and Applications*. July 8–10, 2013, Leuven, Belgium. ftp://inmmic.org/sista/kdebaban/13-21.pdf

[4] T. M. Cover and P. E. Hart. Nearest neighbor pattern classification. *IEEE Transactions on Information Theory*, IT-13:21–27, 1967.

[5] L. Devroye, P. G. Ferrario, L. Györfi, and H. Walk. Strong universal consistent estimate of the minimum mean squared error. In *Empirical Inference - Festschrift in Honor of Vladimir N. Vapnik*, ed. by Bernhard Schölkopf, Zhiyuan Luo, and Vladimir Vovk, Springer, Heidelberg.

[6] L. Devroye, L. Györfi, and G. Lugosi. *A Probabilistic Theory of Pattern Recognition*. Springer–Verlag, New York, 1996.

[7] L. Devroye, D. Schäfer, L. Györfi, and H. Walk. The estimation problem of minimum mean squared error. *Statistics and Decisions*, 21:15–28, 2003.

[8] S. Dudoit and M.J. van der Laan. Asymptotics of cross-validated risk estimation in estimator selection and performance assessment. *Statistical Methodology*, 2:131–154, 2005.

[9] B. Efron and C. Stein. The jackknife estimate of variance. *Annals of Statistics*, 9:586–596, 1981.

[10] P.G. Ferrario and H. Walk. Nonparametric partitioning estimation of residual and local variance based on first and second nearest neighbors. *Journal of Nonparametric Statistics*, 24:1019–1039, 2012.

[11] L. Györfi, M. Kohler, A. Krzyżak, and H. Walk. *A Distribution-Free Theory of Nonparametric Regression*. Springer, New York, 2002.

[12] H. Koepke and M. Bilenko. Fast prediction of new feature utility. http://arxiv.org/ftp/arxiv/papers/1206/1206.4680.pdf

[13] M. Kohler. Nonparametric regression with additional measurement errors in the dependent variable. *Journal of Statistical Planning and Inference*, 136:3339–3361, 2006.

[14] M. Kohler and A. Krżyzak. On the rate of convergence of local averaging plug-in classification rules under a margin condition. *IEEE Transactions on Information Theory*, 53:1735–1742, 2007.

[15] E. Liitiäinen, F. Corona, and A. Lendasse. On nonparametric residual variance estimation. *Neural Processing Letters*, 28:155–167, 2009.

[16] E. Liitiäinen, F. Corona, and A. Lendasse. Residual variance estimation using a nearest neighbor statistic. *Journal of Multivariate Analysis*, 101:811–823, 2010.

[17] E. Liitiäinen, M. Verleysen, F. Corona, and A. Lendasse. Residual variance estimation in machine learning. *Neurocomputing*, 72:3692–3703, 2009.

[18] E. Mammen and A. B. Tsybakov. Smooth discrimination analysis. *Annals of Statistics*, 27:1808–1829, 1999.

[19] H.-G. Müller and U. Stadtmüller. Estimation of heteroscedasticity in regression analysis. *Annals of Statistics*, 15:610–625, 1987.

[20] U. Müller, A. Schick, and W. Wefelmeyer. Estimating the error variance in nonparametric regression by a covariate-matched U-statistic. *Statistics*, 37:179–188, 2003.

[21] M.-H. Neumann. Fully data-driven nonparametric variance estimators. *Statistics*, 25:189–212, 1994.

[22] N. Neumeyer and I. Van Keilegom. Estimating the error distribution in nonparametric multiple regression with applications to model testing. *Journal of Multivariate Analysis*, 101:1067–1078, 2010.

[23] K. Pelckmans, J. De Brabanter, J.A.K. Suykens, and B. De Moor. The differogram: Non-parametric noise variance estimation and its use for model selection. *Neurocomputing*, 69:100–122, 2005.

[24] U. Stadtmüller and A. Tsybakov. Nonparametric recursive variance estimation. *Statistics*, 27:55–63, 1995.

[25] C.J. Stone. Consistent nonparametric regression. *Annals of Statistics*, 5:595–645, 1977.

[26] A. B. Tsybakov. Optimal aggregation of classifiers in statistical learning. *Annals of Statistics*, 32:135–166, 2004.

[27] N. C. Weber. A martingale approach to central limit theorems for exchangeable random variables. *Journal of Applied Probability*, 17:662–673, 1980.

Chapter 9

Quadratic Basis Pursuit

Henrik Ohlsson

Department of Electrical Engineering and Computer Sciences, University of California, Berkeley

Allen Y. Yang

Department of Electrical Engineering and Computer Sciences, University of California, Berkeley

Roy Dong

Department of Electrical Engineering and Computer Sciences, University of California, Berkeley

Michel Verhaegen

Delft Center for Systems and Control, Delft University

S. Shankar Sastry

Department of Electrical Engineering and Computer Sciences, University of California, Berkeley

9.1   Introduction

Consider the problem of finding the sparsest signal $\mathbf{x}$ satisfying a system of linear equations:

$$
\begin{aligned}
\min_{\mathbf{x}\in\Re^n} \quad & \|\mathbf{x}\|_0 \\
\text{subj. to} \quad & y_i = \boldsymbol{b}_i^{\mathsf{T}}\mathbf{x}, \quad y_i \in \Re,\ \boldsymbol{b}_i \in \Re^n,\ i = 1,\dots,N.
\end{aligned}
\tag{9.1}
$$

This problem is known to be combinatorial and NP-hard [45]. A number of approaches to approximate its solution have been proposed. One of the most well known approaches is to relax the zero norm and replace it with the ℓ_1-norm:

$$
\min_{\mathbf{x}\in\Re^n} \|\mathbf{x}\|_1 \quad \text{subj. to} \quad y_i = \boldsymbol{b}_i^{\mathsf{T}}\mathbf{x}, \quad i = 1,\dots,N.
\tag{9.2}
$$

This approach is often referred to as *basis pursuit* (BP) [19].

The ability to recover the optimal solution to (9.1) is essential in the theory of *compressive sensing* (CS) [16, 24]. A tremendous amount of work has been dedicated to solving and analyzing the solution of (9.1) and (9.2) in the last decade. Today CS is regarded as a powerful tool in signal processing and widely used in many applications. For a detailed review of the literature, the reader is referred to several recent publications such as [10, 27].

It has recently been shown that CS can be extended to nonlinear models. More specifically, the relatively new topic of *nonlinear compressive sensing* (NLCS) deals with a more general problem of finding the sparsest signal $\mathbf{x}$ to a nonlinear set of equations:

$$
\begin{aligned}
\min_{\mathbf{x}\in\Re^n} \quad & \|\mathbf{x}\|_0 \\
\text{subj. to} \quad & y_i = f_i(\mathbf{x}), \quad y_i \in \Re,\ i = 1,\dots,N,
\end{aligned}
\tag{9.3}
$$

where each $f_i : \Re^n \to \Re$ is a continuously differentiable function. Compared to CS, the literature on NLCS is still very limited. The interested reader is referred to [3, 7] and references therein.

In this chapter, we will restrict our attention from general nonlinear systems. Instead, we focus on nonlinearities that depends quadratically on the unknown $\mathbf{x}$. More specifically, we consider the following problem formulated in the complex domain:

$$
\begin{aligned}
\min_{\mathbf{x}\in\mathbb{C}^n} \quad & \|\mathbf{x}\|_0 \\
\text{subj. to} \quad & y_i = a_i + \boldsymbol{b}_i^{\mathsf{H}}\mathbf{x} + \mathbf{x}^{\mathsf{H}}\boldsymbol{c}_i + \mathbf{x}^{\mathsf{H}}\boldsymbol{Q}_i\mathbf{x}, \\
& i = 1,\dots,N,
\end{aligned}
\tag{9.4}
$$

where $a_i \in \mathbb{C}$, $\boldsymbol{b}_i, \boldsymbol{c}_i \in \mathbb{C}^n$, $y_i \in \mathbb{C}$, $\boldsymbol{Q}_i \in \mathbb{C}^{n\times n}$, $i = 1,\dots,N$, and H denotes the conjugate transpose. In a sense, being able to solve (9.4) would make it possible to apply the principles of CS to a second-order Taylor expansion of the nonlinear relationship in (9.3), while traditional CS mainly considers its linear approximation or first-order Taylor expansion. In particular, in the most simple case, when a second order Taylor expansion is taken around zero (i.e., a Maclaurin expansion), let $a_i = f_i(0)$, $b_i = c_i = \nabla_{\mathbf{x}}^{\mathsf{T}} f_i(0)/2$ and $\mathbf{Q}_i = \nabla_{\mathbf{x}}^2 f_i(0)/2$, $i = 1, \ldots, N$, with $\nabla_{\mathbf{x}}$ and $\nabla_{\mathbf{x}}^2$ denoting the gradient and Hessian with respect to $\mathbf{x}$. In this case, $\mathbf{Q}$ is a Hermitian matrix. Nevertheless, we note that our derivations in the chapter do not depend on the matrix $\mathbf{Q}$ to be symmetric in the real domain or Hermitian in the complex domain.

In another motivating example, we consider the well-known phase retrieval problem in x-ray crystallography; see for instance [36, 33, 31, 29, 41, 2]. The underlying principle of x-ray crystallography is that the information about the crystal structure can be obtained from its diffraction pattern by hitting the crystal by an x-ray beam. Due to physical limitations, typically only the intensity of the diffraction pattern can be measured but not its phase. This leads to a nonlinear relation

$$y_i = |\mathbf{a}_i^{\mathsf{H}} \mathbf{x}|^2 = \mathbf{x}^{\mathsf{H}} \mathbf{a}_i \mathbf{a}_i^{\mathsf{H}} \mathbf{x}, \quad i = 1, \ldots, N, \tag{9.5}$$

between the measurements $y_1, \ldots, y_N \in \Re$ and the structural information contained in $\mathbf{x} \in \mathbb{C}^n$. The complex vectors $\mathbf{a}_1, \ldots, \mathbf{a}_N \in \mathbb{C}^n$ are known. The mathematical problem of recovering $\mathbf{x}$ from $y_1, \ldots, y_N$, and $\mathbf{a}_1, \ldots, \mathbf{a}_N$ is referred to as the *phase retrieval* (PR) problem. The traditional phase retrieval problem is known to be combinatorial [17].

If $\mathbf{x}$ is sparse under an appropriate basis in (9.5) and the measurements are under-sampled, the problem is referred to as *compressive phase retrieval* (CPR) in [44, 49] or *quadratic compressed sensing* (QCS) in [53]. Compressive phase retrieval is of relevance to several important imaging applications, such as diffraction imaging [11], astronomy [21, 30], optics [57], x-ray tomography [23], microscopy [43, 1, 55], and quantum mechanics [20], to mention a few. As we will later show, our solution as a convex relaxation of (9.4), called *quadratic basis pursuit* (QBP), can be readily applied to solve this type of problem, namely, let $a_i = \mathbf{b}_i = \mathbf{c}_i = 0$, $\mathbf{Q}_i = \mathbf{a}_i \mathbf{a}_i^{\mathsf{H}}$, $i = 1, \ldots, N$.

9.1.1 Literature Review

Overall, the literature on nonlinear sparse problems and NLCS is very limited. One of the first papers discussing these topics is [8]. The authors presented a greedy gradient based algorithm for estimating the sparsest solution to a general nonlinear equation system. A greedy approach was also proposed in [38] for the estimation of sparse solutions of nonlinear equation systems. The work of [3] proposed several iterative hard-thresholding and sparse simplex pursuit algorithms. As the algorithms are nonconvex greedy solutions, the analysis of the theoretical convergence only concerns their local behavior. In [7] and [26], the authors also considered a generalization of the *restricted isom-*

etry property (RIP) to support the use of similar iterative hard-thresholding algorithms and orthogonal least squares for solving general NLCS problems.

The work presented in this chapter is inspired by several recent papers on CS applied to the phase retrieval problem [44, 42, 18, 53, 49, 50, 55, 35, 51, 52]. In particular, the generalization of compressive sensing to phase retrieval was first proposed in [44]. In [53], the problem was also referred to as QCS. These methods typically do not consider a general quadratic constraint as in (9.4) but a pure quadratic form (i.e., $a_i = b_i = c_i = 0$, $i = 1, \ldots, N$, in (9.4)).

In terms of the numerical algorithms that solve the CPR problem, most of the existing methods are greedy algorithms, where a solution to the underlying non-convex problem is sought by a sequence of local decisions [44, 42, 53, 50, 55, 52]. In particular, the QCS algorithm in [53] used a *lifting* technique similar to that in [54, 40, 46, 32] and *iterative rank minimization* resulting in a series of semidefinite programs (SDPs) that would converge to a local optimum.

The first work that applied the lifting technique to the PR and CPR problems was presented in [18]. Extensions of similar techniques were also studied in [39, 35]. The methods presented in our previous publications [48, 49] were also based on the lifting technique. It is important to note that the algorithms proposed in [18, 48, 49] are non-greedy global solutions, which are different from the previous local solutions [44, 53]. The technique presented in this chapter was inspired by the solutions to phase retrieval via low-rank approximation in [18, 17, 13]. Given an oversampled phase retrieval problem, a lifting technique was used to relax the nonconvex problem with an SDP. The authors of [17, 13] also derived an upper-bound for the sampling rate that guarantees exact recovery in the noise-free case and stable recovery in the noisy case. Nevertheless, the work in [17, 13] only addressed the oversampled phase retrieval problem but not CPR or NLCS. The only similarities between our work and theirs are the lifting technique and convex relaxation. This lifting technique has also been used in other topics to convert nonconvex quadratic problems to SDPs; see for instance [56, 35]. The work presented in [18] and our previous contributions [48, 49] only discussed the CPR problem.

Finally, in [51], a *message passing* algorithm similar to that in CS was proposed to solve the compressive phase retrieval problem. The work in [28, 47] further considered stability and uniqueness in real phase retrieval problems. CPR has also been shown useful in practice and we refer the interested reader to [44, 55] for two very nice contributions. We find the work presented in [55] especially fascinating where the authors show how CPR can be used to facilitate sub-wavelength imaging in microscopy.

9.1.2   Notation

In this chapter, we will use boldface to denote vectors and matrices and normal font for scalars. $\boldsymbol{X}_{i,j}$ is used to denote the (i, j)th element, $\boldsymbol{X}_{i,:}$ the ith row, and $\boldsymbol{X}_{:,j}$ the jth column of a matrix $\boldsymbol{X}$, respectively. We will use the notation $\boldsymbol{X}_{i_1:i_2,j_1:j_2}$ to denote a submatrix constructed from rows i_1 to

i_2 and columns j_1 to j_2 of X. Given two matrices X and Y, we use the following fact that their product in the trace function commutes, namely, $\text{Tr}(XY) = \text{Tr}(YX)$, under the assumption that the dimensions match. $\|\cdot\|_0$ counts the number of nonzero elements in a vector or matrix; similarly, $\|\cdot\|_1$ denotes the element-wise ℓ_1-norm of a vector or matrix, i.e., the sum of the magnitudes of the elements; whereas $\|\cdot\|$ represents the ℓ_2-norm for vectors and the Frobenius norm for matrices. For a quadratic matrix X, $X \succeq 0$ is used to denote that X is positive semi-definite.

9.2 Quadratic Basis Pursuit

9.2.1 Convex Relaxation via Lifting

As optimizing the ℓ_0-norm function in (9.4) is known to be a combinatorial problem, in this section, we first introduce a convex relaxation of (9.4).

It is easy to see that the general quadratic constraint of (9.4) can be rewritten as the quadratic form:

$$y_i = \begin{bmatrix} 1 & \mathbf{x}^H \end{bmatrix} \begin{bmatrix} a_i & b_i^H \\ c_i & Q_i \end{bmatrix} \begin{bmatrix} 1 \\ \mathbf{x} \end{bmatrix} \in \mathbb{C}, \quad i = 1, \dots, N. \tag{9.6}$$

Since each y_i is a scalar, we further have

$$y_i = \text{Tr}\left(\begin{bmatrix} 1 & \mathbf{x}^H \end{bmatrix} \begin{bmatrix} a_i & b_i^H \\ c_i & Q_i \end{bmatrix} \begin{bmatrix} 1 \\ \mathbf{x} \end{bmatrix} \right) \tag{9.7}$$

$$= \text{Tr}\left(\begin{bmatrix} a_i & b_i^H \\ c_i & Q_i \end{bmatrix} \begin{bmatrix} 1 \\ \mathbf{x} \end{bmatrix} \begin{bmatrix} 1 & \mathbf{x}^H \end{bmatrix} \right). \tag{9.8}$$

Define $\boldsymbol{\Phi}_i = \begin{bmatrix} a_i & b_i^H \\ c_i & Q_i \end{bmatrix}$ and $X = \begin{bmatrix} 1 \\ \mathbf{x} \end{bmatrix} \begin{bmatrix} 1 & \mathbf{x}^H \end{bmatrix}$, both matrices of dimensions $(n+1) \times (n+1)$. The operation that constructs X from the vector $\begin{bmatrix} 1 \\ \mathbf{x} \end{bmatrix}$ is known as the *lifting* operator [54, 40, 46, 32]. By definition, X is a Hermitian matrix, and it satisfies the constraints that $X_{1,1} = 1$ and $\text{rank}(X) = 1$. Hence, (9.4) can be rewritten as

$$\begin{aligned} \min_X \quad & \|X\|_0 \\ \text{subj. to} \quad & y_i = \text{Tr}(\boldsymbol{\Phi}_i X), \quad i = 1, \dots, N, \\ & \text{rank}(X) = 1, X_{1,1} = 1, X \succeq 0. \end{aligned} \tag{9.9}$$

When the optimal solution X^* is found, the unknown $\mathbf{x}$ can be obtained by the rank-1 decomposition of X^* via *singular value decomposition* (SVD).

The above problem is still non-convex and combinatorial. Therefore, solving it for any moderate size of n is impractical. Inspired by recent literature on matrix completion [15, 18, 17, 13] and sparse PCA [22], we relax the problem into the following convex non-smooth *semidefinite program* (SDP):

$$\begin{aligned} \min_{\boldsymbol{X}} \quad & \mathrm{Tr}(\boldsymbol{X}) + \lambda \|\boldsymbol{X}\|_1 \\ \text{subj. to} \quad & y_i = \mathrm{Tr}(\boldsymbol{\Phi}_i \boldsymbol{X}), \quad i = 1, \dots, N, \\ & \boldsymbol{X}_{1,1} = 1, \ \boldsymbol{X} \succeq 0, \end{aligned} \tag{9.10}$$

where $\lambda \geq 0$ is a design parameter. In particular, the trace of $\boldsymbol{X}$ is a convex surrogate of the low-rank condition and $\|\boldsymbol{X}\|_1$ is the well-known convex surrogate for $\|\boldsymbol{X}\|_0$ in (9.9). We refer to the approach as *quadratic basis pursuit* (QBP).

One can further consider a noisy counterpart of the QBP problem, where some deviation between the measurements and the estimates is allowed. More specifically, we propose the following *quadratic basis pursuit denoising* (QBPD) problem:

$$\begin{aligned} \min_{\boldsymbol{X}} \quad & \mathrm{Tr}(\boldsymbol{X}) + \lambda \|\boldsymbol{X}\|_1 \\ \text{subj. to} \quad & \sum_i^N \left(y_i - \mathrm{Tr}(\boldsymbol{\Phi}_i \boldsymbol{X}) \right)^2 \leq \varepsilon, \\ & \boldsymbol{X}_{1,1} = 1, \ \boldsymbol{X} \succeq 0, \end{aligned} \tag{9.11}$$

for some $\varepsilon > 0$.

9.2.2   Theoretical Analysis

In this section, we highlight some theoretical results derived for QBP. The analysis follows that of CS, and is inspired by derivations given in [17, 16, 18, 24, 12, 4, 10]. For further analysis on special cases of QBP and its noisy counterpart QBPD, please refer to [49].

First, it is convenient to introduce a linear operator B:

$$B : \boldsymbol{X} \in \mathbb{C}^{n \times n} \mapsto \{\mathrm{Tr}(\boldsymbol{\Phi}_i \boldsymbol{X})\}_{1 \leq i \leq N} \in \mathbb{C}^N. \tag{9.12}$$

We consider a generalization of the *restricted isometry property* (RIP) of the linear operator B.

Definition 9.1 (RIP). *A linear operator $B(\cdot)$ as defined in (9.12) is (ϵ, k)-RIP if*

$$\left| \frac{\|B(\boldsymbol{X})\|^2}{\|\boldsymbol{X}\|^2} - 1 \right| < \epsilon \tag{9.13}$$

for all $\|\boldsymbol{X}\|_0 \leq k$ and $\boldsymbol{X} \neq 0$.

We can now state the following theorem:

Theorem 9.1 (Recoverability/Uniqueness). *Let $\bar{\mathbf{x}} \in \mathbb{C}^n$ be a solution to (9.4). If $\boldsymbol{X}^* \in \mathbb{C}^{(n+1) \times (n+1)}$ satisfies $\mathbf{y} = B(\boldsymbol{X}^*)$, $\boldsymbol{X}^* \succeq 0$, $\mathrm{rank}(\boldsymbol{X}^*) = 1$, $\boldsymbol{X}_{1,1}^* = 1$ and if $B(\cdot)$ is a $(\epsilon, 2\|\boldsymbol{X}^*\|_0)$-RIP linear operator with $\epsilon < 1$ then $\boldsymbol{X}^*$ and $\bar{\mathbf{x}}$ are unique and $\boldsymbol{X}_{2:n+1,1}^* = \bar{\mathbf{x}}$.*

Proof. Assume the contrary, i.e., $X^*_{2:n+1,1} \neq \bar{\mathbf{x}}$ and hence that $X^* \neq \begin{bmatrix} 1 \\ \bar{\mathbf{x}} \end{bmatrix} \begin{bmatrix} 1 & \bar{\mathbf{x}}^H \end{bmatrix}$. It is clear that $\left\| \begin{bmatrix} 1 \\ \bar{\mathbf{x}} \end{bmatrix} \begin{bmatrix} 1 & \bar{\mathbf{x}}^H \end{bmatrix} \right\|_0 \leq \|X^*\|_0$ and hence $\left\| \begin{bmatrix} 1 \\ \bar{\mathbf{x}} \end{bmatrix} \begin{bmatrix} 1 & \bar{\mathbf{x}}^H \end{bmatrix} - X^* \right\|_0 \leq 2\|X^*\|_0$. Since $\left\| \begin{bmatrix} 1 \\ \bar{\mathbf{x}} \end{bmatrix} \begin{bmatrix} 1 & \bar{\mathbf{x}}^H \end{bmatrix} - X^* \right\|_0 \leq 2\|X^*\|_0$, we can apply the RIP inequality (9.13) to $\begin{bmatrix} 1 \\ \bar{\mathbf{x}} \end{bmatrix} \begin{bmatrix} 1 & \bar{\mathbf{x}}^H \end{bmatrix} - X^*$. If we use that $\mathbf{y} = B(X^*) = B\left(\begin{bmatrix} 1 \\ \bar{\mathbf{x}} \end{bmatrix} \begin{bmatrix} 1 & \bar{\mathbf{x}}^H \end{bmatrix} \right)$ and hence $B\left(\begin{bmatrix} 1 \\ \bar{\mathbf{x}} \end{bmatrix} \begin{bmatrix} 1 & \bar{\mathbf{x}}^H \end{bmatrix} - X^* \right) = 0$, we are led to the contradiction $1 < \epsilon$. We therefore conclude that $X^* = \begin{bmatrix} 1 \\ \bar{\mathbf{x}} \end{bmatrix} \begin{bmatrix} 1 & \bar{\mathbf{x}}^H \end{bmatrix}$, $X^*_{2:n+1,1} = \bar{\mathbf{x}}$, and that X^* and $\bar{\mathbf{x}}$ are unique. □

We can also give a bound on the sparsity of $\bar{\mathbf{x}}$:

Theorem 9.2 (Bound on $\|\bar{\mathbf{x}}\|_0$ from above). *Let $\bar{\mathbf{x}}$ be the sparsest solution (or one of the solutions if the sparsest solution is not unique) to (9.4) and let $\tilde{X}$ be a solution of QBP (9.10). If $\tilde{X}$ has rank 1 then $\|\tilde{X}_{2:n+1,1}\|_0 \geq \|\bar{\mathbf{x}}\|_0$.*

Proof. Let $\tilde{X}$ be a rank-1 solution of QBP (9.10). By contradiction, assume $\|\tilde{X}_{2:n+1,1}\|_0 < \|\bar{\mathbf{x}}\|_0$. Since $\tilde{X}_{2:n+1,1}$ satisfies the constraints of (9.4), it is a feasible solution of (9.4). As assumed, $\tilde{X}_{2:n+1,1}$ also gives a lower objective value than $\bar{\mathbf{x}}$ in (9.4). This is a contradiction since $\bar{\mathbf{x}}$ was assumed to be a solution of (9.4). Hence we must have that $\|\tilde{X}_{2:n+1,1}\|_0 \geq \|\bar{\mathbf{x}}\|_0$. □

The following result now holds trivially:

Corollary 9.1 (Guaranteed recovery using RIP). *Let $\bar{\mathbf{x}}$ be the sparsest solution to (9.4). The solution of QBP $\tilde{X}$ is equal to $\begin{bmatrix} 1 \\ \bar{\mathbf{x}} \end{bmatrix} \begin{bmatrix} 1 & \bar{\mathbf{x}}^H \end{bmatrix}$ if it has rank 1 and $B(\cdot)$ is $(\epsilon, 2\|\tilde{X}\|_0)$-RIP with $\epsilon < 1$.*

Proof. This follows trivially from Theorem 9.1 by realizing that $\tilde{X}$ satisfy all properties of X^*. □

Given the RIP analysis, it may be that the linear operator $B(\cdot)$ does not satisfy the RIP property defined in Definition 9.1 with a small enough ϵ, as pointed out in [17]. In these cases, RIP-1 may be considered:

Definition 9.2 (RIP-1). *A linear operator $B(\cdot)$ is (ϵ, k)-RIP-1 if*

$$\left| \frac{\|B(X)\|_1}{\|X\|_1} - 1 \right| < \epsilon \tag{9.14}$$

for all matrices $X \neq 0$ and $\|X\|_0 \leq k$.

Theorems 9.1–9.2 and Corollary 9.1 all hold with RIP replaced by RIP-1 and will not be restated in detail here. Instead, we summarize the most important property in the following theorem:

Theorem 9.3 (Upper bound and recoverability using RIP-1). *Let $\bar{\mathbf{x}}$ be the sparsest solution to (9.4). The solution of QBP (9.10), $\tilde{\mathbf{X}}$, is equal to $\begin{bmatrix} 1 \\ \bar{\mathbf{x}} \end{bmatrix} \begin{bmatrix} 1 & \bar{\mathbf{x}}^H \end{bmatrix}$ if it has rank 1 and $B(\cdot)$ is $(\epsilon, 2\|\tilde{\mathbf{X}}\|_0)$-RIP-1 with $\epsilon < 1$.*

Proof. The proof follows trivially from the proof of Theorem 9.1.   □

The RIP-type argument may be difficult to check for a given matrix and is more useful for claiming results for classes of matrices/linear operators. For instance, it has been shown that random Gaussian matrices satisfy the RIP with high probability. However, given realization of a random Gaussian matrix, it is indeed difficult to check if it actually satisfies the RIP. Two alternative arguments are the *spark condition* [19] and the *mutual coherence* [25, 14]. The spark condition usually gives tighter bounds but is known to be difficult to compute as well. On the other hand, mutual coherence may give less tight bounds, but is more tractable. We will focus on mutual coherence, which is defined as:

Definition 9.3 (Mutual coherence). *For a matrix $\mathbf{A}$, define the mutual coherence as*

$$\mu(\mathbf{A}) = \max_{1 \leq i,j \leq n, i \neq j} \frac{|\mathbf{A}_{:,i}^H \mathbf{A}_{:,j}|}{\|\mathbf{A}_{:,i}\| \|\mathbf{A}_{:,j}\|}. \tag{9.15}$$

Let $\mathbf{B}$ be the matrix satisfying $\mathbf{y} = \mathbf{B}\mathbf{X}^s = B(\mathbf{X})$ with $\mathbf{X}^s$ being the vectorized version of $\mathbf{X}$. We are now ready to state the following theorem:

Theorem 9.4 (Recovery using mutual coherence). *Let $\bar{\mathbf{x}}$ be the sparsest solution to (9.4). The solution of QBP (9.10), $\tilde{\mathbf{X}}$, is equal to $\begin{bmatrix} 1 \\ \bar{\mathbf{x}} \end{bmatrix} \begin{bmatrix} 1 & \bar{\mathbf{x}}^H \end{bmatrix}$ if it has rank 1 and $\|\tilde{\mathbf{X}}\|_0 < 0.5(1 + 1/\mu(\mathbf{B}))$.*

Proof. It follows from [25] [10, Thm. 5] that if

$$\|\tilde{\mathbf{X}}\|_0 < \frac{1}{2}\left(1 + \frac{1}{\mu(\mathbf{B})}\right) \tag{9.16}$$

then $\tilde{\mathbf{X}}$ is the sparsest solution to $\mathbf{y} = B(\mathbf{X})$. Since $\begin{bmatrix} 1 \\ \bar{\mathbf{x}} \end{bmatrix} \begin{bmatrix} 1 & \bar{\mathbf{x}}^H \end{bmatrix}$ is by definition the sparsest rank 1 solution to $\mathbf{y} = B(\mathbf{X})$, it follows that $\tilde{\mathbf{X}} = \begin{bmatrix} 1 \\ \bar{\mathbf{x}} \end{bmatrix} \begin{bmatrix} 1 & \bar{\mathbf{x}}^H \end{bmatrix}$.   □

9.3   Numerical Algorithms

In addition to the above analysis of guaranteed recovery properties, a critical issue for practitioners is the efficiency of numerical solvers that can handle moderate-sized SDP problems. Several numerical solvers used in CS may be applied to solve nonsmooth SDPs, which include interior-point methods, e.g., those used in CVX [34], gradient projection methods [5], and augmented Lagrangian methods (ALM) [5]. However, interior-point methods are known to scale badly to moderate-sized convex problems in general. Gradient projection methods also fail to meaningfully accelerate QBP due to the complexity of the projection operator. Alternatively, nonsmooth SDPs can be solved by ALM. However, the augmented primal and dual objective functions are still SDPs, which are equally expensive to solve in each iteration. There also exist a family of iterative approaches, often referred to as *outer approximation methods*, that successively approximate the solution of an SDP by solving a sequence of linear programs (see [37]). These methods approximate the positive semidefinite cone by a set of linear constraints and refine the approximation in each iteration by adding a new set of linear constraints. However, we have experienced slow convergence using these types of methods. In summary, QBP as a nonsmooth SDP is categorically more expensive to solve compared to the linear programs in CS, and the task exceeds the capability of many popular sparse optimization techniques.

In this chapter, we propose an effective solver to the nonsmooth SDP underlying QBP via the *alternating directions method of multipliers* (ADMM, see for instance [9] and [6, Sec. 3.4]) technique. The motivation to use ADMM is two-fold:

1. It scales well to large data sets.

2. It is known for its fast convergence.

There are also a number of strong convergence results that further motivate the choice [9].

To set the stage for ADMM, let n denote the dimension of $\mathbf{x}$, and let N denote the number of measurements. Then, rewrite (9.10) to the equivalent SDP

$$\min_{\mathbf{X}_1,\mathbf{X}_2,\mathbf{Z}} \quad f_1(\mathbf{X}_1) + f_2(\mathbf{X}_2) + g(\mathbf{Z}),$$
$$\text{subj. to} \quad \mathbf{X}_1 - \mathbf{Z} = 0, \quad \mathbf{X}_2 - \mathbf{Z} = 0, \tag{9.17}$$

where $\mathbf{X}_1 = \mathbf{X}_1^{\mathsf{H}} \in \mathbb{C}^{(n+1)\times(n+1)}$, $\mathbf{X}_2 = \mathbf{X}_2^{\mathsf{H}} \in \mathbb{C}^{(n+1)\times(n+1)}$, $\mathbf{Z} = \mathbf{Z}^{\mathsf{H}} \in \mathbb{C}^{(n+1)\times(n+1)}$, and

$$f_1(\mathbf{X}) \triangleq \begin{cases} \mathrm{Tr}(\mathbf{X}) & \text{if } y_i = \mathrm{Tr}(\mathbf{\Phi}_i \mathbf{X}), \ i = 1,\dots,N \\ & \text{and } \mathbf{X}_{1,1} = 1 \\ \infty & \text{otherwise} \end{cases}$$

$$f_2(\mathbf{X}) \triangleq \begin{cases} 0 & \text{if } \mathbf{X} \succeq 0 \\ \infty & \text{otherwise} \end{cases}$$

$$g(\mathbf{Z}) \triangleq \lambda \|\mathbf{Z}\|_1.$$

Define two matrices $\mathbf{Y}_1$ and $\mathbf{Y}_2$ as the Lagrange multipliers of the two equality constraints in (9.17), respectively. Then the update rules of ADMM lead to the following:

$$\begin{aligned}
\boldsymbol{X}_i^{l+1} &= \arg\min_{\boldsymbol{X}=\boldsymbol{X}^{\mathsf{H}}} f_i(\boldsymbol{X}) + \mathrm{Tr}(\boldsymbol{Y}_i^l(\boldsymbol{X}-\boldsymbol{Z}^l)) \\
&\quad + \tfrac{\rho}{2}\|\boldsymbol{X}-\boldsymbol{Z}^l\|^2, \\
\boldsymbol{Z}^{l+1} &= \arg\min_{\boldsymbol{Z}=\boldsymbol{Z}^{\mathsf{H}}} g(\boldsymbol{Z}) + \sum_{i=1}^2 \mathrm{Tr}(\boldsymbol{Y}_i^l \boldsymbol{Z}) \\
&\quad + \tfrac{\rho}{2}\|\boldsymbol{X}_i^{l+1}-\boldsymbol{Z}\|^2, \\
\boldsymbol{Y}_i^{l+1} &= \boldsymbol{Y}_i^l + \rho(\boldsymbol{X}_i^{l+1}-\boldsymbol{Z}^{l+1}),
\end{aligned} \tag{9.18}$$

for $i = 1, 2$, where $\rho \geq 0$ is a parameter that enforces consensus between $\boldsymbol{X}_1$, $\boldsymbol{X}_2$, and $\boldsymbol{Z}$. Each of these steps has a tractable calculation. After some simple manipulations, we have:

$$\begin{aligned}
\boldsymbol{X}_1^{l+1} = \mathrm{argmin}_{\boldsymbol{X}=\boldsymbol{X}^{\mathsf{H}}} \quad & \|\boldsymbol{X} - (\boldsymbol{Z}^l - \tfrac{\boldsymbol{I}+\boldsymbol{Y}_1^l}{\rho})\|, \\
\text{subj. to} \quad & y_i = \mathrm{Tr}(\boldsymbol{\Phi}_i \boldsymbol{X}), \quad i = 1, \ldots, N, \\
& X_{1,1} = 1.
\end{aligned} \tag{9.19}$$

Let $\tilde{B} : \mathbb{C}^{(n+1)\times(n+1)} \to \mathbb{C}^{N+1}$ be the augmented linear operator such that $\tilde{B}(\boldsymbol{X}) = \begin{bmatrix} B(\boldsymbol{X}) \\ X_{1,1} \end{bmatrix}$, where B is the linear operator defined by (9.12). Assuming that a feasible solution exists, and defining $\Pi_{\tilde{B}}$ as the orthogonal projection onto the convex set given by the linear constraints, i.e., $\begin{bmatrix} \mathbf{y} \\ 1 \end{bmatrix} = \tilde{B}(\boldsymbol{X})$, the solution is: $\boldsymbol{X}_1^{l+1} = \Pi_{\tilde{B}}(\boldsymbol{Z}^l - \tfrac{\boldsymbol{I}+\boldsymbol{Y}_1^l}{\rho})$. This matrix-valued problem can be solved by converting the linear constraint on Hermitian matrices into an equivalent constraint on real-valued vectors.

Next,

$$\boldsymbol{X}_2^{l+1} = \mathrm{argmin}_{\boldsymbol{X}\succeq 0} \left\| \boldsymbol{X} - \left(\boldsymbol{Z}^l - \tfrac{\boldsymbol{Y}_2^l}{\rho}\right) \right\| = \Pi_{PSD}\left(\boldsymbol{Z}^l - \tfrac{\boldsymbol{Y}_2^l}{\rho}\right), \tag{9.20}$$

where Π_{PSD} denotes the orthogonal projection onto the positive-semidefinite cone, which can easily be obtained via eigenvalue decomposition.

Finally, let $\overline{\boldsymbol{X}}^{l+1} = \tfrac{1}{2}\sum_{i=1}^2 \boldsymbol{X}_i^{l+1}$ and similarly $\overline{\boldsymbol{Y}}^l$. Then, the $\boldsymbol{Z}$ update rule can be written:

$$\begin{aligned}
\boldsymbol{Z}^{l+1} &= \mathrm{argmin}_{\boldsymbol{Z}=\boldsymbol{Z}^{\mathsf{T}}} \lambda\|\boldsymbol{Z}\|_1 + \rho\|\boldsymbol{Z} - (\overline{\boldsymbol{X}}^{l+1} + \tfrac{\overline{\boldsymbol{Y}}^l}{\rho})\|^2 \\
&= \mathrm{soft}(\overline{\boldsymbol{X}}^{l+1} + \tfrac{\overline{\boldsymbol{Y}}^l}{\rho}, \tfrac{\lambda}{2\rho})
\end{aligned} \tag{9.21}$$

where $\mathrm{soft}(\cdot)$ in the complex domain is defined with respect to a positive real scalar q as:

$$\mathrm{soft}(x, q) = \begin{cases} 0 & \text{if } |x| \leq q, \\ \tfrac{|x|-q}{|x|}x & \text{otherwise.} \end{cases} \tag{9.22}$$

Note that if the first argument is a complex value, the soft operator is defined in terms of the magnitude rather than the sign and if it is a matrix, the soft operator acts element-wise.

Setting $l = 1, X_1^l = X_2^l = Z^l = I$, where I denotes the identity matrix, and $\rho^l = 1$, setting $l = 0$, the Hermitian matrices $X_i^{l+1}, Z_i^{l+1}, Y_i^{l+1}$ can now be iteratively computed using the ADMM iterations (9.18). The stopping criterion of the algorithm is given by (see for instance [9]):

$$\|r^l\| \leq n\epsilon^{abs} + \epsilon^{rel} \max(\|\overline{X}^l\|, \|Z^l\|), \tag{9.23}$$

$$\|s^l\| \leq n\epsilon^{abs} + \epsilon^{rel}\|\overline{Y}^l\|, \tag{9.24}$$

where $\epsilon^{abs}, \epsilon^{rel}$ are algorithm parameters set to 10^{-3} and r^l and s^l are the primal and dual residuals, respectively, as:

$$r^l = \begin{bmatrix} X_1^l - Z^l & X_2^l - Z^l \end{bmatrix}, \tag{9.25}$$

$$s^l = -\rho \begin{bmatrix} Z^l - Z^{l-1} & Z^l - Z^{l-1} \end{bmatrix}. \tag{9.26}$$

We also update ρ according to the rule discussed in [9]:

$$\rho^{l+1} = \begin{cases} \tau_{incr}\rho^l & \text{if } \|r^l\| > \mu\|s^l\|, \\ \rho^l/\tau_{decr} & \text{if } \|s^l\| > \mu\|r^l\|, \\ \rho^l & \text{otherwise}, \end{cases} \tag{9.27}$$

where τ_{incr}, τ_{decr}, and μ are algorithm parameters. Values commonly used are $\mu = 10$ and $\tau_{incr} = \tau_{decr} = 2$.

In terms of the computational complexity of the ADMM algorithm, its inner loop calculates the updates of X_i, Z, and Y_i, $i = 1, 2$. It is easy to see that its complexity is dominated by (9.19) and (9.20), which is bounded by $\mathcal{O}(n^2N^2 + n^3)$, while the calculation of Z and Y_i is linear with respect to the number of their elements.

9.4 Experiments

In this section, we provide comprehensive experiments to validate the efficacy of the QBP algorithms in solving several representative nonlinear CS problems that depend quadratically on the unknown. We compare their performance primarily with two existing algorithms. As we mentioned in Section 9.1, if an underdetermined nonlinear system is approximated up to the first order, the classical sparse solver in CS is basis pursuit. In NLCS literature, several greedy algorithms have been proposed for nonlinear systems. In this section, we choose to compare them with the *iterative hard thresholding* (IHT) algorithm in [3] in Section 9.4.1 and another greedy algorithm demonstrated in [55] in Section 9.4.3. Besides the comparisons shown here, we have also compared them to a number of CPR algorithms [44, 52]. Not surprisingly, they performed badly on the general quadratic problems since they do not account for the linear term.

9.4.1 Nonlinear Compressive Sensing in Real Domain

In this experiment, we illustrate the concept of nonlinear compressive sensing. Assume that there is a cost associated with sampling and that we would like to recover $\mathbf{z}_0 \in \Re^m$, related to our samples $y_i \in \Re$, $i = 1, \ldots, N$, via

$$y_i = f_i(\mathbf{z}_0), \quad i = 1, \ldots, N, \tag{9.28}$$

using as few samples as possible. Also, assume that there is a sparsifying basis $\mathbf{D} \in \Re^{m \times n}$, possibly overcomplete, such that

$$\mathbf{z}_0 = \mathbf{D}\mathbf{x}_0, \quad \text{with } \mathbf{x}_0 \text{ sparse.} \tag{9.29}$$

Hence, we have

$$y_i = f_i(\mathbf{D}\mathbf{x}_0), \quad i = 1, \ldots, N, \tag{9.30}$$

with $\mathbf{x}_0$ a sparse vector. If we approximate the nonlinear equation system (9.30) using a second order Maclaurin expansion we end up with a set of quadratic equations,

$$y_i = f_i(0) + \nabla f_i(0)\mathbf{D}\mathbf{x}_0 + \mathbf{x}_0^\mathsf{T}\mathbf{D}^\mathsf{T}\frac{\nabla^2 f_i(0)}{2}\mathbf{D}\mathbf{x}_0, \ i = 1, \ldots, N. \tag{9.31}$$

Hence, we can use QBP to recover $\mathbf{x}_0$ given $\{f_i(\mathbf{x}), y_i\}_{i=1}^N$ and $\mathbf{D}$.

In particular, let $\mathbf{D} = \mathbf{I}$, $n = m = 20$, $N = 25$, $f_i(\mathbf{x}) = a_i + \mathbf{b}_i^\mathsf{T}\mathbf{x} + \mathbf{x}^\mathsf{T}\mathbf{Q}_i\mathbf{x}$, $i = 1, \ldots, N$, and generate $\{y_i\}_{i=1}^N$ by sampling $\{a_i, \mathbf{b}_i, \mathbf{Q}_i\}_{i=1}^N$ from a unitary Gaussian distribution. Let $\mathbf{x}_0$ be a binary vector with three elements different than zero. Given $\{y_i, a_i, \mathbf{b}_i, \mathbf{Q}_i\}_{i=1}^N$, the task is now to recover $\mathbf{x}_0$. The results of this simulation are shown in Figure 9.1.

First, as the noiseless measurements are generated by a quadratic system of equations, it is not surprising that QBP perfectly recovers the sparse signal $\mathbf{x}_0$ when $\lambda = 50$. One may wonder whether in the 20-D ambient space, the solution $\mathbf{x}_0$ is unique. To show that the solution is not unique, we let $\lambda = 0$ and again apply QBP. As shown in Figure 9.1 (c), the solution is dense and it also satisfies the quadratic constraints. Therefore, we have verified that the system is underdetermined and there exist multiple solutions.

Second, in Figure 9.1 (d), we approximate (9.31) only up to the first order and set $\mathbf{Q}_i = 0, i = 1, \ldots, N$. The approximation enables us to employ the classical basis pursuit algorithm in CS to seek the best 3-sparse estimate $\mathbf{x}$. As expected, the approximation is not accurate enough, and the estimate is far from the ground truth.

Third, we implement the iterative hard thresholding (IHT) algorithm in

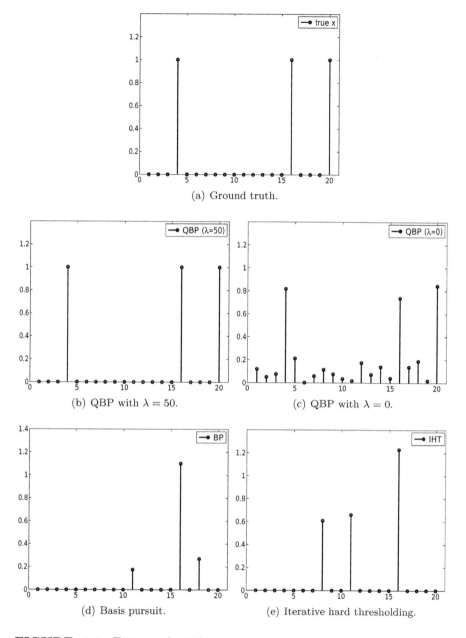

(a) Ground truth.

(b) QBP with $\lambda = 50$.

(c) QBP with $\lambda = 0$.

(d) Basis pursuit.

(e) Iterative hard thresholding.

FIGURE 9.1: Estimated 20-D sparse signals measured in a simulated quadratic system of equations. The QBP solution perfectly recovers the ground truth with $\lambda = 50$, while the remaining algorithms fail to recover the correct sparse coefficients.

[3], and the correct number of nonzero coefficients in $\mathbf{x}_0$ is also provided to the algorithm. Its estimate is given in Figure 9.1 (e). As IHT is a greedy algorithm, its performance is affected by the initialization. In Figure 9.1 (e), the initial value is set by $\mathbf{x} = 0$, and the estimate is incorrect.

Finally, we note that the advantage of using general CS theory is that fewer samples are needed to recover a source signal from its observations. This remains true for NLCS presented in this chapter. However, as (9.28) and (9.31) are nonlinear equation systems, typically $N \gg m$ measurements are required for recovering a unique solution. In the same simulation shown in Figure 9.1, one could ignore the sparsity constraint (i.e., by letting $\lambda = 0$ in Figure 9.1 (c)), and it would require $N' = 40$ observations for QBP to recover the unique solution, which is exactly the ground-truth signal.

Clearly, Figure 9.1 is only able to illustrate one set of simulation results. To more systematically demonstrate the accuracy of the four algorithms in probability, a Monte Carlo simulation is performed that repeats the above simulation but with different randomly generated $\mathbf{x}_0$ and $\{a_i, b_i, Q_i\}$. Table 9.1 shows the rates of successful recovery. We can see QBP achieves the highest success rate, which is followed by IHT. BP and the dense QBP solution basically fail to return enough good results. $\lambda = 50$ was used in all trials.

TABLE 9.1: The percentage of correctly recovering $\mathbf{x}_0$ in 100 trials.

Method	QBP ($\lambda = 50$)	QBP ($\lambda = 0$)	BP	IHT
Success rate	79%	5%	3%	54%

9.4.2   The Shepp-Logan Phantom

In this experiment, we consider recovery of images from random samples. More specifically, we formulate an example of the CPR problem in the QBP framework using the Shepp-Logan phantom. Our goal is to show that using the QBPD algorithm provides approximate solutions that are visually close to the ground-truth images.

Consider the ground-truth image in Figure 9.2. This 30×30 Shepp-Logan phantom has a 2D Fourier transform with 100 nonzero complex coefficients. We generate N linear combinations of pixels, and then measure the square of the measurements. This relationship can be written as:

$$\mathbf{y} = |\mathbf{Ax}|^2 = \{\mathbf{x}^H a_i a_i^H \mathbf{x}\}_{1 \leq i \leq N}, \tag{9.32}$$

where $\mathbf{A} = \mathbf{RF}$ is the concatenation of a random matrix $\mathbf{R}$ and the Fourier basis $\mathbf{F}$, and the image $\mathbf{Fx}$ is represented as a stacked vector in the 900-D complex domain. The CPR problem minimizes the following objective function:

$$\min_{\mathbf{x}} \|\mathbf{x}\|_1 \quad \text{subj. to} \quad \mathbf{y} = |\mathbf{Ax}|^2 \in \Re^N. \tag{9.33}$$

Previously, an SDP solution to the nonsparse phase retrieval problem was proposed in [17], which is called *PhaseLift*. In a sense, PhaseLift can be viewed as a special case of the QBP solution in (9.10) where $\lambda = 0$, namely, the sparsity constraint is not enforced. In Figure 9.2 (b), the recovered result using PhaseLift is shown with $N = 2400$.

To compare visually the performance of the QBP solution when the sparsity constraint is properly enforced, two recovered results are shown in Figure 9.2 (c) and (d) with $N = 2400$ and 1500, respectively. Note that the number of measurements with respect to the sparsity in $\mathbf{x}$ is too low for both QBP and PhaseLift to perfectly recover $\mathbf{x}$. Therefore, in this case, we employ the noisy version of the algorithm QBPD to recover the image. We can clearly see from the illustrations that QBPD provides a much better approximation and outperforms PhaseLift visually even though it uses considerably fewer measurements.

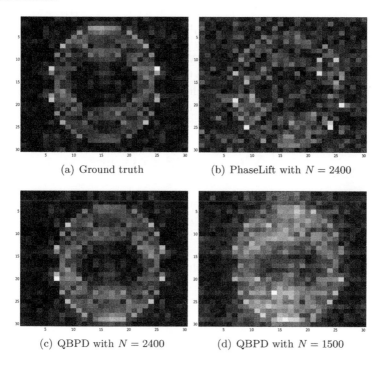

(a) Ground truth   (b) PhaseLift with $N = 2400$

(c) QBPD with $N = 2400$ (d) QBPD with $N = 1500$

FIGURE 9.2: Recovery of a Shepp-Logan image by PhaseLift and QBPD.

9.4.3 Subwavelength Imaging

In this example, we present an example in sub-wavelength coherent diffractive imaging. The experiment and the data collection were conducted by [55].

Let y_i, $i = 1, \ldots, N$, be intensity samples of a 2D diffraction pattern. The diffraction pattern is the result of a 532 nm laser beam passing through an arrangement of holes made on an opaque piece of glass. The task is to decide the location of the holes out of a number of possible locations.

It can be shown that the relation between the intensity measurements and the arrangements of holes is of the following type:

$$y_i = |a_i^H x|^2, \quad i = 1, \ldots, N, \tag{9.34}$$

where $y_i \in \mathfrak{R}$, $i = 1, \ldots, N$, are intensity measurements, $a_i \in \mathbb{C}^n$, $i = 1, \ldots, N$, are known complex vectors, and $x \in \mathfrak{R}^n$ is the sought entity, each element giving the likelihood of a hole at a given location.

We use QBPD with $\varepsilon = 0.0012$ and $\lambda = 100$. 89 measurements were selected by taking every 200th intensity measurement from the dataset of [55]. The quantity x is from the setup of the experiment known to be real and $a_i = b_i = c_i = 0$. We hence have

$$y_i = x^T Q_i x = |a_i^H x|^2, \quad i = 1, \ldots, N, \tag{9.35}$$

with $Q_i = a_i a_i^H \in \mathbb{C}^{n \times n}$, $i = 1, \ldots, N$, and $x \in \mathfrak{R}^n$.

The resulting estimate is given to the left in Figure 9.3. The result deviates from the ground truth and the result presented in [55] (shown in Figure 9.3 right), and it actually finds a more sparse pattern. It is interesting to note that both estimates are, however, within the noise level estimated in [55]:

$$\frac{1}{N} \sum_i^N (y_i - |a_i^H x|^2)^2 \leq 1.8 \times 10^{-6}. \tag{9.36}$$

Therefore, under the same noise assumptions, the two solutions are equally likely to lead to the same observations y. However, knowing that there is a solution within the noise level that is indeed sparser than the ground-truth pattern, it should *not* be the optimal solution to have recovered the ground truth, since there exists a sparser solution.

9.5   Conclusion

Classical compressive sensing assumes a linear relation between samples and the unknowns. The ability to more accurately characterize nonlinear models has the potential to improve the results in both existing compressive sensing applications and those where a linear approximation does not suffice, e.g., phase retrieval.

This chapter presents an extension of classical compressive sensing to quadratic relations or second order Taylor expansions of the nonlinearity relating measurements and the unknowns. The novel extension is based on lifting

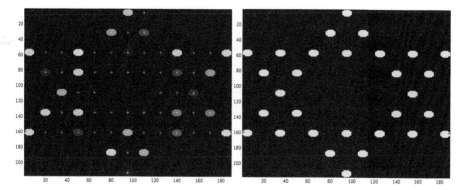

FIGURE 9.3: The estimated sparse vector **x**. The crosses mark possible positions for holes, while the dots represent the recovered nonzero coefficients. **Left:** Recovered pattern by QBPD. Note that this estimate is sparser than the ground truth but within the estimated noise level. **Right:** Recovered pattern by the compressive phase retrieval method used in [55].

and convex relaxations and the final formulation takes the form of a SDP. The proposed method, quadratic basis pursuit, inherits properties of basis pursuit and classical compressive sensing, and conditions for perfect recovery, etc., are derived. We also give an efficient numerical implementation.

Acknowledgments

The authors would like to acknowledge useful discussions and inputs from Yonina C. Eldar, Mordechai Segev, Laura Waller, Filipe Maia, Stefano Marchesini, and Michael Lustig. We also want to acknowledge the authors of [55] for kindly sharing their data with us.

Ohlsson is partially supported by the Swedish Research Council in the Linnaeus center CADICS, the European Research Council under the advanced grant LEARN, contract 267381, by a postdoctoral grant from the Sweden-America Foundation, donated by ASEA's Fellowship Fund, and by a postdoctoral grant from the Swedish Research Council. Yang is supported in part by ARO 63092-MA-II and DARPA FA8650-11-1-7153. Dong is supported by an NSF Graduate Research Fellowship under grant DGE 1106400 TRUST (Team for Research in Ubiquitous Secure Technology) which receives support from NSF (award number CCF-0424422).

Bibliography

[1] J. Antonello, M. Verhaegen, R. Fraanje, T. van Werkhoven, H. C. Gerritsen, and C. U. Keller. Semidefinite programming for model-based sensorless adaptive optics. *J. Opt. Soc. Am. A*, 29(11):2428–2438, November 2012.

[2] R. Balan, P. Casazza, and D. Edidin. On signal reconstruction without phase. *Applied and Computational Harmonic Analysis*, 20:345–356, 2006.

[3] A. Beck and Y. C. Eldar. Sparsity constrained nonlinear optimization: Optimality conditions and algorithms. Technical Report arXiv:1203.4580, 2012.

[4] R. Berinde, A. Gilbert, P. Indyk, H. Karloff, and M. Strauss. Combining geometry and combinatorics: A unified approach to sparse signal recovery. In *Communication, Control, and Computing, 2008 46th Annual Allerton Conference on*, pages 798–805, September 2008.

[5] D. P. Bertsekas. *Nonlinear Programming*. Athena Scientific, 1999.

[6] D. P. Bertsekas and J. N. Tsitsiklis. *Parallel and Distributed Computation: Numerical Methods*. Athena Scientific, 1997.

[7] T. Blumensath. Compressed sensing with nonlinear observations and related nonlinear optimization problems. Technical Report arXiv:1205.1650, 2012.

[8] T. Blumensath and M. E. Davies. Gradient pursuit for non-linear sparse signal modelling. In *European Signal Processing Conference*, Lausanne, Switzerland, April 2008.

[9] S. Boyd, N. Parikh, E. Chu, B. Peleato, and J. Eckstein. Distributed optimization and statistical learning via the alternating direction method of multipliers. *Foundations and Trends in Machine Learning*, 3(1), January 2011.

[10] A. Bruckstein, D. Donoho, and M. Elad. From sparse solutions of systems of equations to sparse modeling of signals and images. *SIAM Review*, 51(1):34–81, 2009.

[11] O. Bunk, A. Diaz, F. Pfeiffer, C. David, B. Schmitt, D. K. Satapathy, and J. F. van der Veen. Diffractive imaging for periodic samples: retrieving one-dimensional concentration profiles across microfluidic channels. *Acta Crystallographica Section A*, 63(4):306–314, July 2007.

[12] E. Candès. The restricted isometry property and its implications for compressed sensing. *Comptes Rendus Mathematique*, 346(9–10):589–592, 2008.

[13] E. Candès, Y. C. Eldar, T. Strohmer, and V. Voroninski. Phase retrieval via matrix completion. Technical Report arXiv:1109.0573, Stanford University, September 2011.

[14] E. Candès, X. Li, Y. Ma, and J. Wright. Robust Principal Component Analysis? *Journal of the ACM*, 58(3), 2011.

[15] E. Candès and B. Recht. Exact matrix completion via convex optimization. *CoRR*, abs/0805.4471, 2008.

[16] E. Candès, J. Romberg, and T. Tao. Robust uncertainty principles: Exact signal reconstruction from highly incomplete frequency information. *IEEE Transactions on Information Theory*, 52:489–509, February 2006.

[17] E. Candès, T. Strohmer, and V. Voroninski. PhaseLift: Exact and stable signal recovery from magnitude measurements via convex programming. Technical Report arXiv:1109.4499, Stanford University, September 2011.

[18] A. Chai, M. Moscoso, and G. Papanicolaou. Array imaging using intensity-only measurements. Technical report, Stanford University, 2010.

[19] S. Chen, D. Donoho, and M. Saunders. Atomic decomposition by basis pursuit. *SIAM Journal on Scientific Computing*, 20(1):33–61, 1998.

[20] J.V. Corbett. The Pauli problem, state reconstruction and quantum-real numbers. *Reports on Mathematical Physics*, 57(1):53–68, 2006.

[21] J.C. Dainty and J.R. Fienup. Phase retrieval and image reconstruction for astronomy. In editor H. Stark, editor, *Image Recovery: Theory and Application*. Academic Press, New York, 1987.

[22] A. d'Aspremont, L. El Ghaoui, M. Jordan, and G. Lanckriet. A direct formulation for Sparse PCA using semidefinite programming. *SIAM Review*, 49(3):434–448, 2007.

[23] M. Dierolf, A. Menzel, P. Thibault, P. Schneider, C. M. Kewish, R. Wepf, O. Bunk, and F. Pfeiffer. Ptychographic x-ray computed tomography at the nanoscale. *Nature*, 467:436–439, 2010.

[24] D. Donoho. Compressed sensing. *IEEE Transactions on Information Theory*, 52(4):1289–1306, April 2006.

[25] D. Donoho and M. Elad. Optimally sparse representation in general (nonorthogonal) dictionaries via ℓ_1-minimization. *PNAS*, 100(5):2197–2202, March 2003.

[26] M. Ehler, M. Fornasier, and J. Sigl. Quasi-Linear Compressed Sensing. *ArXiv e-prints*, November 2013.

[27] Y. C. Eldar and G. Kutyniok. *Compresed Sensing: Theory and Applications*. Cambridge University Press, 2012.

[28] Y. C. Eldar and S. Mendelson. Phase Retrieval: Stability and Recovery Guarantees. *ArXiv e-prints*, November 2012.

[29] J. Fienup. Phase retrieval algorithms: a comparison. *Applied Optics*, 21(15):2758–2769, 1982.

[30] J. R. Fienup, J. C. Marron, T. J. Schulz, and J. H. Seldin. Hubble space telescope characterized by using phase-retrieval algorithms. *Applied Optics*, 32(10):1747–1767, Apr 1993.

[31] R. Gerchberg and W. Saxton. A practical algorithm for the determination of phase from image and diffraction plane pictures. *Optik*, 35:237–246, 1972.

[32] M. X. Goemans and D. P. Williamson. Improved approximation algorithms for maximum cut and satisfiability problems using semidefinite programming. *J. ACM*, 42(6):1115–1145, November 1995.

[33] R. Gonsalves. Phase retrieval from modulus data. *Journal of Optical Society of America*, 66(9):961–964, 1976.

[34] M. Grant and S. Boyd. CVX: Matlab software for disciplined convex programming, version 1.21. `http://cvxr.com/cvx`, August 2010.

[35] K. Jaganathan, S. Oymak, and B. Hassibi. Recovery of Sparse 1-D Signals from the Magnitudes of their Fourier Transform. *ArXiv e-prints*, June 2012.

[36] D. Kohler and L. Mandel. Source reconstruction from the modulus of the correlation function: a practical approach to the phase problem of optical coherence theory. *Journal of the Optical Society of America*, 63(2):126–134, 1973.

[37] H. Konno, J. Gotoh, T. Uno, and A. Yuki. A cutting plane algorithm for semi-definite programming problems with applications to failure discriminant analysis. *Journal of Computational and Applied Mathematics*, 146(1):141–154, 2002.

[38] L. Li and B. Jafarpour. An iteratively reweighted algorithm for sparse reconstruction of subsurface flow properties from nonlinear dynamic data. *CoRR*, abs/0911.2270, 2009.

[39] X. Li and V. Voroninski. Sparse Signal Recovery from Quadratic Measurements via Convex Programming. *ArXiv e-prints*, September 2012.

[40] L. Lovász and A. Schrijver. Cones of matrices and set-functions and 0-1 optimization. *SIAM Journal on Optimization*, 1:166–190, 1991.

[41] S. Marchesini. Phase retrieval and saddle-point optimization. *Journal of the Optical Society of America A*, 24(10):3289–3296, 2007.

[42] S. Marchesini. Ab Initio Undersampled Phase Retrieval. *Microscopy and Microanalysis*, 15, July 2009.

[43] J. Miao, T. Ishikawa, Q. Shen, and T. Earnest. Extending x-ray crystallography to allow the imaging of noncrystalline materials, cells, and single protein complexes. *Annual Review of Physical Chemistry*, 59(1):387–410, 2008.

[44] M. Moravec, J. Romberg, and R. Baraniuk. Compressive phase retrieval. In *SPIE International Symposium on Optical Science and Technology*, 2007.

[45] B. K. Natarajan. Sparse approximate solutions to linear systems. *SIAM Journal on Computing*, 24(2):227–234, 1995.

[46] Y. Nesterov. Semidefinite relaxation and nonconvex quadratic optimization. *Optimization Methods & Software*, 9:141–160, 1998.

[47] H. Ohlsson and Y. C. Eldar. On conditions for uniqueness in sparse phase retrieval. *CoRR*, abs/1308.5447, 2013.

[48] H. Ohlsson, A. Yang, R. Dong, and S. S. Sastry. CPRL — an extension of compressive sensing to the phase retrieval problem. In P. Bartlett, F.C.N. Pereira, C.J.C. Burges, L. Bottou, and K.Q. Weinberger, editors, *Advances in Neural Information Processing Systems 25*, pages 1376–1384. 2012.

[49] H. Ohlsson, A. Y. Yang, R. Dong, and S. Sastry. Compressive Phase Retrieval From Squared Output Measurements Via Semidefinite Programming. Technical Report arXiv:1111.6323, University of California, Berkeley, November 2011.

[50] E. Osherovich, Y. Shechtman, A. Szameit, P. Sidorenko, E. Bullkich, S. Gazit, S. Shoham, E.B. Kley, M. Zibulevsky, I. Yavneh, Y.C. Eldar, O. Cohen, and M. Segev. Sparsity-based single-shot subwavelength coherent diffractive imaging. In *2012 Conference on Lasers and Electro-Optics (CLEO)*, San Jose, CA, USA, May 2012.

[51] P. Schniter and S. Rangan. Compressive phase retrieval via generalized approximate message passing. In *Proceedings of Allerton Conference on Communication, Control, and Computing*, Monticello, IL, USA, October 2012.

[52] Y. Shechtman, A. Beck, and Y. C. Eldar. GESPAR: Efficient Phase Retrieval of Sparse Signals. *ArXiv e-prints*, January 2013.

[53] Y. Shechtman, Y. C. Eldar, A. Szameit, and M. Segev. Sparsity based sub-wavelength imaging with partially incoherent light via quadratic compressed sensing. *Opt. Express*, 19(16):14807–14822, Aug 2011.

[54] N.Z. Shor. Quadratic optimization problems. *Soviet Journal of Computer and Systems Sciences*, 25:1–11, 1987.

[55] A. Szameit, Y. Shechtman, E. Osherovich, E. Bullkich, P. Sidorenko, H. Dana, S. Steiner, E. B. Kley, S. Gazit, T. Cohen-Hyams, S. Shoham, M. Zibulevsky, I. Yavneh, Y. C. Eldar, O. Cohen, and M. Segev. Sparsity-based single-shot subwavelength coherent diffractive imaging. *Nature Materials*, 11(5):455–459, May 2012.

[56] I. Waldspurger, A. d'Aspremont, and S. Mallat. Phase Recovery, MaxCut and Complex Semidefinite Programming. *ArXiv e-prints*, June 2012.

[57] A. Walther. The question of phase retrieval in optics. *Optica Acta*, 10:41–49, 1963.

Chapter 10

Robust Compressive Sensing

Esa Ollila

Aalto University

Hyon-Jung Kim

Aalto University

Visa Koivunen

Aalto University

10.1 Introduction

The *compressed sensing (CS)* or *sparse signal reconstruction/approximation (SSR)* is a signal processing technique that exploits the fact that acquired data can have a sparse representation in some basis. It allows for solving underdetermined systems of equations. The problem can be formulated as follows [8, 7, 5]. Let $\mathbf{y} = (y_1, \ldots, y_M)^\top$ denote the observed data vector (*measurements*) modelled as

$$\mathbf{y} = \mathbf{\Phi}\mathbf{x} + \boldsymbol{\varepsilon} \tag{10.1}$$

$$\text{i.e., } y_i = \boldsymbol{\phi}_i^\top \mathbf{x} + \varepsilon_i, \quad i = 1, \ldots, M \tag{10.2}$$

where $\mathbf{\Phi} = \begin{pmatrix} \boldsymbol{\phi}_1 & \cdots & \boldsymbol{\phi}_M \end{pmatrix}^\top$ is $M \times N$ *measurement matrix* with more column vectors than row vectors $\boldsymbol{\phi}_i$ (i.e., $N > M$), $\mathbf{x} = (x_1, \ldots, x_N)^\top$ is the unobserved *signal vector*, and $\boldsymbol{\varepsilon} = (\varepsilon_1, \ldots, \varepsilon_M)^\top$ is the (unobserved) *random noise* vector. It is assumed that the signal vector $\mathbf{x}$ is *K-sparse* (i.e., it has K

non-zero elements) or is *compressible* (i.e., it has a representation whose entries decay rapidly when sorted in order of decreasing magnitude). Note that compressible signals are well approximated by K-sparse signals and typically in many applications $K \ll N$. Let us denote the signal *support* (i.e., the locations of non-zero elements) as $\Gamma = \text{supp}(\mathbf{x}) = \{j : x_j \neq 0\}$. The objective is then to reconstruct or approximate the signal vector $\mathbf{x}$ by K-sparse representation knowing only the acquired vector $\mathbf{y}$, the measurement matrix $\boldsymbol{\Phi}$, and the sparsity K.

One approach for finding a K-sparse estimate of $\mathbf{x}$ is by solving the following optimization problem

$$\hat{\mathbf{x}}^* = \arg \min_{\mathbf{x}} \|\mathbf{y} - \boldsymbol{\Phi}\mathbf{x}\|_2^2 \quad \text{subject to} \quad \|\mathbf{x}\|_0 \leq K \qquad (10.3)$$

where $\| \cdot \|_0$ denotes the ℓ_0 pseudo-norm, $\|\mathbf{x}\|_0 = \#\{j : x_j \neq 0\}$, and $\| \cdot \|_p$ for $p \geq 1$ denotes the usual ℓ_p-norm, $\|\mathbf{x}\|_p = \left(\sum_{i=1}^N |x_i|^p\right)^{1/p}$. This optimization problem is known to be NP-hard and hence suboptimal strategies have been under active research; see [8] for a review. These methods, such as *Compressive Sampling Matching Pursuit (CoSaMP)* [14] or *Iterative Hard Thresholding (IHT)* [2, 3] are guaranteed to perform very well provided that suitable conditions (e.g., restricted isometry property on $\boldsymbol{\Phi}$ and non impulsive noise conditions) are met. It is important to notice that the derived recovery bounds depend linearly on $\|\boldsymbol{\varepsilon}\|_2$ and thus the methods are not guaranteed to provide accurate reconstruction/approximation under heavy-tailed (spiky) non-Gaussian noise. An alternative approach to enforce sparsity of the solution is to deploy ℓ_1 constraint on the signal, i.e., $\|\mathbf{x}\|_1 \leq \delta$, as is done in the celebrated *LASSO* method [18].

Despite the vast interest in CS/SSR during the past decade, *sparse and robust signal reconstruction methods*, i.e., methods that can provide accurate sparse reconstruction even in heavy-tailed non-Gaussian noise conditions or in the face of outliers (gross errors), have been considered in the literature only recently. In [6], the authors robustify the iterative hard thresholding (IHT) approach for K-sparse signal reconstruction by replacing the ℓ_2 norm on the residuals (reconstruction error) by the Lorentzian pseudo-norm; this leads to a slightly modified algorithm referred to as *Lorentzian IHT (LIHT)* hereafter. In [16], the authors use the ℓ_0-regularized least absolute deviation (LAD) regression model and propose an approximate algorithm that utilizes weighted median regression. This method is computationally demanding and also involves tuning parameters whose selection is not an easy task. In our earlier work in [17], robust versions of the CoSaMP, IHT and orthogonal matching pursuit (OMP) [19] algorithms were proposed. These methods were based on replacing the non-robust ℓ_2-norm of the residuals by robust loss functions (i.e., which downweights large residuals) that are commonly used in *robust statistics* [11, 13], leading to modifications of the respective algorithms. See also our earlier work in [12] in which we proposed robust and sparse ℓ_1-regularized estimation of tensor decompositions and devised an alternating LS-type algorithm for their computations.

In this chapter, we propose a new and novel IHT method based on joint estimation of the unknown parameters of the system model, namely the K-sparse or compressible signal $\mathbf{x}$ and the scale parameter σ of the error distribution. The chapter is organized as follows. First, Section 10.2 provides a review of robust M-estimation approach to regression with particular emphasis put on different robust loss and objective functions. We shall then adopt these approaches for obtaining (constrained) sparse and robust estimates of $\mathbf{x}$ in the CS system model using the IHT technique. Section 10.3 describes the IHT method [2, 3] and Section 10.4 its robustification proposed in our earlier paper [17] that is based on preliminary estimate of the scale. This method, referred to as generalized IHT, utilizes robust objective functions arising from M-estimation of regression reviewed in Section 10.2. The disadvantage of this method is that it requires a preliminary (auxiliary) robust estimate of the scale parameter σ of the error distribution. Section 10.4 provides the main contribution of this chapter, describing in detail the new robust IHT method that is based on joint estimation of signal $\mathbf{x}$ and scale σ. This method is also described in a recent unpublished manuscript [15]. Section 10.4 provides extensive simulation studies illustrating the effectiveness and usefulness of the proposed method in reconstructing a K-sparse signal in various noise conditions and SNR settings.

Notations: For a vector $\mathbf{a} \in \mathbb{R}^m$, a matrix $\mathbf{A} \in \mathbb{R}^{n \times m}$, and an index set $\Gamma = (\gamma_1, \ldots, \gamma_p)$, a_i denotes the ith component of $\mathbf{a}$, $\mathbf{a}_i$ denotes the ith column vector of $\mathbf{A}$, and $\mathbf{a}_\Gamma$ denotes the p-vector of $\mathbf{a}$ with elements a_{γ_i} selected according to the support set Γ. Similarly $\mathbf{A}_\Gamma = (\mathbf{a}_{\gamma_1} \cdots \mathbf{a}_{\gamma_p})$ is an $n \times p$ matrix whose columns are selected from the columns of $\mathbf{A}$ according to the index set Γ. Notation $=_d$ is read "has the same distributions as".

10.2 Robust Regression and Robust Loss Functions

We assume that the noise terms ε_i are independent and identically distributed (i.i.d.) random variables from a continuous symmetric distribution and let $\sigma > 0$ denote the *scale parameter* of the error distribution. The density of ε_i is then $f(e) = (1/\sigma)f_0(e/\sigma)$, where $f_0(\cdot)$ denotes the standard form of the density, e.g., $f_0(e) = (1/\sqrt{2\pi}) \exp(-\frac{1}{2}e^2)$ in case of normal (Gaussian) error distribution. Let us denote residuals for a given (candidate) signal vector $\mathbf{x}$ as

$$e_i \equiv e_i(\mathbf{x}) = y_i - \boldsymbol{\phi}_i^\top \mathbf{x}$$

and write $\mathbf{e} \equiv \mathbf{e}(\mathbf{x}) = (e_1, \ldots, e_M)^\top = \mathbf{y} - \boldsymbol{\Phi}\mathbf{x}$ for the vector residuals. Let us first point out that if $M > N$ and sparse approximation of $\mathbf{x}$ is not looked for (unconstrained problem), then (10.1) is just a conventional *regression model*. In this section we review robust regression approach (namely, M-estimation) with particular emphasis put on different robust loss and objective functions and their properties. See [11, 13] for a more detailed overview of M-estimation of regression parameters.

Recall that the least squares (LS-) estimator of regression minimizes the sum of squares of the residuals. A common approach to obtaining a robust estimator of regression parameters is then to replace the LS (ℓ_2-)loss function $\rho(e) = \frac{1}{2}e^2$ with a *robust loss function* that downweights large residuals, thus reducing the influence (effect, impact) of gross errors in the obtained solution. Suppose for a moment that we have a preliminary estimate of the scale parameter σ or that the scale parameter is known; in both cases, we denote this (estimated or known) value of σ by $\hat{\sigma}$. Then the *objective function approach* for obtaining a robust M-estimator $\hat{\mathbf{x}}$ of $\mathbf{x}$ is to solve the optimization problem

$$\hat{\mathbf{x}} = \arg\min_{\mathbf{x}} \sum_{i=1}^{M} \rho\left(\frac{y_i - \boldsymbol{\phi}_i^\top \mathbf{x}}{\hat{\sigma}}\right) \tag{10.4}$$

where ρ is continuous, even function and increasing for $e \geq 0$. An *M-estimating equation approach* to robust regression is to find $\hat{\mathbf{x}}$ that solves

$$\sum_{i=1}^{M} \psi\left(\frac{y_i - \boldsymbol{\phi}_i^\top \mathbf{x}}{\hat{\sigma}}\right) \boldsymbol{\phi}_i = \mathbf{0} \tag{10.5}$$

where ψ is continuous and odd function ($\psi(-e) = -\psi(e)$). Naturally when $\psi = \rho'$, a stationary point of the objective function in (10.4) is a solution to (10.5). Moreover, if ρ is a convex function, then solving (10.4) is equivalent to solving (10.5) and vice versa, i.e., they are equivalent characterizations of the problem. If the distribution of the errors is known, then $\rho(e) = -\log f_0(e) + c$ is the *optimal loss function*, where $f_0(e)$ denotes the standard form of the error density. In this case, i.e., in the terminology of maximum likelihood (ML-) estimation, ψ-function is called as score function; we adopt this terminology and refer to ψ function as the *score function*. Below we review two commonly used robust loss functions, namely Huber's and Cauchy loss functions. We note that in both cases there also exists a noise distribution for which these functions are also optimal loss functions.

Huber's loss function combines ℓ_2 and ℓ_1 loss functions. Recall that the optimal loss function when errors follow a Gaussian distribution, i.e., $\varepsilon_i \sim \mathcal{N}(0, \sigma^2)$, is the ℓ_2 loss function $\rho(r) = \frac{1}{2}e^2$ whereas the optimal loss function for Laplacian (double exponential) errors, $\varepsilon_i \sim Lap(0, \sigma)$, is the ℓ_1-loss function $\rho(e) = |e|$. The corresponding $\psi = \rho'$ functions are $\psi(e) = e$ and $\psi(e) = \text{sign}(e)$ (the latter being "pseudo-derivative" since $|e|$ is not differentiable at $e = 0$). Huber's loss function is then defined as

$$\rho_{\mathrm{H}}(e) = \begin{cases} \frac{1}{2}e^2, & \text{for } |e| \leq c \\ c|e| - \frac{1}{2}c^2, & \text{for } |e| > c \end{cases} \tag{10.6}$$

where c is a user-defined *tuning constant* that affects robustness and efficiency of the method. For example, the following choices of constant c,

$$c_1 = 1.345 \quad \text{and} \quad c_2 = 0.732 \tag{10.7}$$

yield 95 and 85 percent (asymptotic) relative efficiency compared to LS-estimator of regression in the case of Gaussian errors. These constants will also be used in the sequel in the simulations for the Huber IHT method introduced in Section 10.5. Huber's loss function is also an optimal loss function in the case that the errors follow the so-called (Huber's) *least favourable distribution*, i.e., a symmetric unimodal distribution with smallest Fisher Information within a "neighborhood" of the normal distribution, which can be characterized as being Gaussian in the middle and double exponential in the tails. The corresponding score function $\psi = \rho'$ is a *winsorizing (clipping, trimming) function*

$$\psi_H(e) = \max[-c, \min(c, e)] = \begin{cases} e, & \text{for } |e| \leq c \\ c\,\text{sign}(e), & \text{for } e > c. \end{cases}$$

Thus the smaller the c, the more severe is the trimming (clipping) of the residuals. Huber's ρ and ψ-functions are depicted in Figure 10.1.

Cauchy loss function is the optimal loss function when the errors follow the Cauchy distribution, $\varepsilon_i \sim Cau(0, \sigma)$; it is given by

$$\rho_C(e) = \tfrac{1}{2}\log(1 + e^2). \tag{10.8}$$

Note that Cauchy loss function is differentiable but non-convex function. Hence solving the optimization problem (10.4) is a non-convex optimization problem, and finding a global solution remains an open problem. The corresponding $\psi = \rho'$ function is

$$\psi_C(e) = \frac{e}{1 + e^2}.$$

The fact the Cauchy loss function is "more robust" than Huber's loss, i.e., downweighting the residuals more rigidly, is attested by the fact that ψ is redescending to zero. Hence the "scores" of very large residuals can be zero or close to zero and hence will not have much effect on the solution.

10.3 Iterative Hard Thresholding (IHT)

Iterative hard thresholding (IHT) has been proposed and extensively studied for sparse signal recovery by Blumensath and Davies [1, 2, 3]. Let the initial value of iteration be $\mathbf{x}^0 = \mathbf{0}$. Then the *IHT algorithm* iterates

$$\mathbf{x}^{n+1} = H_K(\mathbf{x}^n + \mu\mathbf{\Phi}^\top\mathbf{e}^n), \tag{10.9}$$

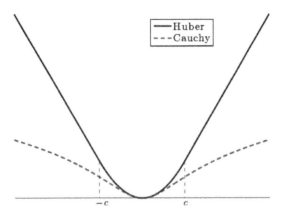

(a) Loss functions $\rho(e)$

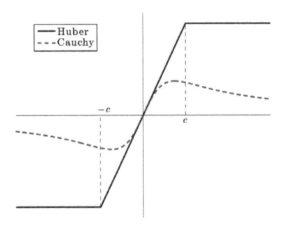

(b) Score functions $\psi(e)$

FIGURE 10.1: Huber's and Cauchy ρ and ψ functions. Unlike the Huber's loss function, the Cauchy loss function is non-convex. Cauchy loss function on the other hand can handle very spiky or large residuals. This can be seen from the Cauchy score function, which redescends to zero.

where $\mathbf{e}^n = \mathbf{y} - \mathbf{\Phi}\mathbf{x}^n$ denote the residual (error) vector at nth iteration, $\mu > 0$ denotes a stepsize, and $H_K(\cdot)$ denotes the *hard thresholding operator* that sets all but the largest (in magnitude) K elements of its vector-valued argument to zero. In [1] it was shown that the iterations converge to a local minimum of the optimization problem (10.3) given that $\|\mathbf{\Phi}\|_2 \le 1$ and $\mathbf{\Phi}$ has full rank. Later in [2], the authors derived performance guarantees of the algorithm based on the *restricted isometry property (RIP)* [4] of matrix $\mathbf{\Phi}$ and showed that the algorithm will reach a neighborhood of the best K-term approximation. Since the recovery bound [2, c.f., Theorems 4 & 5] is linearly dependent on the ℓ_2-norm of the noise vector $\|\boldsymbol{\varepsilon}\|_2$, it is clear that the IHT method will not be robust or provide accurate reconstruction in heavy-tailed noise (in such cases $\|\mathbf{e}_2\|$ will be very large). Later in [2], the authors proposed an improvement of the method that computes an optimal stepsize update (in terms of reduction in squared approximation error) at each iteration. As its benefits, the resulting *normalized IHT* algorithm has a faster convergence rate. Another important advantage is that the requirement $\|\mathbf{\Phi}\|_2 \le 1$ is avoided.

Stepsize selection. Let Γ^n denote the support set of $\mathbf{x}^n$ at nth iteration. The stepsize is updated as

$$\mu^n = \frac{\mathbf{g}_{\Gamma^n}^\top \mathbf{g}_{\Gamma^n}}{\mathbf{g}_{\Gamma^n}^\top \mathbf{\Phi}_{\Gamma^n}^\top \mathbf{\Phi}_{\Gamma^n} \mathbf{g}_{\Gamma^n}}, \qquad (10.10)$$

where $\mathbf{g} = \mathbf{\Phi}^\top \mathbf{e}^n$ is the gradient update at nth iteration. If $\Gamma^{n+1} \ne \Gamma^n$, then this stepsize might not be optimal and does not guarantee convergence. For guaranteed convergence, one requires that

$$\mu^n \le (1-c)\frac{\|\mathbf{x}^{n+1} - \mathbf{x}^n\|_2^2}{\|\mathbf{\Phi}(\mathbf{x}^{n+1} - \mathbf{x}^n)\|_2^2} = \omega^n$$

for some $c > 0$. Then, if $\Gamma^{n+1} \ne \Gamma^n$, one calculates ω^n and if $\mu^n > \omega^n$, one sets $\mu^n \leftarrow \mu^n/((1-c)\kappa)$ for some $\kappa > 1/(1-c)$. Then a new proposal is calculated using (10.9) and the value of ω^n is updated. The process terminates when $\mu^n < \omega^n$ in which case the latest proposal is accepted and one continues with the next iteration.

10.4 Robust IHT Based on Preliminary Estimate of Scale

We now describe the robust IHT method proposed in our earlier work [17]. Let $\hat{\sigma}$ denote a preliminary estimate of the scale parameter of the error distribution. Instead of using the LS-loss on the residuals, we look for the minimum of the problem

$$\hat{\sigma}^2 \sum_{i=1}^{M} \rho\left(\frac{y_i - \boldsymbol{\phi}_i^\top \mathbf{x}}{\hat{\sigma}}\right) \quad \text{subject to} \quad \|\mathbf{x}\|_0 \le K \qquad (10.11)$$

where the multiplying factor σ^2 is used so that the optimization problem (10.11) coincides with (10.3) when ρ is the LS-loss function $\rho(e) = \frac{1}{2}e^2$. As earlier, let us denote the residual vector by $\mathbf{e} = \mathbf{y} - \mathbf{\Phi x}$ and let $J(\mathbf{x}) = \hat{\sigma}^2 \sum_{i=1}^{M} \rho\left(\frac{y_i - \boldsymbol{\phi}_i^\top \mathbf{x}}{\hat{\sigma}}\right)$ denote the objective function in (10.11). Now the gradient descent update is

$$\mathbf{g} = -\nabla J(\mathbf{x}) = \hat{\sigma} \sum_{i=!}^{M} \psi\left(\frac{e_i}{\hat{\sigma}}\right)\boldsymbol{\phi}_i = \mathbf{\Phi}^\top \mathbf{e}_\psi$$

where $\psi = \rho'$ and

$$\mathbf{e}_\psi = \psi\left(\frac{\mathbf{e}}{\hat{\sigma}}\right)\hat{\sigma} \tag{10.12}$$

denotes the vector of *pseudo-residuals*. Above the notation is such that for a vector $\mathbf{e}$, $\psi(\mathbf{e}) = (\psi(e_i), \ldots, \psi(e_M))^\top$, i.e., a co-ordinatewise application of the function ψ on the elements of vector $\mathbf{e}$. As an example, when the LS loss function $\rho(e) = \frac{1}{2}e^2$ is used, then $\mathbf{e}_\psi = \mathbf{e}$. On the other hand, when Huber's loss function is used, $\mathbf{e}_\psi$ is simply a vector of Winsorized (clipped, trimmed) residuals. Again let the initial value of iteration be $\mathbf{x}^0 = \mathbf{0}$. Then the *generalized IHT* algorithm iterates

$$\mathbf{x}^{n+1} = H_K(\mathbf{x}^n + \mu\mathbf{\Phi}^\top \mathbf{e}_\psi^n), \tag{10.13}$$

where μ is the stepsize and $\mathbf{e}_\psi^n = \psi(\mathbf{e}^n/\hat{\sigma})\hat{\sigma}$ denotes the pseudo-residual vector at nth iteration. Finally, we note that one can use a similar idea as in normalized IHT to find an approximate value of the optimal stepsize at each iteration. The stepsize calculation of the robust IHT will be discussed in detail in Section 10.5.

Computation of the preliminary scale parameter $\hat{\sigma}$. If robust loss functions are used then a preliminary scale estimate $\hat{\sigma}$ is needed. Even if one chooses an optimal loss function for the noise distribution (e.g., Huber loss function when the errors follow the least favorable distribution), the performance of the method deteriorates if the scale parameter is badly estimated. For example, over-estimation of $\hat{\sigma}$ can result in the case that pseudo-residuals in (10.12) are not Winsorized (clipped) at all, or, Winsorized too much if $\hat{\sigma}$ is severely under-estimated. Obtaining a robust estimate of scale $\hat{\sigma}$ is a very difficult problem. A first step is to find a robust initial K-sparse estimate $\mathbf{x}^0$ of $\mathbf{x}$ after which $\hat{\sigma}$ can be computed by some robust scale statistic (e.g., *median absolute deviation (MAD)*) of the obtained residuals $\mathbf{e}^0 = \mathbf{y} - \mathbf{\Phi x}^0$. A robust initial estimate $\mathbf{x}^0$ can be, e.g., the robust OMP estimate proposed in [17]. Naturally, it can be questioned if this makes sense, as the resulting robust IHT estimate may not provide a better (more robust or accurate) estimate than the initial OMP estimator $\mathbf{x}^0$. To avoid computation of an initial estimate $\mathbf{x}^0$, one approach is to recompute $\hat{\sigma}$ at each iteration by MAD of the current residuals as in [17]. The problem with this approach is that the obtained estimate (given that the iterations converged) is not a solution to an optimization problem (10.13) and hence any performance guarantees are difficult to provide. To overcome the above problems, we shall propose in Section 10.5 a sound approach that estimates $\mathbf{x}$ and σ simultaneously. Moreover, the proposed IHT algorithm converges to a local minimum of the proposed optimization problem.

Lorentzian IHT (LIHT) proposed in [6] is essentially a special case of the generalized IHT method described above based on Cauchy loss function (10.8). In this case, the objective function in (10.11) (if one discards the multiplier term $(\hat{\sigma}/2)$) can be given as

$$\sum_{i=1}^{M} \log(1 + e_i^2/\hat{\sigma}^2),$$

which authors denote as $\|\mathbf{e}\|_{LL}$ and refer to as *Lorentzian* (pseudo-)norm. Note that $\|\cdot\|_{LL}$ is pseudo-norm as it does not verify axioms of norms such as the triangle-inequality. The authors propose to estimate $\hat{\sigma}$ (denoted as γ in their paper) by

$$\hat{\sigma} = (y_{(0.875)} - y_{(0.125)}) \qquad (10.14)$$

where $y_{(a)}$ denotes ath empirical quantile of the measurements $y_1, \ldots, y_M$. Thus $\hat{\sigma}$ is the range between .875th and .125th sample quantiles of the measurements. This choice has no guarantees of estimating the scale parameter of the error distribution and thus the performance of the LIHT method with this choice of $\hat{\sigma}$ is varying heavily with the signal-to-noise ratio. This is evident in the simulations in Section 10.6. Therein, it is observed that for all noise distributions and all considered SNR ranges, the LIHT method is performing much worse than the Huber IHT method proposed in Section 10.5, which estimates the signal and scale simultaneously.

10.5 Robust IHT Based on Joint Estimation of Signal and Scale

The main problem with the generalized IHT method described in the previous section was in obtaining accurate and robust preliminary scale estimate $\hat{\sigma}$. To circumvent the above problem, we propose to estimate $\mathbf{x}$ and σ simultaneously. To do this elegantly, we propose to minimize

$$Q(\mathbf{x}, \sigma) = \sigma \sum_{i=1}^{M} \rho\left(\frac{y_i - \boldsymbol{\phi}_i^\top \mathbf{x}}{\sigma}\right) + (M - K)\alpha\sigma \qquad (10.15)$$

$$\text{subject to} \quad \|\mathbf{x}\|_0 \leq K,$$

where ρ is a convex loss function, $\rho(0) = 0$, which should verify

$$\lim_{|x| \to \infty} \frac{\rho(x)}{|x|} = c \leq \infty$$

and $\alpha > 0$ is a scaling factor chosen so that the solution $\hat{\sigma}$ is Fisher-consistent for σ when the errors follow the normal distribution, i.e., $\varepsilon_i \sim \mathcal{N}(0, \sigma^2)$. This is achieved by setting $\alpha = \mathbb{E}[\chi(u)]$, where $u \sim \mathcal{N}(0, 1)$ and

$$\chi(e) = \psi(e)e - \rho(e). \tag{10.16}$$

Note also that a multiplier $(M - K)$ is used in the second term of (10.15) instead of M in order to reduce the bias of the obtained scale estimate $\hat{\sigma}$ at small sample lengths. The objective function Q in (10.15) was proposed for joint estimation of location and scale and regression and scale parameters by Huber (1964, 1973) in [9, 10] and is often referred to as *"Huber's proposal 2"*. Note that $Q(\mathbf{x}, \sigma)$ is a convex function of $(\mathbf{x}, \sigma)$ [11, p. 177–178] which enables to derive a simple convergence proof of an iterative algorithm to compute the solution $(\hat{\mathbf{x}}, \hat{\sigma})$.

Let us now choose Huber's loss function $\rho_H(e)$ in Equation (10.6) as our choice of ρ function. In this case χ in Equation (10.16) becomes $\chi_H(e) = \frac{1}{2}\psi_H^2(e)$ and the scaling factor $\alpha = \beta/2$ can be computed as

$$\beta = 2\{c^2(1 - F_G(c)) + F_G(c) - 1/2 - c f_G(c)\}, \tag{10.17}$$

where F_G and f_G denote the cumulative distribution function (c.d.f.) and the probability density function (p.d.f.) of $\mathcal{N}(0, 1)$ distribution, respectively, and c is the trimming threshold of Huber's loss function that controls the trade-off between robustness and efficiency of the method. Thus the choice of c determines $\beta \equiv \beta(c)$. For example, for the choices in (10.7), the respective β constants are $\beta_1 = 0.7102$ and $\beta_2 = 0.3378$. The algorithm for finding the solution to (10.15) then proceeds as follows:

Huber IHT algorithm

Initialization Let $\mathbf{x}^0 = \mathbf{0}$ and $\sigma^0 = 1$. For a given trimming threshold c, compute the scaling factor $\beta = \beta(c)$ in (10.17).

0. *Compute the initial signal support* $\Gamma^0 = \text{supp}\big(H_K(\mathbf{\Phi}^\top \mathbf{y}_\psi)\big)$, where $\mathbf{y}_\psi = \psi_H(\mathbf{y})$.

   For $n = 0, 1, \ldots$, iterate the steps

1. *Compute the residuals* $\mathbf{e}^n = \mathbf{y} - \mathbf{\Phi}\mathbf{x}^n$

2. *Update the value of the scale:*

$$(\sigma^{n+1})^2 = \frac{(\sigma^n)^2}{(M-K)\beta} \sum_{i=1}^{M} \psi_H^2\left(\frac{e_i^n}{\sigma^n}\right)$$

3. *Compute the pseudo-residuals and the gradient update*

$$\mathbf{e}_\psi^n = \psi_H\left(\frac{\mathbf{e}^n}{\sigma^{n+1}}\right)\sigma^{n+1}$$
$$\mathbf{g} = \mathbf{\Phi}^\top \mathbf{e}_\psi^n$$

4. *Compute the stepsize* μ^n *using (10.20) if* $n = 0$ *and (10.22) otherwise*

5. *Update the value of the signal vector and the signal support*

$$\mathbf{x}^{n+1} = H_K(\mathbf{x}^n + \mu^n \mathbf{g})$$
$$\Gamma^{n+1} = \text{supp}(\mathbf{x}^{n+1})$$

6. *Approve the updates* $(\mathbf{x}^{n+1}, \Gamma^n)$ *or recompute them (discussed later).*

7. *Terminate iteration if*

$$\frac{\|\mathbf{x}^{n+1} - \mathbf{x}^n\|^2}{\|\mathbf{x}^n\|^2} < \delta,$$

where δ is a predetermined tolerance/accuracy level (e.g., $\delta = 1.0^{-6}$).

Relation to IHT algorithm. Let us consider the special that we let the trimming threshold c be arbitrarily large ($c \to \infty$). In this case, the Huber's loss function coincides with LS loss function and hence $\psi_H(e) = e$. Also note that when $c \to \infty$, then the scaling factor $\beta = \beta(c)$ in (10.17) converges to 1 ($\beta \to 1$). It is now easy to verify that in this special case, the Huber IHT algorithm above coincides with the IHT algorithm [2, 3]. In this case Step 2 can be discarded as it does not have any effect on Step 3 because $\mathbf{e}_\psi^n = \mathbf{e}^n$. Furthermore, now $\mathbf{V}^0 = \mathbf{I}$ in (10.20) and $\mathbf{W}^n = \mathbf{I}$ in (10.22), and hence Step 4 reduces to

4. *Compute the stepsize μ^n using (10.10)*

which equals with the optimal stepsize update used in the normalized IHT algorithm [3]. Step 6 of the normalized IHT is computed as follows

6. if $\Gamma^{n+1} \neq \Gamma^n$ and $\mu^n > \omega^n$, then recompute new values for $\mathbf{x}^{n+1}$ and Γ^{n+1} using the procedure described in Section 10.3.

Computing the stepsize μ^n in Step 4. Recall the following notations: at nth iteration, Γ^n denotes the current signal support, $\mathbf{e}^n$ denotes the residual vector computed in Step 1, σ^{n+1} denotes the updated value of the scale computed in Step 2, and $\mathbf{g}$ denotes the gradient update $\mathbf{g} = \mathbf{\Phi}^\top \mathbf{e}_\psi^n$ computed in Step 3. Assuming that we have identified the correct signal support, an optimal step size can be found in gradient ascent direction $\mathbf{x}_{\Gamma^n} + \mu^n \mathbf{g}_{\Gamma^n}$ by solving

$$\mu_{opt}^n = \arg\min_\mu \sum_{i=1}^M \rho\left(\frac{y_i - [\phi_i]_{\Gamma^n}^\top (\mathbf{x}_{\Gamma^n}^n + \mu \mathbf{g}_{\Gamma^n})}{\sigma^{n+1}}\right). \tag{10.18}$$

Since closed-form solution can not be derived for μ_{opt}^n above, we aim at finding a good approximation in closed-form. By writing $v(e) = \rho(e)/e^2$, we can express the optimization problem above in an equivalent form

$$\mu_{opt}^n = \arg\min_\mu \sum_{i=1}^M v_i^n(\mu)\left(y_i - [\phi_i]_{\Gamma^n}^\top (\mathbf{x}_{\Gamma^n}^n + \mu \mathbf{g}_{\Gamma^n})\right)^2 \tag{10.19}$$

where the "weights", defined as

$$v_i^n(\mu) = v\left(\frac{y_i - [\phi_i]_{\Gamma^n}^\top (\mathbf{x}_{\Gamma^n}^n + \mu \mathbf{g}_{\Gamma^n})}{\sigma^{n+1}}\right),$$

depend on μ. If we replace $v_i^n(\mu)$ in the above optimization problem by their approximation $v_i^n = v_i^n(0)$, then we can calculate stepsize (i.e., an approximation of μ_{opt}^n) in closed-form by elementary calculus. Hence, when the iteration starts at $n = 0$, we calculate the stepsize μ^0 in Step 4 as

$$\mu^0 = \frac{(\mathbf{e}^0)^\top \mathbf{V}^0 \mathbf{\Phi}_{\Gamma^0} \mathbf{g}_{\Gamma^0}}{\mathbf{g}_{\Gamma^0}^\top \mathbf{\Phi}_{\Gamma^0}^\top \mathbf{V}^0 \mathbf{\Phi}_{\Gamma^0} \mathbf{g}_{\Gamma^0}}, \tag{10.20}$$

where $\mathbf{V}^0 = \mathrm{diag}(v_1^0, \dots, v_M^0)$. When iteration proceeds (for $n = 1, 2, \dots,$) the current support Γ^n and the signal update $\mathbf{x}^n$ are more accurate estimates of the true signal support Γ and the K-sparse signal $\mathbf{x}$. Hence, when $n \geq 1$, we find an approximation of the optimal stepsize μ_{opt}^n by solving

$$\mu^n = \arg\min_\mu \sum_{i=1}^n w_i^n\left(y_i - [\phi_i]_{\Gamma^n}^\top (\mathbf{x}_{\Gamma^n}^n + \mu \mathbf{g}_{\Gamma^n})\right)^2 \tag{10.21}$$

where the "weights" w_i^n are defined as

$$w_i^n = w\left(\frac{y_i - [\phi_i]_{\Gamma^n}^\top \mathbf{x}_{\Gamma^n}^n}{\sigma^{n+1}}\right),$$

and $w(\cdot)$ is the weight function, defined as $w_{\mathrm{H}}(e) = \psi_{\mathrm{H}}(e)/e$. The solution to
(10.21) can be given in closed form as

$$\mu^n = \frac{\mathbf{g}_{\Gamma^n}^\top \mathbf{g}_{\Gamma^n}}{\mathbf{g}_{\Gamma^n}^\top \mathbf{\Phi}_{\Gamma^n}^\top \mathbf{W}^n \mathbf{\Phi}_{\Gamma^n} \mathbf{g}_{\Gamma^n}}, \tag{10.22}$$

where $\mathbf{W}^n = \mathrm{diag}(w_1^n, \ldots, w_M^n)$. The idea of (10.21) is based on the property that if Γ^n correctly identifies the signal support, then the signal elements can be found solving the M-estimation objective function $Q(\mathbf{x}_{\Gamma^n}, \sigma^{n+1})$ (where σ^{n+1} is considered fixed) in the regression model $\mathbf{y} = \mathbf{\Phi}_{\Gamma^n}\mathbf{x}_{\Gamma^n} + \varepsilon$. Recall that a solution to an M-estimate of regression can be found by an iteratively reweighted least squares (IRLS) algorithm, which iteratively solves $\sum_{i=1}^n w_i(y_i - [\phi_i]_{\Gamma^n}^\top \mathbf{x})^2$ with weights dependent on the current residuals $w_i = w(e_i/\sigma^{n+1})$. These iterations are guaranteed to decrease the objective function.

Approving or recomputing the updates $(\mathbf{x}^{n+1}, \Gamma^{n+1})$ in Step 6. We accept the updates if the value of the new objective function is smaller than the old objective function, i.e., $Q(\mathbf{x}^{n+1}, \sigma^{n+1}) < Q(\mathbf{x}^n, \sigma^n)$, otherwise we set $\mu^n \leftarrow \mu^n/2$ and step back to Step 5 and recompute new updates.

10.6 Simulation Studies

Next, extensive simulation studies are used to illustrate the validity and usefulness of the proposed methods in a variety of noise environments and SNR levels. The considered methods and their acronyms in the simulations are

- **IHT** corresponds to the normalized IHT method, implemented as described in [3] and in Section 10.3.

- **LIHT** corresponds to the Lorentzian IHT method, implemented as described in [6]. See also Section 10.4 for discussions about this approach.

- **Hub IHT** (c_i), $i \in \{1, 2\}$, corresponds to the Huber IHT method (estimating $\mathbf{x}$ and σ jointly) using the trimming threshold c_1 (or c_2) in (10.7). The method is implemented as described in detail in Section 10.5.

Description of the setup and performance measures. For all experiments, the elements of the measurement matrix $\mathbf{\Phi}$ are drawn from $\mathcal{N}(0, 1)$

distribution after which the columns are normalized to have unit norm. Furthermore, the K nonzero coefficients of $\mathbf{x}$ are set to have equal amplitude $\sigma_s = |x_i| = 10$ for all $i \in \Gamma$, equiprobable signs (i.e,. $\pm 1 = \operatorname{sign}(x_i)$ with equal probability $\frac{1}{2}$), and the signal support set $\Gamma = \operatorname{supp}(\mathbf{x})$ is randomly chosen from $\{1, \dots, N\}$ without replacement for each trial. In all of our experiments, the noise random vector $\boldsymbol{\varepsilon}$ consists of i.i.d. elements ε_i from a continuous symmetric distribution F_ε with p.d.f. $f_\varepsilon(e) = (1/\sigma)f_0(e/\sigma)$, where $\sigma > 0$ denotes the scale parameter and $f_0(e)$ the standard form of the p.d.f. Then the *signal to noise ratio (SNR)* can be defined

$$\mathrm{SNR}(\sigma) = 20\log_{10}\frac{\sigma_s}{\sigma}$$

and thus depends on the used scale parameter σ of the error distribution. When the underlying noise distribution is not particularly heavy-tailed in nature, a conventional scale parameter is the *standard deviation* (SD) $\sigma_{\mathrm{SD}} = \sqrt{\mathbb{E}[|\varepsilon|^2]}$ (e.g., for Gaussian noise distribution). When SD is employed as the scale parameter, $\mathrm{SNR}(\sigma_{\mathrm{SD}})$ is denoted shortly as SNR. In some cases, the *mean absolute deviation* (MeAD) $\sigma_{\mathrm{MeAD}} = \mathbb{E}[|\varepsilon|]$ is the most natural scale parameter of the distribution. This is the case for the Laplace (double exponential) distribution. However for noise distribution families with possibly infinite variance such as Student's t_ν-distribution for degrees of freedom (d.o.f.) $\nu \leq 2$, we choose the *median absolute deviation* (MAD) $\sigma_{\mathrm{MAD}} = \mathrm{Med}(|\varepsilon_i|)$ as the scale parameter.

As performance measures of sparse signal recovery, we use both the (observed) *mean squared error*

$$\mathrm{MSE}(\hat{\mathbf{x}}) = \frac{1}{Q}\sum_{q=1}^{Q}\|\hat{\mathbf{x}}^{[q]} - \mathbf{x}^{[q]}\|_2^2$$

and the (observed) *probability of exact recovery*

$$\mathrm{PER} \triangleq \frac{1}{Q}\sum_{q=1}^{Q}\mathrm{I}(\hat{\Gamma}^{[q]} = \Gamma^{[q]})$$

where $\mathrm{I}(\cdot)$ denotes the indicator function, $\hat{\mathbf{x}}^{[q]}$ and $\hat{\Gamma}^{[q]} = \operatorname{supp}(\hat{\mathbf{x}}^{[q]})$ denote the estimate of the K-sparse signal $\mathbf{x}^{[q]}$ and the signal support $\Gamma^{[q]}$ for the qth Monte Carlo (MC) trial, respectively. In all simulation settings described below, the number of MC trials is $Q = 2000$, the length of the signal is $M = 512$, the number of measurements is $N = 256$ and the sparsity level is $K = 8$.

Experiment I: Gaussian noise. The noise distribution F_ε is Gaussian $\mathcal{N}(0, \sigma^2)$ distribution with conventional scale parameter $\sigma = \sigma_{\mathrm{SD}}$. Figure 10.2(a) gives the signal reconstruction performance as a function of SNR. As expected, the IHT has the best performance, but Huber IHT with c_1 suffers a negligible 0.2 dB performance loss compared to IHT whereas Huber IHT

with c_2 has 0.65 dB performance loss. All three methods had a full PER rate
($= 1$) for all SNR. We note that the Lorentzian IHT was also included in the
study but the method experienced convergence problems and hence was left
out. These problems may be due to the choice of the preliminary scale estimate $\hat{\sigma}$ in (10.14), which seems appropriate only for heavy-tailed distributions
at low SNR regimes.

Experiment II: Laplacian noise. The noise distribution F_ε is Laplace
distribution $Lap(0, \sigma^2)$ with conventional scale parameter $\sigma = \sigma_{\mathrm{MeAD}}$. Recall
that the Laplace distribution is a heavy-tailed distribution with kurtosis equal
to 3. Figure 10.2(b) gives the signal reconstruction performance as a function
of $\mathrm{SNR}(\sigma_{\mathrm{MeAD}})$. As expected, the Huber IHT with c_2 has the best performance,
next comes Huber IHT with c_1, whereas LIHT has only slightly better perfor-
mance compared to IHT in the high SNR regime [30, 40] dB. The performance
loss of IHT as compared to Huber IHT with c_2 is 1.9 dB on average in SNR
[22, 40], but jumps to 2.5 dB at SNR 20 dB. Huber IHT with c_1 has 0.59 dB
performance loss compared to Huber IHT with c_2. Note also that LIHT has
the worst performance at low SNR regime. In terms of PER rates, Huber IHT
methods had again the best performance and they attained full PER ($= 1$)
rate at all SNR levels considered. The PER rates of LIHT decayed to 0.99,
0.93, and 0.7 ad SNR $= 24, 22$, and 20 dB, respectively. The PER rate of IHT
was full until SNR 20 dB at which it decayed to 0.97.

Experiment III: t_ν-distributed noise. The noise distribution F_ε is symmetric Student's t-distribution, $t_\nu(0, \sigma)$, with scale parameter $\sigma = \sigma_{\mathrm{MAD}}$. Recall that t-distribution with $\nu > 0$ degrees of freedom (d.o.f.) is heavy-tailed
distribution with $\nu = 1$ corresponding to Cauchy distribution and $\nu \to \infty$
corresponding to Gaussian distribution. Figure 10.3 depicts signal recovery
performance in terms of MSE as a function of d.o.f. ν and for three different $\mathrm{SNR}(\sigma_{\mathrm{MAD}})$ levels. Note that the right hand side (RHS) plots provide a
zoom of the lower right corner of the LHS figure in the $\nu > 2$ regime (i.e, for
less heavy-tailed distributions). First we wish to point out that the proposed
Huber's IHT methods are outperforming the competing methods in all cases.
Note especially that at high SNR 40dB in Figure 10.3(a), the Huber IHT
method with c_2 tuning constant is able to retain a steady MSE around -6.5
dB for all $\nu = 1, \ldots, 5$. The Huber IHT method with c_1 is (as expected) less
robust with slightly worse performance; namely, its MSE is increasing mildly
as the distributions get heavier tailed (i.e., as ν decreases). IHT is already
performing poorly at $\nu = 5$ and its performance deteriorates at a rapid rate
with decreasing ν. Note that the performance decay of LIHT is much milder
than that of IHT, yet it also has a rapid decay when it is compared to the
Huber IHT method. The PER rates given in Table 10.1 illustrate the remarkable performance of Huber's IHT methods, which are able to maintain full
recovery rates even at Cauchy distribution ($\nu = 1$) for SNR 30 and 40 dB.
At low SNR 20 dB, only the Huber's IHT methods are able to maintain good
PER rates whereas the other methods, IHT or LIHT, provide estimates that
are completely corrupted.

TABLE 10.1: Probability of exact recovery (PER) rates for different methods under $t_\nu(0, \sigma_{\mathrm{MAD}})$ distributed noise at varying $\mathrm{SNR}(\sigma_{\mathrm{MAD}})$ and d.o.f. ν. System parameters were $(M, N, K) = (512, 256, 8)$ and results are averages over 2000 trials As can be seen, the Huber IHT methods are able to maintain full recovery rates even at Cauchy distribution ($\nu = 1$) for SNR 30 and 40 dB. At low SNR 20 dB, the performance of Huber IHT methods are remarkably better than those of IHT or LIHT, which are breaking down completely.

SNR(σ_{MAD})	Method	Degrees of freedom ν							
		1	1.25	1.5	1.75	2	3	4	5
40 dB	IHT	.51	.86	.95	.99	0.99	1.0	1.0	1.0
	LIHT	1.0	1.0	1.0	1.0	1.0	1.0	1.0	1.0
	Hub IHT (c_1)	1.0	1.0	1.0	1.0	1.0	1.0	1.0	1.0
	Hub IHT (c_2)	1.0	1.0	1.0	1.0	1.0	1.0	1.0	1.0
30 dB	IHT	.06	.33	.66	.85	0.93	1.0	1.0	1.0
	LIHT	0.98	1.0	1.0	1.0	1.0	1.0	1.0	1.0
	Hub IHT (c_1)	1.0	1.0	1.0	1.0	1.0	1.0	1.0	1.0
	Hub IHT (c_2)	1.0	1.0	1.0	1.0	1.0	1.0	1.0	1.0
20 dB	IHT	0	0	0	.01	.04	.31	.57	.72
	LIHT	.09	.10	.13	.13	.17	.22	.24	.24
	Hub IHT (c_1)	.46	.61	.70	.77	.81	.90	.92	.93
	Hub IHT (c_2)	.61	.66	.72	.73	.74	.80	.82	.82

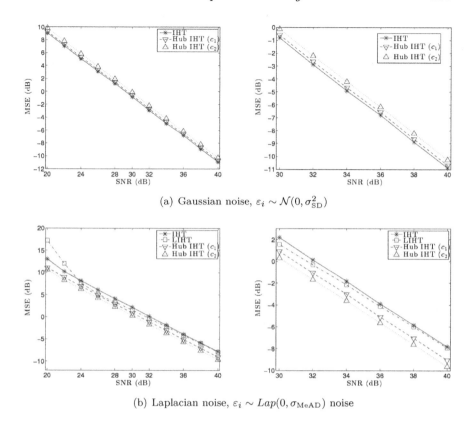

(a) Gaussian noise, $\varepsilon_i \sim \mathcal{N}(0, \sigma_{\mathrm{SD}}^2)$

(b) Laplacian noise, $\varepsilon_i \sim Lap(0, \sigma_{\mathrm{MeAD}})$ noise

FIGURE 10.2: Average MSE as a function of $\mathrm{SNR}(\sigma)$ in reconstructing a K-sparse signal using IHT, LIHT, and Huber IHT methods under $\mathcal{N}(0, \sigma_{\mathrm{SD}}^2)$ and $Lap(0, \sigma_{\mathrm{MeAD}})$ distributed noise. The RHS plot provides a zoom of the lower right corner of the LHS plot in high SNR regime [30, 40] dB. System parameters were $(M, N, K) = (256, 512, 8)$ and number of trials is $Q = 2000$.

10.7 Conclusions

In this chapter, we have provided an overview of recent robust approaches to compressive sensing. The Huber IHT method proposed here (and in the submitted conference paper [15]) avoids the computation of the preliminary estimate of scale that is needed in the generalized IHT method [17] and in the Lorentzian IHT [6]. The Huber IHT method provides excellent signal reconstruction performance under various noise distributions and SNR levels. Under non-Gaussian noise (e.g., Laplace and Cauchy) it outperforms the normalized IHT [3] and the Lorentzian IHT by orders of magnitude, yet it has a negligible performance loss compared to the normalized IHT under the nominal Gaus-

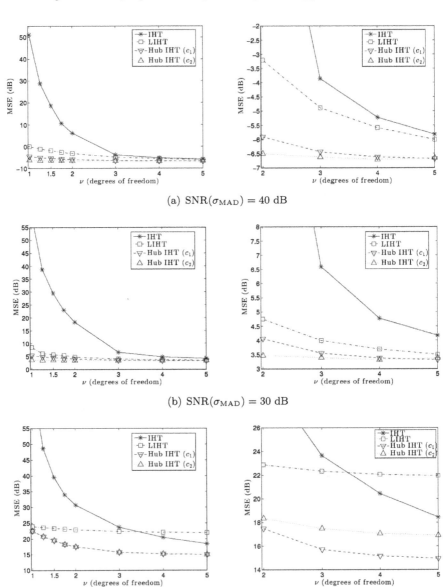

(a) SNR(σ_{MAD}) = 40 dB

(b) SNR(σ_{MAD}) = 30 dB

(c) SNR(σ_{MAD}) = 20 dB

FIGURE 10.3: Average MSE in reconstructing a K-sparse signal using IHT, LIHT, and the Huber IHT using c_1 or c_2 under $t_\nu(0, \sigma_{\mathrm{MAD}})$ distributed noise as a function of d.o.f. ν. Figures (a)-(c) illustrate the results at SNR(σ_{MAD}) = 40, 30, 20 dB, respectively. Note that the RHS plot provides a zoom of the lower right corner of the LHS figure in the $\nu > 2$ regime (i.e, for less heavy-tailed distributions). System parameters were $(M, N, K) = (256, 512, 8)$ and the number of trials is $Q = 2000$.

sian noise. A plan for future work is to derive exact theoretical performance guarantees of the proposed Huber IHT method, which will be reported in a separate paper.

Bibliography

[1] T. Blumensath and M. E. Davies. Iterative thresholding for sparse approximations. *Journal of Fourier Analysis and Applications*, 14(5–6):629–654, 2008.

[2] T. Blumensath and M. E. Davies. Iterative hard thresholding for compressed sensing. *Applied and Computational Harmonic Analysis*, 27(3):265–274, 2009.

[3] T. Blumensath and M. E. Davies. Normalized iterative hard thresholding: guaranteed stability and performance. *IEEE Journal of Selected Topics in Signal Processing*, 4(2):298–309, 2010.

[4] E. Candés, J. Romberg, and T. Tao. Robust uncertainty principles: Exact signal reconstruction from highly incomplete frequency information. *IEEE Trans. Inform. Theory*, 52(4):1289–1306, 2006.

[5] E. J. Candes and M. B Wakin. An introduction to compressive sampling. *IEEE Signal Proc. Mag.*, 25(2):21–30, 2008.

[6] R. E. Carrillo and K. E. Barner. Lorentzian based iterative hard thresholding for compressed sensing. In *Proc. IEEE Int. Conf. on Acoustics, Speech, and Signal Processing (ICASSP'11)*, pages 3664–3667, Prague, Czech Republic, May 22–27, 2011.

[7] D. Donoho. Compressive sensing. *IEEE Trans. Inform. Theory*, 52(2):5406–5425, 2006.

[8] M. Elad. *Sparse and redundant representations*. Springer, New York, 2010.

[9] P. J. Huber. Robust estimation of a location parameter. *Ann. Math. Statist.*, 35:73–101, 1964.

[10] P. J. Huber. Robust regression: Asymptotics, conjectures and Monte Carlo. *Ann. Statist.*, 1:799–821, 1973.

[11] Peter J. Huber. *Robust Statistics*. Wiley, New York, 1981.

[12] H.-J. Kim, E. Ollila, and V. Koivunen. Robust and sparse estimation of tensor decompositions. In *Proc. IEEE Global Conference on Signal and Information Processing (GlobalSIP'13)*, pages 1–5, Austin, Texas, USA, December 3–5, 2013.

[13] R. A. Maronna, R. D. Martin, and V. J. Yohai. *Robust Statistics: Theory and Methods*. Wiley, New York, 2006.

[14] D. Needell and J. A. Tropp. CoSaMP: Iterative signal recovery from incomplete and inaccurate samples. *Applied and Computational Harmonic Analysis*, 26(3):301–321, 2009.

[15] E. Ollila, H.-J. Kim, and V. Koivunen. Robust iterative hard thresholding for compressed sensing. In *Proc. 6th International Symposium on Communications, Control, and Signal Processing (ISCCSP'14)*, Athens, Greece, May 21–24, 2014.

[16] J. L. Paredes and G. R. Arce. Compressive sensing signal reconstruction by weighted median regression estimates. *IEEE Trans. Signal Processing*, 59(6):2585–2601, 2011.

[17] S. A. Razavi, E. Ollila, and V. Koivunen. Robust greedy algorithms for compressed sensing. In *Proc. 20th European Signal Processing Conference (EUSIPCO'12)*, pages 969–973, Bucharest, Romania, 2012.

[18] R. Tibshirani. Regression shrinkage and selection via the lasso. *J. Royal Stat. Soc., Ser. B*, 58:267–288, 1996.

[19] J. A. Tropp and A. C. Gilbert. Signal recovery from random measurements via orthogonal matching pursuit. *IEEE Trans. Inform. Theory*, 53(12):4655–4666, 2007.

Chapter 11

Regularized Robust Portfolio Estimation

Theodoros Evgeniou

Decision Sciences and Technology Management INSEAD

Massimiliano Pontil

Department of Computer Science, University College London

Diomidis Spinellis

Department of Management Science and Technology, Athens University of Economics and Business

Nick Nassuphis

31 St. Martin's Lane, London

11.1   Introduction

Given a vector-valued time series, we study the problem of learning the weights of a linear combination of the series' components (e.g., a portfolio), which has large autocorrelation, and discuss the extension to the problem of learning two combinations, which have large cross-correlation. Both problems have been studied from different perspectives in various areas, ranging from computational neuroscience [27], to computer vision [20, 14], to information retrieval [15], among others. In this chapter, we address these problems from the point of view of robust optimization (see, e.g., [5, 8] and references therein) and regularization, and highlight their application to the context of financial time series analysis; see, e.g., [25].

The autocorrelation (or cross-correlation) function is a quantity difficult to measure, as it depends on the lag-1 autocovariance matrix of the time series, which is typically unstable. To mitigate this problem, we propose a robust optimization approach that leads to regularized versions of the autocorrelation (or cross-correlation) function. We describe various forms of regularizers derived from different constraints on the uncertainty region of the lag-1 autocovariance matrix, which in particular induce ℓ_2 or ℓ_1 regularized portfolio estimation methods. We present an optimization algorithm to solve the ℓ_1 regularization problem, which is inspired by recent work on sparse principal component analysis [16, 17], also linking this work to the broader literature on sparsity regularization. We then apply the proposed methods to estimate high lag-1 autocorrelation portfolios for financial time series. On the way, we link specific instances of our method to portfolio creation strategies previously considered in the finance literature. Finally, we extend the proposed approach to the setting of canonical correlation analysis [2, 13], an older statistical technique, which has seen revived interest in machine learning and statistics; see, e.g., [15, 28].

In summary, the key contributions of this chapter are both methodological, namely developing novel regularization methodologies and optimization based estimation algorithms, as well as theoretical, namely establishing a link between robust optimization and regularization in the context of portfolio estimation. Although the methods are more broadly applicable, we study them in the context of portfolio creation for financial time series. This type of data is among the most challenging ones to develop predictive models for. Another important contribution of this chapter is a demonstration of the proposed approach application potential on financial time series. Although developing portfolio estimation methods for such series that can be used in practice (e.g., for trading [23]) is beyond the scope of this chapter, we discuss potential future research that can lead to such methods building on the approach we develop. Over the past few years there has been a rising interest within the financial industry in employing machine learning techniques [1]; this work also builds in that direction.

This chapter is organized as follows. In Section 11.2, we introduce the portfolio learning problem. In Section 11.3, we address the issue of implementing the learning method numerically. In Section 11.4, we extend our regularization approach to the setting of canonical correlation analysis. Finally, in Section 11.5 we report experiments with the proposed methods in the context of financial time series.

11.2 Finding Robust Autocorrelation Portfolios

We start with the relation between a robust optimization formulation of the maximally autocorrelated portfolio estimation problem and regularization. As the experimental focus of the chapter is on financial time series, we first introduce some notation from that context.

11.2.1 Financial Time Series: Notation and Definitions

Let $r_1, \ldots, r_T \in \mathbb{R}^n$ be the realization of a vector-valued time series over T consecutive time frames (e.g., days). In the experiments, r_t represents the vector of log-returns on day t of n assets (e.g., stocks). A common goal in practice is to learn a weight vector $x \in \mathbb{R}^n$ that maximizes some investment performance, such as the cumulative return or Sharpe ratio. The former quantity is defined as the sum of the daily returns, that is, $\sum_{t=2}^{T} f(x^\top r_{t-1}) \, x^\top r_t$, where the function f can, for example, be $\mathrm{sign}(\cdot)$ or $-\mathrm{sign}(\cdot)$, depending on whether the portfolio follows a momentum or mean reversion strategy; we refer to [7] for background. The Sharpe ratio is defined as the ratio between the average daily returns and the standard deviation of the daily returns. Both quantities are difficult to optimize in x, i.e., they are not differentiable because of the sign function in the numerator. Therefore in this chapter we use as a surrogate function the lag-1 autocorrelation of the portfolio. Intuitively, a positive (respectively negative) autocorrelated portfolio will favor a momentum (respectively mean reversion) strategy: we buy (respectively sell) the portfolio on day t if it had a positive (respectively negative) return on the previous day.

11.2.2 Regularization Problem

The vector x gives rise to the scalar time series $p_t := x^\top r_t, t = 1, \ldots, T$, called the portfolio series [11]. Our goal is to find a portfolio that has maximal lag-1 autocorrelation, which is defined as the correlation between p_{t-1} and p_t.

If the series is stationary[1] this quantity simplifies to

$$\nu(x) = \frac{x^\top \Theta x}{x^\top \Gamma x} \tag{11.1}$$

where Θ and Γ are the lag-1 autocovariance and the covariance of the time series, respectively. The latter matrix is assumed to be invertible[2]. In order to emphasize the fact that the autocorrelation ν depends on matrix Θ, we will sometimes use the notation $\nu(\cdot|\Theta)$. We then solve the problem

$$\max_{x \in \mathbb{R}^n} \nu(x). \tag{11.2}$$

This is a generalized eigenvalue problem [12], whose solution is given by $x = \Gamma^{-\frac{1}{2}} u$, where u is the leading eigenvector of the matrix $\Gamma^{-\frac{1}{2}} \Theta \Gamma^{-\frac{1}{2}}$.

In practice, matrices Θ and Γ are estimated from historical data, the most common estimates being $\hat{\Theta} = \frac{1}{T} \sum_{t=2}^{T} r_t r_{t-1}^\top$ and $\hat{\Gamma} = \frac{1}{T} \sum_{t=1}^{T} r_t r_t^\top$ (see, e.g., [25], where for simplicity we assumed that the series has mean zero. Often, these estimates are inaccurate and it therefore can be useful to introduce robust versions of Problem (11.2), in which we suppose that the matrix Θ and/or Γ is known to belong to some uncertainty set. For simplicity here we only address the robustness of Θ, which is typically the main problem of concern since the covariance is in practice often more stable than the lag-1 autocovariance [7, 25]. In particular, we prescribe an *uncertainty set* $\mathcal{A}$ and consider the problem

$$\max_{x \in \mathbb{R}^n} \min_{\Theta \in \mathcal{A}} \nu(x|\Theta). \tag{11.3}$$

That is, we maximize the worst autocorrelation obtained when varying matrix Θ in the set $\mathcal{A}$.

A natural choice for the uncertainty set is a ball centered at the empirical estimate $\hat{\Theta}$, namely we choose

$$\mathcal{A} = \{\Theta : \|\Theta - \hat{\Theta}\|_p \le \epsilon\},$$

where $\|\cdot\|_p$ is the elementwise ℓ_p norm of the matrix (or vector) elements, $p \in [1, \infty]$. The associated dual norm is $\|\cdot\|_q$, where $q \in [1, \infty]$ is defined by the equation $1/p + 1/q = 1$.

Lemma 11.1. *It holds that*

$$\min_{\Theta \in \mathcal{A}} \{x^\top \Theta y : \Theta \in \mathbb{R}^{n \times m}\} = x^\top \hat{\Theta} y - \epsilon \|x\|_q \|y\|_q.$$

---

[1] Relaxing this assumption within our framework is left for future research.
[2] If this assumption is violated, adding a small positive value to the diagonal elements of Γ will ensure that it is invertible. Invertibility was not an issue in the numerical experiments presented in Section 11.5 below.

Proof. We write $\Theta = \hat{\Theta} + \Delta$ for some $\|\Delta\|_p \leq \epsilon$. Using Hölder's inequality, we obtain that

$$x^\top \Theta y = x^\top \hat{\Theta} y + x^\top \Delta y \geq x^\top \hat{\Theta} y - \|\Delta\|_p \|xy^\top\|_q.$$

This inequality is tight for $\Delta = \epsilon \delta \gamma^\top$, where $\delta \in \mathbb{R}^n$ satisfies $\|\delta\|_p = 1$ and $\delta^\top x = \|x\|_q$, and $\gamma \in \mathbb{R}^m$ satisfies $\|\gamma\|_p = 1$ and $\gamma^\top y = \|y\|_q$. The result follows by noting that $\|xy^\top\|_q = \|x\|_q \|y\|_q$. $\square$

Using this lemma with $x = y$, we see that $\min\{x^\top \Theta x : \Theta \in \mathcal{A}\} = x^\top \hat{\Theta} x - \epsilon \|x\|_q^2$. Consequently, Problem (11.3) becomes

$$\max_{x \in \mathbb{R}^n} \left\{ \frac{x^\top \hat{\Theta} x - \epsilon \|x\|_q^2}{x^\top \Gamma x} \right\}. \tag{11.4}$$

Parameter ϵ is related to the size of the uncertainty set $\mathcal{A}$, and, as we adopt the regularization framework below, for simplicity we call it hereafter the regularization parameter. In this chapter we consider the cases $q = 2$ and $q = 1$. In the first case, Problem (11.4) is still a generalized eigenvalue problem of the form (11.1) with matrix Θ replaced by $\hat{\Theta} - \epsilon I$. In the second case, the problem becomes a nonlinear one, for which we present an optimization method in the next section.

The above robust analysis can be applied in a similar way to the problem of finding the most negative autocorrelated portfolio, namely $\min\{\nu(x|\Theta) : x \in \mathbb{R}^n\}$. This merely requires replacing Θ by $-\Theta$ in Problem (11.4). Moreover, the above analysis can be extended to take into account uncertainty in both matrices Θ and Γ, with not much additional difficulty. Specifically, if we choose the set

$$\mathcal{A} = \{\Theta : \|\hat{\Theta} - \Theta\|_p \leq \epsilon\} \times \{\Gamma : \|\hat{\Gamma} - \Gamma\|_p \leq \lambda\}$$

then we obtain a robust optimization problem similar to (11.4) but with the denominator replaced by $x^\top \Gamma x + \lambda \|x\|_q^2$. As we have already noted, in financial time series Θ is much more unstable than Γ and, so, we do not explore robustness with respect to both Θ and Γ further in this chapter. We also refer to [26] for related ideas in the case of ℓ_2.

11.2.3 Interpretation of the Case $\epsilon \to \infty$

We note that, in the case of financial data, when $\epsilon \to \infty$ the solutions of Problem (11.4) are related to portfolios studied before in the finance literature. Specifically, for $q = 2$ we recover the leading eigenvector of Γ, which for financial time series is close to the "market portfolio" [3]. If $q = 1$, and the series components all have the same unit variance, then the solution of Problem (11.4) is (up to a nonzero multiplicative constant) given by $x = e_j$, where j is the most positively autocorrelated series component [19], that is $\Theta_{jj} = \max_{i=1}^n \Theta_{ii}$. To see this, note that Problem (11.4) is equivalent to

$\max\{x^\top \Theta x - \epsilon \|x\|_1^2 : x^\top \Gamma x = 1\}$. If ϵ is large enough there is an advantage in choosing x such that $\|x\|_1$ is as small as possible provided that $x^\top \Gamma x = 1$. Since we assumed that the series all have the same unit variance, we have that

$$\begin{aligned} 1 &= x^\top \Gamma x = \sum_{i=1}^n x_i^2 + \sum_{i \neq j} \Gamma_{ij} x_i x_j \\ &= \|x\|_1^2 + \sum_{i \neq j} |x_i x_j| \left(\Gamma_{ij} \operatorname{sign}(x_i x_j) - 1 \right). \end{aligned}$$

Since $|\Gamma_{ij}| < 1$ if $i \neq j$ (otherwise Γ would not be strictly positive definite) we conclude that

$$\|x\|_1^2 = 1 + \sum_{i \neq j} |x_i x_j| \left(1 - \Gamma_{ij} \operatorname{sign}(x_i x_j) \right) \geq 1$$

and $\|x\|_1 = 1$ if and only if $x \in \{e_1, \dots, e_n\}$.

11.2.4   Connection to Slow Feature Analysis

We end this section by noting a connection between Problem (11.2) and a method of unsupervised learning. Using the identity $2(x^\top r_{t-1})(x^\top r_t) = (x^\top r_{t-1})^2 + (x^\top r_t)^2 - (x^\top (r_t - r_{t-1}))^2$, we can rewrite twice the numerator in (11.1) as $x^\top \Gamma_{t-1} x + x^\top \Gamma_t x - x^\top V x$, where $\Gamma_t = \mathbb{E} r_t r_t^\top$ and $V_t = \mathbb{E} \delta_t \delta_t^\top$ is the covariance of the "velocity" process $\delta_t = r_t - r_{t-1}$. If the process is stationary, so Γ_t and V_t are time invariant, then $\nu(x) := 1 - \frac{1}{2}\psi(x)$, where

$$\psi(x) := \frac{x^\top V x}{x^\top \Gamma x}.$$

Thus, maximizing ν is the same as minimizing ψ. The latter optimization problem is very similar to the method of slow feature analysis [20, 22, 27], an unsupervised learning technique that was originally designed to extract invariant representations from time varying visual signals [6]. The robust version (11.4) of Problem (11.2) could likewise be interpreted as a robust version of slow feature analysis, which could lead to interesting applications to that context.

11.3   Optimization Method

In this section, we address the issue of implementing the learning method (11.4) in the case $q = 1$.[3] We begin by rewriting Problem (11.4) as that of

---

[3] Similar observations apply to the general case $q \in (1, \infty]$.

Algorithm 10 ℓ_1-Regularized Autocorrelation

Choose a starting point $x^0 \in \mathbb{R}^n$ and tolerance parameter *tol*.
for $k = 0, 1, \ldots$ **do**
$\quad$ Let $x^{k+1} = \operatorname{argmin}\{\phi(x|x^k) : x \in \mathbb{R}^n\}$
$\quad$ If $|\eta(x^{k+1}) - \eta(x^k)| \leq tol$ terminate.
end for

---

minimizing the function

$$\eta(x) = \frac{x^\top N x - x^\top P x + \epsilon\|x\|_1^2}{x^\top \Gamma x}$$

where P and N are symmetric positive definite matrices such that $P - N = S := (\hat{\Theta} + \hat{\Theta}^\top)/2$. These matrices can be obtained via the eigenvalue decomposition of S as $P = (S)_+$ and $N = P - S$, where $(\cdot)_+$ is a spectral function that acts on the eigenvalues $\lambda \in \mathbb{R}$ as $(\lambda)_+ = \max(\lambda, 0)$.

Fix a point x^0 and let $\phi(\cdot|x^0)$ be the function, defined, for every $x \in \mathbb{R}^n$, as

$$\begin{aligned}
\phi(x|x^0) &= x^\top N x - (x^0)^\top P x^0 - 2(x - x^0)^\top P x^0 \\
&\quad + \epsilon\|x\|_1^2 - \eta(x^0)\gamma(x|x^0)
\end{aligned} \tag{11.5}$$

where

$$\gamma(x|x^0) = \begin{cases} (x^0)^\top \Gamma x^0 + 2(x - x^0)\Gamma x^0 & \text{if } \eta(x^0) > 0, \\ x^\top \Gamma x & \text{otherwise.} \end{cases}$$

We then consider the convex optimization problem

$$\min_x \phi(x|x^0). \tag{11.6}$$

The following lemma provides a rationale behind this problem.

Lemma 11.2. *If $\phi(x|x^0) < 0$ then $\eta(x) < \eta(x^0)$.*

Proof. This result follows from the inequality $x^\top N x - x^\top P x + \epsilon\|x\|_1^2 - \eta(x^0)x^\top \Gamma x \leq \phi(x|x^0)$. $\qquad\square$

This observation leads to the descent Algorithm 10, which is inspired by a method outlined in [16] for sparse principal component analysis.

Algorithm 10 iteratively solves a sequence of problems of the form (11.6). Lemma 11.2 guarantees that the algorithm produces a sequence of points $\{x^k : k \in \mathbb{N}\}$ such that the corresponding sequence of function values $\{\eta(x^k) : k \in \mathbb{N}\}$ is strictly monotonically decreasing or the algorithm terminates. In practice, we terminate the algorithm when $|\eta(x^{k+1}) - \eta(x^k)|$ is less than some tolerance parameter, e.g., 10^{-4}.

It remains to show how to solve Problem (11.6). This problem is of the form $\min \|Ax - b\|_2^2 + \epsilon\|x\|_1^2$, for an appropriate choice of the $n \times n$ matrix A and vector $b \in \mathbb{R}^n$. Hence, it is equivalent to the Lasso method and can be solved up to numerical precision by proximal gradient methods; see, e.g., [4, 24, 9, 10]. In our numerical experiments below we have found that we do not need to solve (11.6) exactly. It is enough to find a point x that strictly decreases the objective. The simplest choice for the update rule is to set

$$x^{k+1} = \text{prox}_{r\|\cdot\|_1^2}\left[\left(I + \frac{1}{\alpha^k}\left(S + \eta(x^k)\Gamma\right)x^k\right]\right] \tag{11.7}$$

where $\alpha^k = |||N|||$ if $\eta(x^k) > 0$ and $\alpha^k = |||N - \eta(x^k)\Gamma|||$ otherwise, with $|||\cdot|||$ the spectral norm. This corresponds to a single step of the proximal gradient method [24]. The function $\text{prox}_{r\|\cdot\|_1^2}$ is the proximity operator of the function $r\|\cdot\|_1^2$, where $r := \frac{\epsilon}{2\alpha^k}$ and it is defined, for every $z \in \mathbb{R}^n$, as

$$\text{prox}_{r\|\cdot\|_1^2}(z) = \underset{x \in \mathbb{R}^n}{\text{argmin}}\left\{\frac{1}{2}\|x - z\|_2^2 + r\|x\|_1^2\right\}. \tag{11.8}$$

To solve problem (11.8), we use the following identity [21, Lemma 26]

$$\|x\|_1^2 = \inf\left\{\sum_{i=1}^n \frac{x_i^2}{\lambda_i} : \lambda > 0, \sum_{i=1}^n \lambda_i = 1\right\}$$

where $\lambda \in \mathbb{R}^n$ denotes the vector $(\lambda_1, \ldots, \lambda_n)$ and $\lambda > 0$ means that all components of λ must be greater than zero. Replacing the above expression in the right hand side of (11.8), fixing λ and minimizing over x we obtain the solution

$$x_i(\lambda) = \frac{\lambda_i z_i}{2r + \lambda_i}. \tag{11.9}$$

Using this equation, we obtain the problem

$$\min\left\{\sum_{i=1}^n \frac{r z_i^2}{2r + \lambda_i} : \lambda \geq 0, \sum_{i=1}^n \lambda_i = 1\right\}.$$

One verifies that the minimizing λ is given by

$$\lambda_i = (\rho|z_i| - 2r)_+$$

where the positive parameter ρ is found by binary search in order to ensure that $\|\lambda\|_1 = 1$. Finally, we replace the obtained value of λ in the right hand side of (11.9) to obtain the solution of (11.8).

11.4   Robust Canonical Correlation Analysis

In this section, we sketch an extension of the proposed approach to the problem of maximizing the correlation between the one-dimensional

---

Algorithm 11 Robust CCA

Choose a starting point $z^0 \in \mathbb{R}^n$ and tolerance parameter *tol*.
for $k = 0, 1, \ldots$ **do**
    Let $(\hat{x}, \hat{y}) = \hat{z}$ where $\hat{z} = \mathrm{argmin}\{\phi(z|z^k) : z \in \mathbb{R}^n\}$
    Set $z^{k+1} = (x^{k+1}, y^{k+1})$ with $x^{k+1} = \hat{x}/\sqrt{\hat{x}^\top \Gamma \hat{x}}$, $y^{k+1} = \hat{y}/\sqrt{\hat{y}^\top \Sigma \hat{y}}$.
    If $|\eta(z^{k+1}) - \eta(z^k)| \leq tol$ terminate.
end for

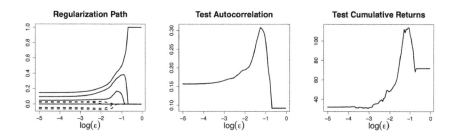

FIGURE 11.1: Synthetic Data: Solution path (left), Test Autocorrelation (center), and Test Cumulative Return (right), as a function of the regularization parameter ϵ, for the ℓ_1-Method.

projections of the time series r_t and a second m-dimensional series, denoted by s_t. Specifically, for every pair of vectors $x \in \mathbb{R}^n$ and $y \in \mathbb{R}^m$, we define the cross-correlation function

$$\rho(x, y) = \frac{x^\top \Theta y}{\sqrt{x^\top \Gamma x}\sqrt{y^\top \Sigma y}} \tag{11.10}$$

where Γ and Σ are estimates of the covariance of r_t and s_t respectively (both assumed to be invertible) and the $n \times m$ matrix Θ is an estimate of the cross-covariance of r_t and s_t. This problem is valuable when additional time series s_t (e.g., commodities, other markets, series formed by technical indicators, etc.), which co-vary with the main stock series r_t are at hand. Notice also that if $s_t = r_{t-1}$ and $x = y$ we recover problem (11.1).

Our goal is to solve the problem

$$\max_{x,y} \rho(x, y) = \max_{x,y} \frac{2x^\top \Theta y}{x^\top \Gamma x + y^\top \Sigma y} \tag{11.11}$$

where the equality follows by the arithmetic-geometric mean inequality and the fact that the solutions are invariant by rescaling; see, e.g., [2, 15]. The right problem is a generalized eigenvalue problem of the form

$$\max\{z^\top A z : z \in \mathbb{R}^{n+m}, \ z^\top B z = 1\},$$

where

$$z = \begin{pmatrix} x \\ y \end{pmatrix}, \quad A = \begin{bmatrix} 0 & \Theta \\ \Theta^\top & 0 \end{bmatrix}, \quad B = \begin{bmatrix} \Gamma & 0 \\ 0 & \Sigma \end{bmatrix}.$$

We can now have an extension of Problem (11.3) to the case of canonical correlation analysis (CCA) where again we consider robustness with respect to perturbations of matrix Θ, that is, we consider the problem

$$\max_{x,y} \min_{\Theta \in \mathcal{A}} \rho(x, y | \Theta) \tag{11.12}$$

for $\mathcal{A} = \{\Theta : \|\Theta - \hat{\Theta}\|_p \leq \epsilon\}$. Using Lemma 11.1 to compute the inner minimum in (11.12), we obtain the problem

$$\max_{x,y} \frac{x^\top \hat{\Theta} y - \epsilon \|x\|_q \|y\|_q}{\sqrt{x^\top \Gamma x} \sqrt{y^\top \Sigma y}}. \tag{11.13}$$

Two choices of interest are $q = 2$ and $q = 1$. In the latter case we obtain again an interpretation of the problem for large values of ϵ, which is equivalent to selecting the pair of most positively correlated series, studied in [19]. This analysis requires that Γ and Σ have all their diagonal elements equal to one. Indeed, if ϵ is sufficiently large the second term in the numerator of (11.13) dominates, hence we want this term to be as small as possible. Thus, it must be the case that $x \in \{e_1, \ldots, e_n\}$ and $y \in \{e_1, \ldots, e_m\}$. We conclude that the solution of (11.13) is given by $x = e_j$, $y = e_k$ such that $\hat{\Theta}_{jk} = \max\{\hat{\Theta}_{i\ell} : i = 1, \ldots, n, \ \ell = 1, \ldots, m\}$. A similar reasoning applies for the most negatively autocorrelated series pair. This is obtained by replacing Θ by $-\Theta$ in the above analysis.

Algorithm 11 presents a method to find a solution of Problem (11.13). The algorithm, which is similar in spirit to Algorithm 10, starts from a vector z^0 and iteratively decreases the objective by solving the convex optimization problem

$$\phi\left(z | z^k\right) = z^\top \left(N + |\eta\left(z^k\right)| B\right) z - \\ \left(z - z^k\right)^\top \left(2Pz^k + \epsilon u + 1_{\{\eta(z^k) > 0\}} v\right) \tag{11.14}$$

where $P = (A)_+$, $N = P - A$, $u \in \partial \|z^k\|_{2,q}^2$, and $v \in \partial (\|\Gamma^{\frac{1}{2}} x^k\|_2 + \|\Sigma^{\frac{1}{2}} y^k\|_2)^2$. Similarly to Lemma 11.2 one can show that function $\phi(\cdot | z^k)$ has the property that if $\phi(z | z^k) < 0$ then $\eta(z) < \eta(z^k)$. Furthermore this function can be optimized by means of proximal gradient methods; see, e.g., [4, 24]. A detailed description of the algorithm in presented in the appendix.

11.5   Experiments

In this section, we present experiments with the proposed methods on a synthetic and a real dataset. As the main goal of this chapter is to study the use of robust optimization through the development of regularization methods for time series prediction, a key aim of the experiments is to explore whether robustness — and the corresponding regularization — has the potential to improve test performance when the data is highly noisy, such as in the case of financial data. Under these circumstances, one would expect that as we "add robustness" by increasing the regularization parameter ϵ, the test performance of the methods improves and, potentially, drops after some point.

11.5.1   Synthetic Data

We used synthetic data to test the ℓ_1 method above. We generated 30 synthetic time series of length 200. Among these, three were governed by a stationary AR(1) process [25], whereas the remaining 27 series were white noise (normally distributed with zero mean and unit variance). The AR(1) time series were centered and normalized in order to have the same mean and variance as the remaining 27 time series. We trained Algorithm 10 using this dataset for 100 values of the regularization parameter ϵ. The left plot in Figure 11.1 shows the weights of the portfolio found by Algorithm 10 as a function of ϵ. It can be seen that for the portfolios of sparsity less than or equal to 3, the AR(1) time series were heavily favored by our method. While the noise time series had non-zero weights for small values of ϵ, these weights were usually smaller than the weights of the AR(1) time series for all values of ϵ, and, as ϵ increased, they tended to be suppressed to zero earlier than the weights of the AR(1) time series. The next two plots in Figure 11.1 show the performance of the portfolios on 800 consecutive test data points, again as a function of ϵ. Both test performances are maximized at approximately the same value of ϵ in this case.

11.5.2   S&P 500 Stock Data

Next, we tested the methods using a dataset that is known to be notoriously challenging: we focused on the problem of constructing portfolios x of n stocks using daily adjusted close prices of stocks in the S&P 500 index.[4] We used data for the past 10 years, from January 1, 2003 until April 12, 2013 (the date of the final data construction). This corresponds to a total of 2586 daily (close to close) returns. We tested the methods by constructing portfolios using only stocks from specific sectors, for a number of different sectors. This was done both in order to perform multiple experiments and because companies in the same sector are known to "co-move" [3], making the proposed methods more applicable. We consider four large sectors defined

---

[4]The data (downloaded from Yahoo!) and the R code used to run the experiments are available from the authors upon request.

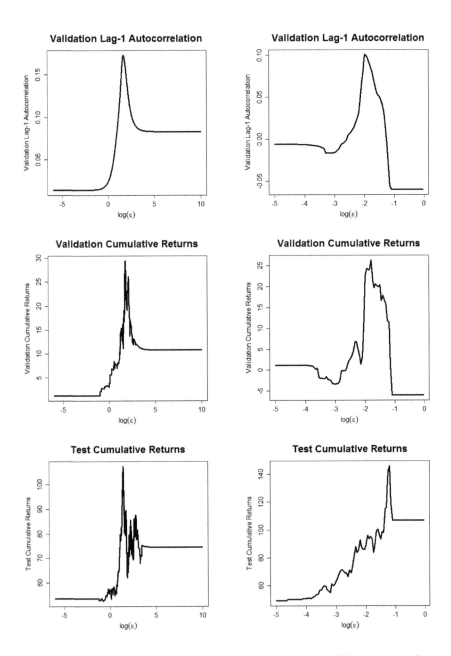

FIGURE 11.2: S&P 500 stock data: validation ν, validation cumulative return, and test cumulative return over 1336 days for the ℓ_2-Method (left) and ℓ_1-Method (right) as functions of the regularization parameter ϵ.

TABLE 11.1: Comparison of methods for the S&P 500 Stock Data: for each of the four sectors we note with MARKET the average of the stocks in that sector, with MARKET AC the selection between momentum and mean reversion of the market, with MAX AC the non-regularized maximum autocorrelation solution ($\epsilon = 0$), with SELECTED L1/L2 the regularization methods with the selected regularization parameter ϵ using the validation data, and with BEST L1/L2 the best ϵ using the test data (hence with hindsight). Cumulative returns during 1336 test days are reported, with annual Sharpe ratio (values above 0.85 are statistically significant, see text) in parentheses. Best performance for each sector, without considering the "best" cases, is indicated in bold.

Method	Healthcare	Financials	Energy	Technology
Market	37.5% (0.47)	2.0% (0.01)	7.8% (0.04)	28.1% (0.19)
Market AC	57.2% (0.47)	353.3% (1.60)	66.25% (0.31)	22.44% (0.15)
Max AC	53.6% (1.52)	115.7% (1.80)	-3.6% (-0.12)	5.1% (0.16)
Selected L2	85.8% (0.77)	300.8 % (1.36)	34.4% (0.17)	**52.7%** (0.34)
Best L2	107.3% (2.13)	346.3% (1.57)	67.7% (0.30)	88.56% (0.71)
Selected L1	**94.0%** (1.02)	**439.6%** (1.43)	**161.1%** (1.03)	-24.7% (-0.47)
Best L1	145.7% (1.04)	446.9% (1.48)	162.9% (0.88)	281.3% (0.83)

based on a standard industry classification:[5] energy, financial, healthcare, and technology (other sectors led to similar conclusions). These consist of, respectively, $n = 30, 30, 24$, and 35 stocks. For each sector we use the first 1000 days for training, the next 250 days for validation, and the remaining 1336 days for testing.[6] The values of the regularization parameter ϵ considered were $\{0, 0.1^k, 10^{10}\}$ for k between -6 and 6 at increments of 0.005 (hence a total of 2403 values) for the ℓ_2 method, and for k between 0.05 and 5 at increments of 0.05 (hence a total of 100 values) for the ℓ_1 method, which led to "complete U-curves" below for both methods. We used fewer values for the ℓ_1 method as it is a computationally more costly one. We report the following performance metrics:

- *Cumulative return*: the sum of the 1336 daily returns of the constructed portfolios. It corresponds to the cumulative returns one would get if one had invested using the method over the period's last 1336 days, investing "one unit" every day.

---

[5]Based on the industry classification at *http://www.nasdaq.com/screening/industries. aspx*, sorted by market capitalization, and using only those companies with market capitalization larger than $10 billion in April 2013.

[6]Although the conclusions are similar for other data splits, the problem (discussed below) that the selected parameter ϵ changes across windows (for some of which the performances of the different portfolios tested are similar) makes the analysis of the "average across windows" effects of ϵ on performance (figures below) noisy. To better present this effect we only report the results for one data split.

- *Yearly Sharpe ratio:* the ratio of the average daily return over the standard deviation of the returns in the 1336 days. We scale it by a factor of $\sqrt{250}$ to get an annual Sharpe ratio as typically reported in practice, e.g., [25]. Note that, as we have 1336 test data, any Sharpe ratio larger than $\frac{1.96}{\sqrt{1336/250}} = 0.85$ is statistically significant.

We report the performances of the two methods (ℓ_1 and ℓ_2) using both the regularization parameter ϵ selected using the cumulative returns in the validation data, and the optimal regularization parameter based on the test data (hence with hindsight). The latter indicates potential space for improvement, as discussed below. We also report the performance of three benchmarks:

(a) MARKET: the buy-and-hold of the average of the stocks during the same test window. This is typically used as a benchmark in practice and beating the market, which is a key challenge, means doing better than this benchmark.

(b) MARKET AC: the portfolio x with equal weights (i.e., $1/n$) for all stocks. This corresponds to doing mean-reversion or momentum of the market of the stocks considered.

(c) MAX AC: the maximally autocorrelated portfolio, corresponding to the non-regularization based solution, which is obtained for $\epsilon = 0$.

For comparison in all cases we normalize the solution x so that it has ℓ_1 norm 1 (hence in all cases "we invest 1 unit every day").

Figure 11.2 shows the validation (minus) autocorrelation, the validation cumulative return, and the test cumulative return as a function of the regularization parameter ϵ, both for the case of the ℓ_2 (top row) and ℓ_1 (bottom row) methods. The plots shown are for mean reversion, hence most negatively autocorrelated portfolio, for the healthcare sector; similar conclusions can be drawn from plots for the other sectors and for momentum/most positively autocorrelated portfolio. The figure illustrates the main experimental finding: using the proposed robust optimization approach improves performance and, more interestingly, the observed inverted U-curve indicates that the proposed methodologies capture "structure" even in the highly unpredictable S&P 500 daily stock returns time series.

Using the validation data, we choose between momentum (maximum positive autocorrelation) and mean reversion (minimum negative autocorrelation), and select the regularization parameter for the proposed methods. We report the performances in Table 11.1. For each case we report the cumulative return and the Sharpe ratio (in parentheses) in the test data. From the values in Table 1 we can make the following observations.

1. For all sectors, the proposed approach leads to portfolios that outperform the sector's MARKET. Given that in practice outperforming the market, particularly for stocks in the S&P 500 index, is considered challenging, the results indicate the potential of the proposed approach.

2. For all sectors, regularization improves performance relative to the case $\epsilon = 0$ (MAX AC).

3. For all sectors the best ϵ with hindsight is, as expected, (much) better than the performance of the selected ϵ. This indicates that there can be further improvements if a better method to select ϵ for this data is developed. Note that Figure 11.2 also illustrates this challenge of selecting ϵ for the specific financial data: the best performing regularization parameter ϵ for the validation data may differ from that in the test data. Although the "inverse U-curves" are observed for different time windows indicating that the proposed methods capture structure in this data, selecting the regularization parameter in a "rolling window" setup can be a challenge in practice as the "U-curve" may shift across time windows (e.g., this structure may be non-stationary).

4. The largest performance improvement is for the financial sector. This indicates that it may be the case that the proposed methods work better for certain groups of time series/stocks.[7] Future work can improve our understanding of the characteristics of such groups of time series and/or lead to other methods for different types of groups of time series.

11.6 Conclusion and Future Research

We proposed an approach to estimate large autocorrelation portfolios using regularization methods derived from a robust optimization formulation. We developed two regularization methods as special cases of the proposed general approach. For one of these methods we developed an iterative optimization learning algorithm that estimates sparse portfolios. We then tested the methods using notoriously noisy financial time series data. The experiments indicate the potential of the proposed approach to uncover structure in time series of daily S&P 500 stock returns. The results also indicate that the proposed method can lead to portfolios that outperform "the market" for this data. Finally, we discussed an extension and a novel algorithm for the more general case of CCA.

A number of future research directions can further improve the proposed approach. One of the key questions is the selection of the regularization parameter for (non-stationary, among others) time series such as the stock data we explored. Another question may be to build on the proposed methods in order to better identify subsets of time series for which the approach performs best.

---

[7]Note however that short selling financial companies may require higher transaction costs, but as noted we do not consider such costs in this chapter.

Yet another direction for research is to further develop the proposed CCA approach and test it using potentially diverse "predictors", e.g., for the financial time series explored. Finally, potentially novel regularization methodologies for time series analysis can be developed based on the robust optimization approach used in this chapter.

11.7   Appendix: Robust CCA Algorithm

We describe the main steps behind Algorithm 11, which solves the robust CCA problem. If f is a convex function we denote by $\text{Lin}_{f,u}(z|z^0) = f(z^0) + u^\top(z - z^0)$ a linear approximation of f at z^0, for some $u \in \partial f(z^0)$. We sometimes omit u and write $\text{Lin}_f(z|z^0)$ to denote a generic linear approximation. This approximation can be visualized as a linear lower bound for f, which touches f at z^0.

We first rewrite Problem (11.13) as that of minimizing the quantity

$$\eta(x, y) = \frac{-x^\top \hat{\Theta} y + \epsilon\|x\|_q\|y\|_q}{\sqrt{x^\top \Gamma x}\sqrt{y^\top \Sigma y}} \tag{11.15}$$

over $x \in \mathbb{R}^n$ and $y \in \mathbb{R}^m$. Recall the notation

$$z = \begin{pmatrix} x \\ y \end{pmatrix}, \quad A = \begin{bmatrix} 0 & \hat{\Theta} \\ \hat{\Theta}^\top & 0 \end{bmatrix}, \quad B = \begin{bmatrix} \Gamma & 0 \\ 0 & \Sigma \end{bmatrix}$$

and note that $-2x^\top \hat{\Theta} y = z^\top A z$. Using the equality $2\|x\|_q\|y\|_q = (\|x\|_q + \|y\|_q)^2 - \|x\|_q^2 - \|y\|_q^2$ and the formula $A = P - N$, for $P = (A)_+$ and $N = P - A$, twice the numerator in (11.15) has the DC decomposition (difference of convex functions)

$$R_1(z) - R_2(z) = \left(z^\top N z + \epsilon\|z\|_{1,q}^2\right) - \left(z^\top P z + \epsilon\|z\|_{2,q}^2\right)$$

where we defined the mixed norms $\|z\|_{1,q} = \|x\|_q + \|y\|_q$ and $\|z\|_{2,q} = \sqrt{\|x\|_q^2 + \|y\|_q^2}$. Using this formula, problem (11.13) can be rewritten as

$$\min_z \left\{ \frac{z^\top N z - z^\top P z + \epsilon\|z\|_{1,q}^2 - \epsilon\|z\|_{2,q}^2}{2\sqrt{x^\top \Gamma x}\sqrt{y^\top \Sigma y}} \right\}. \tag{11.16}$$

For every vector $z^0 \in \mathbb{R}^{n+m}$ we shall construct a function $\phi(\cdot|z^0)$ that has the property that if $\phi(z|z^0) < 0$ then $\eta(z) < \eta(z^0)$. We distinguish between two cases.

Case 1: $\eta(z^0) \leq 0$. We replace the denominator in (11.16) by the quadratic form $z^\top B z$, which provides a simplification of the problem. Indeed, if the objective is negative by the arithmetic-geometric mean inequality, we obtain that

$$\eta(z) \le h(z) := \frac{R(z)}{z^\top B z}$$

and equality holds if and only if $x^\top \Gamma x = y^\top \Sigma y$.

Fix $z^0 \in \mathbb{R}^{n+m}$ and let $\phi(\cdot|z^0)$ be the convex function defined, for every $z \in \mathbb{R}^{n+m}$, as

$$\phi(z|z^0) = R_1(z) - \mathrm{Lin}_{R_2}(z, z^0) - \eta(z^0)z^\top B z. \tag{11.17}$$

Note that $\phi(z^0|z^0) = 0$. If z is a point such that $\phi(z|z^0) < 0$, we conclude that $h(z) < h(z^0)$. Indeed,

$$0 > \phi(z|z^0) \ge R_1(z) - R_2(z) - h(z^0)z^\top B z.$$

Furthermore, if we rescale z so that $x^\top \Gamma x = y^\top \Sigma y$, then $\eta(z) = h(z)$.

Case 2: $\eta(z^0) > 0$. In this case we need to work with the original denominator, $\sqrt{x^\top \Gamma x}\sqrt{y^\top \Sigma y}$. We rewrite twice this quantity as the DC decomposition

$$S_1(z) - S_2(z) = \left(\|\Gamma^{\frac{1}{2}}x\|_2 + \|\Sigma^{\frac{1}{2}}y\|_2 \right)^2 - x^\top \Gamma x - y^\top \Sigma y.$$

Now, we choose

$$\phi(z|z^0) = R_1(z) - \mathrm{Lin}_{R_2}(z, z^0) - \eta(z^0)\left(\mathrm{Lin}_{S_1}(z|z^0) - S_2(z) \right). \tag{11.18}$$

This case is slightly more difficult to handle since we also need to compute a subgradient of the function S_1. Combining [18, Chapter VI, Theorems 4.2.1 and 4.3.1], we obtain that

$$\partial \left(\|\Gamma^{\frac{1}{2}}x\|_2 + \|\Sigma^{\frac{1}{2}}y\|_2 \right)^2 =$$
$$2(\|\Gamma^{\frac{1}{2}}x\|_2 + \|\Sigma^{\frac{1}{2}}y\|_2)\left\{ (\Gamma^{\frac{1}{2}}\alpha, \Sigma^{\frac{1}{2}}\beta) : \alpha \in \partial\|\Gamma^{\frac{1}{2}}x\|_2, \beta \in \partial\|\Sigma^{\frac{1}{2}}y\|_2 \right\}.$$

We are now ready to summarize the formula for $\phi(z|z^0)$. Combining equations (11.17) and (11.18) we have

$$\phi(z|z^0) = z^\top(N + |\eta(z^0)|B)z - (z - z^0)^\top(2Pz^0 + \epsilon u + 1_{\{\eta(z^0)>0\}}v)$$

where $u \in \partial\|z^0\|_{2,q}^2$ and $v \in \partial S_1(z^0)$.

Algorithm 11 solves a sequence of convex optimization problems of the form

$$\min\{\phi(z|z^k) : z \in \mathbb{R}^{n+m}\}.$$

In practice it is sufficient to solve this problem approximately, finding a point z such that $\phi(z|z^k) < 0$. Similarly to Lemma 11.2 it is easily seen that if $\phi(z|z^k) < 0$ then $\eta(z) < \eta(z^0)$. The simplest updating rule is provided by one step of the proximal gradient method [4, 24]

$$z^{k+1} = \mathrm{prox}_{r\|\cdot\|_{1,q}^2}\left[z^k - \frac{1}{\|\|N + |\eta^k|B\|\|}\left((A + |\eta^k|B)z^k - \frac{\epsilon u + (\eta^k)_+v}{2}\right) \right]$$

where, recall, $|||\cdot|||$ denotes the spectral norm of a matrix, $r = \frac{\epsilon}{2|||N+|\eta(z^0)|B|||}$, and we have defined $\eta^k = \eta(z^k)$.

Next, we discuss how to compute a subgradient of $\|z\|_{2,q}^2$ and the proximity operator when $q \in \{1,2\}$. For this purpose, we recall that if $\|\cdot\|$ is a norm, then by [18, S4.3] the subdifferential of $\|\cdot\|^2$ at z is equal to $2\|z\|$ times the subdifferential of $\|\cdot\|$ at z, that is $\partial\|z\|^2 = 2\|z\|\partial\|z\|$. In particular, we obtain that

$$\partial\|z\|_{2,1}^2 = \{2(\|x\|_1\alpha, \|y\|_1\beta) : \alpha \in \partial\|x\|_1, \ \beta \in \partial\|y\|_1\}.$$

On the other hand, if $q = 2$ then $\|z\|_{2,2}^2$ is just the square ℓ_2 norm of z and its gradient is equal to $2z$.

It remains to obtain the formula for the proximity operator defined above. The case $q = 1$ is conceptually identical to the derivation of the proximity operator presented at the end of Section 11.3, with the understanding that n is replaced by $n + m$ and x by the vector z. The case $q = 2$ is derived along the same lines and we only sketch the main points here. We obtain, for every $z = (x, y) \in \mathbb{R}^{n+m}$, that

$$\mathrm{prox}_{r\|z\|_{1,2}^2}(z) = \left(\frac{\lambda x}{2r + \lambda}, \frac{\tau y}{2r + \tau}\right)$$

where $\lambda = (\rho\|x\|_2 - 2r)_+$, $\tau = (\rho\|y\|_2 - 2r)_+$ and the positive parameter ρ is determined by binary search in order to ensure that $\lambda + \tau = 1$.

Bibliography

[1] A. d'Aspremont. Identifying small mean-reverting portfolios. *Quantitative Finance*, 11(3):351–364, 2011.

[2] T.M. Anderson. *An Introduction to Multivariate Statistical Analysis.* John Wiley & Sons, second edition, 1984.

[3] M. Avellaneda and J.H. Lee. Statistical arbitrage in the U.S. equities market. *Quantitative Finance*, 10:761-782, 2010.

[4] A. Beck and M. Teboulle. A fast iterative shrinkage-thresholding algorithm for linear inverse problems. *SIAM Journal of Imaging Sciences*, 2 (1):183–202, 2009.

[5] A. Ben-Tal, L. El Ghaoui, and A. Nemirovski. *Robust Optimization*, Princeton University Press, 2009.

[6] M. Borga. *Learning Multidimensional Signal Processing.* Linkping Studies in Science and Technology. Dissertations No. 531, Linköping University, Sweden, 1998.

[7] J.Y. Campbell, A.W.C. Lo, and A.C. MacKinlay. *The Econometrics of Financial Markets.* Princeton University Press, 1997.

[8] G. Cornuejols and R. Tütüncü. *Optimization Methods in Finance.* Cambridge University Press, 2011.

[9] K.-C. Chang. Variational methods for non-differentiable functionals and their applications to partial differential equations. *Journal of Mathematical Analysis and Applications,* 80:102-129, 1981.

[10] F.H. Clarke. *Optimization and Nonsmooth Analysis.* Wiley, 1983.

[11] V. DeMiguel and F.J. Nogales. Portfolio selection with robust estimation. *Operations Research,* 57(3):560–577, 2009.

[12] G.H. Golub and C.F. Van Loan. *Matrix Computations.* John Hopkins University Press, 1996.

[13] H. Hotelling. Relations between two sets of variates. *Biometrika,* 28:321–377, 1936.

[14] T.-K. Kim, J. Kittler, and R. Cipolla. Discriminative learning and recognition of image set classes using canonical correlations. *IEEE Trans. Pattern Analysis and Machine Intelligence,* 29(6):1005–1018, 2007.

[15] D.R. Hardoon, S. Szedmak, and J. Shawe-Taylor. Canonical correlation analysis: An overview with application to learning methods. *Neural Computation,* 16(12):2639–2664, 2004.

[16] M. Hein and T. Buehler. An inverse power method for nonlinear eigenproblems with applications in 1-spectral clustering and sparse PCA. *Advances in Neural Information Processing Systems 23,* pages 847–855, 2010.

[17] M. Hein and S. Setzer, Beyond spectral clustering - tight relaxations of balanced graph cuts. *Advances in Neural Information Processing Systems 24,* pages 2366–2374, 2011.

[18] J.-B. Hiriart-Urruty and C. Lemaréchal. *Convex Analysis and Minimization Algorithms, Part I.* Springer, 1996.

[19] A.W. Lo and A.C. MacKinlay. When are contrarian profits due to stock market overreaction? *Review of Financial studies,* 1990.

[20] A. Maurer. Unsupervised slow subspace-learning from stationary processes. *Theoretical Computer Science,* 405(3):237–255, 2008.

[21] C.A. Micchelli and M. Pontil. Learning the kernel function via regularization. *J. Machine Learning Research,* 6:1099-1125, 2005.

[22] H.Q. Minh and L. Wiskott. Multivariate slow feature analysis and decorrelation filtering for blind source separation. *IEEE Transactions on Image Processing*, 22(7):2737–2750, 2013.

[23] J. Moody and M. Saffell. Learning to trade via direct reinforcement. *Neural Networks, IEEE Transactions on*, 12(4):875–889, 2001.

[24] Y. Nesterov. Gradient methods for minimizing composite objective functions. *ECORE Discussion Paper*, 2007/96, 2007.

[25] R.S. Tsay. *Analysis of Financial Time Series*. John Wiley & Sons, 2002.

[26] P. Xanthopoulos, M.R. Guarracino, and P.M. Pardalos. Robust generalized eigenvalue classifier with ellipsoidal uncertainty. *Annals of Operations Research*, 2013.

[27] L. Wiskott and T.J. Sejnowski. Slow feature analysis: unsupervised learning of invariances. *Neural Computation*, 14:715–770, 2002.

[28] D.M. Witten, R. Tibshirani, and T. Hastie. A penalized matrix decomposition, with applications to sparse principal components and canonical correlation analysis. *Biostatistics*, 10(3):515–534, 2009.

Chapter 12

The Why and How of Nonnegative Matrix Factorization

Nicolas Gillis*

Department of Mathematics and Operational Research, Faculté Polytechnique,
Université de Mons

---

*nicolas.gillis@umons.ac.be

12.1   Summary

Nonnegative matrix factorization (NMF) has become a widely used tool for the analysis of high-dimensional data as it automatically extracts sparse and meaningful features from a set of nonnegative data vectors. We first illustrate this property of NMF on three applications, in image processing, text mining, and hyperspectral imaging — this is the why. Then we address the problem of solving NMF, which is NP-hard in general. We review some standard NMF algorithms, and also present a recent subclass of NMF problems, referred to as near-separable NMF, that can be solved efficiently (that is, in polynomial time), even in the presence of noise — this is the how. Finally, we briefly describe some problems in mathematics and computer science closely related to NMF via the nonnegative rank.

12.2   Introduction

Linear dimensionality reduction (LDR) techniques are a key tool in data analysis, and are widely used: for example, for compression, visualization, feature selection, and noise filtering. Given a set of data points $x_j \in \mathbb{R}^p$ for $1 \leq j \leq n$ and a dimension $r < \min(p, n)$, LDR amounts to computing a set of r basis elements $w_k \in \mathbb{R}^p$ for $1 \leq k \leq r$ such that the linear space spanned by the w_k's approximates the data points as closely as possible, that is, such that we have for all j

$$x_j \approx \sum_{k=1}^{r} w_k h_j(k), \qquad \text{for some weights } h_j \in \mathbb{R}^r. \qquad (12.1)$$

In other words, the p-dimensional data points are represented in an r-dimensional linear subspace spanned by the basis elements w_k's and whose coordinates are given by the vectors h_j's. LDR is equivalent to low-rank matrix approximation: in fact, constructing

- the matrix $X \in \mathbb{R}^{p \times n}$ such that each column is a data point, that is, $X(:, j) = x_j$ for $1 \leq j \leq n$,

- the matrix $W \in \mathbb{R}^{p \times r}$ such that each column is a basis element, that is, $W(:, k) = w_k$ for $1 \leq k \leq r$, and

- the matrix $H \in \mathbb{R}^{r \times n}$ such that each column of H gives the coordinates of a data point $X(:, j)$ in the basis W, that is, $H(:, j) = h_j$ for $1 \leq j \leq n$, the above LDR model (12.1) is equivalent to $X \approx WH$, that is, to approximate the data matrix X with a low-rank matrix WH.

A first key aspect of LDR is the choice of the measure to assess the quality of the approximation. It should be chosen depending on the *noise model*. The most widely used measure is the Frobenius norm of the error, that is, $||X - WH||_F^2 = \sum_{i,j}(X - WH)_{ij}^2$. The reason for the popularity of the Frobenius norm is two-fold. First, it implicitly assumes the noise N present in the matrix $X = WH + N$ to be Gaussian, which is reasonable in many practical situations (see also the introduction of Section 12.4). Second, an optimal approximation can be computed efficiently through the truncated singular value decomposition (SVD); see [57] and the references therein. Note that the SVD is equivalent to principal component analysis (PCA) after mean centering of the data points (that is, after shifting all data points so that their mean is on the origin).

A second key aspect of LDR is the assumption on the structure of the factors W and H. The truncated SVD and PCA do not make any assumption on W and H. For example, assuming independence of the columns of W leads to independent component analysis (ICA) [29], or assuming sparsity of W (and/or H) leads to sparse low-rank matrix decompositions, such as sparse PCA [32]. Nonnegative matrix factorization (NMF) is an LDR where both the basis elements w_k's and the weights h_j's are assumed to be *component-wise nonnegative*. Hence NMF aims at decomposing a given nonnegative data matrix X as $X \approx WH$ where $W \geq 0$ and $H \geq 0$ (meaning that W and H are component-wise nonnegative). NMF was first introduced in 1994 by Paatero and Tapper [97] and gathered more and more interest after an article by Lee and Seung [79] in 1999.

In this paper, we explain *why* NMF has been so popular in different data mining applications, and *how* one can compute NMF's. The aim of this paper is not to give a comprehensive overview of all NMF applications and algorithms (and we apologize for not mentioning many relevant contributions) but rather to serve as an introduction to NMF, describing three applications and several standard algorithms.

12.3   The Why — NMF Generates Sparse and Meaningful Features

The reason why NMF has become so popular is because of its ability to automatically extract sparse and easily interpretable factors. In this section, we illustrate this property of NMF through three applications, in image processing, text mining, and hyperspectral imaging. Other applications include air emission control [97], computational biology [34], blind source separation [22], single-channel source separation [82], clustering [35], music analysis [42], collaborative filtering [92], and community detection [106].

12.3.1   Image Processing — Facial Feature Extraction

Let each column of the data matrix $X \in \mathbb{R}_+^{p \times n}$ be a vectorized gray-level image of a face, with the (i,j)th entry of matrix X being the intensity of the ith pixel in the jth face. NMF generates two factors (W, H) so that each image $X(:,j)$ is approximated using a linear combination of the columns of W; see Equation (12.1), and Figure 12.1 for an illustration. Since W is

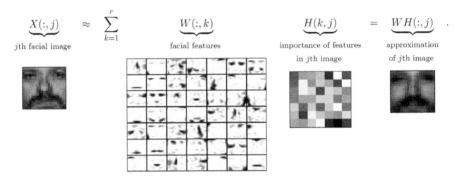

$$\underbrace{X(:,j)}_{j\text{th facial image}} \approx \sum_{k=1}^{r} \underbrace{W(:,k)}_{\text{facial features}} \underbrace{H(k,j)}_{\substack{\text{importance of features} \\ \text{in } j\text{th image}}} = \underbrace{WH(:,j)}_{\substack{\text{approximation} \\ \text{of } j\text{th image}}} .$$

FIGURE 12.1: Decomposition of the CBCL face database, MIT Center for Biological and Computation Learning (2429 gray-level 19-by-19 pixels images) using $r = 49$ as in [79].

nonnegative, the columns of W can be interpreted as images (that is, vectors of pixel intensities), which we refer to as the basis images. As the weights in the linear combinations are nonnegative ($H \geq 0$), these basis images can only be summed up to reconstruct each original image. Moreover, the large number of images in the data set must be reconstructed approximately with only a few basis images (in fact, r is in general much smaller than n), hence the latter should be localized features (hence sparse) found simultaneously in several images. In the case of facial images, the basis images are features such as eyes, noses, mustaches, and lips (see Figure 12.1) while the columns of H indicate which feature is present in which image (see also [79, 61]).

A potential application of NMF is in face recognition. It has for example been observed that NMF is more robust to occlusion than PCA (which generates dense factors): in fact, if a new occluded face (e.g., with sunglasses) has to be mapped into the NMF basis, the non-occluded parts (e.g., the mustache or the lips) can still be well approximated [61].

12.3.2 Text Mining — Topic Recovery and Document Classification

Let each column of the nonnegative data matrix X correspond to a document and each row to a word. The (i, j)th entry of the matrix X could for example be equal to the number of times the ith word appears in the jth document in which case each column of X is the vector of word counts of a document; in practice, more sophisticated constructions are used, e.g., the term frequency-inverse document frequency (tf-idf). This is the so-called bag-of-words model: each document is associated with a set of words with different weights, while the ordering of the words in the documents is not taken into account (see, e.g., the survey [10] for a discussion). Note that such a matrix X is in general rather sparse as most documents only use a small subset of the dictionary. Given such a matrix X and a factorization rank r, NMF generates two factors (W, H) such that, for all $1 \leq j \leq n$, we have

$$\underbrace{X(:, j)}_{j\text{th document}} \approx \sum_{k=1}^{r} \underbrace{W(:, k)}_{k\text{th topic}} \underbrace{H(k, j)}_{\substack{\text{importance of } k\text{th topic} \\ \text{in } j\text{th document}}}, \quad (12.2)$$

where $W \geq 0$ and $H \geq 0$. This decomposition can be interpreted as follows (see, also, e.g., [79, 101, 3]):

- Because W is nonnegative, each column of W can be interpreted as a document, that is, as a bag of words.

- Because the weights in the linear combinations are nonnegative ($H \geq 0$), one can only take the union of the sets of words defined by the columns of W to reconstruct all the original documents.

- Moreover, because the number of documents in the data set is much larger than the number of basis elements (that is, the number of columns of W), the latter should be a set of words found simultaneously in several documents. Hence the basis elements can be interpreted as *topics*, that is, sets of words found simultaneously in different documents, while the weights in the linear combinations (that is, the matrix H) assign the documents to the different topics, that is, identify which document discusses which topic.

Therefore, given a set of documents, *NMF identifies topics and simultaneously classifies the documents among these different topics*. Note that NMF is closely related to existing topic models, in particular probabilistic latent semantic analysis and indexing (PLSA and PLSI) [45, 37].

12.3.3   Hyperspectral Unmixing — Identify Endmembers and Classify Pixels

Let the columns of the nonnegative data matrix X be the spectral signatures of the pixels in a scene being imaged. The spectral signature of a pixel is the fraction of incident light being reflected by that pixel at different wavelengths, and is therefore nonnegative. For a hyperspectral image, there are usually between 100 and 200 wavelength-indexed bands, observed in much broader spectrum than the visible light. This allows for more accurate analysis of the scene under study.

Given a hyperspectral image (see Figure 12.2 for an illustration), the goal

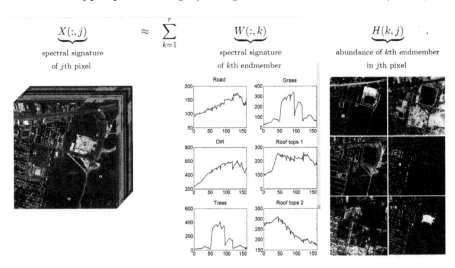

$$\underbrace{X(:,j)}_{\substack{\text{spectral signature} \\ \text{of } j\text{th pixel}}} \quad \approx \quad \sum_{k=1}^{r} \quad \underbrace{W(:,k)}_{\substack{\text{spectral signature} \\ \text{of } k\text{th endmember}}} \quad \underbrace{H(k,j)}_{\substack{\text{abundance of } k\text{th endmember} \\ \text{in } j\text{th pixel}}} .$$

FIGURE 12.2: Decomposition of the urban hyperspectral image from http://www.agc.army.mil/, constituted mainly of six endmembers ($r = 6$). Each column of the matrix W is the spectral signature of an endmember, while each row of the matrix H is the abundance map of the corresponding endmember, that is, it contains the abundance of all pixels for that endmember. (Note that to obtain this decomposition, we used a sparse prior on the matrix H; see Section 12.4.)

of blind hyperspectral unmixing (blind HU) is two-fold:

1. Identify the constitutive materials present in the image; for example, it could be grass, roads, or metallic surfaces. These are referred to as the *endmembers.*

2. Classify the pixels, that is, identify which pixel contains which endmember and in which proportion. (In fact, pixels are in general mixture of several endmembers, due for example to low spatial resolution or mixed materials.)

The simplest and most popular model used to address this problem is the *linear mixing model*. It assumes that the spectral signature of a pixel results from the linear combination of the spectral signature of the endmembers it contains. The weights in the linear combination correspond to the abundances of these endmembers in that pixel. For example, if a pixel contains 30% of grass and 70% of road surface, then, under the linear mixing model, its spectral signature will be 0.3 times the spectral signature of the grass plus 0.7 times the spectral signature of the road surface. This is exactly the NMF model: the spectral signatures of the endmembers are the basis elements, that is, the columns of W, while the abundances of the endmembers in each pixel are the weights, that is, the columns of H. Note that the factorization rank r corresponds to the number of endmembers in the hyperspectral image. Figure 12.2 illustrates such a decomposition.

Therefore, given a hyperspectral image, *NMF is able to compute the spectral signatures of the endmembers and simultaneously the abundance of each endmember in each pixel*. We refer the reader to [8, 90] for recent surveys on blind HU techniques.

12.4   The How — Some Algorithms

We have seen in the previous section that NMF is a useful LDR technique for nonnegative data. The question is now: can we compute such factorizations? In this paper, we focus on the following optimization problem

$$\min_{W \in \mathbb{R}^{p \times r}, H \in \mathbb{R}^{r \times n}} ||X - WH||_F^2 \quad \text{such that} \quad W \geq 0 \text{ and } H \geq 0. \quad (12.3)$$

Hence we implicitly assume Gaussian noise on the data; see Introduction. Although this NMF model is arguably the most popular, it is not always reasonable to assume Gaussian noise for nonnegative data, especially for sparse matrices such as document data sets; see the discussion in [23]. In fact, many other objective functions are used in practice, e.g., the (generalized) Kullback-Leibler divergence for text mining [23], the Itakura-Saito distance for music analysis [42], the ℓ_1 norm to improve robustness against outliers [71], and the earth mover's distance for computer vision tasks [100]. Other NMF models are motivated by statistical considerations; we refer the reader to the recent survey [102].

There are many issues when using NMF in practice. In particular,

- **NMF is NP-hard**. Unfortunately, as opposed to the unconstrained problem that can be solved efficiently using the SVD, NMF is NP-hard in general [105]. Hence, in practice, most algorithms are applications of standard nonlinear optimization methods and may only be guaranteed to converge to stationary points; see Section 12.4.1. However, these heuristics have been proved to be successful in many applications. More recently, Arora et al. [4] described a subclass of nonnegative matrices for which NMF can be solved efficiently. These are the near-separable matrices that will be addressed in Section 12.4.2. Note that Arora et al. [4] also described an algorithmic approach for exact NMF[1] requiring $\mathcal{O}\big((pn)^{2^r r^2}\big)$ operations (later improved to $\mathcal{O}\big((pn)^{r^2}\big)$ by Moitra [94]) hence polynomial in the dimensions p and n for r fixed. Although r is usually small in practice, this approach cannot be used to solve real-world problems because of its high computational cost (in contrast, most heuristic NMF algorithms run in $\mathcal{O}(pnr)$ operations; see Section 12.4.1).

- **NMF is ill-posed.** Given an NMF (W, H) of X, there usually exist equivalent NMF's (W', H') with $W'H' = WH$. In particular, any matrix Q satisfying $WQ \geq 0$ and $Q^{-1}H \geq 0$ generates such an equivalent factorization. The matrix Q can always be chosen as the permutation of a diagonal matrix with positive diagonal elements (that is, as a monomial matrix) and this amounts to the scaling and permutation of the rank-one factors $W(:, k)H(k, :)$ for $1 \leq k \leq r$; this is not an issue in practice. The issue is when there exist non-monomial matrices Q satisfying the above conditions. In that case, such equivalent factorizations generate different interpretations: for example, in text mining, they would lead to different topics and classifications; see the discussion in [48]. Here is a simple example

$$\begin{pmatrix} 0 & 1 & 1 & 1 \\ 1 & 0 & 1 & 1 \\ 1 & 1 & 0 & 1 \end{pmatrix} = \begin{pmatrix} 0 & 1 & 1 \\ 1 & 0 & 1 \\ 1 & 1 & 0 \end{pmatrix} \begin{pmatrix} 1 & 0 & 0 & 0.5 \\ 0 & 1 & 0 & 0.5 \\ 0 & 0 & 1 & 0.5 \end{pmatrix}$$

$$= \begin{pmatrix} 1 & 0 & 0 \\ 0 & 1 & 0 \\ 0 & 0 & 1 \end{pmatrix} \begin{pmatrix} 0 & 1 & 1 & 1 \\ 1 & 0 & 1 & 1 \\ 1 & 1 & 0 & 1 \end{pmatrix}.$$

We refer the reader to [66] and the references therein for recent results on non-uniqueness of NMF.

In practice, this issue is tackled using other priors on the factors W and H and adding proper regularization terms in the objective function. The most popular prior is sparsity, which can be tackled with projections [64] or with ℓ_1-norm penalty [72, 48]. For example, in blind HU (Section 12.3.3), the abundance maps (that is, the rows of matrix H) are usually very sparse (most pixels contain only a few endmembers) and applying plain NMF (12.3) usually gives poor results for these data sets. Other priors for blind HU include piece-wise smoothness of the spectral signatures or spatial coherence (neighboring pixels are more likely to contain the same materials), which are usually tackled with TV-like

---

[1] Exact NMF refers to the NMF problem where an exact factorization is sought: $X = WH$ with $W \geq 0$ and $H \geq 0$.

regularizations (that is, ℓ_1 norm of the difference between neighboring values to preserve the edges in the image); see, e.g., [68, 67], and the references therein. Note that the design and algorithmic implementation of refined NMF models for various applications is a very active area of research, e.g., graph regularized NMF [16], orthogonal NMF [24], tri-NMF [38, 85], semi and convex NMF [36], projective NMF [110], minimum volume NMF [93], and hierarchical NMF [86], to cite only a few.

- **Choice of r.** The choice of the factorization rank r, that is, the problem of order model selection, is usually rather tricky. Several popular approaches are: trial and error (that is, try different values of r and pick the one performing best for the application at hand), estimation using the SVD (that is, look at the decay of the singular values of the input data matrix), and the use of experts' insights (e.g., in blind HU, experts might have a good guess for the number of endmembers present in a scene); see also [7, 104, 70] and the references therein.

In this section, we focus on the first issue. In Section 12.4.1, we present several standard algorithms for the general problem (12.3). In Section 12.4.2, we describe the near-separable NMF problem and several recent algorithms.

12.4.1 Standard NMF Algorithms

Almost all NMF algorithms designed for (12.3) use a two-block coordinate descent scheme (exact or inexact; see below), that is, they optimize alternatively over one of the two factors, W or H, while keeping the other fixed. The reason is that the subproblem in one factor is convex. More precisely, it is a nonnegative least squares problem (NNLS): for example, for H fixed, we have to solve $\min_{W \geq 0} ||X - WH||_F^2$. Note that this problem has a particular structure as it can be decomposed into p independent NNLS in r variables since

$$||X - WH||_F^2 = \sum_{i=1}^{p} ||X_{i:} - W_{i:}H||_2^2$$

$$= \sum_{i=1}^{p} W_{i:} \left(HH^T \right) W_{i:}^T - 2W_{i:} \left(HX_{i:}^T \right) + ||X_{i:}||_2^2. \quad (12.4)$$

Many algorithms exist to solve the NNLS problem, and NMF algorithms based on two-block coordinate descent differ by which NNLS algorithm is used; see also, e.g., the discussion in [74]. It is interesting to notice that the problem is symmetric in W and H since $||X - WH||_F^2 = ||X^T - H^T W^T||_F^2$. Therefore, we can focus on the update of only one factor and, in fact, most NMF algorithms use the same update for W and H, and therefore adhere to the framework described in Algorithm 12.

Algorithm 12 Two-Block Coordinate Descent–Framework of Most NMF Algorithms

---

Input: Input nonnegative matrix $X \in \mathbb{R}_+^{p \times n}$ and factorization rank r.
Output: $(W, H) \geq 0$: A rank-r NMF of $X \approx WH$.

1: Generate some initial matrices $W^{(0)} \geq 0$ and $H^{(0)} \geq 0$; see Section 12.4.1.8.
2: **for** $t = 1, 2, \ldots$ † **do**
3:     $W^{(t)} = \text{update}\big(X, H^{(t-1)}, W^{(t-1)}\big)$.
4:     $H^{(t)T} = \text{update}\big(X^T, W^{(t)T}, H^{(t-1)T}\big)$.
5: **end for**
    †See Section 12.4.1.7 for stopping criteria.

The update in steps 3 and 4 of Algorithm 12 usually guarantees the objective function to decrease. In this section, we describe the most widely used updates, that is, we describe several standard and widely used NMF algorithms, and compare them in Section 12.4.1.6. But first we address an important tool to designing NMF algorithms: the optimality conditions. To simplify notations, we will drop the iteration index t.

12.4.1.1   First-Order Optimality Conditions

Given X, let us denote $F(W, H) = \frac{1}{2}\|X - WH\|_F^2$. The first-order optimality conditions for (12.3) are

$$W \geq 0, \quad \nabla_W F = WHH^T - XH^T \geq 0, \quad W \circ \nabla_W F = 0, \quad (12.5)$$
$$H \geq 0, \quad \nabla_H F = W^TWH - W^TX \geq 0, \quad H \circ \nabla_H F = 0,$$

where $\circ$ is the component-wise product of two matrices. Any (W, H) satisfying these conditions is a stationary point of (12.3).

It is interesting to observe that these conditions give a more formal explanation of why NMF naturally generates sparse solutions [51]: in fact, any stationary point of (12.3) is expected to have zero entries because of the conditions $W \circ \nabla_W F = 0$ and $H \circ \nabla_H F = 0$, that is, the conditions that for all i, k either W_{ik} is equal to zero or the partial derivative of F with respect to W_{ik} is, and similarly for H.

12.4.1.2   Multiplicative Updates

Given X, W, and H, the multiplicative updates (MU) modify W as follows

$$W \leftarrow W \circ \frac{[XH^T]}{[WHH^T]} \qquad (12.6)$$

where $\dfrac{}{}$ denotes the component-wise division between two matrices. The MU were first developed in [33] for solving NNLS problems, and later rediscovered and used for NMF in [80]. The MU are based on the majorization-minimization framework. In fact, (12.6) is the global minimizer of a quadratic function majorizing F, that is, a function that is larger than F everywhere and is equal to F at the current iterate [33, 80]. (Note that the majorizing function is separable, that is, its Hessian is diagonal which explains why its global minimizer over the nonnegative orthant can be written in closed form, which is not possible for F.) Hence minimizing that function guarantees F to decrease and therefore leads to an algorithm for which F monotonically decreases. The MU can also be interpreted as a rescaled gradient method: in fact,

$$W \circ \frac{[XH^T]}{[WHH^T]} = W - \frac{[W]}{[WHH^T]} \circ \nabla_W F. \qquad (12.7)$$

Another more intuitive interpretation is as follows: we have that

$$\frac{[XH^T]_{ik}}{[WHH^T]_{ik}} \geq 1 \quad \Longleftrightarrow \quad (\nabla_W F)_{ik} \leq 0.$$

Therefore, in order to satisfy (12.5), for each entry of W, the MU either (i) increase it if its partial derivative is negative, (ii) decrease it if its partial derivative is positive, or (iii) leave it unchanged if its partial derivative is equal to zero.

If an entry of W is equal to zero, the MU cannot modify it, hence it may occur that an entry of W is equal to zero while its partial derivative is negative, which would not satisfy (12.5). Therefore, the MU are not guaranteed to converge to a stationary point[2]. There are several ways to fix this issue, e.g., rewriting the MU as a rescaled gradient descent method, see Equation (12.7): only entries in the same row interact, and modifying the step length [87], or using a small positive lower bound for the entries of W and H [52, 103]; see also [5]. A simpler and nice way to guarantee convergence of the MU to a stationary point is proposed in [23]: use the original updates (12.6) while reinitializing zero entries of W to a small positive constant when their partial derivatives become negative.

The MU became extremely popular mainly because (i) they are simple to implement[3], (ii) they scale well and are applicable to sparse matrices[4], and (iii) they were proposed in the paper of Lee and Seung [79], which launched the research on NMF. However, the MU converge relatively slowly; see, e.g., [62] for a theoretical analysis, and Section 12.4.1.6 for some numerical experiments. Note that the original MU only update W once before updating H. They can be significantly accelerated using a more effective alternation strategy [52]: the

---

[2]If the initial matrices are chosen positive, some entries can first converge to zero while their partial derivative eventually becomes negative or zero (when strict complementarity is not met), which is numerically unstable; see [52] for some numerical experiments.

[3]For example, in Matlab: `W = W.*(X*H')./(W*(H*H'))`.

[4]When computing the denominator WHH^T in the MU, it is crucial to compute HH^T first in order to have the lowest computational cost, and make the MU scalable for sparse matrices; see, e.g., footnote 3.

idea is to update W several times before updating H because the products HH^T and XH^T do not need to be recomputed.

12.4.1.3   Alternating Least Squares

The alternating least squares method (ALS) first computes the optimal solution of the unconstrained least squares problem $\min_W ||X - WH||_F$ and then projects the solution onto the nonnegative orthant:

$$W \leftarrow \max\left(\operatorname{argmin}_{Z \in \mathbb{R}^{p \times r}} ||X - ZH||_F, 0\right),$$

where the max is taken component-wise. The method has the advantage to be relatively cheap, and easy to implement[5]. ALS usually does not converge: the objective function of (12.3) might oscillate under the ALS updates (especially for dense input matrices X; see Section 12.4.1.6). It is interesting to notice that, because of the projection, the solution generated by ALS is not scaled properly. In fact, the error can be reduced (sometimes drastically) by multiplying the current solution WH by the constant

$$\alpha^* = \operatorname{argmin}_{\alpha \geq 0} ||X - \alpha WH||_F = \frac{\langle X, WH \rangle}{\langle WH, WH \rangle} = \frac{\langle XH^T, W \rangle}{\langle W^T W, HH^T \rangle}. \quad (12.8)$$

Although it is in general not recommended to use ALS because of the convergence issues, ALS can be rather powerful for initialization purposes (that is, perform a few steps of ALS and then switch to another NMF algorithm), especially for sparse matrices [28].

12.4.1.4   Alternating Nonnegative Least Squares

Alternating nonnegative least squares (ANLS) is a class of methods where the subproblems in W and H are solved exactly, that is, the update for W is given by

$$W \leftarrow \operatorname{argmin}_{W \geq 0} ||X - WH||_F.$$

Many methods can be used to solve the NNLS $\operatorname{argmin}_{W \geq 0} ||X - WH||_F$, and dedicated active-set methods have been shown to perform very well in practice[6]; see [72, 73, 75]. Other methods are based, for example, on projected gradients [88], Quasi-Newton [26], or fast gradient methods [60]. ANLS is guaranteed to converge to a stationary point [59]. Since each iteration of ANLS computes an optimal solution of the NNLS subproblem, each iteration of ANLS decreases the error the most among NMF algorithms following the framework described in Algorithm 12. However, each iteration is computationally more expensive, and more difficult to implement.

Note that, because usually the initial guess WH is a poor approximation

---

[5]For example, in Matlab: `W = max(0,(X*H')/(H*H'))`.
[6]In particular, the Matlab function `lsqnonneg` implements an active-set method from [78].

of X, it does not make much sense to solve the NNLS subproblems exactly at the first steps of Algorithm 12, and therefore it might be profitable to use ANLS rather in a refinement step of a cheaper NMF algorithm (such as the MU or ALS).

12.4.1.5   Hierarchical Alternating Least Squares

Hierarchical alternating least squares (HALS) solves the NNLS subproblem using an exact coordinate descent method, updating one column of W at a time. The optimal solutions of the corresponding subproblems can be written in closed form. In fact, the entries of a column of W do not interact (see Equation (12.4)). Hence the corresponding problem can be decoupled into p quadratic problems with a single nonnegative variable. HALS updates W as follows. For $\ell = 1, 2, \ldots, r$:

$$W(:,\ell) \leftarrow \mathrm{argmin}_{W(:,\ell) \geq 0} \left\| X - \sum_{k \neq \ell} W(:,k)H(k,:) - W(:,\ell)H(\ell,:) \right\|_F$$

$$\leftarrow \max\left(0, \frac{XH(\ell,:)^T - \sum_{k \neq \ell} W(:,k)\left(H(k,:)H(\ell,:)^T\right)}{\|H(\ell,:)\|_2^2}\right).$$

HALS has been rediscovered several times, originally in [27] (see also [25]), then as the rank-one residue iteration (RRI) in [63], as FastNMF in [84], and also in [89]. Actually, HALS was first described in Rasmus Bro's thesis [14, pp. 161–170] (although it was not investigated thoroughly):

> ...to solve for a column vector w of W it is only necessary to solve the unconstrained problem and subsequently set negative values to zero. Though the algorithm for imposing non-negativity is thus simple and may be advantageous in some situations, it is not pursued here. Since it optimizes a smaller subset of parameters than the other approaches it may be unstable in difficult situations.

HALS was observed to converge much faster than the MU (see [47, p. 131] for a theoretical explanation, and Section 12.4.1.6 for a comparison) while having almost the same computational cost; see [52] for a detailed account of the flops needed per iteration. Moreover, HALS is, under some mild assumptions, guaranteed to converge to a stationary point; see the discussion in [52]. Note that one should be particularly careful when initializing HALS, otherwise the algorithm could set some columns of W to zero initially (e.g., if WH is badly scaled with $WH \gg X$), hence it is recommended to initially scale (W, H) according to (12.8); see the discussion in [47, p. 72].

In the original HALS, each column of W is updated only once before updating H. However, as for the MU, it can be sped up by updating W several times before updating H [52], or selecting the entries of W to update following a Gauss-Southwell-type rule [65]. HALS can also be generalized to other cost functions using Taylor expansion [83].

12.4.1.6   Comparison

Figure 12.3 displays the evolution of the objective function of (12.3) for the algorithms described in the previous section: on the top, the dense CBCL data set, and, on the bottom, the sparse Classic document data set.

As anticipated in the description of the different algorithms in the previous sections, we observe that:

- The MU converge rather slowly.

- ALS oscillates for the dense matrix (CBCL data set) and performs quite poorly while, for the sparse matrix (Classic data set), it converges initially very fast but then stabilizes and cannot compute a solution with small objective function value.

- ANLS performs rather well for the dense matrix and is the second best after HALS. However, it performs rather poorly for the sparse matrix.

- HALS performs the best as it generates the best solutions within the allotted time.

For other comparisons of NMF algorithms and more numerical experiments, we refer the reader to the book [28], the theses [63, 47], the survey [6], and the references therein.

Further research on NMF includes the design of more efficient algorithms, in particular for regularized problems; see, e.g., [98] for a recent example of imposing sparsity in a more robust and stable way. We conclude this section with some comments about stopping criteria and initializations of NMF algorithms.

12.4.1.7   Stopping Criterion

There are several approaches for the stopping criterion of NMF algorithms, as in usual non-linear optimization schemes, e.g., based on the evolution of the objective function, on the optimality conditions (12.5), or on the difference between consecutive iterates. These criteria are typically combined with either a maximum number of iterations or a time limit to ensure termination; see, e.g., the discussion in [47]. In this section, we would like to point out an issue that is sometimes overlooked in the literature when using the optimality conditions to assess the convergence of NMF algorithms. A criterion based on the optimality conditions is, for example, $C(W, H) = C_W(W) + C_H(H)$ where

$$C_W(W) = \underbrace{|| \min(W, 0)||_F}_{W \geq 0} + \underbrace{|| \min(\nabla_W F, 0)||_F}_{\nabla_W F \geq 0} + \underbrace{||W \circ \nabla_W F||_F}_{W \circ \nabla_W F = 0}, \quad (12.9)$$

and similarly $C_H(H)$ for H, so that $C(W, H) = 0 \iff (W, H)$ is a stationary point of (12.3). There are several problems to using $C(W, H)$ (and other

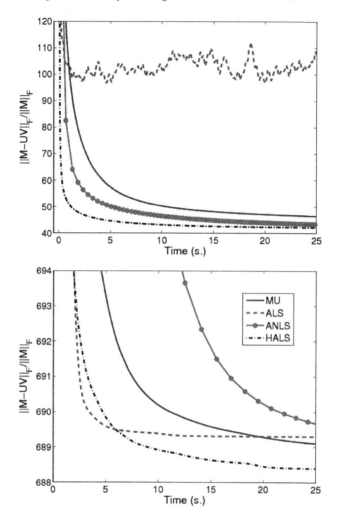

FIGURE 12.3: Comparison of MU, ALS, ANLS, and HALS on the CBCL facial images with $r = 49$, same data set as in Figure 12.1 (top), and the Classic document data set with $m = 7094$, $n = 41681$, and $r = 20$; see, e.g., [113] (bottom). The figure displays the average error using the same ten initial matrices W and H for all algorithms, randomly generated with the **rand** function of MATLAB. All tests were performed using MATLAB on a laptop Intel CORE i5-3210M CPU @2.5GHz 2.5GHz 6Go RAM. Note that, for ALS, we display the error after scaling; see Equation (12.8). For MU and HALS, we used the implementation from https://sites.google.com/site/nicolasgillis/, for ANLS from http://www.cc.gatech.edu/~hpark/nmfsoftware.php, and ALS was implemented following footnote 5.

similar variants) as a stopping criterion and for comparing the convergence of different algorithms:

- It is sensitive to scaling. For $\alpha > 0$ and $\alpha \neq 1$, we will have in general that

$$C_W(W) + C_H(H) = C(W, H) \neq C(\alpha W, \alpha^{-1}H),$$

  since the first two terms in (12.9) are sensitive to scaling. For example, if one solves the subproblem in W exactly and obtains $C_W(W) = 0$ (this will be the case for ANLS; see Section 12.4.1.4), then $\nabla_H F$ can be made arbitrarily small by multiplying W by a small constant and dividing H by the same constant (while, if $H \geq 0$, it will not influence the first term, which is equal to zero). This issue can be handled with proper normalization, e.g., imposing $||W(:, k)||_2 = ||H(k, :)||_2$ for all k; see [63].

- The value of $C(W, H)$ after the update of W can be very different from the value after an update of H (in particular, if the scaling is bad or if $|m - n| \gg 0$). Therefore, one should be very careful when using this type of criterion to compare ANLS-type methods with other algorithms such as the MU or HALS as the evolution of $C(W, H)$ can be misleading (in fact, an algorithm that monotonically decreases the objective function, such as the MU or HALS, is not guaranteed to monotonically decrease $C(W, H)$.) A potential fix would be to scale the columns of W and the rows of H so that $C_W(W)$ after the update of H and $C_H(H)$ after the update of W have the same order of magnitude.

12.4.1.8   Initialization

A simple way to initialize W and H is to generate them randomly (e.g., generating all entries uniformly at random in the interval $[0,1]$). Several more sophisticated initialization strategies have been developed in order to have better initial estimates in the hope to (i) obtain a good factorization with fewer iterations, and (ii) converge to a better stationary point. However, most initialization strategies come with no theoretical guarantee (e.g., a bound on the distance of the initial point to optimality), which can be explained in part by the complexity of the problem (in fact, NMF is NP-hard in general; see the introduction of this section). This could be an interesting direction for further research. We list some initialization strategies here; they are based on:

- *Clustering techniques.* Use the centroids computed with some clustering method, e.g., with k-means or spherical k-means, to initialize the columns of W, and initialize H as a proper scaling of the cluster indicator matrix (that is, $H_{kj} \neq 0 \iff X(:, j)$ belongs to the kth cluster) [107, 109]; see also [18] and the references therein for some recent results.

- *The SVD.* Let $\sum_{k=1}^{r} u_k v_k^T$ be the best rank-r approximation of X (which can be computed, e.g., using the SVD; see Introduction). Each rank-one factor $u_k v_k^T$ might contain positive and negative entries (except for the first one, by the Perron-Frobenius theorem[7]). However, denoting $[x]_+ = \max(x, 0)$, we have

$$u_k v_k^T = [u_k]_+ [v_k^T]_+ + [-u_k]_+ [-v_k^T]_+ - [-u_k]_+ [v_k^T]_+ - [u_k]_+ [-v_k^T]_+,$$

and the first two rank-one factors in this decomposition are nonnegative. Boutsidis et al. [11] proposed to replace each rank-one factor in $\sum_{k=1}^{r} u_k v_k^T$ with either $[u_k]_+ [v_k^T]_+$ or $[-u_k]_+ [-v_k^T]_+$, selecting the one with larger norm and scaling it properly.

- *Column subset selection.* It is possible to initialize the columns of W using data points, that is, initialize $W = X(:, \mathcal{K})$ for some set $\mathcal{K}$ with cardinality r; see [21, 112] and Section 12.4.2.

In practice, one may use several initializations, and keep the best solution obtained; see, e.g., the discussion in [28].

12.4.2   Near-Separable NMF

A matrix X is r-separable if there exists an index set $\mathcal{K}$ of cardinality r such that

$$X = X(:, \mathcal{K})H \qquad \text{for some } H \geq 0.$$

In other words, there exists a subset of r columns of X that generates a convex cone containing all columns of X. Hence, given a separable matrix, the goal of separable NMF is to identify the subset of columns $\mathcal{K}$ that allows us to reconstruct all columns of X (in fact, given $X(:, \mathcal{K})$, H can be computed by solving a convex optimization program; see Section 12.4.1). The separability assumption makes sense in several applications: for example,

- In text mining (see Section 12.3.2), separability of the word-by-document matrix requires that for each topic, there exists a document only on that topic. Note that we can also assume separability of the transpose of X (that is, of the document-by-word matrix), i.e., for each topic there exists one word used only by that topic (referred to as an 'anchor' word). In fact, the latter is considered a more reasonable assumption in practice; see [77, 3, 39] and also the thesis [46] for more details.

- In hyperspectral unmixing (see Section 12.3.3), separability of the wavelength-by-pixel matrix requires that for each endmember there exists a pixel containing only that endmember. This is the so-called pure-pixel assumption, and makes sense for relatively high spatial resolution hyperspectral images; see [8, 90] and the references therein.

---

[7]Actually, the first factor could contain negative entries if the input matrix is reducible and its first two singular values are equal to one another; see, e.g., [47, p. 16].

Separability has also been used successfully in blind source separation [95, 22], video summarization and image classification [40], and foreground-background separation in computer vision [76]. Note that for facial feature extraction described in Section 12.3.1, separability does not make sense since we cannot expect features to be present in the data set.

It is important to point out that separable NMF is closely related to several problems, including

- Column subset selection, which is a long-standing problem in numerical linear algebra (see [12] and the references therein).

- Pure-pixel search in hyperspectral unmixing, which has been addressed long before NMF was introduced; see [90] for a historical note.

- The problem of identifying a few important data points in a data set (see [40] and the references therein).

- Convex NMF [36], and the CUR decomposition [91].

Therefore, it is difficult to pinpoint the roots of separable NMF, and a comprehensive overview of all methods related to separable NMF is out of the scope of this paper. However, to the best of our knowledge, it is only very recently that provably efficient algorithms for separable NMF have been proposed. This new direction of research was launched by a paper by Arora et al. [4] which shows that NMF of separable matrices can be computed efficiently (that is, in polynomial time), even in the presence of noise (the error can be bounded in terms of the noise level; see below). We focus in this section on these provably efficient algorithms for separable NMF.

In the noiseless case, separable NMF reduces to identifying the extreme rays of the cone spanned by the columns of X. If the columns of the input matrix X are normalized so that their entries sum to one, that is, $X(:,j) \leftarrow ||X(:,j)||_1^{-1} X(:,j)$ for all j (and discarding the zero columns of X), then the problem reduces to identifying the vertices of the convex hull of the columns of X. In fact, since the entries of each column of X sum to one and $X = X(:,\mathcal{K})H$, the entries of each column of H must also sum to one: as X and H are nonnegative, we have for all j

$$1 = ||X(:,j)||_1 = ||X(:,\mathcal{K})H(:,j)||_1$$
$$= \sum_k ||X(:,\mathcal{K}(k))||_1 H(k,j) = \sum_k H(k,j) = ||H(:,j)||_1.$$

Therefore, the columns of X are convex combinations (that is, linear combinations with nonnegative weights summing to one) of the columns of $X(:,\mathcal{K})$.

In the presence of noise, the problem is referred to as near-separable NMF, and can be formulated as follows:

(Near-Separable NMF) *Given a noisy r-separable matrix* $\tilde{X} = X + N$ *with* $X = W[I_r, H']\Pi$ *where* W *and* H' *are nonnegative matrices,* Π *is a permutation matrix, and* N *is the noise, find a set* $\mathcal{K}$ *of* r *indices such that* $\tilde{X}(:, \mathcal{K}) \approx W$.

In the following, we describe some algorithms for near-separable NMF; they are classified in two categories: algorithms based on self-dictionary and sparse regression (Section 12.4.2.1) and geometric algorithms (Section 12.4.2.2).

12.4.2.1 Self-Dictionary and Sparse Regression Framework

In the noiseless case, separable NMF can be formulated as follows

$$\min_{Y \in \mathbb{R}^{n \times n}} ||Y||_{\text{row},0} \quad \text{such that} \quad X = XY \text{ and } Y \geq 0, \quad (12.10)$$

where $||Y||_{\text{row},0}$ is the number of non-zero rows of Y. In fact, if all the entries of a row of Y are equal to zero, then the corresponding column of X is not needed to reconstruct the other columns of X. Therefore, minimizing the number of rows of Y different from zero is equivalent to minimizing the number of columns of X used to reconstruct all the other columns of X, which solves the separable NMF problem. In particular, given an optimal solution Y^* of (12.10) and denoting $\mathcal{K} = \{i | Y^*(i, :) \neq 0\}$, we have $X = WY^*(\mathcal{K}, :)$ where $W = X(:, \mathcal{K})$.

In the presence of noise, the constraints $X = XY$ are usually reformulated as $||X - XY|| \leq \delta$ for some $\delta > 0$ or put as a penalty $\lambda ||X - XY||$ in the objective function for some penalty parameter $\lambda > 0$. In [40, 41], $||Y||_{\text{row},0}$ is replaced using ℓ_1-norm type relaxation:

$$||Y||_{q,1} = \sum_j ||Y(i, :)||_q,$$

where $q > 1$ so that $||Y||_{q,1}$ is convex and (12.10) becomes a convex optimization problem. Note that this idea is closely related to compressive sensing where ℓ_1-norm relaxation is used to find the sparsest solution to an underdetermined linear system. This relaxation is exact given that the matrix involved in the linear system satisfies some incoherence properties. In separable NMF, the columns and rows of matrix X are usually highly correlated, hence it is not clear how to extend the results from the compressive sensing literature to this separable NMF model; see, e.g., the discussion in [90].

A potential problem in using convex relaxations of (12.10) is that it cannot distinguish duplicates of the columns of W. In fact, if a column of W is present twice in the data matrix X, the corresponding rows of Y can both be non-zero, hence both columns of W can potentially be extracted (this is because of the convexity and the symmetry of the objective function)–in [40], k-means is used as a pre-processing in order to remove duplicates. Moreover, although this model was successfully used to solve real-world problems, no robustness results have been developed so far: so it is not clear how this model behaves in the presence of noise (only asymptotic results were proved, that is, when the noise level goes to zero and when no duplicates are present [40]).

A rather different approach to enforce row sparsity was suggested in [9], and later improved in [49, 54]. Row sparsity of Y is enforced by: (i) minimizing a weighted sum of the diagonal entries of Y hence enforcing diag(Y) to be sparse (in fact, this is nothing but a weighted ℓ_1 norm since Y is nonnegative), and (ii) imposing all entries in a row of Y to be smaller than the corresponding diagonal entry of Y (we assume here that the columns of X are normalized). The second condition implies that if diag(Y) is sparse then Y is row sparse. The corresponding near-separable NMF model is:

$$\min_{Y \in \mathbb{R}^{n \times n}} p^T \text{diag}(Y) \ \text{ such that } \ ||X - XY||_1 \leq \delta \ \text{ and } \ 0 \leq Y_{ij} \leq Y_{ii} \leq 1,$$

$$(12.11)$$

for some positive vector $p \in \mathbb{R}^n$ with distinct entries (this breaks the symmetry so that the model can distinguish duplicates). This model has been shown to be robust: defining the parameter[8] α as

$$\alpha(W) = \min_{1 \leq j \leq r} \min_{x \in \mathbb{R}_+^{r-1}} ||W(:,j) - W(:,\mathcal{J}_j)x||_1, \ \text{ where } \mathcal{J}_j = \{1, 2, \ldots, r\} \setminus \{j\},$$

and for a near-separable matrix $\tilde{X} = W[I_r, H']\Pi + N$ (see above) with $\epsilon = \max_j ||N(:,j)||_1 \leq \mathcal{O}\left(\frac{\alpha^2(W)}{r}\right)$, the model (12.11) can be used to identify the columns of W with ℓ_1 error proportional to $\mathcal{O}\left(\frac{r\epsilon}{\alpha(W)}\right)$, that is, the identified index set $\mathcal{K}$ satisfies $\max_j \min_{k \in \mathcal{K}} ||\tilde{X}(:,k) - W(:,j)||_1 \leq \mathcal{O}\left(\frac{r\epsilon}{\alpha(W)}\right)$; see [54, Th.7] for more details.

Finally, a drawback of the approaches based on self-dictionary and sparse regression is that they are computationally expensive as they require us to tackle optimization problems with n^2 variables.

12.4.2.2 Geometric Algorithms

Another class of near-separable algorithms are based on geometric insights and in particular on the fact that the columns of W are the vertices of the convex hull of the normalized columns of X. Many geometric algorithms can be found in the remote sensing literature (they are referred to as endmember extraction algorithms or pure-pixel search algorithms); see [90] for a historical note and [8] for a comprehensive survey. Because of the large body of literature, we do not aim at surveying all algorithms but rather focus on a single algorithm that is particularly simple while being rather effective in practice: the successive projection algorithm (SPA). Moreover, the ideas behind SPA

---

[8]The larger the parameter α is, the less sensitive the data to noise. For example, it can be easily checked that $\epsilon = \max_j ||N(:,j)||_1 < \frac{\alpha}{2}$ is a necessary condition to being able to distinguish the columns of W [49].

are at the heart of many geometric-based near-separable NMF algorithms (see below).

SPA looks for the vertices of the convex hull of the columns of the input data matrix X and works as follows: at each step, it selects the column of X with maximum ℓ_2 norm and then updates X by projecting each column onto the orthogonal complement of the extracted column; see Algorithm 13. SPA is extremely fast as it can be implemented in $2pnr + \mathcal{O}(pr^2)$ operations, using the formula $||(I - uu^T)v||_2^2 = ||v||_2^2 - (u^Tv)^2$, for any $u, v \in \mathbb{R}^m$ with $||u||_2 = 1$ [55]. Moreover, if r is unknown, it can be estimated using the norm of the residual R.

Algorithm 13 Successive Projection Algorithm [2]

Input: Near-separable matrix $\tilde{X} = W[I_r, H']\Pi + N$ where W is full rank, $H' \geq 0$, the entries of each column of H' sum to at most one, Π is a permutation, and N is the noise, and the number r of columns of W.
Output: Set of r indices $\mathcal{K}$ such that $\tilde{X}(:,\mathcal{K}) \approx W$ (up to permutation).

1: Let $R = \tilde{X}$, $\mathcal{K} = \{\}$.
2: **for** $k = 1 : r$ **do**
3: $\quad p = \text{argmax}_j ||R_{:j}||_2$.
4: $\quad R = \left(I - \frac{R_{:p} R_{:p}^T}{||R_{:p}||_2^2}\right) R$.
5: $\quad \mathcal{K} = \mathcal{K} \cup \{p\}$.
6: **end for**

Let us prove the correctness of SPA in the noiseless case using induction, and assuming W is full rank (this is a necessary and sufficient condition) and assuming the entries of each column of H' sum to at most one (this can be achieved through normalization; see above). At the first step, SPA identifies a column of W because the ℓ_2 norm can only be maximized at a vertex of the convex hull of a set of points; see Figure 12.4 for an illustration. In fact, for all $1 \leq j \leq n$,

$$||X(:,j)||_2 = ||WH(:,j)||_2 \leq \sum_{k=1}^{r} H(k,j)||W(:,k)||_2 \leq \max_{1 \leq k \leq r} ||W(:,k)||_2.$$

The first inequality follows from the triangle inequality, and the second since $H(k,j) \geq 0$ and $\sum_k H(k,j) \leq 1$. Moreover, by strong convexity of the ℓ_2 norm and the full rank assumption on W, the first inequality is strict unless $H(:,k)$ is a column of the identity matrix, that is, unless $X(:,j) = W(:,k)$ for some k. For the induction step, assume without loss of generality that SPA has extracted the first ℓ columns of W, and let $W_\ell = W(:,1{:}\ell)$ and $P_{W_\ell}^\perp$ be the projection onto the orthogonal complement of the columns of W_ℓ so that

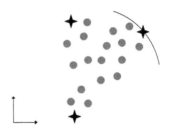

FIGURE 12.4: Illustration of SPA.

$P_{W_\ell}^\perp W_\ell = 0$. We have, for all $1 \le j \le n$,

$$||P_{W_\ell}^\perp X(:,j)||_2 = ||P_{W_\ell}^\perp W H(:,j)||_2 \le \sum_{k=1}^{r} H(k,j)||P_{W_\ell}^\perp W(:,k)||_2$$

$$\le \max_{\ell+1 \le k \le r} ||P_{W_\ell}^\perp W(:,k)||_2,$$

where $P_{W_\ell}^\perp W(:,k) \ne 0$ for $\ell+1 \le k \le r$ since W is full rank. Hence, using the same reasoning as above, SPA will identify a column of W not extracted yet, which concludes the proof.

Moreover, SPA is robust to noise: given a near-separable matrix $\tilde{X} = W[I_r, H']\Pi + N$ with W full rank, H' nonnegative with $||H'(:,j)||_1 \le 1 \; \forall j$, and $\epsilon = \max_j ||N(:,j)||_2 \le \mathcal{O}\left(\frac{\sigma_{\min}(W)}{\sqrt{r}\kappa^2(W)}\right)$, SPA identifies the columns of W up to ℓ_2 error proportional to $\mathcal{O}\left(\epsilon \kappa^2(W)\right)$, where $\kappa(W) = \frac{\sigma_{\max}(W)}{\sigma_{\min}(W)}$ [55, Th.3]. These bounds can be improved using post-processing (see below), which reduces the error to $\mathcal{O}\left(\epsilon \kappa(W)\right)$ [3], or preconditioning, which significantly increases the upper bound on the noise level, to $\epsilon \le \mathcal{O}\left(\frac{\sigma_{\min}(W)}{r\sqrt{r}}\right)$, and reduces the error to $\mathcal{O}\left(\epsilon \kappa(W)\right)$ [56].

It is interesting to note that SPA has been developed and used for rather different purposes in various fields:

- *Numerical linear algebra.* SPA is closely related to the modified Gram-Schmidt algorithm with column pivoting, used for example to solve linear least squares problems [15].

- *Chemistry* (and in particular spectroscopy). SPA is used for variable selection in spectroscopic multicomponent analysis; in fact, the name SPA comes from [2].

- *Hyperspectral imaging.* SPA is closely related to several endmember extraction algorithms; in particular N-FINDR [108] and its variants, the automatic target generation process (ATGP) [99], and the successive volume maximization algorithm (SVMAX) [21]; see the discussion in [90] for more details. The motivation behind all these approaches is to identify an index set $\mathcal{K}$ that maximizes the volume of the convex hull of the columns of $X(:, \mathcal{K})$. Note that most endmember extraction algorithms use an LDR (such as the SVD) as a pre-processing step for noise filtering, and SPA can be combined with an LDR to improve performance.

- *Text mining.* Arora et al. [3] proposed FastAnchorWords whose differences with SPA are that (i) the projection is made onto the affine hull of the columns extracted so far (instead of the linear span), and (ii) the index set extracted is refined using the following post-processing step: let $\mathcal{K}$ be the extracted index set by SPA, for each $k \in \mathcal{K}$:

  - Compute the projection R of X into the orthogonal complement of $X(:, \mathcal{K} \backslash \{k\})$.

  - Replace k with the index corresponding to the column of R with maximum ℓ_2 norm.

- *Theoretical computer science.* SPA was proved to be a good heuristic to identify a subset of columns of a given matrix whose convex hull has maximum volume [19, 20] (in the sense that no polynomial-time algorithm can achieve better performance up to some logarithmic factors).

- *Sparse regression with self-dictionary.* SPA is closely related to orthogonal matching pursuit and can be interpreted as a greedy method to solve the sparse regression problem with self-dictionary (12.10); see [44] and the references therein.

Moreover, there exist many geometric algorithms that are variants of SPA, e.g., vertex component analysis (VCA) using linear functions instead of the ℓ_2-norm [96], ℓ_p-norm based pure pixel algorithm (TRI-P) using p-norms [1], FastSepNMF using strongly convex functions [55], the successive nonnegative projection algorithm (SNPA) [50], and the fast conical hull algorithm (XRAY) [77] using nonnegativity constraints for the projection step.

Further research on near-separable NMF includes the design of faster and/or provably more robust algorithms.

12.5 Connections with Problems in Mathematics and Computer Science

In this section, we briefly mention several connections between NMF and problems outside data mining and machine learning. The minimum r such that an exact NMF of a nonnegative matrix X exists is the nonnegative rank of X, denoted $\mathrm{rank}_+(X)$. More precisely, given $X \in \mathbb{R}_+^{p \times n}$, $\mathrm{rank}_+(X)$ is the minimum r such that there exist $W \in \mathbb{R}_+^{p \times r}$ and $H \in \mathbb{R}_+^{r \times n}$ with $X = WH$. The nonnegative rank has tight connections with several problems in mathematics and computer science:

- *Graph Theory.* Let $G(X) = (V_1 \cup V_2, E)$ be the bipartite graph induced by X (that is, $(i,j) \in E \iff X_{ij} \neq 0$). The minimum biclique cover $\mathrm{bc}(G(X))$ of $G(X)$ is the minimum number of complete bipartite subgraphs needed to cover $G(X)$. It can be checked easily that for any $(W,H) \geq 0$ such that $X = WH = \sum_{k=1}^r W_{:k} H_{k:}$, we have

$$G(X) = \cup_{k=1}^r G(W_{:k} H_{k:}),$$

  where $G(W_{:k} H_{k:})$ are complete bipartite subgraphs, hence $\mathrm{bc}(G(W_{:k} H_{k:})) = 1 \,\forall k$. Therefore,

$$\mathrm{bc}(G(X)) \leq \mathrm{rank}_+(X).$$

  This lower bound on the nonnegative rank is referred to as the rectangle covering bound [43].

- *Extended Formulations.* Given a polytope P, an extended formulation (or lift, or extension) is a higher dimensional polyhedron Q and a linear projection π such that $\pi(Q) = P$. When the polytope P has exponentially many facets, finding extended formulations of polynomial size is of great importance since it allows us to solve linear programs (LP) over P in polynomial time. It turns out that the minimum number of facets of an extended formulation Q of a polytope P is equal to the nonnegative rank of its slack matrix [111], defined as $X(i,j) = a_i^T v_j - b_i$ where v_j is the jth vertex of P and $\{x \in \mathbb{R}^n \mid a_i^T x - b_i \geq 0\}$ its ith facet with $a_i \in \mathbb{R}^n$ and $b_i \in \mathbb{R}$, that is, X is a facet-by-vertex matrix and $X(i,j)$ is the slack of the jth vertex with respect to ith facet; see the surveys [30, 69] and the references therein. These ideas can be generalized to approximate extended formulations, directly related to approximate factorizations (hence NMF) [13, 58].

- *Probability.* Let $X^{(k)} \in \{1, \ldots, p\}$ and $Y^{(k)} \in \{1, \ldots, n\}$ be two independent variables for each $1 \leq k \leq r$, and $P^{(k)}$ be the joint distribution with

$$P_{ij}^{(k)} = \mathbb{P}\left(X^{(k)} = i, Y^{(k)} = j\right) = \mathbb{P}\left(X^{(k)} = i\right) \mathbb{P}\left(Y^{(k)} = j\right).$$

  Each distribution $P^{(k)}$ corresponds to a nonnegative rank-one matrix. Let us define the mixture P of these k independent models as follows:

  - Choose the distribution $P^{(k)}$ with probability α_k ($\sum_{k=1}^r \alpha_k = 1$).
  - Draw X and Y from the distribution $P^{(k)}$.

We have that $P = \sum_{k=1}^{r} \alpha_k P^{(k)}$ is the sum of r rank-one nonnegative matrices. In practice, only P is observed and computing its nonnegative rank and a corresponding factorization amounts to explaining the distribution P with as few independent variables as possible; see [17] and the references therein.

- *Communication Complexity.* In its simplest variant, communication complexity addresses the following problem: Alice and Bob have to compute the following function

$$f : \{0,1\}^m \times \{0,1\}^n \to \{0,1\} : (x,y) \to f(x,y).$$

Alice only knows x and Bob y, and the aim is to minimize the number of bits exchanged between Alice and Bob to compute f exactly. Nondeterministic communication complexity (NCC) is a variant where Bob and Alice first receive a message before starting their communication; see [81, Ch.3] and the references therein for more details. The communication matrix $X \in \{0,1\}^{2^n \times 2^m}$ is equal to the function f for all possible combinations of inputs. Yannakakis [111] showed that the NCC for computing f is upper bounded by the logarithm of the nonnegative rank of the communication matrix (this result is closely related to the rectangle covering bound described above: in fact, $\lceil \log(\mathrm{bc}(G(X))) \rceil$ equals the NCC of f).

- *Computational Geometry.* Computing the nonnegative rank is closely related to the problem of finding a polytope with minimum number of vertices nested between two given polytopes [53]. This is a well-known problem is computational geometry, referred to as the nested polytopes problem; see [31] and the references therein.

12.6 Conclusion

NMF is an easily interpretable linear dimensionality reduction technique for nonnegative data. It is a rather versatile technique with many applications, and brings together a broad range of researchers. In the context of "Big Data" science, which is becoming an increasingly important topic, we believe NMF has a bright future; see Figure 12.5 for an illustration of the number of publications related to NMF since the publication of the Lee and Seung paper [79].

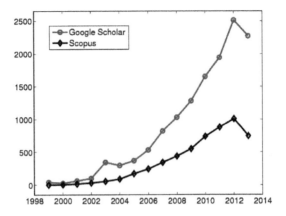

FIGURE 12.5: Number of search results for papers containing either "non-negative matrix factorization" or "non-negative matrix factorization" on Google Scholar and Scopus (as of December 12, 2013).

Acknowledgments

The author would like to thank Rafal Zdunek, Wing-Kin Ma, Marc Pirlot, and the editors of this book for insightful comments that helped improve the chapter.

Bibliography

[1] A. Ambikapathi, T.-H. Chan, C.-Y. Chi, and K. Keizer. Hyperspectral data geometry based estimation of number of endmembers using p-norm based pure pixel identification. *IEEE Trans. on Geoscience and Remote Sensing*, 51(5):2753–2769, 2013.

[2] U.M.C. Araújo, B.T.C. Saldanha, R.K.H. Galvão, T. Yoneyama, H.C. Chame, and V. Visani. The successive projections algorithm for variable selection in spectroscopic multicomponent analysis. *Chemometrics and Intelligent Laboratory Systems*, 57(2):65–73, 2001.

[3] S. Arora, R. Ge, Y. Halpern, D. Mimno, A. Moitra, D. Sontag, Y. Wu, and M. Zhu. A practical algorithm for topic modeling with provable guarantees. In *Int. Conf. on Machine Learning (ICML '13)*, volume 28, pages 280–288. 2013.

[4] S. Arora, R. Ge, R. Kannan, and A. Moitra. Computing a nonnegative matrix factorization – provably. In *Proc. of the 44th Symp. on Theory of Computing (STOC '12)*, pages 145–162, 2012.

[5] R. Badeau, N. Bertin, and E. Vincent. Stability analysis of multiplicative update algorithms and application to nonnegative matrix factorization. *IEEE Trans. on Neural Networks*, 21(12):1869–1881, 2010.

[6] M.W. Berry, M. Browne, A. Langville, V.P. Pauca, and R.J. Plemmons. Algorithms and Applications for Approximate Nonnegative Matrix Factorization. *Computational Statistics & Data Analysis*, 52:155–173, 2007.

[7] J.M. Bioucas-Dias and J.M.P. Nascimento. Estimation of signal subspace on hyperspectral data. In *Remote Sensing*, page 59820L. International Society for Optics and Photonics, 2005.

[8] J.M. Bioucas-Dias, A. Plaza, N. Dobigeon, M. Parente, Q. Du, P. Gader, and J. Chanussot. Hyperspectral unmixing overview: Geometrical, statistical, and sparse regression-based approaches. *IEEE J. of Selected Topics in Applied Earth Observations and Remote Sensing*, 5(2):354–379, 2012.

[9] V. Bittorf, B. Recht, E. Ré, and J.A. Tropp. Factoring nonnegative matrices with linear programs. In *Advances in Neural Information Processing Systems (NIPS '12)*, pages 1223–1231, 2012.

[10] D.M. Blei. Probabilistic topic models. *Communications of the ACM*, 55(4):77–84, 2012.

[11] C. Boutsidis and E. Gallopoulos. SVD based initialization: A head start for nonnegative matrix factorization. *Pattern Recognition*, 41:1350–1362, 2008.

[12] C. Boutsidis, M.W. Mahoney, and P. Drineas. An improved approximation algorithm for the column subset selection problem. In *Proc. of the 20th Annual ACM-SIAM Symp. on Discrete Algorithms (SODA '09)*, pages 968–977, 2009.

[13] G. Braun, S. Fiorini, S. Pokutta, and D. Steurer. Approximation limits of linear programs (beyond hierarchies). In *Proc. of the 53rd Annual IEEE Symp. on Foundations of Computer Science (FOCS' 12)*, pages 480–489, 2012.

[14] R. Bro. *Multi-Way Analysis in the Food Industry: Models, Algorithms, and Applications*. PhD thesis, University of Copenhagen, 1998. `http://curis.ku.dk/ws/files/13035961/Rasmus_Bro.pdf`.

[15] P. Businger and G.H. Golub. Linear least squares solutions by householder transformations. *Numerische Mathematik*, 7:269–276, 1965.

[16] D. Cai, X. He, J. Han, and T.S. Huang. Graph regularized nonnegative matrix factorization for data representation. *IEEE Trans. on Pattern Analysis and Machine Intelligence*, 33(8):1548–1560, 2011.

[17] E. Carlini and F. Rapallo. Probability matrices, non-negative rank, and parameterization of mixture models. *Linear Algebra and its Applications*, 433:424–432, 2010.

[18] G. Casalino, N. Del Buono, and C. Mencar. Subtractive clustering for seeding non-negative matrix factorizations. *Information Sciences*, (257):369–387, 2013.

[19] A. Çivril and M. Magdon-Ismail. On selecting a maximum volume submatrix of a matrix and related problems. *Theoretical Computer Science*, 410(47-49):4801–4811, 2009.

[20] A. Çivril and M. Magdon-Ismail. Exponential inapproximability of selecting a maximum volume sub-matrix. *Algorithmica*, 65(1):159–176, 2013.

[21] T.-H. Chan, W.-K. Ma, A. Ambikapathi, and C.-Y. Chi. A simplex volume maximization framework for hyperspectral endmember extraction. *IEEE Trans. on Geoscience and Remote Sensing*, 49(11):4177–4193, 2011.

[22] T.-H. Chan, W.-K. Ma, C.-Y. Chi, and Y. Wang. A convex analysis framework for blind separation of non-negative sources. *IEEE Trans. on Signal Processing*, 56(10):5120–5134, 2008.

[23] E.C. Chi and T.G. Kolda. On tensors, sparsity, and nonnegative factorizations. *SIAM J. on Matrix Analysis and Applications*, 33(4):1272–1299, 2012.

[24] S. Choi. Algorithms for orthogonal nonnegative matrix factorization. In *Proc. of the Int. Joint Conf. on Neural Networks*, pages 1828–1832, 2008.

[25] A. Cichocki and A.H. Phan. Fast local algorithms for large scale Nonnegative Matrix and Tensor Factorizations. *IEICE Trans. on Fundamentals of Electronics*, Vol. E92-A No.3:708–721, 2009.

[26] A. Cichocki, R. Zdunek, and S.-I. Amari. Non-negative Matrix Factorization with Quasi-Newton Optimization. In *Lecture Notes in Artificial Intelligence, Springer*, volume 4029, pages 870–879, 2006.

[27] A. Cichocki, R. Zdunek, and S.-I. Amari. Hierarchical ALS Algorithms for Nonnegative Matrix and 3D Tensor Factorization. In *Lecture Notes in Computer Science, Vol. 4666, Springer*, pages 169–176, 2007.

[28] A. Cichocki, R. Zdunek, A.H. Phan, and S.-I. Amari. *Nonnegative Matrix and Tensor Factorizations: Applications to Exploratory Multi-way Data Analysis and Blind Source Separation.* Wiley-Blackwell, 2009.

[29] P. Comon. Independent component analysis: A new concept? *Signal Processing*, 36:287–314, 1994.

[30] M. Conforti, G. Cornuéjols, and G. Zambelli. Extended formulations in combinatorial optimization. *4OR: A Quarterly Journal of Operations Research*, 10(1):1–48, 2010.

[31] G. Das and D.A. Joseph. The Complexity of Minimum Convex Nested Polyhedra. In *Proc. of the 2nd Canadian Conf. on Computational Geometry*, pages 296–301, 1990.

[32] A. d'Aspremont, L. El Ghaoui, M.I. Jordan, and G.R.G. Lanckriet. A Direct Formulation for Sparse PCA Using Semidefinite Programming. *SIAM Review*, 49(3):434–448, 2007.

[33] M.E. Daube-Witherspoon and G. Muehllehner. An iterative image space reconstruction algorithm suitable for volume ECT. *IEEE Trans. on Medical Imaging*, 5:61–66, 1986.

[34] K. Devarajan. Nonnegative Matrix Factorization: An Analytical and Interpretive Tool in Computational Biology. *PLoS Computational Biology*, 4(7):e1000029, 2008.

[35] C. Ding, X. He, and H.D. Simon. On the Equivalence of Nonnegative Matrix Factorization and Spectral Clustering. In *SIAM Int. Conf. Data Mining (SDM'05)*, pages 606–610, 2005.

[36] C. Ding, T. Li, and M.I. Jordan. Convex and semi-nonnegative matrix factorizations. *IEEE Trans. on Pattern Analysis and Machine Intelligence*, 32(1):45–55, 2010.

[37] C. Ding, T. Li, and W. Peng. On the equivalence between non-negative matrix factorization and probabilistic latent semantic indexing. *Computational Statistics & Data Analysis*, 52(8):3913–3927, 2008.

[38] C. Ding, T. Li, W. Peng, and H. Park. Orthogonal nonnegative matrix t-factorizations for clustering. In *Proc. of the 12th ACM SIGKDD Int. Conf. on Knowledge Discovery and Data Mining*, pages 126–135, 2006.

[39] W. Ding, M.H. Rohban, P. Ishwar, and V. Saligrama. Topic discovery through data dependent and random projections. In *Int. Conf. on Machine Learning (ICML '13)*, volume 28, pages 471–479. 2013.

[40] E. Elhamifar, G. Sapiro, and R. Vidal. See all by looking at a few: Sparse modeling for finding representative objects. In *IEEE Conf. on Computer Vision and Pattern Recognition (CVPR '12)*, 2012.

[41] E. Esser, M. Moller, S. Osher, G. Sapiro, and J. Xin. A convex model for nonnegative matrix factorization and dimensionality reduction on physical space. *IEEE Trans. on Image Processing*, 21(7):3239–3252, 2012.

[42] C. Févotte, N. Bertin, and J.-L. Durrieu. Nonnegative matrix factorization with the Itakura-Saito divergence: With application to music analysis. *Neural Computation*, 21(3):793–830, 2009.

[43] S. Fiorini, V. Kaibel, K. Pashkovich, and D.O. Theis. Combinatorial bounds on nonnegative rank and extended formulations. *Discrete Mathematics*, 313(1):67–83, 2013.

[44] X. Fu, W.-K. Ma, T.-H. Chan, J.M. Bioucas-Dias, and M.-D. Iordache. Greedy algorithms for pure pixels identification in hyperspectral unmixing: A multiple-measurement vector viewpoint. In *Proc. of 21st European Signal Processing Conf. (EUSIPCO '13)*, 2013.

[45] E. Gaussier and C. Goutte. Relation between PLSA and NMF and implications. In *Proc. of the 28th Annual Int. ACM SIGIR Conf. on Research and Development in Information Retrieval*, pages 601–602, 2005.

[46] R. Ge. *Provable Algorithms for Machine Learning Problems*. PhD thesis, Princeton University, 2013. `http://dataspace.princeton.edu/jspui/bitstream/88435/dsp019k41zd62n/1/Ge_princeton_0181D_10819.pdf`.

[47] N. Gillis. *Nonnegative Matrix Factorization: Complexity, Algorithms and Applications*. PhD thesis, Université catholique de Louvain, 2011. `https://sites.google.com/site/nicolasgillis/`.

[48] N. Gillis. Sparse and unique nonnegative matrix factorization through data preprocessing. *Journal of Machine Learning Research*, 13(Nov):3349–3386, 2012.

[49] N. Gillis. Robustness analysis of Hottopixx, a linear programming model for factoring nonnegative matrices. *SIAM J. on Matrix Analysis and Applications*, 34(3):1189–1212, 2013.

[50] N. Gillis. Successive nonnegative projection algorithm for robust nonnegative blind source separation. arXiv:1310.7529, 2013.

[51] N. Gillis and F. Glineur. Using underapproximations for sparse nonnegative matrix factorization. *Pattern Recognition*, 43(4):1676–1687, 2010.

[52] N. Gillis and F. Glineur. Accelerated multiplicative updates and hierarchical ALS algorithms for nonnegative matrix factorization. *Neural Computation*, 24(4):1085–1105, 2012.

[53] N. Gillis and F. Glineur. On the geometric interpretation of the nonnegative rank. *Linear Algebra and its Applications*, 437(11):2685–2712, 2012.

[54] N. Gillis and R. Luce. Robust near-separable nonnegative matrix factorization using linear optimization. *Journal of Machine Learning Research*, 2014. to appear.

[55] N. Gillis and S.A. Vavasis. Fast and robust recursive algorithms for separable nonnegative matrix factorization. *IEEE Trans. Pattern Anal. Mach. Intell.*, 2013. doi:10.1109/TPAMI.2013.226.

[56] N. Gillis and S.A. Vavasis. Semidefinite programming based preconditioning for more robust near-separable nonnegative matrix factorization. arXiv:1310.2273, 2013.

[57] G.H. Golub and C.F. Van Loan. *Matrix Computation, 3rd Edition.* The Johns Hopkins University Press Baltimore, 1996.

[58] J. Gouveia, P.A. Parrilo, and R.R. Thomas. Approximate cone factorizations and lifts of polytopes. arXiv:1308.2162, 2013.

[59] L. Grippo and M. Sciandrone. On the convergence of the block nonlinear Gauss-Seidel method under convex constraints. *Operations Research Letters*, 26:127–136, 2000.

[60] N. Guan, D. Tao, Z. Luo, and B. Yuan. NeNMF: an optimal gradient method for nonnegative matrix factorization. *IEEE Trans. on Signal Processing*, 60(6):2882–2898, 2012.

[61] D. Guillamet and J. Vitrià. Non-negative matrix factorization for face recognition. In *Lecture Notes in Artificial Intelligence*, pages 336–344. Springer, 2002.

[62] J. Han, L. Han, M. Neumann, and U. Prasad. On the rate of convergence of the image space reconstruction algorithm. *Operators and Matrices*, 3(1):41–58, 2009.

[63] N.-D. Ho. *Nonnegative Matrix Factorization - Algorithms and Applications.* PhD thesis, Université Catholique de Louvain, 2008.

[64] P.O. Hoyer. Nonnegative matrix factorization with sparseness constraints. *Journal of Machine Learning Research*, 5:1457–1469, 2004.

[65] C.-J. Hsieh and I.S. Dhillon. Fast coordinate descent methods with variable selection for non-negative matrix factorization. In *Proc. of the 17th ACM SIGKDD Int. Conf. on Knowledge Discovery and Data Mining*, pages 1064–1072, 2011.

[66] K. Huang, N.D. Sidiropoulos, and A. Swami. Non-negative matrix factorization revisited: Uniqueness and algorithm for symmetric decomposition. *IEEE Trans. on Signal Processing*, 62(1):211–224, 2014.

[67] M.-D. Iordache, J.M. Bioucas-Dias, and A. Plaza. Total variation spatial regularization for sparse hyperspectral unmixing. *IEEE Trans. on Geoscience and Remote Sensing*, 50(11):4484–4502, 2012.

[68] S. Jia and Y. Qian. Constrained nonnegative matrix factorization for hyperspectral unmixing. *IEEE Trans. on Geoscience and Remote Sensing*, 47(1):161–173, 2009.

[69] V. Kaibel. Extended Formulations in Combinatorial Optimization. *Optima*, 85:2–7, 2011.

[70] B. Kanagal and V. Sindhwani. Rank selection in low-rank matrix approximations. In *Advances in Neural Information Processing Systems (NIPS '10)*, 2010.

[71] Q. Ke and T. Kanade. Robust L_1 norm factorization in the presence of outliers and missing data by alternative convex programming. In *IEEE Conf. on Computer Vision and Pattern Recognition (CVPR '05)*, pages 739–746, 2005.

[72] H. Kim and H. Park. Sparse non-negative matrix factorizations via alternating non-negativity-constrained least squares for microarray data analysis. *Bioinformatics*, 23(12):1495–1502, 2007.

[73] H. Kim and H. Park. Non-negative Matrix Factorization Based on Alternating Non-negativity Constrained Least Squares and Active Set Method. *SIAM J. on Matrix Analysis and Applications*, 30(2):713–730, 2008.

[74] J. Kim, Y. He, and H. Park. Algorithms for nonnegative matrix and tensor factorizations: A unified view based on block coordinate descent framework. *Journal of Global Optimization*, 2013. doi:10.1007/s10898-013-0035-4.

[75] J. Kim and H. Park. Fast nonnegative matrix factorization: An active-set-like method and comparisons. *SIAM J. on Scientific Computing*, 33(6):3261–3281, 2011.

[76] A. Kumar and V. Sindhwani. Near-separable non-negative matrix factorization with ℓ_1- and Bregman loss functions. arXiv:1312.7167, 2013.

[77] A. Kumar, V. Sindhwani, and P. Kambadur. Fast conical hull algorithms for near-separable non-negative matrix factorization. In *Int. Conf. on Machine Learning (ICML '13)*, volume 28, pages 231–239. 2013.

[78] C.L. Lawson and R.J. Hanson. *Solving Least Squares Problems.* Prentice-Hall, 1974.

[79] D.D. Lee and H.S. Seung. Learning the Parts of Objects by Nonnegative Matrix Factorization. *Nature*, 401:788–791, 1999.

[80] D.D. Lee and H.S. Seung. Algorithms for Non-negative Matrix Factorization. *In Advances in Neural Information Processing (NIPS '01)*, 13, 2001.

[81] T. Lee and A. Shraibman. *Lower bounds in communication complexity.* Now Publishers Inc., 2009.

[82] A. Lefèvre. *Dictionary learning methods for single-channel source separations.* PhD thesis, Ecole Normale Supérieure de Cachan, 2012.

[83] L. Li, G. Lebanon, and H. Park. Fast Bregman divergence NMF using Taylor expansion and coordinate descent. In *Proc. of the 18th ACM SIGKDD Int. Conf. on Knowledge Discovery and Data Mining*, pages 307–315, 2012.

[84] L. Li and Y.-J. Zhang. FastNMF: highly efficient monotonic fixed-point nonnegative matrix factorization algorithm with good applicability. *J. Electron. Imaging*, Vol. 18(033004), 2009.

[85] T. Li, Y. Zhang, and V. Sindhwani. A non-negative matrix tri-factorization approach to sentiment classification with lexical prior knowledge. In *Association of Computational Lingustics*, pages 244–252, 2009.

[86] Y. Li, D.M. Sima, S. Van Cauter, A.R. Croitor Sava, U. Himmelreich, Y. Pi, and S. Van Huffel. Hierarchical non-negative matrix factorization (hNMF): a tissue pattern differentiation method for glioblastoma multiforme diagnosis using MRSI. *NMR in Biomedicine*, 26(3):307–319, 2013.

[87] C.-J. Lin. On the convergence of multiplicative update algorithms for nonnegative matrix factorization. *IEEE Trans. on Neural Networks*, 18(6):1589–1596, 2007.

[88] C.-J. Lin. Projected Gradient Methods for Nonnegative Matrix Factorization. *Neural Computation*, 19:2756–2779, 2007.

[89] J. Liu, J. Liu, P. Wonka, and J. Ye. Sparse non-negative tensor factorization using columnwise coordinate descent. *Pattern Recognition*, 45(1):649–656, 2012.

[90] W.-K. Ma, J.M. Bioucas-Dias, T.-H. Chan, N. Gillis, P. Gader, A. Plaza, A. Ambikapathi, and C.-Y. Chi. A Signal Processing Perspective on Hyperspectral Unmixing. *IEEE Signal Processing Magazine*, 31(1):67–81, 2014.

[91] M.W. Mahoney and P. Drineas. CUR matrix decompositions for improved data analysis. *Proc. of the National Academy of Sciences*, 106(3): 697–702, 2009.

[92] P. Melville and V. Sindhwani. Recommender systems. *Encyclopedia of machine learning*, 1:829–838, 2010.

[93] L. Miao and H. Qi. Endmember extraction from highly mixed data using minimum volume constrained nonnegative matrix factorization. *IEEE Trans. on Geoscience and Remote Sensing*, 45(3):765–777, 2007.

[94] A. Moitra. An almost optimal algorithm for computing nonnegative rank. In *Proc. of the 24th Annual ACM-SIAM Symp. on Discrete Algorithms (SODA '13)*, pages 1454–1464, 2013.

[95] W. Naanaa and J.-M. Nuzillard. Blind source separation of positive and partially correlated data. *Signal Processing*, 85(9):1711–1722, 2005.

[96] J.M.P. Nascimento and J.M. Bioucas-Dias. Vertex component analysis: a fast algorithm to unmix hyperspectral data. *IEEE Trans. on Geoscience and Remote Sensing*, 43(4):898–910, 2005.

[97] P. Paatero and U. Tapper. Positive matrix factorization: a non-negative factor model with optimal utilization of error estimates of data values. *Environmetrics*, 5:111–126, 1994.

[98] J. Rapin, J. Bobin, A. Larue, and J.-L. Starck. Sparse and non-negative BSS for noisy data. *IEEE Trans. on Signal Processing*, 61(22):5620–5632, 2013.

[99] H. Ren and C.-I. Chang. Automatic spectral target recognition in hyperspectral imagery. *IEEE Trans. on Aerospace and Electronic Systems*, 39(4):1232–1249, 2003.

[100] R. Sandler and M. Lindenbaum. Nonnegative matrix factorization with earth mover's distance metric. In *IEEE Conf. on Computer Vision and Pattern Recognition (CVPR '09)*, pages 1873–1880, 2009.

[101] F. Shahnaz, M.W. Berry, V.P. Pauca, and R.J. Plemmons. Document clustering using nonnegative matrix factorization. *Information Processing and Management*, 42:373–386, 2006.

[102] P. Smaragdis, C. Févotte, G.J. Mysore, N. Mohammadiha, and M. Hoffman. A Unified View of Static and Dynamic Source Separation Using Non-Negative Factorizations. *IEEE Signal Processing Magazine*, 2014.

[103] N. Takahashi and R. Hibi. Global convergence of modified multiplicative updates for nonnegative matrix factorization. *Computational Optimization and Applications*, 2013. doi:10.1007/s10589-013-9593-0.

[104] V.Y.F. Tan and C. Févotte. Automatic relevance determination in non-negative matrix factorization. In *Signal Processing with Adaptive Sparse Structured Representations (SPARS '09)*, 2009.

[105] S.A. Vavasis. On the complexity of nonnegative matrix factorization. *SIAM J. on Optimization*, 20(3):1364–1377, 2009.

[106] F. Wang, T. Li, X. Wang, S. Zhu, and C. Ding. Community discovery using nonnegative matrix factorization. *Data Min. Knowl. Disc.*, 22(3):493–521, 2011.

[107] S. Wild, J. Curry, and A. Dougherty. Improving non-negative matrix factorizations through structured initialization. *Pattern Recognition*, 37(11):2217–2232, 2004.

[108] M. Winter. N-FINDR: an algorithm for fast autonomous spectral end-member determination in hyperspectral data. In *Proc. SPIE Conf. on Imaging Spectrometry V*, pages 266–275, 1999.

[109] Y. Xue, C.S. Tong, Y. Chen, and W.-S. Chen. Clustering-based initialization for non-negative matrix factorization. *Applied Mathematics and Computation*, 205(2):525–536, 2008.

[110] Z. Yang and E. Oja. Linear and nonlinear projective nonnegative matrix factorization. *IEEE Trans. on Neural Networks*, 21(5):734–749, 2010.

[111] M. Yannakakis. Expressing Combinatorial Optimization Problems by Linear Programs. *Journal of Computer and System Sciences*, 43(3):441–466, 1991.

[112] R. Zdunek. Initialization of nonnegative matrix factorization with vertices of convex polytope. In *Artificial Intelligence and Soft Computing*, volume 7267 of *Lecture Notes in Computer Science*, pages 448–455. Springer Berlin Heidelberg, 2012.

[113] S. Zhong and J. Ghosh. Generative model-based document clustering: a comparative study. *Knowledge and Information Systems*, 8(3):374–384, 2005.

Chapter 13

Rank Constrained Optimization Problems in Computer Vision

Ivan Markovsky

Department ELEC, Vrije Universiteit Brussel

13.1 Introduction

The claim that

"Behind every data modeling problem there is a (hidden) low rank
approximation problem"   [13]

is demonstrated in this book chapter via four problems in computer vision:

- multidimensional scaling,

- conic section fitting,

- fundamental matrix estimation, and

- least squares contour alignment.

293

A matrix constructed from exact data is rank deficient. The corresponding data fitting problem in the case of noisy data is a rank constraint optimization problem. In general, rank constrained optimization is a hard nonconvex problem, for which application specific heuristics are proposed. In this chapter, I do not describe solution methods for rank constrained optimization but refer the reader to the literature.

Our main contribution is the analytic solution of the contour alignment problem presented in Section 13.5.1. This problem is also nonconvex in the original problem variables; however, a nonlinear change of variables renders the problem convex in the transformed variables. The link to low-rank approximation (the theme of the chapter) is presented in Section 13.5.4, where the problem is shown to be equivalent to the orthogonal Procrustes problem, which is a constrained low-rank approximation problem [12].

13.2   Multidimensional Scaling

Consider N points $\{x_1, \ldots, x_N\}$ in an n-dimensional real space and let d_{ij} be the squared Euclidean distances between x_i and x_j. The matrix $D = [d_{ij}] \in \mathbb{R}^{N \times N}$ of the pair-wise squared distances is symmetric, element-wise nonnegative, and has zero diagonal elements. Moreover, since

$$d_{ij} := (x_i - x_j)^\top (x_i - x_j) = x_i^\top x_i - 2 x_i^\top x_j + x_j^\top x_j,$$

D has the following structure

$$D = \begin{bmatrix} x_1^\top x_1 \\ \vdots \\ x_N^\top x_N \end{bmatrix} \begin{bmatrix} 1 & \cdots & 1 \end{bmatrix} - 2 \begin{bmatrix} x_1^\top \\ \vdots \\ x_N^\top \end{bmatrix} \begin{bmatrix} x_1 & \cdots & x_N \end{bmatrix} + \begin{bmatrix} 1 \\ \vdots \\ 1 \end{bmatrix} \begin{bmatrix} x_1^\top x_1 & \cdots & x_N^\top x_N \end{bmatrix},$$

or

$$D = \operatorname{diag}(X^\top X) 1_N^\top - 2 X^\top X + 1_N \operatorname{diag}^\top (X^\top X) =: \mathcal{S}(X), \tag{13.1}$$

where

$$X := \begin{bmatrix} x_1 & \cdots & x_N \end{bmatrix} \quad \text{and} \quad 1_N = \begin{bmatrix} 1 & \cdots & 1 \end{bmatrix}^\top \in \mathbb{R}^N.$$

In particular, from (13.1) it can be seen that D is rank deficient:

$$\operatorname{rank}(D) \leq n + 2. \tag{13.2}$$

The image of the function $\mathcal{S} : X \mapsto D$ is referred to as the set of *element-wise-squared-distance matrices*. The inverse of $\mathcal{S}$ is a set valued function

$$\mathcal{S}^{-1}(D) := \{ X \mid (13.1) \text{ holds} \}.$$

If D is a distance matrix of a set of points X, $\mathcal{S}^{-1}(D)$ consists of all rigid transformations (translation, rotation, and reflection) of X. In other words the nonuniqueness in finding X, given D, is up to a rigid transformation.

Theorem 13.1. *Let D be a distance matrix and let $\bar{X}$ be a particular solution of the Equation (13.1). Then*

$$\mathcal{S}^{-1}(D) = \{\, R\bar{X} + c\mathbf{1}_N^\top \mid c \in \mathbb{R}^n \text{ and } R \in \mathbb{R}^{n\times n}, \text{ such that } RR^\top = I \,\}.$$

The considered problem is defined informally as follows:

> Given noisy and incomplete information about the pair-wise squared-distances d_{ij} among the points $\{\, x_1, \dots, x_N \,\}$ and the dimension n of the ambient space, find estimates of the points $\{\, x_1, \dots, x_N \,\}$, up to a rigid transformation.

With exact data, the problem can be posed and solved as a rank revealing factorization problem (see the Appendix). With noisy measurements, however, the matrix D is generically full rank. In this case, the relative (up to rigid transformation) point locations can be *estimated* by approximating D by a rank-$(n + 2)$ matrix $\hat{D}$. In order to be a valid distance matrix, however, $\hat{D}$ must have the structure $\hat{D} = \mathcal{S}(\hat{X})$, for some $\hat{X} = \begin{bmatrix} \hat{x}_1 & \cdots & \hat{x}_N \end{bmatrix}$, i.e., the estimation problem is a *bilinearly structured low-rank approximation problem*:

$$\text{minimize over } \hat{D} \in \mathbb{R}^{N\times N} \text{ and } \hat{X} \in \mathbb{R}^{n\times N} \left\| D - \hat{D} \right\|_{\mathrm{F}} \text{ subject to } \hat{D} = \mathcal{S}(\hat{X}),$$

where $\|\cdot\|_{\mathrm{F}}$ is the Frobenius norm. Note that the rank constraint (13.2) is automatically satisfied by the structure constraint (13.1).

For comprehensive treatment of applications and solution methods for multidimensional scaling, the reader is refered to the books [4, 2].

13.3 Conic Section Fitting

A conic section is a static quadratic model. In this section, I show that the conic section fitting problem can be formulated as a low-rank approximation of an extended data matrix. The mapping from the original data to the extended data is called in the machine learning literature the *feature map*. In the application at hand, the feature map is naturally defined by the conic model, i.e., it is a quadratic function.

Let

$$\{\, d_1, \dots, d_N \,\} \subset \mathbb{R}^2, \qquad \text{where } d_j = \begin{bmatrix} x_j \\ y_j \end{bmatrix},$$

be the given data. A conic section is a set defined by a second order equation

$$\mathcal{B}(A, b, c) := \{\, d \in \mathbb{R}^2 \mid d^\top A d + b^\top d + c = 0 \,\}. \tag{13.3}$$

Here A is a 2×2 symmetric matrix, b is a 2×1 vector, and c is a scalar. A, b, and c are the parameters of the conic section. In order to avoid a trivial solution $\mathcal{B} = \mathbb{R}^2$, it is assumed that at least one of the parameters A, b, or c is nonzero. The representation (13.3) is an implicit representation of the conic section, because it imposes a relation (implicit function) on the elements x and y of d. In special cases, it is possible to use explicit representations defined by a function from x to y or from y to x; however, this approach is restrictive as it does not cover all conic sections (e.g., an ellipse cannot be represented by a map from one variable to the other).

Defining the parameter vector

$$\theta := \begin{bmatrix} a_{11} & 2a_{12} & b_1 & a_{22} & b_2 & c \end{bmatrix},$$

and the extended data vector

$$d_{\text{ext}} := \begin{bmatrix} x^2 & xy & x & y^2 & y & 1 \end{bmatrix}^\top, \tag{13.4}$$

we have that

$$d \in \mathcal{B}(\theta) = \mathcal{B}(A, b, c) \qquad \Longleftrightarrow \qquad \theta d_{\text{ext}} = 0.$$

(The map $d \mapsto d_{\text{ext}}$, defined by (13.4), is the feature map for the conic section model.) Consequently, all data points $d_1, \ldots, d_N$ are fitted by the model if

$$\theta \underbrace{\begin{bmatrix} d_{\text{ext},1} & \cdots & d_{\text{ext},N} \end{bmatrix}}_{D_{\text{ext}}} = 0 \qquad \Longleftrightarrow \qquad \text{rank}(D_{\text{ext}}) \leq 5. \tag{13.5}$$

Indeed, for $\theta \neq 0$, the left-hand-side of the equivalence states that D_{ext} has a nontrivial left kernel. Since D_{ext} has 6 rows (see (13.4)), its rank is at most 5. The mapping $D \mapsto D_{\text{ext}}$ is denoted by $\mathcal{S}$.

In the presence of noise, generically, $\text{rank}(D_{\text{ext}}) > 5$. Then, the aim is to

> approximate the data points $d_1, \ldots, d_N$ by nearby points $\hat{d}_1, \ldots, \hat{d}_N$ that lie exactly on a conic section.

Minimizing the sum of squares of the orthogonal distances from the data points to their approximations leads to the structured low-rank approximation problem

$$\text{minimize} \quad \text{over } \hat{D} \in \mathbb{R}^{2 \times N} \quad \left\| D - \hat{D} \right\|_F \quad \text{subject to} \quad \text{rank}\left(\mathcal{S}(\hat{D}) \right) \leq 5,$$

where

$$D := \begin{bmatrix} d_1 & \cdots & d_N \end{bmatrix}, \qquad \hat{D} := \begin{bmatrix} \hat{d}_1 & \cdots & \hat{d}_N \end{bmatrix}$$

are the data matrix and the approximating matrix, respectively.

In the computer vision literature, see, e.g., the tutorial paper [23], conic section fitting by orthogonal projections is called *geometric fitting*. As shown above, the corresponding computational problem is a quadratically structured low-rank approximation problem. The problem is intuitively appealing; however, it is nonconvex and, moreover, leads to an inconsistent estimator. This has motivated work on easier to compute methods [1, 6, 8, 5, 10, 14, 19] that also reduce or even eliminate the bias.

13.4   Fundamental Matrix Estimation

In two-dimensional motion analysis [11] a scene is captured by two cameras at fixed locations (stereo vision) and N matching pairs of points

$$\{ u_1, \ldots, u_N \} \subset \mathbb{R}^2 \qquad \text{and} \qquad \{ v_1, \ldots, v_N \} \subset \mathbb{R}^2 \qquad (13.6)$$

are located in the resulting images. The corresponding points u and v in the two images satisfy what is called an *epipolar constraint*

$$\begin{bmatrix} v^\top & 1 \end{bmatrix} F \begin{bmatrix} u \\ 1 \end{bmatrix} = 0, \qquad \text{for some } F \in \mathbb{R}^{3 \times 3}, \text{ with } \operatorname{rank}(F) = 2. \qquad (13.7)$$

The 3×3 matrix $F \neq 0$, called the *fundamental matrix*, characterizes the relative position and orientation of the cameras and does not depend on the selected pairs of points. Estimation of F from data is a necessary calibration step in many computer vision methods.

The epipolar constraint (13.7) is linear in F. Indeed, defining

$$d_{\text{ext}} := \begin{bmatrix} u_x v_x & u_x v_y & u_x & u_y v_x & u_y v_y & u_y & v_x & v_y & 1 \end{bmatrix}^\top \in \mathbb{R}^9, \qquad (13.8)$$

where $u = \begin{bmatrix} u_x \\ u_y \end{bmatrix}$ and $v = \begin{bmatrix} v_x \\ v_y \end{bmatrix}$, (13.7) can be written as

$$\operatorname{vec}^\top(F) d_{\text{ext}} = 0.$$

Note that, as in the application for conic section fitting, the original data (u, v) is mapped to an extended data vector d_{ext} via a nonlinear function (a feature map). In this case, however, the function is *bilinear*.

Taking into account the epipolar constraints for all data points, we obtain the matrix equation

$$\operatorname{vec}^\top(F) \underbrace{\begin{bmatrix} d_{\text{ext},1} & \cdots & d_{\text{ext},N} \end{bmatrix}}_{D_{\text{ext}}} = 0. \qquad (13.9)$$

The rank constraint imposed on F implies that F is a nonzero matrix. Therefore, by (13.9) D_{ext} has a nontrivial left kernel and since D_{ext} is $9 \times N$

$$\text{rank}(D_{ext}) \leq 8.$$

It can be concluded that for $N \geq 8$ data points, D_{ext} is not full row rank. Moreover, if the left kernel is one dimensional, the fundamental matrix F can be reconstructed up to a scaling factor from the data.

In the case of noisy data,

> the aim is to perturb as little as possible the data (13.6), so that the perturbed data satisfies exactly the epipolar constraints for some $\hat{F}$ with $\text{rank}(\hat{F}) = 2$.

The resulting estimation problem is a *bilinearly structured low-rank approximation* with an additional rank constraint. This problem defines a maximum-likelihood estimator for the true parameter value. As in the conic section fitting problem, the maximum-likelihood estimator is a nonconvex optimization problem and is inconsistent in the measurement errors or *errors-in-variables* setup. These facts motivated the development of methods that are convex and unbiased; see [21, 3, 9, 14] and the references therein. Closely related to the estimation of the fundamental matrix problem in *two-view computer vision* is the shape from motion problem [20].

13.5   Least Squares Contour Alignment

Let $\mathcal{R}_\theta$ be the operator in $\mathbb{R}^2$ that rotates its argument by θ rad (positive angle corresponding to anticlockwise rotation) and let $\mathcal{R}'_\theta$ be the operator that reflects its argument about a line, passing through the origin, at $\theta/2$ rad with respect to the first basis vector (see Figure 13.1).

It can be shown that $\mathcal{R}_\theta$ and $\mathcal{R}'_\theta$ have matrix representations

$$\mathcal{R}_\theta(p) = \begin{bmatrix} \cos\theta & -\sin\theta \\ \sin\theta & \cos\theta \end{bmatrix} p = R_\theta p$$

and

$$\mathcal{R}'_\theta(p) = \begin{bmatrix} \cos\theta & \sin\theta \\ \sin\theta & -\cos\theta \end{bmatrix} p = R'_\theta p.$$

In [15], the authors considered transformation by rotation, scaling, and translation, i.e.,

$$\mathcal{A}_{a,\theta,s}(p) = s\mathcal{R}_\theta(p) + a, \tag{13.10}$$

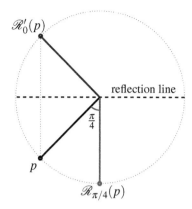

FIGURE 13.1: Rotation $\mathcal{R}_{\theta_1}$ and reflection $\mathcal{R}'_{\theta_2}$ of a point p.

where $s > 0$ is the scaling factor and $a \in \mathbb{R}^2$ is the translation parameter. The problem of determining the parameters θ, s, and a of a transformation $\mathcal{A}_{a,\theta,s}(p)$ that best, in a least squares sense, matches one set of points $p^{(1)}, \ldots, p^{(N)}$ to another set of points $q^{(1)}, \ldots, q^{(N)}$ can be used to align two explicitly represented contours, specified by corresponding points. Although this alignment problem is a nonlinear least squares problem in the parameters θ, s, and a, it is shown in [15] that the change of variables

$$b = \begin{bmatrix} b_1 \\ b_2 \end{bmatrix} = s \begin{bmatrix} \cos \theta \\ \sin \theta \end{bmatrix} \quad , \quad \begin{bmatrix} \theta \\ s \end{bmatrix} = \begin{bmatrix} \sin^{-1}(b_2/\|b\|) \\ \|b\| \end{bmatrix} \tag{13.11}$$

results in an equivalent linear least squares problem in the parameters a and b. This fact allowed efficient solution of image registration problems (see, e.g., [17, 16]) with a large number of corresponding points. The invariance to rigid transformation appears also in learning with linear functionals on reproducing kernel Hilbert space; see [22].

It is well known, however, that dilation and rigid transformation involves reflection, in addition to rotation, scaling, and translation. Therefore, a problem occurs regarding how to optimally align in a least squares sense the set of points

$$\{ p^{(1)}, \ldots, p^{(N)} \} \quad \text{and} \quad \{ q^{(1)}, \ldots, q^{(N)} \}$$

under reflection, rotation, scaling, and translation, i.e., transformation of the type

$$\mathcal{A}_{a,\theta_1,\theta_2,s}(p) = s\mathcal{R}_{\theta_1}\left(\mathcal{R}'_{\theta_2}(p)\right) + a. \tag{13.12}$$

In order to solve the problem of alignment by dilation and rigid transformation, first consider alignment by reflection, scaling, and translation, i.e., transformation of the type

$$\mathcal{A}'_{a,\theta,s}(p) = s\mathcal{R}'_{\theta}(p) + a. \tag{13.13}$$

The solution of this latter problem, given in Section 13.5.1, also uses the change of variables (13.11) to convert the original nonlinear least squares problem to a linear one. The derivation given in Section 13.5.1, however, is different from the derivation in [15] and reveals a link between the alignment problems by rotation and reflection.

The solution to the general least squares alignment problem by rigid transformation is given in Section 13.5.2. Since a transformation (13.12) is either rotation, scaling, and translation, or reflection, scaling, and translation, the alignment problem (13.12) reduces to solving problems (13.10) and (13.13) separately, and choosing the solution that corresponds to the better fit.

In Section 13.5.4, I show that least squares alignment by rotation and reflection is equivalent to the orthogonal Procrustes problem [7, p. 601]. An extension of the orthogonal Procrustes problem to alignment by (13.12), presented in [18], gives an alternative solution method for contour alignment by dilation and rigid transformation. An advantage of the approach based on the orthogonal Procrustes problem is that the solution is applicable to data in higher dimensional space; however, the method requires singular value decomposition of a matrix computed from the data, which may be computationally more expensive than solving an ordinary linear least squares problem.

13.5.1 Alignment by Reflection, Scaling, and Translation

Let C_1 and C_2 be the matrices of the points $p^{(1)}, \ldots, p^{(N)}$ stacked next to each other, and $q^{(1)}, \ldots, q^{(N)}$, respectively, i.e.,

$$C_1 := \begin{bmatrix} p^{(1)} & \cdots & p^{(N)} \end{bmatrix} \quad \text{and} \quad C_2 := \begin{bmatrix} q^{(1)} & \cdots & q^{(N)} \end{bmatrix},$$

and let $\|\cdot\|_{\mathrm{F}}$ be the Frobenius norm, defined as

$$\|C_1\|_{\mathrm{F}} := \sqrt{\sum_{i=1}^{N} \left\| p^{(i)} \right\|_2^2}.$$

The problem considered in this section is least squares alignment by reflection:

$$\begin{aligned} \text{minimize} \quad & \left\| C_1 - \mathcal{A}'_{a,\theta,s}(C_2) \right\|_{\mathrm{F}} \\ \text{over} \quad & a \in \mathbb{R}^2, \ s > 0, \ \theta \in [-\pi, \pi). \end{aligned} \tag{13.14}$$

Similarly to the alignment by rotation problem

$$\begin{aligned} \text{minimize} \quad & \left\| C_1 - \mathcal{A}_{a,\theta,s}(C_2) \right\|_{\mathrm{F}} \\ \text{over} \quad & a \in \mathbb{R}^2, \ s > 0, \ \theta \in [-\pi, \pi), \end{aligned} \tag{13.15}$$

(13.14) is a nonlinear least squares problem in the parameters θ, s, and a. The change of variables (13.11), however, also transforms problem (13.14) into a linear least squares problem.

Theorem 13.2 (Alignment by reflection, scaling, and translation). *Problem (13.14) is equivalent to the linear least squares problem*

$$\text{minimize} \quad \text{over } a,b \in \mathbb{R}^2 \quad \left\| \text{vec}(C_1) - \left[(C_2^\top \otimes I_2)E \quad \mathbf{1}_N \otimes I_2 \right] \begin{bmatrix} b \\ a \end{bmatrix} \right\|_2$$
(13.16)

where $\text{vec}(\cdot)$ *is the column-wise matrix vectorization operator,* $\otimes$ *is the Kronecker product,*

$$\mathbf{1}_N := \begin{bmatrix} 1 \\ \vdots \\ 1 \end{bmatrix} \in \mathbb{R}^N, \quad E := \begin{bmatrix} 1 & 0 \\ 0 & 1 \\ 0 & 1 \\ -1 & 0 \end{bmatrix}, \quad \text{and } I_2 := \begin{bmatrix} 1 & 0 \\ 0 & 1 \end{bmatrix}. \tag{13.17}$$

The one-to-one relation between the parameters θ, s *and* b_1, b_2 *is given by (13.11).*

Proof. Note that

$$\mathcal{A}'_{a,\theta,s}(C_2) = sR'_\theta C_2 - a\mathbf{1}_N^\top.$$

Using the identity,

$$\text{vec}(AXB) = (B^\top \otimes A)\,\text{vec}(X),$$

we rewrite the cost function of (13.14) as

$$\left\| C_1 - I_2 \begin{bmatrix} s\cos\theta & s\sin\theta \\ s\sin\theta & -s\cos\theta \end{bmatrix} C_2 - a \right\|_F = \left\| \text{vec}(C_1) - (C_2^\top \otimes I_2) \begin{bmatrix} s\cos\theta \\ s\sin\theta \\ s\sin\theta \\ -s\cos\theta \end{bmatrix} - a \right\|_2$$

$$= \left\| \text{vec}(C_1) - (C_2^\top \otimes I_2) \begin{bmatrix} 1 & 0 \\ 0 & 1 \\ 0 & 1 \\ -1 & 0 \end{bmatrix} \begin{bmatrix} s\cos\theta \\ s\sin\theta \end{bmatrix} - a \right\|_2.$$

Problem (13.14) and the relation (13.11) follows by setting

$$b_1 := s\cos\theta \quad \text{and} \quad b_2 := s\sin\theta.$$

$\square$

Note 13.1 (Alignment by rotation, scaling, and translation). *The above solution of Problem (13.14) can be modified easily for the corresponding alignment problem with rotation (13.15), giving an alternative shorter proof to Theorem 1 in [15]. Indeed, the only necessary modification is to replace the matrix* E *in (13.17) by*

$$E = \begin{bmatrix} 1 & 0 \\ 0 & 1 \\ 0 & -1 \\ 1 & 0 \end{bmatrix}.$$

Example 13.1. *As an illustration of the presented alignment procedure, consider the contours shown in Figure 13.2. The optimal alignment by rotation, scaling, and translation is shown in Figure 13.3, right, and the optimal alignment by reflection, scaling, and translation is shown in Figure 13.3, left.*

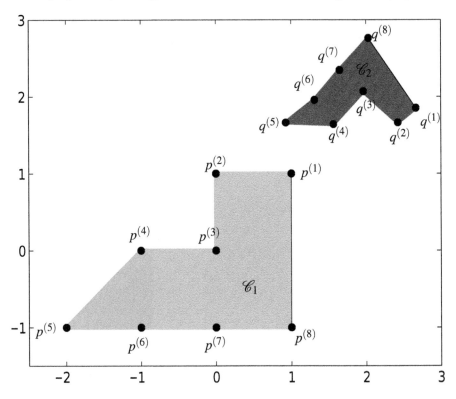

FIGURE 13.2: Example of contour alignment problem (13.14): given contours C_1 and C_2 with corresponding points $p^{(i)} \leftrightarrow q^{(i)}$, find a transformation $\mathcal{A}'_{a,\theta,s}$ that minimizes the distance between C_1 and the transformed contour $\mathcal{A}'_{a,\theta,s}(C_2)$.

13.5.2   Alignment by Rigid Transformation

The problem considered in this section is:

$$\text{minimize} \quad \|C_1 - \mathcal{A}_{a,\theta_1,\theta_2,s}(C_2)\|_{\mathrm{F}}$$
$$\text{over} \quad a \in \mathbb{R}^2, \; s > 0, \; \theta_1, \theta_2 \in [-\pi, \pi) \tag{13.18}$$

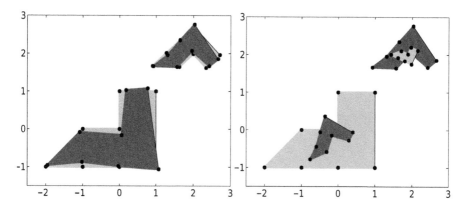

FIGURE 13.3: Left: optimal alignment of $\mathcal{C}_2$ to $\mathcal{C}_1$ and $\mathcal{C}_1$ to $\mathcal{C}_2$ by $\mathcal{A}'_{a,\theta,s}$ (reflection), Right: optimal alignment of $\mathcal{C}_2$ to $\mathcal{C}_1$ and $\mathcal{C}_1$ to $\mathcal{C}_2$ by $\mathcal{A}_{a,\theta,s}$ (rotation).

The following fact allows us to reduce Problem (13.18) to the already studied Problems (13.14) and (13.15).

Proposition 13.1. *A transformation by rotation and reflection, $R_{\theta_1}\left(R'_{\theta_2}(p)\right)$, is equivalent to a transformation by an orthogonal matrix Qp. Moreover,*

$$Qp = R_\theta(p), \quad \text{if } \det(Q) = 1,$$

and

$$Qp = R'_\theta(p), \quad \text{if } \det(Q) = -1,$$

where, in either case,

$$\theta = \cos^{-1}(q_{11}). \tag{13.19}$$

Proof. The matrix $R_{\theta_1} R'_{\theta_2}$ is orthogonal, because R_{θ_1} and R'_{θ_2} are orthogonal matrices and the product of orthogonal matrices is an orthogonal matrix. Next, I show that an orthogonal matrix Q is either a rotation matrix R_θ, for some $\theta \in [-\pi, \pi)$, or a reflection matrix R'_θ, for some $\theta \in [-\pi, \pi)$.

Since Q is orthogonal

$$\begin{bmatrix} q_{11} & q_{12} \\ q_{21} & q_{22} \end{bmatrix} \begin{bmatrix} q_{11} & q_{21} \\ q_{12} & q_{22} \end{bmatrix} = \begin{bmatrix} 1 & 0 \\ 0 & 1 \end{bmatrix}.$$

Without loss of generality we can choose

$$q_{11} = \cos\theta \quad \text{and} \quad q_{12} = \sin\theta.$$

Then, there are two possibilities for q_{21} and q_{22}

$$q_{21} = \cos(\theta + \pi/2) \quad \text{and} \quad q_{22} = \sin(\theta + \pi/2)$$

or

$$q_{21} = \cos(\theta - \pi/2) \quad \text{and} \quad q_{22} = \sin(\theta - \pi/2).$$

In the first case, Q is a rotation matrix and in the second case Q is a reflection matrix. Therefore,

$$Q = R_\theta \quad \text{or} \quad Q = R'_\theta,$$

where

$$\theta = \cos^{-1}(q_{11}).$$

It is easy to check that

$$\det(R_\theta) = 1 \quad \text{and} \quad \det(R'_\theta) = -1, \qquad \text{for any } \theta.$$

$\square$

The result of Proposition 13.1 shows that Problem (13.18) can be solved by the following procedure.

1. Solve the alignment problem by reflection (13.14).

2. Solve the alignment problem by rotation (13.15).

3. Select the solution of the problem that gives smaller cost function value.

Since Problems (13.14) and (13.15) are already solved in Section 13.5.1, we have a complete solution to (13.18).

13.5.3   Invariance Properties and a Distance Measure

It turns out that the minimum value of (13.18)

$$\text{dist}'(C_1, C_2) := \min_{a \in \mathbb{R}^2, \ s>0, \ \theta_1, \theta_2 \in [-\pi, \pi)} \|C_1 - \mathcal{A}_{a, \theta_1, \theta_2, s}(C_2)\|_{\mathrm{F}}$$

is not a proper distance measure. (A counter example is given in Example 13.3.) Proposition 4 (invariance) and Theorem 7 (distance measure) were stated and proved in [15] for alignment by (13.10). However, they hold for the more general problem of alignment by dilation and rigid transformation.

Proposition 13.2. *If the contours C_1 and C_2, defined by the sets of corresponding points $\{p^{(i)}\}$ and $\{p^{(i)}\}$, are centered (i.e., $C_1 1_N = C_2 1_N = 0$), $\text{dist}'(C_1, C_2)$ is invariant to a rigid transformation, i.e.,*

$$\begin{aligned}
\text{dist}'(C_1, C_2) &= \text{dist}' \left(\mathcal{R}_\theta(C_1), \mathcal{R}_\theta(C_2) \right) \\
&= \text{dist}' \left(\mathcal{R}'_\theta(C_1), \mathcal{R}'_\theta(C_2) \right), \quad \text{for any } \theta \in [-\pi, \pi).
\end{aligned} \tag{13.20}$$

If, in addition, C_1 and C_2 are normalized by $\|C_1\|_{\mathrm{F}} = \|C_2\|_{\mathrm{F}} = 1$,

$$\text{dist}'(C_1, C_2) = \text{dist}'(C_2, C_1). \tag{13.21}$$

Example 13.2. *Consider again the contours from Example 13.1. The points $p^{(i)}$ and $q^{(i)}$ are preprocessed, so that the resulting contours, say $C_{1,c}$ and $C_{2,c}$, are centered. As a numerical verification of (13.20), we have*

$$
\begin{aligned}
\text{dist}'(C_{1,c}, C_{2,c}) &= \text{dist}'\left(\mathcal{R}_{0.3}(C_{1,c}), \mathcal{R}_{0.3}(C_{2,c})\right) \\
&= \text{dist}'\left(\mathcal{R}'_{0.3}(C_{1,c}), \mathcal{R}'_{0.3}(C_{2,c})\right) = 0.40640.
\end{aligned}
$$

Let, in addition, the points $p^{(i)}$ and $q^{(i)}$ be preprocessed, so that the resulting contours, say $C_{1,cn}$ and $C_{2,cn}$, are centered and normalized. As a numerical verification of (13.21), we have

$$
\text{dist}'(C_{1,cn}, C_{2,cn}) = \text{dist}'(C_{2,cn}, C_{1,cn}) = 0.11271.
$$

As in the case of the transformation (13.10), treated in [15, Section III], the following definition gives a distance measure.

Definition 13.1 (2-norm distance between contours modulo rigid transformation).

$$
\text{dist}(C_1, C_2) := \frac{1}{\left\| C_1 - \frac{1}{N} C_1 1_N 1_N^\top \right\|_{\text{F}}} \times
$$
$$
\min_{a \in \mathbb{R}^2, \ s > 0, \ \theta_1, \theta_2 \in [-\pi, \pi)} \left\| C_1 - \mathcal{A}_{a, \theta_1, \theta_2, s}(C_2) \right\|_{\text{F}}. \quad (13.22)
$$

Theorem 13.3. *The distance measure $\text{dist}(C_1, C_2)$ is symmetric and invariant to dilation and a rigid transformation, i.e.,*

$$
\text{dist}(C_1, C_2) = \text{dist}(C_2, C_1) = \text{dist}\left(\mathcal{A}_{a, \theta_1, \theta_2, s}(C_1), \mathcal{A}_{a, \theta_1, \theta_2, s}(C_2)\right),
$$
$$
\text{for all } a \in \mathbb{R}^2, \ \theta_1, \theta_2 \in [-\pi, \pi), \ \text{and } s > 0. \quad (13.23)
$$

Example 13.3. *For the contours in Example 13.1, we have*

$$
\text{dist}'(C_1, C_2) = 0.40640 \qquad \text{and} \qquad \text{dist}'(C_2, C_1) = 0.20748,
$$

while

$$
\text{dist}(C_1, C_2) = \text{dist}(C_2, C_1) = 0.11271.
$$

13.5.4   Contour Alignment as an Orthogonal Procrustes Problem

As a consequence of Proposition 13.1, we have that

Problem (13.18) is equivalent to

$$
\begin{aligned}
\text{minimize} \quad & \| C_1 - sQC_2 - a \|_{\text{F}} \\
\text{subject to} \quad & Q^\top Q = I_2 \\
\text{over} \quad & a \in \mathbb{R}^2, \ s > 0, \ Q \in \mathbb{R}^{2 \times 2}.
\end{aligned} \quad (13.24)
$$

In turn, Problem (13.24) is related to the orthogonal Procrustes problem in numerical linear algebra.

Problem 13.1 (Orthogonal Procrustes problem). *Given $q \times N$ real matrices C_1 and C_2,*

$$\text{minimize} \quad \text{over } Q \quad \|C_1 - QC_2\|_{\mathrm{F}} \quad \text{subject to} \quad Q^\top Q = I_q.$$

The classical solution of the orthogonal Procrustes problem is given by

$$Q = UV^\top,$$

where $U\Sigma V^\top$ is the singular value decomposition (SVD) of $C_1^\top C_2$, see [7, p. 601].

The orthogonal Procrustes problem does not involve scaling and translation. The extension of the problem to alignment by dilation and rigid transformation is done in [18]. The resulting procedure is summarized in Algorithm 14. It presents an alternative solution approach for solving Problem (13.18). Compared to the solution proposed in Section 13.5.4, Algorithm 14 has the advantage of being applicable to data of any dimension ($C_1, C_2 \in \mathbb{R}^{q \times N}$, for any natural number q), i.e., the solution based on the orthogonal Procrustes problem is applicable to contours in spaces of dimension higher than 2.

The solution based on the orthogonal Procrustes problem, however, uses the singular value decomposition, while the solution proposed in Section 13.5.4 involves two ordinary least least squares problems. Therefore, an advantage of the proposed solution is its conceptual simplicity. In particular, exploiting the Kronecker structure of the coefficients matrix in (13.16) one can derive an efficient algorithm for alignment of contours specified by a large number of corresponding points. Furthermore, in the case of sequential but not necessarily corresponding points (see [15, Section IV]), N alignment problems are solved, which makes the computational efficiency an important factor.

Algorithm 14 Algorithm for least-squares contour alignment, based on the orthogonal Procrustes problem.

Input: Contours with corresponding points, specified by matrices C_1 and C_2.

1: Centering of the contours:

$$C_{i,c} := C_i - a^{(i)} 1_N^\top, \qquad \text{where } a^{(i)} := \frac{1}{N} C_i 1_N.$$

2: Alignment of the centered data by orthogonal transformation:

$$Q := UV^\top, \qquad \text{where } U\Sigma V^\top \text{ is the SVD of } C_{2,c}^\top C_{1,c}.$$

3: Computation of the scaling parameter:

$$s := \frac{\text{trace}(QC_2 C_1^\top)}{\|C_{2,c}\|_F^2}$$

4: Rigid transformation of C_2 to fit C_1:

$$\hat{C}_1 := sQ(C_2 - a^{(2)} 1_N^\top) + a^{(1)} 1_N^\top.$$

Output: Rigid transformation parameters:

- $a^{(1)} - sQa^{(2)}$ — translation,

- Q — orthogonal transformation, and

- s — scaling.

13.6   Conclusions

This chapter illustrated the claim that every data modeling problem is related to a (structured) low-rank approximation problem for a matrix obtained from the data via a nonlinear transformation (feature map) by four specific examples in computer vision: multidimensional scaling, conic section fitting, fundamental matrix estimation, and contour alignment. In multidimensional scaling, the data is the squared distances between a set of points and the structure of the low-rank approximation problem is given by (13.1). This structure automatically makes the constructed matrix rank deficient, so that the low-rank approximation problem has no additional rank constraint. In the conic section fitting problem, the feature map is a quadratic function and, in the fundamental matrix estimation problem, the feature map is a bilinear function. Finally the contour alignment problem was reduced to the orthogonal Procrustes problem, which is a low-rank approximation problem with an additional orthogonality constraint. A summary of the application is given in Table 13.1.

TABLE 13.1: Summary of applications, matrix structures, and rank constraints.

application	data	data matrix	structure	rank =
multidim. scaling	distances d_{ij} pair-wise	$\begin{bmatrix} d_{ij} \end{bmatrix}$	(13.1)	$\dim(x) + 2$
conic section fitting	points d_i	(13.4), (13.5)	quadratic	5
fundamental matrix estimation	corresponding points u_j, v_j	(13.8), (13.9)	bilinear	8
contour alignment	corresponding points C_1, C_2	$\begin{bmatrix} C_1 \\ C_2 \end{bmatrix}$	unstructured	2

Acknowledgments

Funding from the European Research Council under the European Union's Seventh Framework Programme (FP7/2007-2013)/ERC Grant agreement number 258581, "Structured low-rank approximation: Theory, algorithms, and applications", is gratefully acknowledged.

Appendix: Position Estimation from Exact and Complete Distances

Consider the change of variables

$$S := X^\top X. \tag{A.1}$$

The inverse transformation $S \mapsto X$ is a set valued function with nonuniqueness described by the orthogonal transformation $X \mapsto RX$ (i.e., rotation or reflection of the set of points X). A particular solution of Equation (A.1), for given symmetric matrix X of rank at most n, can be computed by the eigenvalue decomposition of X. Let

$$S = V\Lambda V^\top = \begin{bmatrix} V_1 & V_2 \end{bmatrix} \begin{bmatrix} \Lambda_1 & \\ & 0 \end{bmatrix} \begin{bmatrix} V_1 & V_2 \end{bmatrix}^\top,$$

where the diagonal elements of Λ_1 are all positive, be the eigenvalue decomposition of X. Then

$$\sqrt{\Lambda_1} V_1^\top = RX,$$

for some orthogonal matrix R.

Equation (13.1) is linear in S. We have,

$$\mathrm{vec}(D) = (\mathbf{1}_N \otimes E + E \otimes \mathbf{1}_N - 2I)\,\mathrm{vec}(S) =: L\,\mathrm{vec}(S). \tag{A.2}$$

Furthermore, taking into account the symmetry of D and S, (A.2) becomes

$$\mathrm{vec_s}(D) = L_\mathrm{s}\,\mathrm{vec_s}(S). \tag{A.3}$$

The matrix L_s is of size $N_\mathrm{s} \times N_\mathrm{s}$, where $N_\mathrm{s} := N(N+1)/2$, and is a submatrix of $L \in \mathbb{R}^{N \times N}$.

The system of linear equations (A.3) has N_s equations and N_s unknowns. The matrix L_s, however, is rank deficient

$$\mathrm{rank}(L_\mathrm{s}) = N_\mathrm{s} - N,$$

so that a solution is nonunique. (Assuming that D is a distance matrix, an exact solution of (A.3) exists.) We are aiming at a solution S of (A.3) of rank at most n, finding such a solution in the affine set of solutions is a hard problem.

A simple transformation avoids the nonuniqueness issue. The translated set of points

$$\bar{X} := X - x_1 \mathbf{1}_N^\top = \begin{bmatrix} 0 & \bar{x}_2 & \cdots & \bar{x}_N \end{bmatrix}$$

has the same distance matrix as X, i.e., $\mathcal{S}(\bar{X}) = D$. The change of variables (A.1) then results in a matrix

$$\bar{S} := \bar{X}^\top \bar{X} = \begin{bmatrix} 0_{1\times 1} & 0_{N-1\times 1} \\ 0_{1\times N-1} & * \end{bmatrix},$$

so that

$$\mathrm{vec}_\mathrm{s}(\bar{S}) = \begin{bmatrix} 0_{N\times 1} \\ * \end{bmatrix}.$$

From (A.3), we have

$$\mathrm{vec}_\mathrm{s}(D) = L_\mathrm{s} \begin{bmatrix} 0_{N\times 1} \\ \bar{s} \end{bmatrix} =: L_\mathrm{s}(:, N+1 :)\bar{s}. \tag{A.4}$$

The submatrix $L_\mathrm{s}(:, N+1 :)$ of L_s is full column rank, which implies that $\bar{s}$ is the unique solution of (A.4).

Bibliography

[1] F. L. Bookstein. Fitting conic sections to scattered data. *Computer Graphics Image Proc.*, 9:59–71, 1979.

[2] I. Borg and P. Groenen. *Modern Multidimensional Scaling: Theory and Applications.* Springer, 2005.

[3] S. Chaudhuri and S. Chatterjee. Recursive estimation of motion parameters. *Computer Vision and Image Understanding*, 64(3):434–442, November 1996.

[4] T. Cox and M. Cox. *Multidimensional Scaling, Second Edition.* CRC Press, 2000.

[5] A. Fitzgibbon, M. Pilu, and R. Fisher. Direct least-squares fitting of ellipses. *IEEE Trans. Pattern Anal. Machine Intelligence*, 21(5):476–480, 1999.

[6] W. Gander, G. Golub, and R. Strebel. Fitting of circles and ellipses: Least squares solution. *BIT*, 34:558–578, 1994.

[7] G. Golub and C. Van Loan. *Matrix Computations*. Johns Hopkins University Press, third edition, 1996.

[8] K. Kanatani. Statistical bias of conic fitting and renormalization. *IEEE Trans. Pattern Anal. Machine Intelligence*, 16(3):320–326, 1994.

[9] A. Kukush, I. Markovsky, and S. Van Huffel. Consistent fundamental matrix estimation in a quadratic measurement error model arising in motion analysis. *Comput. Statist. Data Anal.*, 41(1):3–18, 2002.

[10] A. Kukush, I. Markovsky, and S. Van Huffel. Consistent estimation in an implicit quadratic measurement error model. *Comput. Statist. Data Anal.*, 47(1):123–147, 2004.

[11] Y. Ma, S. Soatto, J. Kosecká, and S. Sastry. *An Invitation to 3-D Vision*, volume 26 of *Interdisciplinary Applied Mathematics*. Springer, 2004.

[12] I. Markovsky. Structured low-rank approximation and its applications. *Automatica*, 44(4):891–909, 2008.

[13] I. Markovsky. *Low Rank Approximation: Algorithms, Implementation, Applications*. Communications and Control Engineering. Springer, 2012.

[14] I. Markovsky, A. Kukush, and S. Van Huffel. Consistent least squares fitting of ellipsoids. *Numerische Mathematik*, 98(1):177–194, 2004.

[15] I. Markovsky and S. Mahmoodi. Least-squares contour alignment. *IEEE Signal Proc. Letters*, 16(1):41–44, 2009.

[16] J. Marques. A fuzzy algorithm for curve and surface alignment. *Pattern Recognition Letters*, 19:797–803, 1998.

[17] J. Marques and A. Abrantes. Shape alignment—optimal initial point and pose estimation. *Pattern Recognition Letters*, 18:49–53, 1997.

[18] P. Schönemann and R. Carroll. Fitting one matrix to another under choice of a central dilation and a rigid motion. *Psychometrika*, 35(2):245–255, 1970.

[19] S. Shklyar, A. Kukush, I. Markovsky, and S. Van Huffel. On the conic section fitting problem. *Journal of Multivariate Analysis*, 98:588–624, 2007.

[20] C. Tomasi and T. Kanade. Shape and motion from image streams: A factorization method. *Proc. Natl. Adadem. Sci. USA*, 90:9795–9802, 1993.

[21] P. Torr and D. Murray. The development and comparison of robust methods for estimating the fundamental matrix. *Int. J. Computer Vision*, 24(3):271–300, 1997.

[22] X. Zhang, W. S. Lee, and Y. W. Teh. Learning with invariance via linear functionals on reproducing kernel Hilbert space. In *Proc. of the Neural Information Processing Systems (NIPS) conference*, pages 2031–2039, 2013.

[23] Z. Zhang. Parameter estimation techniques: A tutorial with application to conic fitting. *Image Vision Comp. J.*, 15(1):59–76, 1997.

Chapter 14

Low-Rank Tensor Denoising and Recovery via Convex Optimization

Ryota Tomioka

Toyota Technological Institute, Chicago

Taiji Suzuki

Tokyo Institute of Technology

Kohei Hayashi

National Institute of Informatics

Hisashi Kashima

University of Tokyo

14.1   Introduction

Low-rank decomposition of tensors (multi-way arrays) naturally arises in many application areas, including signal processing, neuroimaging, bioinformatics, recommender systems and other relational data analysis [29, 35, 20].

This chapter reviews convex-optimization-based algorithms for tensor decomposition. There are several reasons to look into convex optimization for tensor decomposition. First, it allows us to prove worst-case-performance guarantees. Although we might be able to give a performance guarantee for an estimator based on a non-convex optimization (see, e.g., [43]), the practical relevance of the bound would be limited if we cannot obtain the optimum efficiently. Second, the convex methods allow us to side-step the tensor rank selection problem; in practice misspecification of tensor rank can significantly deteriorate the performance, whereas choosing a continuous regularization parameter can be considered an easier task. Third, it allows us to use various efficient techniques developed in the mathematical programming communities, such as proximity operation, alternating direction method of multipliers (ADMM), and duality gap monitoring, which enable us to apply these algorithms to a variety of settings reliably. The norms we propose can be used for both denoising of a fully observed noisy tensor and reconstruction of a low-rank tensor from incomplete measurements. Of course there are limitations to what we can achieve with convex optimization, which we will discuss in Section 14.6. Nevertheless we hope that the methods we discuss here serve to connect tensor decomposition with statistics and (convex) optimization, which have been largely disconnected until recently, and contribute to the better understanding of the hardness and challenges of this area.

This chapter is structured as follows: in the next section, we introduce different notions of tensor ranks and present two norms that induce low-rank tensors, namely the overlapped Schatten 1-norm and latent Schatten 1-norm. In Section 14.3, we present denoising and recovery bounds for the two norms. The proofs of the theorems can be found in original papers [54, 53]. In Section 14.4, we propose optimization algorithms for the two norms based on primal and dual ADMM, respectively. Although ADMM has become a standard practice these days, our implementation allows us to deal with the noisy case and the exact case in the same framework (no need for continuation). We also discuss the choice of the penalty parameter η. Section 14.5 consists of some simple demonstrations of the implication of the theorems. Full quantitative evaluation of the bounds can be found in original papers [54, 53]. We discuss various extensions and related work in Section 14.6. We conclude this chapter with possible future directions.

14.2   Ranks and Norms

Let $\mathcal{W} \in \mathbb{R}^{n_1 \times n_2 \times \cdots \times n_K}$ be a K-way tensor. We denote the total number of entries in $\mathcal{W}$ by $N = \prod_{k=1}^{K} n_k$.

14.2.1   Rank and Multilinear Rank

A tensor $\mathcal{W}$ is *rank one* if it can be expressed as an outer product of K vectors as

$$\mathcal{W} = \boldsymbol{u}^{(1)} \circ \boldsymbol{u}^{(2)} \circ \cdots \circ \boldsymbol{u}^{(K)},$$

which can be written element-wise as follows:

$$W_{i_1 i_2 \cdots i_K} = u_{i_1}^{(1)} u_{i_2}^{(2)} \cdots u_{i_K}^{(K)}, \quad (1 \leq i_k \leq n_k, k = 1, \ldots, K).$$

It is easy to verify that a tensor is rank one.

The rank of a tensor $\mathcal{W}$ is the smallest number r such that $\mathcal{W}$ can be expressed as the sum of r rank-one tensors as follows:

$$\mathcal{W} = \sum_{j=1}^{r} \boldsymbol{u}_j^{(1)} \circ \boldsymbol{u}_j^{(2)} \circ \cdots \circ \boldsymbol{u}_j^{(K)}. \tag{14.1}$$

The above decomposition is known as the canonical polyadic (CP) decomposition [22]. It is known that finding the rank r or computing the best rank r approximation (even for $r = 1$) is an NP hard problem [23, 21].

The multilinear rank of $\mathcal{W}$ is the K tuple $(r_1, \ldots, r_K)$ such that r_k is the dimension of the space spanned by the mode-k fibers [13, 29]; here mode-k fibers are the n_k dimensional vectors obtained by fixing all but the kth index. If $\mathcal{W}$ admits decomposition (14.1), r_k is at most r, in which case the multilinear rank of $\mathcal{W}$ is at most $(r, \ldots, r)$.

In contrast to the rank, the multilinear rank $(r_1, \ldots, r_K)$ can be computed efficiently. To this end, it is convenient to define the mode-k unfolding operation. The mode-k unfolding $\boldsymbol{W}_{(k)}$ is a $n_k \times N/n_k$ matrix obtained by concatenating the mode-k fibers along columns. Then r_k is the matrix rank of the mode-k unfolding $\boldsymbol{W}_{(k)}$.

The notion of multilinear rank is connected to another decomposition known as the Tucker decomposition [56] or the higher-order SVD [13, 14]

$$\mathcal{W} = \mathcal{C} \times_1 \boldsymbol{U}_1 \times_2 \boldsymbol{U}_2 \cdots \times_K \boldsymbol{U}_K, \tag{14.2}$$

where $\times_k$ denotes the mode-k product [29].

Computation of Decompositions (14.1) and (14.2) from large noisy tensor with possibly missing entries is a challenging task. Alternating least squares

(ALS) [11] and higher-order orthogonal iteration (HOOI) [14] are well known and many extensions of them are proposed [29]. However, they typically come with no theoretical guarantee either about global optimality of the obtained solution or the statistical performance of the estimator. Kannan and Vempala (see Chapter 8) [27] proposed a sampling based algorithm with a performance bound, which requires knowledge of the Frobenius norms of the slices.

14.2.2   Convex Relaxations

Recently, motivated by the success of the Schatten 1-norm (also known as the trace norm and nuclear norm) for the recovery of low-rank matrices [16, 50, 42, 10, 43, 38], several authors have proposed norms that induce low-rank tensors.

These approaches solve convex problems of the following form:

$$\underset{\mathcal{W}}{\text{minimize}} \quad L(\mathcal{W}) + \lambda \left\| \mathcal{W} \right\|_\star, \tag{14.3}$$

where $L : \mathbb{R}^{n_1 \times \cdots \times n_K} \to \mathbb{R}$ is a convex loss function that measures how well $\mathcal{W}$ fits the data, $\left\| \mathcal{W} \right\|_\star$ is a norm (we discuss in detail below), and $\lambda > 0$ is a regularization parameter.

For example, let's assume that the measurements $\boldsymbol{y} = (y_i)_{i=1}^M$ are generated as

$$y_i = \langle \mathcal{X}_i, \mathcal{W}^* \rangle + \epsilon_i, \tag{14.4}$$

where $\langle \mathcal{X}, \mathcal{W} \rangle$ denotes the inner product between two tensors viewed as vectors in $\mathbb{R}^N$; more precisely, $\langle \mathcal{X}, \mathcal{W} \rangle = \sum_{i_1, \ldots, i_K} X_{i_1 \ldots i_K} W_{i_1 \ldots i_K}$. Then the loss function can be defined as the sum of squared residuals

$$L(\mathcal{W}) = \frac{1}{2} \| \boldsymbol{y} - \mathfrak{X}(\mathcal{W}) \|_2^2,$$

where $\mathfrak{X}(\mathcal{W}) := (\langle \mathcal{X}_i, \mathcal{W} \rangle)_{i=1}^M$.

The minimization problem (14.3) minimizes the loss function also keeping the norm small. The difference from the conventional optimization based approaches for tensor decomposition is that instead of constraining the (multilinear) rank of the decomposition, it only constrains the complexity of the solution measured by a particular norm.

In the case of matrices, it is well known that the Schatten 1-norm

$$\| \boldsymbol{W} \|_{S_1} = \sum_{j=1}^r \sigma_j(\boldsymbol{W}),$$

where $\sigma_j(\boldsymbol{W})$ is the jth singular value of $\boldsymbol{W}$ and r is the rank of $\boldsymbol{W}$ promotes the solution of (14.3) to be low-rank; see, e.g., [15]. Intuitively, this can be understood analogous to the sparsity inducing property of the ℓ_1 norm; it promotes the spectrum of $\boldsymbol{W}$ to be sparse, i.e., a spectral version of lasso [51].

It is known that the Schatten 1-norm of a rank r matrix $\boldsymbol{W}$ can be related to its Frobenius norm as follows [50]:

$$\|\boldsymbol{W}\|_{S_1} \leq \sqrt{r}\|\boldsymbol{W}\|_F.$$

Thus a low-rank matrix has a small Schatten 1-norm relative to its Frobenius norm.

The following norm has been proposed by several authors [47, 18, 33, 52]:

$$\|\mathcal{W}\|_{S_1/1} = \sum_{k=1}^{K} \|\boldsymbol{W}_{(k)}\|_{S_1}. \tag{14.5}$$

We call the norm (14.5) *overlapped Schatten 1-norm*. Intuitively, it penalizes the Schatten 1-norms of the K unfoldings, and minimizing the norm promotes $\mathcal{W}$ to have low-multilinear rank. In fact, it is easy to show (see [54]) the inequality

$$\|\mathcal{W}\|_{S_1/1} \leq \sum_{k=1}^{K} \sqrt{r_k}\|\mathcal{W}\|_F, \tag{14.6}$$

where $\|\mathcal{W}\|_F$ is the Frobenius norm $\|\mathcal{W}\|_F = \sqrt{\langle \mathcal{W}, \mathcal{W}\rangle}$. Thus, tensors that have low multilinear rank (on average) have low overlapped Schatten 1-norm relative to the Frobenius norm.

Another norm proposed in [52, 53] is the latent Schatten 1-norm

$$\|\mathcal{W}\|_{\overline{S_1/1}} = \inf_{(\mathcal{W}^{(1)}+\cdots+\mathcal{W}^{(K)})=\mathcal{W}} \sum_{k=1}^{K} \|\boldsymbol{W}_{(k)}^{(k)}\|_{S_1}. \tag{14.7}$$

Here the norm is defined as the infimum over all tuple of K tensors that sums to the original tensor $\mathcal{W}$. It is also easy to relate the latent Schatten 1-norm to the multilinear rank of $\mathcal{W}$. In [53], it was shown that

$$\|\mathcal{W}\|_{\overline{S_1/1}} \leq \min_{k} \sqrt{r_k}\|\mathcal{W}\|_F. \tag{14.8}$$

Note that the sum in inequality (14.6) is replaced by the minimum in inequality (14.8). Therefore, the latent Schatten 1-norm is small when the minimum mode-k rank of $\mathcal{W}$ is small.

14.3   Statistical Guarantees

In this section we present statistical performance guarantee for the estimators defined by the overlapped and latent Schatten 1-norms.

14.3.1   Denoising Bounds

The first two theorems concern the denoising performance of the two norms.

Suppose that the observation $\mathcal{Y} \in \mathbb{R}^{n_1 \times \cdots \times n_K}$ is obtained as follows:

$$\mathcal{Y} = \mathcal{W}^* + \mathcal{E},$$

where $\mathcal{W}^*$ is the true low-rank tensor with multilinear rank $(r_1, \ldots, r_K)$ and $\mathcal{E} \in \mathbb{R}^{n_1 \times \cdots \times n_K}$ is the noise tensor whose entries are independently identically distributed zero-mean Gaussian random variables with variance σ^2.

Define the estimator $\hat{\mathcal{W}}$ by

$$\hat{\mathcal{W}} = \underset{\mathcal{W}}{\operatorname{argmin}} \left(\frac{1}{2} \|\mathcal{Y} - \mathcal{W}\|_F^2 + \lambda \|\mathcal{W}\|_{\underline{S_1/1}} \right), \qquad (14.9)$$

where $\lambda > 0$ is a regularization parameter.

Then we have the following denoising performance guarantee.

Theorem 14.1 (Denoising via the overlapped Schatten 1-norm [54]). *There are universal constants $c_i > 0$ ($i = 0, 1$) such that any minimizer of (14.9) with $\lambda = c_0 \frac{\sigma}{K} \sum_{k=1}^K (\sqrt{N/n_k} + \sqrt{n_k})$ satisfies the following bound*

$$\frac{1}{N} \|\hat{\mathcal{W}} - \mathcal{W}^*\|_F^2 \le c_1 \sigma^2 \left(\frac{1}{K} \sum_{k=1}^K (\sqrt{1/n_k} + \sqrt{n_k/N}) \right)^2 \left(\frac{1}{K} \sum_{k=1}^K \sqrt{r_k} \right)^2 .$$

with probability at least $1 - \exp(-(\frac{1}{K} \sum_{k=1}^K (\sqrt{N/n_k} + \sqrt{n_k}))^2)$.

In particular, if $n_k = n$, the above bound implies the following:

$$\frac{1}{N} \|\hat{\mathcal{W}} - \mathcal{W}^*\|_F^2 \le O_p \left(\sigma^2 \frac{\|r\|_{1/2}}{n} \right), \qquad (14.10)$$

where $\|r\|_{1/2} := (\frac{1}{K} \sum_{k=1}^K \sqrt{r_k})^2$.

In order to state a bound for the latent Schatten 1-norm, we need additional assumptions. Suppose the following observation model

$$\mathcal{Y} = \mathcal{W}^* + \mathcal{E} = \sum_{k=1}^K \mathcal{W}^{*(k)} + \mathcal{E},$$

where $\mathcal{W}^* = \sum_{k=1}^K \mathcal{W}^{*(k)}$ is the true tensor composed of factors $\mathcal{W}^{*(k)}$ that each are low-rank in the corresponding mode, i.e., $\operatorname{rank}(\mathbf{W}_{(k)}^{*(k)}) = \bar{r}_k$. Note that generally $\bar{r}_k$ is different from the mode-k rank of $\mathcal{W}^*$ denoted by r_k. The entries of the noise tensor $\mathcal{E}$ are distributed according to the Gaussian distribution $\mathcal{N}(0, \sigma^2)$ as above. In addition, we assume that the spectral norm of a factor $\mathcal{W}^{*(k)}$ is bounded when unfolded at a different mode as follows:

$$\|\mathbf{W}_{(k')}^{*(k)}\|_{S_\infty} \le \frac{\alpha}{K} \sqrt{N/n_{k'}} \quad (k \ne k'). \qquad (14.11)$$

In other words, we assume that the spectral norm of the kth factor unfolded at the k'th mode is comparable to that of a random matrix for $k' \neq k$; note that the spectral norm of a random $m \times n$ matrix whose entries are independently distributed centered random variables with finite fourth moment scales as $O(\sqrt{m} + \sqrt{n})$ [57]. This means that we want the kth factor $\mathcal{W}^{(k)}$ to look only low-rank in the kth mode as the spectral norm of a low-rank matrix would be larger than a random full rank matrix.

Now let's consider the estimator

$$\hat{\mathcal{W}} = \operatorname*{argmin}_{\mathcal{W}} \left(\frac{1}{2} \|\mathcal{Y} - \mathcal{W}\|_F^2 + \lambda \|\mathcal{W}\|_{\overline{S_1/1}} \right.$$

$$\left. \text{s.t. } \mathcal{W} = \sum_{k=1}^K \mathcal{W}^{(k)}, \ \|\boldsymbol{W}_{(k')}^{(k)}\|_{S_\infty} \leq \frac{\alpha}{K} \sqrt{N/n_{k'}}, \quad \forall k \neq k' \right). \tag{14.12}$$

The following theorem states the denoising performance of the latent Schatten 1-norm.

Theorem 14.2 (Denoising via the latent Schatten 1-norm [53]). *There are universal constants $c_i > 0$ $(i = 0, 1)$ such that any solution of the minimization problem (14.12) with regularization constant $\lambda = c_0 \sigma \max_k(\sqrt{N/n_k} + \sqrt{n_k})$ satisfies*

$$\frac{1}{N} \sum_{k=1}^K \|\hat{\mathcal{W}}^{(k)} - \mathcal{W}^{*(k)}\|_F^2 \leq c_1 \sigma^2 \left(\max_k (1/\sqrt{n_k} + \sqrt{n_k/N}) \right)^2 \sum_{k=1}^K \bar{r}_k, \tag{14.13}$$

with probability at least $1 - K \exp(-(\max_k(\sqrt{N/n_k} + \sqrt{n_k}))^2)$. Moreover, the total error $\hat{\mathcal{W}} - \mathcal{W}^*$ can be bounded as follows:

$$\frac{1}{N} \|\hat{\mathcal{W}} - \mathcal{W}^*\|_F^2 \leq c_1 \sigma^2 \left(\max_k (1/\sqrt{n_k} + \sqrt{n_k/N}) \right)^2 \min_k r_k, \tag{14.14}$$

with the same probability as above.

In particular, if $n_k = n$, the above bound implies the following:

$$\frac{1}{N} \|\hat{\mathcal{W}} - \mathcal{W}^*\|_F^2 \leq O_p \left(\sigma^2 \frac{\min_k r_k}{n} \right). \tag{14.15}$$

Comparing Inequalities (14.10) and (14.15), we can see that the bound for the latent approach scales by the minimum mode-k rank, whereas that for the overlap approach scales by the average (square-root) of the mode-k ranks; see [53] for more details.

14.3.2   Tensor Recovery Guarantee

The next theorem concerns the problem of recovering a low-rank tensor from a small number of linear measurements. Suppose that the observations $\boldsymbol{y} = (y_i)_{i=1}^M$ are obtained as in (14.4) with $\epsilon_i \sim \mathcal{N}(0, \sigma^2)$. In addition, we assume that the entries of the observation operator $\mathfrak{X}$ are drawn independently and identically from standard Gaussian distribution.

Now consider the estimator

$$\hat{\mathcal{W}} = \underset{\mathcal{W}}{\text{argmin}} \left(\frac{1}{2M} \|\boldsymbol{y} - \mathfrak{X}(\mathcal{W})\|_2^2 + \lambda_M \|\mathcal{W}\|_{S_{1/1}} \right). \tag{14.16}$$

The following theorem gives a bound for tensor reconstruction from a small number of noisy measurements.

Theorem 14.3 (Tensor recovery with the overlapped Schatten 1-norm [54]). *There are universal constants $c_i > 0$ ($i = 0, 1, 2, 3, 4$) such that for a sample size $M \geq c_1 (\frac{1}{K} \sum_{k=1}^K (\sqrt{N/n_k} + \sqrt{n_k}))^2 (\frac{1}{K} \sum_{k=1}^K \sqrt{r_k})^2$, any solution $\hat{\mathcal{W}}$ of the minimization problem (14.16) with the regularization constant $\lambda_M = c_0 \sigma (\frac{1}{K} \sum_{k=1}^K (\sqrt{N/n_k} + \sqrt{n_k}))/\sqrt{M}$ satisfies the following bound:*

$$\|\hat{\mathcal{W}} - \mathcal{W}^*\|_F^2 \leq c_2 \frac{\sigma^2 \left(\frac{1}{K} \sum_{k=1}^K (\sqrt{n_k} + \sqrt{N/n_k}) \right)^2 (\frac{1}{K} \sum_{k=1}^K \sqrt{r_k})^2}{M},$$

with probability at least $1 - c_3 e^{-c_4 M} - \exp(-(\frac{1}{K} \sum_{k=1}^K (\sqrt{n_k} + \sqrt{N/n_k}))^2)$.

In particular, if $n_k = n$ the above bound implies the following:

$$\|\hat{\mathcal{W}} - \mathcal{W}^*\|_F^2 \leq O_p \left(\frac{\sigma^2 \|r\|_{1/2} n^{K-1}}{M} \right), \tag{14.17}$$

where $\|r\|_{1/2} := (\frac{1}{K} \sum_{k=1}^K \sqrt{r_k})^2$.

The above theorem tells us that the number of samples that we need scales as $O(\|r\|_{1/2} n^{K-1})$. This is rather disappointing because it is only better by a factor $\|r\|_{1/2}/n$ compared to not assuming any low-rank-ness of the underlying truth. This motivates some of the extensions we discuss in Section 14.6.

14.4   Optimization

In this section, we discuss optimization algorithms for overlapped Schatten 1-norm (14.5) and latent Schatten 1-norm (14.7) based on the alternating direction method of multipliers (ADMM) [17].

ADMM is a general technique that can be used whenever splitting makes the problem easier to solve; see [8, 55].

14.4.1 ADMM for the Overlapped Schatten 1-Norm Regularization

We reformulate the overlapped Schatten 1-norm based tensor recovery problem as follows:

$$\underset{\mathcal{W}, \boldsymbol{Z}_1, \ldots, \boldsymbol{Z}_K}{\text{minimize}} \quad \frac{1}{2\lambda} \|\boldsymbol{y} - \boldsymbol{X}\boldsymbol{w}\|_2^2 + \sum_{k=1}^{K} \|\boldsymbol{Z}_k\|_{S_1}, \tag{14.18}$$

$$\text{subject to} \quad \boldsymbol{P}_k \boldsymbol{w} = \boldsymbol{z}_k \quad (k = 1, \ldots, K). \tag{14.19}$$

Here $\boldsymbol{Z}_k \in \mathbb{R}^{n_k \times N/n_k}$ $(k = 1, \ldots, K)$ are auxiliary variables and $\boldsymbol{z}_k$ is the vectorization of $\boldsymbol{Z}_k$. We also denote the vectorization of $\mathcal{W}$ by $\boldsymbol{w}$ and $\boldsymbol{X}\boldsymbol{w} = \mathfrak{X}(\mathcal{W})$. $\boldsymbol{P}_k$ denotes the mode-k unfolding operation; i.e., $\text{vec}(\boldsymbol{W}_{(k)}) = \boldsymbol{P}_k\boldsymbol{w}$. Note that the regularization parameter λ is in the denominator of the loss term. Although dividing the objective by λ does not change the minimizer, it keeps the regularization term from becoming negligible in the limit $\lambda \to 0$; this is useful for dealing with the noiseless case as we explain below.

The augmented Lagrangian function for optimization problem (14.18) can be defined as

$$\mathcal{L}(\boldsymbol{w}, (\boldsymbol{z}_k)_{k=1}^{K}, (\boldsymbol{\alpha}_k)_{k=1}^{K}) = \frac{1}{2\lambda} \|\boldsymbol{y} - \boldsymbol{X}\boldsymbol{w}\|_2^2 + \sum_{k=1}^{K} \|\boldsymbol{Z}_k\|_{S_1}$$
$$+ \eta \sum_{k=1}^{K} \left(\boldsymbol{\alpha}_k^{\top}(\boldsymbol{z}_k - \boldsymbol{P}_k\boldsymbol{w}) + \frac{1}{2}\|\boldsymbol{z}_k - \boldsymbol{P}_k\boldsymbol{w}\|_2^2 \right),$$

where $\boldsymbol{\alpha}_k$ is the Lagrange multiplier vector corresponding to the equality constraint $\boldsymbol{z}_k = \boldsymbol{P}_k\boldsymbol{w}$.

The basic idea of ADMM is to minimize the augmented Lagrangian function with respect to $\boldsymbol{w}$ and $(\boldsymbol{z}_k)$ while maximizing it with respect to $(\boldsymbol{\alpha}_k)$. Following a standard derivation (see [8, 55]), we obtain the following iterations (see [52] for the derivation):

$$\begin{cases} \boldsymbol{w}^{t+1} = \left(\boldsymbol{X}^{\top}\boldsymbol{X} + \lambda\eta K\boldsymbol{I}\right)^{-1} \left(\boldsymbol{X}^{\top}\boldsymbol{y} + \lambda\eta \sum_{k=1}^{K} \boldsymbol{P}_k^{\top}(\boldsymbol{z}_k^t + \boldsymbol{\alpha}_k^t)\right), \\ \boldsymbol{z}_k^{t+1} = \text{prox}_{1/\eta}\left(\boldsymbol{P}_k\boldsymbol{w}^{t+1} - \boldsymbol{\alpha}_k^t\right) \quad (k = 1, \ldots, K), \\ \boldsymbol{\alpha}_k^{t+1} = \boldsymbol{\alpha}_k^t + (\boldsymbol{z}_k^{t+1} - \boldsymbol{P}_k\boldsymbol{w}^{t+1}) \quad (k = 1, \ldots, K). \end{cases}$$

Here $\text{prox}_{1/\eta}$ is the proximity operator with respect to Schatten 1-norm and is defined as follows:

$$\text{prox}_{\theta}(\boldsymbol{z}) = \text{vec}\left(\boldsymbol{U} \max(\boldsymbol{S} - \theta, 0)\boldsymbol{V}^{\top}\right), \tag{14.20}$$

where $\boldsymbol{Z} = \boldsymbol{U}\boldsymbol{S}\boldsymbol{V}^{\top}$ is the singular-value decomposition (SVD) of $\boldsymbol{Z}$, $\boldsymbol{z}$ is the vectorization of $\boldsymbol{Z}$, and $\theta \geq 0$ is a nonnegative parameter.

The first step can be carried out efficiently, for example, by precomputing the Cholesky factorization of $(\boldsymbol{X}^\top \boldsymbol{X} + \lambda \eta K \boldsymbol{I})$ or linearization (see [60]). Note that assuming $M \leq N$ and $\operatorname{rank}(\boldsymbol{X}) = M$, we can express the limit of the first step as $\lambda \to 0$ as follows:

$$\boldsymbol{w}^{t+1} = \boldsymbol{X}^+ \boldsymbol{y} + (\boldsymbol{I} - \boldsymbol{X}^+ \boldsymbol{X}) \frac{1}{K} \sum_{k=1}^{K} \boldsymbol{P}_k^\top (\boldsymbol{z}_k^t + \boldsymbol{\alpha}_k^t),$$

where $\boldsymbol{X}^+ := \boldsymbol{X}^\top (\boldsymbol{X} \boldsymbol{X}^\top)^{-1}$ is the pseudo inverse of $\boldsymbol{X}$. Taking the limit $\lambda \to 0$ corresponds to solving the noise-free problem

$$\operatorname*{minimize}_{\mathcal{W}} \quad \sum_{k=1}^{K} \|\boldsymbol{W}_{(k)}\|_{S_1} \quad \text{subject to} \quad \boldsymbol{y} = \mathfrak{X}(\mathcal{W}).$$

Putting $1/\lambda$ in front of the loss term allows us to deal with the two problems in the same framework.

In particular, in the case of tensor completion, $\boldsymbol{X}$ is a zero-or-one matrix that has one non-zero entry in every row corresponding to the observed position. In this case, the update can be further simplified as follows:

$$w_i^{t+1} = \begin{cases} (\boldsymbol{X}^\top \boldsymbol{y})_i & \text{(if position } i \text{ is observed)}, \\ (\frac{1}{K} \sum_{k=1}^{K} \boldsymbol{P}_k^\top (\boldsymbol{z}_k^t + \boldsymbol{\alpha}_k^t))_i & \text{(otherwise)}. \end{cases}$$

Although careful tuning of the parameter η is not essential for the convergence of the above algorithm, in practice the speed of convergence can be quite different. Here we suggest the following heuristic choice. Consider scaling the truth $\mathcal{W}^*$ and the noise ϵ by a constant c as $\mathcal{W}'^* = c\mathcal{W}^*$ and $\epsilon' = c\epsilon$. Using $\lambda' = c\lambda$, we get the original solution multiplied by the same constant. Now we require that the process of optimization should also be essentially the same. To this end, we need to scale η inversely as $1/c$ so that all the terms appearing in the augmented Lagrangian function scales linearly against c. Therefore, we choose η as $\eta = \eta_0/\operatorname{std}(\boldsymbol{y})$ where η_0 is a constant and $\operatorname{std}(\boldsymbol{y})$ is the standard deviation of $\boldsymbol{y}$.

As a stopping criterion we use the primal-dual gap; see [52] for details.

14.4.2   ADMM for Latent Schatten 1-Norm Regularization

In this section, we present the ADMM for solving the dual of the latent Schatten 1-norm regularized least squares regression problem:

$$\operatorname*{minimize}_{\mathcal{W}} \quad \frac{1}{2\lambda} \|\boldsymbol{y} - \mathfrak{X}(\sum_{k=1}^{K} \mathcal{W}^{(k)})\|_2^2 + \sum_{k=1}^{K} \|\boldsymbol{W}_{(k)}^{(k)}\|_{S_1}. \tag{14.21}$$

The dual problem can be written as follows:

$$\operatorname*{minimize}_{\boldsymbol{\alpha}, \boldsymbol{Z}_1, \dots, \boldsymbol{Z}_K} \quad \frac{\lambda}{2} \|\boldsymbol{\alpha}\|_2^2 - \boldsymbol{\alpha}^\top \boldsymbol{y} + \sum_{k=1}^{K} \delta_{S_\infty}(\boldsymbol{Z}_k),$$

$$\text{subject to} \quad \boldsymbol{z}_k = \boldsymbol{P}_k \boldsymbol{X}^\top \boldsymbol{\alpha} \quad (k = 1, \dots, K), \tag{14.22}$$

where δ_{S_∞} is the indicator function of the unit spectral norm ball, i.e.,

$$\delta_{S_\infty}(\boldsymbol{Z}) = \begin{cases} 0 & (\text{if } \|\boldsymbol{Z}\|_{S_\infty} \leq 1), \\ +\infty & (\text{otherwise}). \end{cases}$$

The augmented Lagrangian function can be written as follows:

$$\mathcal{L}_\eta\left(\boldsymbol{\alpha}, (\boldsymbol{Z}_k), (\boldsymbol{W}_k)\right) = \frac{\lambda}{2}\|\boldsymbol{\alpha}\|_2^2 - \boldsymbol{\alpha}^\top\boldsymbol{y} + \sum_{k=1}^K \delta_{S_\infty}(\boldsymbol{Z}_k)$$

$$+ \sum_{k=1}^K \left(\boldsymbol{w}_k^\top(\boldsymbol{P}_k\boldsymbol{X}^\top\boldsymbol{\alpha} - \boldsymbol{z}_k) + \frac{\eta}{2}\|\boldsymbol{P}_k\boldsymbol{X}^\top\boldsymbol{\alpha} - \boldsymbol{z}_k\|_2^2\right),$$

where $\boldsymbol{W}_k$ $(k = 1,\ldots,K)$ is the Lagrange multiplier vector corresponding to the equality constraint (14.22) and equals the mode-k unfolding of primal variable $\mathcal{W}^{(k)}$ at the optimality.

The iterations can be derived as follows (see [52] for details):

$$\begin{cases} \boldsymbol{w}_k^{t+1} = \text{prox}_\eta\left(\boldsymbol{w}_k^t + \eta\boldsymbol{P}_k\boldsymbol{X}^\top\boldsymbol{\alpha}^t\right), \\ \boldsymbol{z}_k^{t+1} = (\boldsymbol{w}_k^t + \eta\boldsymbol{P}_k\boldsymbol{X}^\top\boldsymbol{\alpha}^t - \boldsymbol{w}_k^{t+1})/\eta, \\ \boldsymbol{\alpha}^{t+1} = \frac{1}{\lambda+\eta K}\left(\boldsymbol{y} + \eta\boldsymbol{X}\sum_{k=1}^K \boldsymbol{P}_k^\top(\boldsymbol{z}_k^{t+1} - \boldsymbol{w}_k^{t+1}/\eta)\right), \end{cases}$$

where prox_η is the proximity operator (14.20).

We can see that the algorithm updates the dual variables $((\boldsymbol{z}_k)$ and $\boldsymbol{\alpha})$ and the primal variables $(\boldsymbol{w}_k)$ alternately. In particular, the update equation for the primal variables $(\boldsymbol{w}_k)$ is a popular proximal-gradient-type update. In fact, $\boldsymbol{P}_k\boldsymbol{X}^\top\boldsymbol{\alpha}^t$ converges to the gradient of the loss term at the optimality.

Note that setting $\lambda = 0$ gives the correct update equations for the noiseless case $\lambda \to 0$ in (14.21).

Consideration on the scale invariance of the algorithm similar to that in the previous subsection suggests that we should scale η linearly as the scale of $\boldsymbol{y}$; thus we set $\eta = \eta_0\text{std}(\boldsymbol{y})$.

14.5   Experiments

14.5.1   Tensor Denoising

We generated synthetic tensor denoising problems as follows. First each entry of the core tensor $\mathcal{C}$ was sampled independently from standard normal distribution. Then orthogonal factors drawn from the uniform (Haar) measure were multiplied to each of its modes to obtain the true tensor $\mathcal{W}^*$. Then the observed tensor $\mathcal{Y}$ was obtained by adding zero-mean Gaussian noise with standard deviation $\sigma = 0.1$ to each entry.

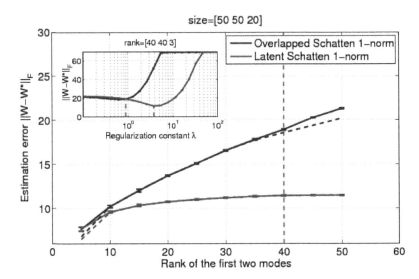

FIGURE 14.1: Estimation of a low-rank $50\times50\times20$ tensor of rank $r \times r \times 3$ from noisy measurements. The noise standard deviation was $\sigma = 0.1$. The estimation errors of overlapped and latent approaches are plotted against the rank r of the first two modes. The solid lines show the error at the fixed regularization constant λ, which was 0.89 for the overlapped approach and 3.79 for the latent approach. The dashed lines show the minimum error over candidates of the regularization constant λ from 0.1 to 100. In the inset, the errors of the two approaches are plotted against the regularization constant λ for rank $r = 40$ (marked with vertical gray dashed line in the outset). The two values (0.89 and 3.79) are marked with vertical dashed lines.

The two approaches (overlap and latent Schatten 1-norms) were applied with different values of the regularization parameter λ ranging from 0.01 to 100. The incoherence parameter α for the latent Schatten 1-norm was set to a sufficiently large constant value so that it had no effect on the solution.

Figure 14.1 shows the result of applying the two approaches to tensors of multilinear rank $(r, r, 3)$ for different r. This experiment was specifically designed to highlight the dependency of the denoising performance of the two methods. The error of the overlapped Schatten 1-norm increases as r increases although the rank of the third mode is constant; this is because the right-hand side of (14.10) depends on the average (square-root) of multilinear ranks. On the other hand, the error of the latent Schatten 1-norm stays almost constant; this is because the minimum multilinear rank 3 is constant; see Theorem 14.2. Of course, this is just one well-constructed example, and we refer the readers to [53] for more results that quantitatively validate Theorem 14.2.

14.5.2   Tensor Completion

A synthetic tensor completion problem was generated as follows. The true tensor $\mathcal{W}^*$ was generated the same way as in the previous subsection. Then we randomly split the entries into training and testing. No observational noise was added.

We trained overlapped and latent Schatten 1-norms using the optimization algorithms discussed in the previous section. The operator $\mathfrak{X}$ was defined as

$$\mathfrak{X}(\mathcal{W}) = (\mathcal{W}_{i_s j_s k_s})_{s=1}^M,$$

where $(i_s, j_s, k_s)_{s=1}^M$ is the set of indices corresponding to the observed positions. Since there is no observational noise, we took the limit $\lambda \to 0$ in the update equations.

The result for $50 \times 50 \times 20$ tensor of multilinear rank (7,8,9) is shown in Figure 14.2. As baselines we included an expectation-maximization-based Tucker decomposition algorithm in [4] with the correct rank (exact) and 20% higher rank (large). We also included matrix completion algorithm that treated a tensor as a matrix by unfolding the tensor at a prespecified mode. This method was implemented by instantiating only one of the auxiliary variables $Z_1, \ldots, Z_K$ in the ADMM for overlapped Schatten 1-norm presented in Section 14.4.1.

The result shows that first, treating tensor as a matrix yields a rather disappointing result, especially when we choose mode 3. This is because the dimensions are not balanced, which is often the case in practice, and unluckily the mode with the smallest dimension (mode 3) has the highest rank. On the other hand, the overlapped Schatten 1-norm can recover this tensor reliably from about 35% of the entries *without any assumption about the low-rank-ness of the modes*.

Second, the reconstruction is exact (up to optimization tolerance) above the sufficient sampling density (35%). This can be predicted from Theorem 14.3 in the following way: first note that the condition for the sample size M does not depend on the noise variance σ^2; second, the right-hand side of the bound is proportional to the noise variance σ^2. Therefore, if we take the limit $\sigma^2 \to 0$, the theorem predicts zero error whenever the condition for the sample size M is satisfied. We would need a lower bound to make this claim more precise, which may be obtained by following the work of [2].

Compared to the overlapped approach, the latent approach recovers the true tensor exactly only around 70% observation. Although we don't have a theory for tensor recovery via the latent approach, it seems to suggest that the number of samples that we need scales faster than the minimum multilinear rank, which appeared in the right-hand side of the denoising bound (14.14).

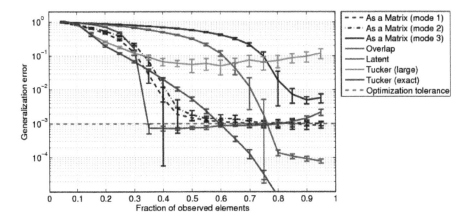

FIGURE 14.2: Comparison of tensor completion performance of overlapped and latent Schatten 1-norm regularization. As baselines, Tucker decomposition with the correct rank (exact) and 20% higher rank (large), and convex optimization based matrix completion (as a matrix) that focuses on a pre-specified mode are included. The size of the tensor is $50 \times 50 \times 20$ and the true multilinear rank is $(7, 8, 9)$. The generalization error is plotted against the fraction of observed elements (M/N) of the underlying low-rank tensor. Also the tolerance of optimization (10^{-3}) is shown.

14.6   Extensions and Related Work

14.6.1   Balanced Unfolding

For a balanced-sized K-way tensor (i.e., $n_k = n$), CP decomposition (14.1) or Tucker decomposition (14.2) has only linearly many parameters in n. Thus we would expect that a reasonable estimator would decrease the error as $O(n/M)$. However, the scaling we see in inequality (14.17) is $O(n^{K-1}/M)$, which is far larger than what we expect.

Looking at the way the bound is derived, we notice (we thank Nam H. Nguyen for pointing this out) that the unbalancedness of the unfolding is the cause. More specifically, the term $\sqrt{n_k} + \sqrt{N/n_k}$ is the spectral norm of a random $n_k \times N/n_k$ matrix with independent centered entries with bounded fourth moment [57]. Thus, we can ask what happens if we unfold the tensor evenly.

Let $\boldsymbol{W}_{(i_1,i_2,\ldots,i_k;j_1,j_2,\ldots,j_l)}$ denote the $\prod_{a=1}^{k} n_{i_a} \times \prod_{b=1}^{l} n_{j_b}$ matrix obtained by concatenating the $\prod_{a=1}^{k} n_{i_a}$ dimensional slices of $\mathcal{W}$ specified by indices in $[n_{j_1}] \times \cdots \times [n_{j_l}]$ along columns. For example, $\boldsymbol{W}_{(1;2,3,4)}$ is the same as $\boldsymbol{W}_{(1)}$ in the original notation defined in Section 14.2. We say that an unfolding is

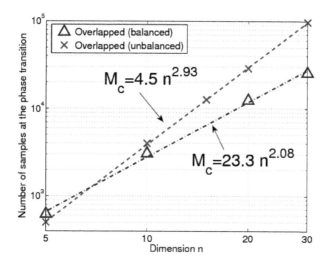

FIGURE 14.3: The number of samples necessary to recover a $n \times n \times n \times n$ tensor of multilinear rank $(2, 2, 2, 2)$. The number of samples at the phase transition M_c was defined as the number of samples at which the empirical probability of obtaining error smaller than 0.01 exceeded $1/2$.

balanced if the number of rows and columns are the same, e.g., $\boldsymbol{W}_{(1,2;3,4)}$ when $n_k = n$.

Figure 14.3 shows the number of samples at the phase transition M_c against n for the completion of 4th order balanced-sized tensors. We compared the original overlapped Schatten 1-norm (14.5) against the following norm based on three balanced unfoldings,

$$\|\mathcal{W}\|_{\text{balanced}} = \|\boldsymbol{W}_{(1,2;3,4)}\|_{S_1} + \|\boldsymbol{W}_{(1,3;2,4)}\|_{S_1} + \|\boldsymbol{W}_{(1,4;2,3)}\|_{S_1}.$$

See Mu et al. [37] for a related approach, though they only considered one of the three possible balanced unfoldings.

The threshold M_c was defined as the number of samples at which the probability that the reconstruction error $\|\hat{\mathcal{W}} - \mathcal{W}^*\|_F$ was smaller than 0.01 exceeded $1/2$. The dashed line corresponds to the original overlapped Schatten 1-norm (one mode against the rest). The dash-dotted line corresponds to the overlapped Schatten 1-norm based on balanced unfoldings.

We can see that the empirical scaling of the balanced version is $n^{2.08}$, whereas that of the ordinary version is $n^{2.93}$. Both of them were close to the theoretically predicted scaling n^2 and n^3, respectively.

However, computationally this approach is more challenging. The major computational cost for optimization is that of SVD. Since SVD scales as $O(m^2 n + m^3)$ for an $m \times n$ matrix with $m \leq n$, the more balanced, the more challenging the computation becomes. Note that the comparison here is made assuming that both approaches use the same ADMM-based optimization algorithm (see Section 14.4.1). Thus there might be another optimization algorithm (see, e.g., Jaggi [25]) that works better in the balanced case.

Recently Mu et al. [37] derived a lower-bound for the overlapped Schatten 1-norm based on the framework developed by Amelunxen et al. [2]. The lower-bound indeed shows that rn^{K-1} samples is unavoidable for the vanilla version of the overlapped Schatten 1-norm. Motivated by the lower bound, they proposed a balanced version (without overlap), which they call the square norm.

14.6.2   Tensor Nuclear Norm

Chandrasekaran et al. [12] discuss a norm for tensors within the framework of *atomic norms*. Let $\mathcal{A}$ be an atomic set that consists of rank one tensors of unit Frobenius norm:

$$\mathcal{A} = \{\boldsymbol{u}_1 \circ \boldsymbol{u}_2 \circ \cdots \circ \boldsymbol{u}_K : \|\boldsymbol{u}_k\| = 1 \quad (k = 1, \dots, K)\}.$$

The nuclear norm for tensor is defined as follows:

$$\|\mathcal{W}\|_{\text{nuc}} = \inf \sum_{a \in \mathcal{A}} c_a \quad \text{s.t.} \quad \mathcal{W} = \sum_{a \in \mathcal{A}} c_a \boldsymbol{u}_1^{(a)} \circ \cdots \circ \boldsymbol{u}_K^{(a)},$$

where with a slight abuse of notation, we use $a \in \mathcal{A}$ as an index for an element in the atomic set.

It can be shown that for a tensor that admits an orthogonal CP decomposition [28] with R terms (Decomposition (14.1) with orthogonality constraints between the components), the nuclear norm can be related to the Frobenius norm as follows:

$$\|\mathcal{W}\|_{\text{nuc}} \leq \sqrt{R} \|\mathcal{W}\|_F.$$

Moreover, the tensor spectral norm

$$\|\mathcal{X}\|_{\text{op}} = \max_{a \in \mathcal{A}} \mathcal{X} \times_1 \boldsymbol{u}_1^{(a)} \times_2 \boldsymbol{u}_2^{(a)} \cdots \times_K \boldsymbol{u}_K^{(a)},$$

which is dual to the nuclear norm, is known to be of order $O(\sqrt{n})$ for a random Gaussian tensor; see [40]. Thus it is natural to hope that we can prove that the nuclear norm would achieve an optimal $O(Rn)$ convex relaxation for tensors. However, computationally, the tensor nuclear norm seems to be intractable for $K \geq 3$. Although it is convex, it involves infinitely many variables. There is no analogue of linear matrix inequality or semidefinite programming for matrices that can be used here to the best of our knowledge.

14.6.3   Interpretation of the Result

That we can bound the error in Frobenius norm as we have presented in Section 14.3 does not mean that our method is useful in practice. In fact, tensor decomposition methods are often used to uncover latent factors and gain insight about the data.

Here we present how such an insight can be gained from the solutions of the two algorithms we presented in Section 14.4.

For the overlapped approach, the factor matrices $U_1, \ldots, U_K$ corresponding to the Tucker decomposition (14.2) can be obtained by computing the left singular vectors of the auxiliary matrices $Z_1, \ldots, Z_K$. The mode-k rank r_k is determined *automatically* by the proximity operator (14.20); importantly, the rank at an optimum does not depend on the choice of η, though the rank during optimization may depend on η. Once the factors are obtained, the core can be obtained as follows:

$$\mathcal{C} = \mathcal{W} \times_1 U_1^\top \times_2 U_2^\top \cdots \times_K U_K^\top.$$

To get the stronger CP decomposition (14.1), one can perform any off-the-shelf CP decomposition algorithm on the *core* $\mathcal{C}$. This post-processing step is easier than applying CP decomposition directly to the original large tensor with noise and missing entries. In other words, this two-step approach allows us to separate the tasks of generalization and interpretation; see [52] for details.

It is less easy to interpret the solution for the latent approach because in general the sum $\mathcal{W} = \sum_{k=1}^{K} \mathcal{W}^{(k)}$ is not low-rank even when each $\mathcal{W}^{(k)}$ is. However, in practice we found that the solution is often singleton, i.e., only one non-zero component $\mathcal{W}^{(k)}$. This corresponds to the intuition that the latent Schatten 1-norm focuses on the low-rank-ness of the mode with the minimum mode-k rank and does not care about the other modes. The fact that the solution is only low-rank in one mode is still disappointing. This could be solved by including more terms in the latent approach, which can be partially low-rank (there are 2^K possibilities to penalize the sum of the Schatten 1-norms of some of the modes) or balanced unfolding. If the resulting solution is a singleton, then the model automatically chose which mode should be low-rank.

14.6.4 Related Work

Liu et al. [33, 34] proposed the overlapped approach in the context of image and video imputation. They used a penalty method to deal with the equality constraints in (14.19). Li et al. [32] extended Liu et al.'s work to sparse+low-rank decomposition of tensors (also known as sparse PCA) and applied to background/shadow removal and face recognition. Li et al. also used a penalty method for the optimization.

Signoretto et al. [47, 49, 48, 46] proposed and extended the overlapped Schatten 1-norm in the context of kernel-based learning, i.e., learning higher-order operators over Hilbert spaces. The use of kernel allows us to incorporate smoothness (or side-information) when we see an entry in a K-way tensor as a representation of a relation among objects from K different domains; see also [1]. They also proposed an optimization algorithm that supports general differentiable loss function L based on an (accelerated) proximal gradient method [39, 6]; the algorithm employs ADMM to compute the proximal operator corresponding to the overlapped Schatten 1-norm.

Gandy et al. [18] proposed ADMM and Douglas-Rachford splitting algorithm for the overlapped approach.

Yang et al. [58] proposed a fast optimization algorithm for the overlapped approach based on a fixed-point iteration combined with continuation.

Goldfarb and Qin [19] studied low-rank+sparse tensor decomposition based on the overlapped and latent Schatten 1-norms. They also proposed adaptive weighting of the terms appearing in the overlapped Schatten 1-norm (14.5) and reported that the adaptive version outperformed other methods in many cases. They also studied the relationship between the normalized rank $\|n^{-1}\|_{1/2}\|r\|_{1/2}$ (the quantity that appears in the condition for M in Theorem 14.3), the necessary sampling density, and the allowable fraction of corrupted entries.

Zhang et al. [61] extended Li et al.'s work [32] on sparse+low-rank decomposition in several interesting ways. They have incorporated transformations that align each image in order to make the spatial low-rank assumption (on the first two modes) as valid as possible (see also [36] for related work), while keeping the sequence of images smooth by also penalizing the Schatten 1-norm for the mode corresponding to the temporal dimension.

On the theoretical side, Nickel and Tresp [41] presented a generalization bound for low multilinear rank tensor in the context of relational data analysis by counting the number of possible sign patterns that low multilinear rank tensors can attain. Although the theory does not lead to a model selection criterion as we cannot provably compute the low-multilinear-rank decomposition at a given rank, it would still be fruitful to study a convex relaxation for the set of low-rank sign tensors; see, e.g., [50].

Romera-Paredes and Pontil [45] proposed a convex relaxation of mode-k rank with respect to the Frobenius norm ball and showed that it is tighter than the overlapped Schatten 1-norm at some points. Although the resulting regularizer is not a continuous function and thus challenging to compute, they proposed a subgradient-based optimization algorithm.

Jiang et al. [26] studied the best rank-one approximation of super symmetric even order tensors. They noticed that for super symmetric (meaning that the tensor is invariant to arbitrary permutation of indices) even order tensors, being rank-one is equivalent to a balanced unfolding (see Section 14.6.1) being rank-one (as a matrix). Then they solved the best rank-one approximation problem with the Schatten 1-norm regularization (which promotes rank-one solution) under linear equality constraints that ensured that the solution was super-symmetric. Empirically the solution was found to be rank one in most of the cases. They also showed many non-symmetric problems can be reduced to the symmetric case.

Krishnamurthy and Singh [30] proposed an adaptive sampling algorithm for tensor completion and showed that it succeeds with high probability with $O(n)$ samples. The number of samples required in their adaptive setting also depends on the true rank r and the coherence parameter μ_0. They also showed a lower bound that scales as $O(r^{K-1}n)$ under the incoherence assumption.

Application of the overlapped approach includes language models [24], hyper-spectral imaging [49], and multi-task learning [46, 44], besides image reconstruction discussed in [33, 34, 32, 61].

14.7   Future Directions

Compared to the overlapped Schatten 1-norm, the behavior of the latent Schatten 1-norm is still unclear in some parts. First, although we have argued that empirically the solution of the optimization problem is *often* a singleton (only one non-zero component), this needs a better explanation. Second, although we believe that the incoherence assumption is necessary to prove the stronger inequality (14.13), it may not be necessary to obtain the weaker one (14.14).

Given that the sample complexities of both the overlapped and latent Schatten 1-norms are far from optimal, it would be extremely interesting to explore the statistics-computation trade-off between what we can provably achieve and how computationally expensive it would be. Balanced unfolding [37], tensor nuclear norm [12], and the new convex relaxation [45] discussed in the previous section are candidates to be evaluated and analyzed. It would also be interesting to study recent work on decomposition of tensors arising from higher order moments of latent variable models [3] in this context.

Finally, nonnegativity [9, 31] and positive semidefiniteness [59] are constraints that are useful to impose on the factors in practice. Generalization of the results for separable nonnegative matrix factorization [5, 7] to tensors would be an interesting direction.

Acknowledgments

The authors would like to thank Franz J. Király, Nati Srebro, Nam H. Nguyen, and Gilles Blanchard for discussions and the editors of this book for comments. This work was partially supported by MEXT Kakenhi 25730013, and JST, CREST.

Bibliography

[1] J. Abernethy, F. Bach, T. Evgeniou, and J.P. Vert. A new approach to collaborative filtering: Operator estimation with spectral regularization. *J. Mach. Learn. Res.*, 10:803–826, 2009.

[2] D. Amelunxen, M. Lotz, M. B. McCoy, and J. A. Tropp. Living on the edge: A geometric theory of phase transitions in convex optimization. Technical report, arXiv:1303.6672, 2013.

[3] A. Anandkumar, R. Ge, D. Hsu, S. M. Kakade, and M. Telgarsky. Tensor decompositions for learning latent variable models. Technical report, arXiv:1210.7559, 2012.

[4] C. A. Andersson and R. Bro. The n-way toolbox for MATLAB. *Chemometr. Intell. Lab.*, 52(1):1–4, 2000. http://www.models.life.ku.dk/source/nwaytoolbox/.

[5] S. Arora, R. Ge, R. Kannan, and A. Moitra. Computing a nonnegative matrix factorization–provably. In *Proceedings of the 44th symposium on Theory of Computing*, pages 145–162, 2012.

[6] A. Beck and M. Teboulle. A fast iterative shrinkage-thresholding algorithm for linear inverse problems. *SIAM J. Imaging Sci.*, 2(1):183–202, 2009.

[7] V. Bittorf, B. Recht, C. Ré, and J. A. Tropp. Factoring nonnegative matrices with linear programs. In *Adv. Neural. Inf. Process. Syst. 25*, pages 1223–1231. 2012.

[8] S. Boyd, N. Parikh, E. Chu, B. Peleato, and J. Eckstein. Distributed optimization and statistical learning via the alternating direction method of multipliers. *Foundations and Trends® in Machine Learning*, 3(1):1–122, 2011.

[9] R. Bro and S. De Jong. A fast non-negativity-constrained least squares algorithm. *J. Chemometr.*, 11(5):393–401, 1997.

[10] E. J. Candès and T. Tao. The power of convex relaxation: Near-optimal matrix completion. *IEEE T. Inform. Theory*, 56(5):2053–2080, 2010.

[11] J.D. Carroll and J.J. Chang. Analysis of individual differences in multidimensional scaling via an n-way generalization of "Eckart-Young" decomposition. *Psychometrika*, 35(3):283–319, 1970.

[12] V. Chandrasekaran, B. Recht, P. A. Parrilo, and A. S. Willsky. The convex geometry of linear inverse problems. *Foundations of Computational Mathematics*, 12(6):805–849, 2012.

[13] L. De Lathauwer, B. De Moor, and J. Vandewalle. A multilinear singular value decomposition. *SIAM J. Matrix Anal. Appl.*, 21(4):1253–1278, 2000.

[14] L. De Lathauwer, B. De Moor, and J. Vandewalle. On the best rank-1 and rank-$(R_1, R_2, \ldots, R_N)$ approximation of higher-order tensors. *SIAM J. Matrix Anal. Appl.*, 21(4):1324–1342, 2000.

[15] M. Fazel. *Matrix rank minimization with applications.* PhD thesis, Stanford University, 2002.

[16] M. Fazel, H. Hindi, and S. P. Boyd. A Rank Minimization Heuristic with Application to Minimum Order System Approximation. In *Proc. of the American Control Conference*, 2001.

[17] D. Gabay and B. Mercier. A dual algorithm for the solution of nonlinear variational problems via finite element approximation. *Comput. Math. Appl.*, 2(1):17–40, 1976.

[18] S. Gandy, B. Recht, and I. Yamada. Tensor completion and low-n-rank tensor recovery via convex optimization. *Inverse Problems*, 27:025010, 2011.

[19] D. Goldfarb and Z. Qin. Robust low-rank tensor recovery: Models and algorithms. Technical report, arXiv:1311.6182, 2013.

[20] R. A. Harshman. Models for analysis of asymmetrical relationships among n objects or stimuli. In *First Joint Meeting of the Psychometric Society and the Society for Mathematical Psychology, McMaster University, Hamilton, Ontario*, 1978.

[21] C. J. Hillar and L.-H. Lim. Most tensor problems are np-hard. *Journal of the ACM*, 60(6):45, 2013.

[22] F. L. Hitchcock. The expression of a tensor or a polyadic as a sum of products. *J. Math. Phys.*, 6(1):164–189, 1927.

[23] J. Håstad. Tensor rank is NP-complete. *Journal of Algorithms*, 11(4):644–654, 1990.

[24] B. Hutchinson, M. Ostendorf, and M. Fazel. Low rank language models for small training sets. *IEEE Signal Proc. Let.*, 18(9):489–492, 2011.

[25] M. Jaggi. Revisiting Frank-Wolfe: Projection-free sparse convex optimization. In *Proceedings of the 30th International Conference on Machine Learning*, pages 427–435, 2013.

[26] B. Jiang, S. Ma, and S. Zhang. Tensor principal component analysis via convex optimization. Technical report, arXiv:1212.2702, 2012.

[27] R. Kannan and S. Vempala. Spectral algorithms. *Foundations and Trends in Theoretical Computer Science*, 4(3–4):157–288, 2008.

[28] T. G. Kolda. Orthogonal tensor decompositions. *SIAM J. Matrix Anal. Appl.*, 23(1):243–255, 2001.

[29] T. G. Kolda and B. W. Bader. Tensor decompositions and applications. *SIAM Review*, 51(3):455–500, 2009.

[30] A. Krishnamurthy and A. Singh. Low-rank matrix and tensor completion via adaptive sampling. In C.J.C. Burges, L. Bottou, M. Welling, Z. Ghahramani, and K.Q. Weinberger, editors, *Adv. Neural. Inf. Process. Syst. 26*, pages 836–844. 2013.

[31] D. D. Lee and H. S. Seung. Learning the parts of objects by non-negative matrix factorization. *Nature*, 401(6755):788–791, 1999.

[32] Y. Li, J. Yan, Y. Zhou, and J. Yang. Optimum subspace learning and error correction for tensors. In *Computer Vision–ECCV 2010*, pages 790–803. Springer, 2010.

[33] J. Liu, P. Musialski, P. Wonka, and J. Ye. Tensor completion for estimating missing values in visual data. In *Proc. ICCV*, 2009.

[34] J. Liu, J. Ye, P. Musialski, and P. Wonka. Tensor completion for estimating missing values in visual data. *IEEE T. Pattern. Anal.*, 35(1):208–220, 2013.

[35] M. Mørup. Applications of tensor (multiway array) factorizations and decompositions in data mining. *Wiley Interdisciplinary Rev.: Data Min. Knowl. Dicov.*, 1(1):24–40, 2011.

[36] M. Mørup, L. K. Hansen, S. M. Arnfred, L.-H. Lim, and K. H. Madsen. Shift-invariant multilinear decomposition of neuroimaging data. *Neuroimage*, 42(4):1439–1450, 2008.

[37] C. Mu, B. Huang, J. Wright, and D. Goldfarb. Square deal: Lower bounds and improved relaxations for tensor recovery. *arXiv preprint arXiv:1307.5870*, 2013.

[38] S. Negahban and M.J. Wainwright. Estimation of (near) low-rank matrices with noise and high-dimensional scaling. *Ann. Statist.*, 39(2):673–1333, 2011.

[39] Y. Nesterov. Gradient methods for minimizing composite objective function. Technical Report 2007/76, Center for Operations Research and Econometrics (CORE), Catholic University of Louvain, 2007.

[40] N. H. Nguyen, P. Drineas, and T. D. Tran. Tensor sparsification via a bound on the spectral norm of random tensors. Technical report, arXiv:1005.4732, 2010.

[41] M. Nickel and V. Tresp. An analysis of tensor models for learning on structured data. In *Machine Learning and Knowledge Discovery in Databases*, pages 272–287. Springer, 2013.

[42] B. Recht, M. Fazel, and P.A. Parrilo. Guaranteed minimum-rank solutions of linear matrix equations via nuclear norm minimization. *SIAM Review*, 52(3):471–501, 2010.

[43] A. Rohde and A. B. Tsybakov. Estimation of high-dimensional low-rank matrices. *Ann. Statist.*, 39(2):887–930, 2011.

[44] B. Romera-Paredes, H. Aung, N. Bianchi-Berthouze, and M. Pontil. Multilinear multitask learning. In *Proceedings of the 30th International Conference on Machine Learning*, pages 1444–1452, 2013.

[45] B. Romera-Paredes and M. Pontil. A new convex relaxation for tensor completion. In *Adv. Neural. Inf. Process. Syst. 26*, pages 2967–2975, 2013.

[46] M. Signoretto, L. De Lathauwer, and J. A. K. Suykens. Learning tensors in reproducing kernel hilbert spaces with multilinear spectral penalties. Technical report, arXiv:1310.4977, 2013.

[47] M. Signoretto, L. De Lathauwer, and J. A. K. Suykens. Nuclear norms for tensors and their use for convex multilinear estimation. Technical Report 10-186, ESAT-SISTA, KU Leuven, 2010.

[48] M. Signoretto, Q. T. Dinh, L. De Lathauwer, and J. A. K. Suykens. Learning with tensors: a framework based on convex optimization and spectral regularization. *Mach. Learn.*, 94(3):303–351, 2014.

[49] M. Signoretto, R. Van de Plas, B. De Moor, and J. A. K. Suykens. Tensor versus matrix completion: a comparison with application to spectral data. *IEEE Signal Proc. Let.*, 18(7):403–406, 2011.

[50] N. Srebro and A. Shraibman. Rank, trace-norm and max-norm. In *Proc. of the 18th Annual Conference on Learning Theory (COLT)*, pages 545–560. Springer, 2005.

[51] R. Tibshirani. Regression shrinkage and selection via the lasso. *J. Roy. Stat. Soc. B*, 58(1):267–288, 1996.

[52] R. Tomioka, K. Hayashi, and H. Kashima. Estimation of low-rank tensors via convex optimization. Technical report, arXiv:1010.0789, 2011.

[53] R. Tomioka and T. Suzuki. Convex tensor decomposition via structured schatten norm regularization. In C.J.C. Burges, L. Bottou, M. Welling, Z. Ghahramani, and K.Q. Weinberger, editors, *Adv. Neural. Inf. Process. Syst. 26*, pages 1331–1339. 2013.

[54] R. Tomioka, T. Suzuki, K. Hayashi, and H. Kashima. Statistical performance of convex tensor decomposition. In *Adv. Neural. Inf. Process. Syst. 24*, pages 972–980. 2011.

[55] R. Tomioka, T. Suzuki, and M. Sugiyama. Augmented lagrangian methods for learning, selecting, and combining features. In Suvrit Sra, Sebastian Nowozin, and Stephen J. Wright, editors, *Optimization for Machine Learning*. MIT Press, 2011.

[56] L. R. Tucker. Some mathematical notes on three-mode factor analysis. *Psychometrika*, 31(3):279–311, 1966.

[57] R. Vershynin. Introduction to the non-asymptotic analysis of random matrices. Technical report, arXiv:1011.3027, 2010.

[58] L. Yang, Z Huang, and X. Shi. A fixed point iterative method for low n-rank tensor pursuit. *IEEE T. Signal Proces.*, 61(11):2952–2962, 2013.

[59] K. Yoshii, R. Tomioka, D. Mochihashi, and M. Goto. Infinite positive semidefinite tensor factorization for source separation of mixture signals. In *Proceedings of the 30th International Conference on Machine Learning*, pages 576–584, 2013.

[60] X. Zhang, M. Burger, and S. Osher. A unified primal-dual algorithm framework based on bregman iteration. *J. Sci. Comput.*, 46(1):20–46, 2010.

[61] X. Zhang, D. Wang, Z. Zhou, and Y. Ma. Simultaneous rectification and alignment via robust recovery of low-rank tensors. In *Adv. Neural. Inf. Process. Syst. 26*, pages 1637–1645, 2013.

Chapter 15

Learning Sets and Subspaces

Alessandro Rudi*

DIBRIS, Università degli Studi di Genova and LCSL, Massachusetts Institute of Technology, and Istituto Italiano di Tecnologia

Guillermo D. Canas*

Massachusetts Institute of Technology

Ernesto De Vito*

DIMA, Università degli Studi di Genova

Lorenzo Rosasco*

DIBRIS, Università degli Studi di Genova and LCSL, Massachusetts Institute of Technology, and Istituto Italiano di Tecnologia

We consider here the classic problem of support estimation, or learning a set from random samples, and propose a natural but novel approach to address it. We do this by investigating its connection with a seemingly distinct problem, namely subspace learning.

---
*alessandro.rudi@unige.it, guilledc@mit.edu, devito@dima.unige.it, lrosasco@mit.edu

The problem of learning the smallest set containing the data distribution is often called support estimation and it is a fundamental problem in statistics and machine learning. As discussed in the following, its applications range from surface estimation, to novelty detection, to name a few. In the following we discuss how a suitable family of positive definite kernels, called separating kernels, allows us to relate the problem of learning a set to the problem of learning an appropriate linear subspace of a Hilbert space. More precisely, we reduce the set learning problem to that of learning the smallest subspace that contains the support of the distribution after a kernel (feature) embedding. This connection between learning sets and learning subspaces allows us on the one hand to design natural spectral estimators for this problem, and on the other hand to use analytic and probabilistic tools to derive generalization guarantees for them.

Besides establishing this novel connection, the goal of this work is to introduce novel sharp sample complexity estimates for subspace and set learning. The theoretical results are illustrated and complemented through some numerical experiments.

The chapter is structured as follows. We begin by briefly discussing some concepts from the statistical analysis of unsupervised learning algorithms (Section 15.1). We then develop our analysis of the subspace learning problem, and discuss set learning in Section 15.3. Finally, we conclude in Section 15.4 with some numerical results.

15.1   Unsupervised Statistical Learning

The present work can be more broadly framed in the context of unsupervised learning, a term typically used to describe the general problem of extracting *patterns* from data [23, 19]. Here, the term pattern refers to some geometric property of the data distribution. Specifically, in the sequel we will be interested in recovering the following: 1) the smallest (closed) set containing the data distribution, and 2) the smallest subspace spanned by the data distribution. As we will discuss, these two problems are indeed tightly connected. After formally describing this connection, our focus will be on introducing a class of spectral estimators for this problem, and deriving sharp generalization error estimates for them.

Given a probability space (X, ρ) from which data X_n are drawn identically and independently, we let $\mathcal{S}$ be a set endowed with a (pseudo) metric d. We view $\mathcal{S}$ as the collection of possible patterns/structures in the data distribution (for instance the set of possible supports of a distribution). In many circumstances, the true distribution ρ identifies an element S_ρ in the space of structures (for instance the true support), and the goal of an (unsupervised) learning algorithm is to estimate an approximation $\hat{S}_n$ given the data. For example, in the context of set learning, $\mathcal{S}$ may be defined as the collection of all closed subsets of X endowed with the Hausdorff distance, and S_ρ to be the support of ρ. In the context of subspace learning, $\mathcal{S}$ is the collection of all linear subspaces of X, with some suitable pseudo-metric such as the reconstruction criterion in (15.3) and S_ρ is the smallest subspace spanned by points drawn from ρ.

Since S_ρ is estimated from random samples, we characterize the learning error of an algorithm through non asymptotic bounds of the form

$$\mathbb{P}\left[d(\hat{S}_n, S_\rho) \leq R_\rho(\delta, n)\right] \geq 1 - \delta \tag{15.1}$$

for $0 < \delta \leq 1$, where the *learning error* $R_\rho(\delta, n)$ typically depends on n and δ, but also on ρ. Once a bound of the form of (15.1) with an asymptotically vanishing learning error R is obtained, almost sure convergence of $d(\hat{S}_n, S_\rho) \to 0$ as $n \to \infty$ follows from the Borel-Cantelli Lemma [29].

15.2   Subspace Learning

Subspace learning is the problem of finding the smallest linear space supporting data drawn from an unknown distribution. It is a classical problem in machine learning and statistics and is at the core of a number of spectral methods for data analysis, most notably PCA [26], but also multidimensional scaling (MDS) [8, 59]. While traditional methods, such as PCA and MDS, perform subspace learning in the original data space, more recent manifold learning methods, such as isomap [51], Hessian eigenmaps [18], maximum-variance unfolding [57, 58, 50], locally-linear embedding [39, 42], and Laplacian eigenmaps [2] (but also kernel PCA [44]), begin by embedding the data in a *feature space*, in which subspace estimation is carried out. As pointed out in [22, 4, 3], all the algorithms in this family have a common structure. They embed the data in a suitable Hilbert space $\mathcal{F}$, and compute a linear subspace that best approximates the embedded data. The local coordinates in this subspace then become the new representation space.

The analysis in this paper applies to learning subspaces both in the data and in a feature space. In the following, we introduce a general formulation of the subspace learning problem and derive novel learning error estimates. Our results rely on natural assumptions on the spectral properties of the covariance operator associated to the data distribution, and hold for a wide class of metrics between subspaces. As a special case, we discuss sharp error estimates for the reconstruction properties of PCA. Key to our analysis is an operator theoretic approach that has broad applicability to the analysis of spectral learning methods.

15.2.1   Problem Definition and Notation

Given a measure ρ with support M in the unit ball of a separable Hilbert space $\mathcal{F}$, we consider in this work the problem of estimating, from n i.i.d. samples $X_n = \{x_i\}_{1 \leq i \leq n}$, the smallest linear subspace $S_\rho := \overline{\text{span}(M)}$ that contains M. In the framework introduced in Section 15.1, the above problem corresponds to a choice of input space $\mathcal{F}$, and the space of candidate structures is the collection of all linear subspaces of $\mathcal{F}$. The target of the learning problem is S_ρ, the smallest linear subspace that contains the support of ρ. As described in Section 15.1, the quality of an estimate $\hat{S}_n$ of S_ρ, for a given metric (or error criterion) d, is characterized in terms of probabilistic bounds of the form of Equation (15.1).

In the following the metric projection operator onto a subspace S is denoted by P_S, where $P_S^2 = P_S^* = P_S$ (every P is idempotent and self-adjoint). We denote by $\|\cdot\|_{\mathcal{F}}$ the norm induced by the dot product $< \cdot, \cdot >_{\mathcal{F}}$ in $\mathcal{F}$, and by $\|A\|p := \sqrt[p]{\text{Tr}(|A|^p)}$ the p-Schatten, or p-class norm of a linear bounded operator A [37, p. 84].

15.2.2   Subspace Estimators

Spectral estimators can be naturally derived from the characterization of S_ρ in terms of the covariance operator C associated to ρ. Indeed, if $C := \mathbb{E}_{x \sim \rho} x \otimes x$ is the (uncentered) covariance operator associated to ρ, it is easy to show that $S_\rho = \overline{\text{Ran} \, C}$. Similarly, given the empirical covariance $C_n := \frac{1}{n} \sum_{i=1}^{n} x \otimes x$, we define the *empirical subspace estimate*,

$$\hat{S}_n := \text{span}(X_n) = \text{Ran} \, C_n,$$

where the closure is not needed because $\hat{S}_n$ is finite-dimensional. We also define the *k-truncated (kernel) PCA subspace estimate* $\hat{S}_n^k := \text{Ran} \, C_n^k$, where C_n^k is obtained from C_n by keeping only its k top eigenvalues; see also Section 15.2.5. Note that, since the PCA estimate $\hat{S}_n^k$ is spanned by the top k eigenvectors of C_n, then clearly $\hat{S}_n^k \subseteq \hat{S}_n^{k'}$ for $k < k'$, and therefore $\{\hat{S}_n^k\}_{k=1}^n$ is a nested family of subspaces (all of which are contained in S_ρ). As discussed in Section 15.2.5, since kernel-PCA reduces to regular PCA in a feature space [44] (and can be computed with knowledge of the kernel alone), the following discussion applies equally to kernel-PCA estimates.

15.2.3   Performance Criteria

We define the pseudo-metric

$$d_{\alpha,p}(U, V) := \|(P_U - P_V)C^\alpha\|_p \tag{15.2}$$

between subspaces U, V, which is a metric over the collection of subspaces contained in S_ρ, for $0 \leq \alpha \leq \frac{1}{2}$ and $1 \leq p \leq \infty$. Note that $d_{\alpha,p}$ depends on

ρ through C but this dependence is omitted in the notation. A number of important performance criteria can be recovered as particular cases of $d_{\alpha,p}$. In particular, the so-called reconstruction error [46, 7],

$$d_R(S_\rho, \hat{S}) := \mathbb{E}_{x \sim \rho} \| P_{S_\rho}(x) - P_{\hat{S}}(x) \|_{\mathcal{F}}^2 \qquad (15.3)$$

is $d_R(S_\rho, \cdot) = d_{1/2,2}(S_\rho, \cdot)^2$. Note that d_R is a natural criterion because a k-truncated PCA estimate minimizes a suitable error d_R over all subspaces of dimension k. Clearly, $d_R(S_\rho, \hat{S})$ vanishes whenever $\hat{S}$ contains S_ρ and, because the family $\{\hat{S}_n^k\}_{k=1}^n$ of PCA estimates is nested, then $d_R(S_\rho, \hat{S}_n^k)$ is non-increasing with k. As shown in [32], a number of unsupervised learning algorithms, including (kernel) PCA, k-means, k-flats, sparse coding, and non-negative matrix factorization, can be written as a minimization of d_R over an algorithm-specific class of sets (e.g., over the set of linear subspaces of a fixed dimension in the case of PCA).

15.2.4   Summary of Results

Our main technical contribution is a bound of the form of Equation (15.1), for the k-truncated PCA estimate $\hat{S}_n^k$ (with the empirical estimate $\hat{S}_n := \hat{S}_n^n$ being a particular case), whose proof is postponed to Section 15.5.

We begin by bounding the distance $d_{\alpha,p}$ between S_ρ and the k-truncated PCA estimate $\hat{S}_n^k$, given a known covariance C.

Theorem 15.1. *Let $\{x_i\}_{1 \leq i \leq n}$ be drawn i.i.d. according to a probability measure ρ supported on the unit ball of a separable Hilbert space $\mathcal{F}$, with covariance C. Assuming $n > 3$, $0 < \delta < 1$, $0 \leq \alpha \leq \frac{1}{2}$, $1 \leq p \leq \infty$, the following holds for each $k \in \{1, \ldots, n\}$:*

$$\mathbb{P}\left[d_{\alpha,p}(S_\rho, \hat{S}_n^k) \leq 3t_k^\alpha \left\| C^\alpha (C + t_k I)^{-\alpha} \right\| p \right] \geq 1 - \delta \qquad (15.4)$$

where $t_k = \max\{\sigma_k, \frac{9}{n} \log \frac{n}{\delta}\}$, and σ_k is the k-th top eigenvalue of C.

We say that C has *eigenvalue decay rate of order r* if there are constants $q, Q > 0$ such that $qj^{-r} \leq \sigma_j \leq Qj^{-r}$, where σ_j are the (decreasingly ordered) eigenvalues of C, and $r > 1$. From Equation (15.2) it is clear that, in order for the subspace learning problem to be well-defined, it must be $\|C^\alpha\| p < \infty$, or alternatively: $\alpha p > 1/r$. Note that this condition is always met for $p = \infty$, and also holds in the reconstruction error case ($\alpha = 1/2, p = 2$), for any decay rate $r > 1$.

Knowledge of an eigenvalue decay rate can be incorporated into Theorem 15.1 to obtain explicit learning rates, as follows.

Theorem 15.2 (Polynomial eigenvalue decay). *Let C have eigenvalue decay rate of order r. Under the assumptions of Theorem 15.1, it holds, with* *probability $1 - \delta$:*

$$d_{\alpha,p}(S_\rho, \hat{S}_n^k) \leq \begin{cases} Q'k^{-r\alpha+\frac{1}{p}} & \text{if } k < k_n^* \quad \text{(polynomial decay)} \\ Q'k_n^{*-r\alpha+\frac{1}{p}} & \text{if } k \geq k_n^* \quad \text{(plateau)} \end{cases} \tag{15.5}$$

where it is $k_n^* = \left(\frac{qn}{9\log(n/\delta)}\right)^{1/r}$, and

$$Q' = 3\left(Q^{\frac{1}{r}}\frac{\Gamma\left(\alpha p - \frac{1}{r}\right)\Gamma\left(1 + \frac{1}{r}\right)}{\Gamma\left(\frac{1}{r}\right)}\right)^{\frac{1}{p}}. \tag{15.6}$$

The above theorem guarantees a decay of $d_{\alpha,p}$ with increasing k, at a rate of $k^{-r\alpha+1/p}$, up to $k = k_n^*$, after which the bound remains constant. The estimated plateau threshold k^* is thus the value of truncation past which the upper bound does not improve. Note that, as described in Section 15.4, this error decay and plateau behavior is observed in practice.

The proofs of Theorems 15.1 and 15.2 rely on recent non-commutative Bernstein-type inequalities on operators [5, 52], and a novel analytical decomposition. Note that classical Bernstein inequalities in Hilbert spaces (e.g.,[34]) could also be used instead of [52]. While this approach would simplify the analysis, it produces looser bounds, as described in Section 15.5.

If we consider an algorithm that produces, for each set of n samples, an estimate $\hat{S}_n^k$ with $k \geq k_n^*$ then, by plugging the definition of k_n^* into Equation 15.5, we obtain an upper bound on $d_{\alpha,p}$ as a function of n.

Corollary 15.1. *Let C have eigenvalue decay rate of order r, and Q', k_n^* be as in Theorem 15.2. Let $\hat{S}_n^*$ be a truncated subspace estimate $\hat{S}_n^k$ with $k \geq k_n^*$. It is, with probability $1 - \delta$,*

$$d_{\alpha,p}(S_\rho, \hat{S}_n^*) \leq Q'\left(\frac{9(\log n - \log \delta)}{qn}\right)^{\alpha - \frac{1}{rp}}$$

Remark 15.1. *Note that, by setting $k = n$, the above corollary also provides guarantees on the rate of convergence of the empirical estimate $\hat{S}_n = \text{span}(X_n)$ to S_ρ, of order*

$$d_{\alpha,p}(S_\rho, \hat{S}_n) = O\left(\left(\frac{\log n - \log \delta}{n}\right)^{\alpha - \frac{1}{rp}}\right).$$

Corollary 15.2 and Remark 15.1 are valid for all n such that $k_n^* \leq n$ (or equivalently such that $n^{r-1}(\log n - \log \delta) \geq q/9$). Note that, because ρ is supported on the unit ball, its covariance has eigenvalues no greater than one, and therefore it must be $q < 1$. It thus suffices to require that $n > 3$ to ensure the condition $k_n^* \leq n$ is to hold.

15.2.5 Kernel PCA and Embedding Methods

One of the main applications of subspace learning is to perform dimensionality reduction. In particular, one may find nested subspaces of dimensions $1 \leq k \leq n$ that minimize the distances from the original to the projected samples. This procedure is known as the Karhunen-Loève, PCA, or Hotelling transform [26], and has been generalized to reproducing-kernel Hilbert spaces (RKHS) [44].

In particular, the above procedure amounts to computing an eigendecomposition of the empirical covariance

$$C_n = \sum_{i=1}^{n} \sigma_i u_i \otimes u_i,$$

where the k-th subspace estimate is $\hat{S}_n^k := \operatorname{Ran} C_n^k = \operatorname{span}\{u_i : 1 \leq i \leq k\}$. In the case of kernel PCA, the samples $\{x_i\}_{1 \leq i \leq n}$ belong to some RKHS $\mathcal{F}$, and we can think of them as the embedding $x_i := \phi(z_i)$ of some original data $(z_1, \ldots, z_n) \in Z^n$, where, e.g., $Z = \mathbb{R}^D$. The measure ρ can be seen as the measure induced by the embedding and the original data distribution. Interestingly, in practice we may only have indirect information about ϕ in the form of a kernel function $K : Z \times Z \to \mathbb{R}$: a symmetric, positive definite function satisfying $K(z, w) = \langle \phi(z), \phi(w) \rangle_{\mathcal{F}}$ [48] (for technical reasons, we also assume K to be continuous). Recall that every such K has a unique associated RKHS, and vice versa [48, p. 120–121], whereas, given K, the embedding ϕ is only unique up to an inner product-preserving transformation. The following reproducing property $f(x) = \langle f, K(z, \cdot) \rangle_{\mathcal{F}}$ holds for all $z \in Z$, $f \in \mathcal{F}$.

If the embedding is defined through a kernel K, it easy to see that the k-truncated kernel PCA can be computed considering the n by n kernel matrix K_n, where $(K_n)_{i,j} = K(x_i, x_j)$ [44]. It is easy to see that the k-truncated kernel PCA subspace $\hat{S}_n^k$ minimizes the empirical reconstruction error $d_R(\hat{S}_n, \hat{S})$, among all subspaces $\hat{S}$ of dimension k. Indeed, it is

$$d_R(\hat{S}_n, \hat{S}) = \mathbb{E}_{x \sim \hat{\rho}} \|x - P_{\hat{S}}(x)\|_{\mathcal{F}}^2 = \mathbb{E}_{x \sim \hat{\rho}} \langle (I - P_{\hat{S}})x, (I - P_{\hat{S}})x \rangle_{\mathcal{F}}$$
$$= \mathbb{E}_{x \sim \hat{\rho}} \langle I - P_{\hat{S}}, x \otimes x \rangle_{HS} = \langle I - P_{\hat{S}}, C_n \rangle_{HS}, \tag{15.7}$$

where $\langle \cdot, \cdot \rangle_{HS}$ is the Hilbert-Schmidt inner product. From this, it clearly follows that the k-dimensional subspace minimizing Equation 15.7 (maximizing $\langle P_{\hat{S}}, C_n \rangle$) is spanned by the k top eigenvectors of C_n. Since we are interested in the expected error $d_R(S_\rho, \hat{S}_n^k)$ of the kernel PCA estimate (rather than the empirical error $d_R(\hat{S}_n, \hat{S})$), we may obtain a learning rate for Equation 15.7 by specializing Theorem 15.2 to the reconstruction error, for all k (Theorem 15.2), and for $k \geq k^*$ with a suitable choice of k^* (Corollary 15.2). In particular, recalling that $d_R(S_\rho, \cdot) = d_{\alpha,p}(S_\rho, \cdot)^2$ with $\alpha = 1/2$ and $p = 2$, and choosing a value of $k \geq k_n^*$ that minimizes the bound of Theorem 15.2, we obtain the following result.

Corollary 15.2 (Performance of PCA/Reconstruction error). *Let C have eigenvalue decay rate of order r, and $\hat{S}_n^*$ be as in Corollary 15.1. Then it holds, with probability $1 - \delta$,*

$$d_R(S_\rho, \hat{S}_n^*) = O\left(\left(\frac{\log n - \log \delta}{n}\right)^{1-1/r}\right).$$

15.2.6   Comparison with Previous Results in the Literature

Figure 15.1 shows a comparison of our learning rates with existing rates in the literature [7, 46]. The plot shows the polynomial decay rate c of the high probability bound $d_R(S_\rho, \hat{S}_n^k) = O(n^{-c})$, as a function of the eigenvalue decay rate r of the covariance C, computed at the best value k_n^* (which minimizes the bound).

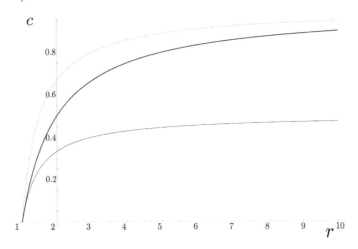

FIGURE 15.1: Known upper bounds for the polynomial decay rate c (for the best choice of k), for the expected distance from a random sample to the empirical k-truncated kernel-PCA estimate, as a function of the covariance eigenvalue decay rate (higher is better). Our bound consistently outperforms previous ones [46]. The top (dashed) line [7], has significantly stronger assumptions, and is only included for completeness.

The learning rate exponent c, under a polynomial eigenvalue decay assumption of the data covariance C, is $c = \frac{s(r-1)}{r-s+sr}$ for [7] and $c = \frac{r-1}{2r-1}$ for [46], where s is related to the fourth moment. Note that, among the two that operate under the same assumptions, our bound is the best by a wide margin. The top, best performing, dashed line [7] is obtained for the best possible fourth-order moment constraint $s = 2r$, and is therefore not a fair comparison. However, it is worth noting that our bounds perform almost as well as the most restrictive one, even when we do not include any fourth-order moment constraints.

Choice of truncation parameter k. Since, as pointed out in Section 15.2.2, the subspace estimates $\hat{S}_n^k$ are nested for increasing k (i.e., $\hat{S}_n^k \subseteq \hat{S}_n^{k'}$ for $k < k'$), the distance $d_{\alpha,p}(S_\rho, \hat{S}_n^k)$, and in particular the reconstruction error $d_R(S_\rho, \hat{S}_n^k)$, is a non-increasing function of k. As discussed [7], this suggests that there is no bias-variance trade-off in the choice of k. Indeed, the fact that the estimates $\hat{S}_n^k$ become increasingly close to S_ρ as k increases indicates that, when minimizing $d_{\alpha,p}(S_\rho, \hat{S}_n^k)$, the best choice is simply $k = n$.

Interestingly, however, both in practice (Section 15.4), and in theory (Section 15.2.4), we observe that a typical behavior for the subspace learning problem in high dimensions (e.g., kernel PCA) is that there is a certain value of $k = k_n^*$, past which performance plateaus. For problems such as spectral embedding methods [51, 18, 58], in which a degree of dimensionality reduction is desirable, producing an estimate $\hat{S}_n^k$ where k is close to the plateau threshold may be a natural parameter choice: it leads to an estimate of the lowest dimension ($k = k_n^*$), whose distance to the true S_ρ is almost as low as the best-performing one ($k = n$).

15.3 Set Learning

The problem of set, or support estimation has received a great deal of attention in the statistics community since the 1960s [36, 21], and since then a number of practical approaches have been proposed to address it [17, 27, 20, 11, 53, 43, 13, 35, 49, 55, 45, 6, 12]. Support estimation is often considered in machine learning in situations in which it is difficult to gather negative examples (as often happens in biological and biomedical problems) or when the negative class is not well defined (as in object detection problems in computer vision), as is the case in one class estimation [43], and novelty and anomaly detection [31, 9].

In this section, we describe an approach that is based on reducing the set learning problem to that of learning a subspace. The results in this section largely draw from [40, 14, 41].

15.3.1 Set Learning via Subspace Learning

We begin by recalling how support estimation can be reduced to subspace learning, and discuss how our results specialize to this setting. From an algorithmic perspective, the approach we discuss is closely related to the one in [25] and has been successfully applied in several practical domains [38, 28, 54, 24, 30, 10, 47].

Central to the connection between set and subspace learning is the notion of *separating kernel* and *separating feature map*, which was introduced in [15].

Let K be a reproducing kernel on some space Z, and $(\phi, \mathcal{F})$ an associated feature map and feature space pair (see Section 15.2.5). For simplicity, we assume that $\|\phi(z)\|_{\mathcal{F}} = 1$ for all $z \in Z$. This assumption is without loss of generality because a kernel with non-zeros in its diagonal can always be normalized. Given a non-empty set $C \subseteq Z$, let $\mathcal{F}_C = \overline{\text{span}}\{\phi(z) \mid z \in C\}$ be the closure of all finite linear combinations of points in the range $\phi(C)$ of C. The distance from any given point $\phi(z)$, with $z \in Z$, to the linear subspace $\mathcal{F}_C$ is

$$d_{\mathcal{F}_C}(\phi(z)) := \inf_{f \in \mathcal{F}_C} \|\phi(z) - f\|_{\mathcal{F}}.$$

The following, key definition is equivalent to the *separating property* in Definition 1 of [56].

Definition 15.1. *We say that a feature map ϕ (and hence the corresponding kernel) separates a set $C \subset Z$ if for all $z \in Z$ it holds:*

$$d_{\mathcal{F}_C}(\phi(z)) = 0 \quad \text{iff} \quad z \in C.$$

An example of separating kernel for $\mathbb{R}^d$ is the exponential kernel $K(x, x') = e^{-\|x-x'\|}$. The proof of this fact, see [14], crucially depends on the fact that for each compact subset of $\mathbb{R}^d$ the associated reproducing kernel Hilbert space contains functions that are zero on the set and non-zero outside. Interestingly, the Gaussian kernel is not separating, because the associated Hilbert space contains only analytic functions, and the only function that is zero on a compact subset (with non-empty interior) is the zero function.

The separating property has a clear geometric interpretation in the feature space: the set $\phi(S_\rho)$ is the intersection of the closed subspace $\mathcal{F}_{S_\rho}$ (the smallest linear subspace containing $\phi(S_\rho)$), and $\phi(Z)$ (see Figure 15.2).

Using the notion of separating kernel, the support S_ρ can be characterized in terms of the subspace $\mathcal{F}_\rho = \overline{\text{span}}\,\phi(S_\rho) \subseteq \mathcal{F}$. More precisely, it can be shown (see the next subsection) that, if the feature map ϕ *separates* S_ρ, then it is

$$S_\rho = \{z \in Z \mid d_{\mathcal{F}_\rho}(\phi(z)) = 0\}.$$

The above discussion naturally leads to an empirical estimate $\hat{S}_n = \{z \in Z \mid d_{\hat{\mathcal{F}}_n}(\phi(z)) \leq \tau\}$ of S_ρ, where $\hat{\mathcal{F}}_n = \overline{\text{span}}\,\phi(Z_n)$, and $\tau > 0$. Given a training set $z_1, \ldots, z_n$, the estimator $\hat{S}_n$ is therefore the set of points $z \in Z$ whose associated distance from $\phi(z)$ to the linear space spanned by $\{\phi(z_1), \ldots, \phi(z_n)\}$ is sufficiently small, according to some tolerance τ. Any point with distance greater than τ will be considered to be *outside of the support* by this estimator.

With the above choice of estimator, it can be shown that almost sure convergence $\lim_{n \to \infty} d_H(S_\rho, \hat{S}_n) = 0$ in the Hausdorff distance [1] is related to the convergence of $\hat{\mathcal{F}}_n$ to $\mathcal{F}_\rho$ [15]. More precisely, if the eigenfunctions of the covariance operator $C = \mathbb{E}_{z \sim \rho}[\phi(z) \otimes \phi(z)]$ are uniformly bounded, then it

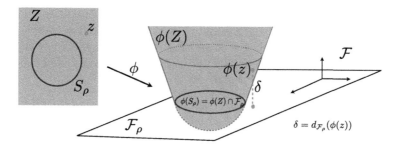

FIGURE 15.2: The input space Z and the support S_ρ are mapped into the feature space $\mathcal{F}$ by the feature map ϕ. Letting $\mathcal{F}_\rho := \mathcal{F}_{S_\rho}$ be the smallest linear subspace containing $\phi(S_\rho)$ then, if the kernel is separable, the image of the support $\phi(S_\rho)$ is given by the intersection between $\phi(Z)$ and $\mathcal{F}_\rho$. By the separating property, a point z belongs to the support if and only the distance between $\phi(z)$ and $\mathcal{F}_\rho$ is zero.

suffices for Hausdorff convergence to bound from above $d_{\frac{r-1}{2r},\infty}$ (where $r > 1$ is the eigenvalue decay rate of C) as shown in Section 15.3.2.

15.3.2 Consistency Results

Before proving the consistency of the set estimator $\hat{S}_n$ we show an improved learning rate for the associated subspace $\hat{\mathcal{F}}_n$. In particular we study the linear subspace $\hat{\mathcal{F}}_n^*$ that is the one spanned by the first k components of the empirical covariance matrix $\hat{C}_n$, where $k \geq k_n^*$ (see Subsection 15.2.5 and Theorem 15.2). Note that $\hat{\mathcal{F}}_n^* = \hat{\mathcal{F}}_n$ when $k = n$. The following result specializes Corollary 15.1 to this setting.

Corollary 15.3 (Performance of KPCA with the set learning metric). *If $0 \leq \alpha \leq \frac{1}{2}$, then it holds, with probability $1 - \delta$,*

$$d_{\alpha,\infty}(\mathcal{F}_\rho, \hat{\mathcal{F}}_n^*) = O\left(\left(\frac{\log n - \log \delta}{n}\right)^\alpha\right)$$

where the constant in the Landau symbol does not depend on δ.

Letting $\alpha = \frac{r-1}{2r}$ above yields a high probability bound of order $O\left(n^{-\frac{r-1}{2r}}\right)$ (up to logarithmic factors), which is considerably sharper than the bound $O\left(n^{-\frac{r-1}{2(3r-1)}}\right)$ found in [16] (Theorem 7). Note that these are upper bounds for the best possible choice of k (which minimizes the bound). While the optima of both bounds vanish with $n \to \infty$, their behavior is qualitatively different. In particular, the bound of [16] is U-shaped, and diverges for $k = n$, while ours is L-shaped (no trade-off), and thus also convergent for $k = n$. Therefore, when compared with [16], our results suggest that no regularization is required from a statistical point of view though, as discussed in the following, it may be required for numerical stability. With the above tools at hand, we are now in a position to prove the consistency of $\hat{S}_n$.

Theorem 15.3 (Consistency of Set Learning). *Let the input space Z be metrizable, K be a kernel on Z with the separating property [15], let the dimension k of the empirical subspace $\hat{\mathcal{F}}_n^*$, be $k_n^* \leq k \leq n$, and the threshold parameter $\tau = \max_{1 \leq i \leq n} d_{\hat{\mathcal{F}}_n^k}(\phi(z_i))$, then*

$$\hat{S}_n^* = \left\{ z \in Z \ \middle| \ d_{\hat{\mathcal{F}}_n^*}(\phi(z)) \leq \tau \right\} \tag{15.8}$$

is a universally consistent unsupervised learning algorithm.

Proof. By Theorem 6 of [56] and our Corollary 15.3, the estimator $\hat{S}_n^*$ satisfies the universal consistency conditions given in Section 15.1, under the given hypotheses.   □

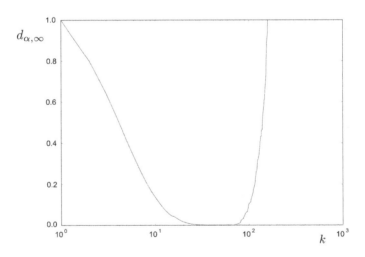

FIGURE 15.3: The experimental behavior of the distance $d_{\alpha,\infty}(\hat{S}^k, S_\rho)$ between the empirical and the actual support subspaces, with respect to the regularization parameter. The setting is the one of Section 15.4. Here the actual subspace is analytically computed, while the empirical one is computed on a dataset with $n = 1000$ and 32bit floating point precision. Note the numerical instability as k tends to 1000.

We note that the above result is an example of how kernel embedding techniques can be used to provably estimate geometric invariants of the the the original data distribution. Note that the considered estimator achieves this without having to explicitly solve a pre-image problem [33].

We end this section noting that, while, as proven in Corollary 15.3, regularization is not needed from a statistical perspective, it can play a role in ensuring numerical stability in practice. Indeed, in order to find $\hat{S}$, we compute $d_{\hat{\mathcal{F}}_n}(\phi(z))$ with $z \in Z$. Using the reproducing property of K, it can be shown that, for $z \in Z$, it is $d_{\hat{\mathcal{F}}_n^k}(\phi(z)) = K(z, z) - \left\langle t_z, (\hat{K}_n^k)^\dagger t_z \right\rangle$ where $(t_z)_i = K(z, z_i)$, $\hat{K}_n$ is the Gram matrix $(\hat{K}_n)_{ij} = K(z_i, z_j)$, $\hat{K}_n^k$ is the rank-k approximation of $\hat{K}_n$, and $(\hat{K}_n^k)^\dagger$ is the pseudo-inverse of $\hat{K}_n^k$. The computation of $\hat{S}$ therefore requires a matrix inversion, which is prone to instability for high condition numbers. Figure 15.3 shows the behavior of the error that results from replacing $\hat{\mathcal{F}}_n$ by its k-truncated approximation $\hat{\mathcal{F}}_n^k$. For large values of k, the small eigenvalues of $\hat{\mathcal{F}}_n$ are used in the inversion, leading to numerical instability.

15.4 Numerical Experiments

In order to validate our analysis empirically, we consider the following experiment. Let ρ be a uniform one-dimensional distribution in the unit interval. We embed ρ into a reproducing-kernel Hilbert space $\mathcal{F}$ using the exponential of the ℓ_1 distance $(k(u, v) = \exp\{-\|u - v\|_1\})$ as kernel. Given n samples drawn from ρ, we compute its empirical covariance in $\mathcal{F}$ (whose spectrum is plotted in Figure 15.4 (top)), and truncate its eigen-decomposition to obtain a subspace estimate $\hat{\mathcal{F}}_n^k$, as described in Section 15.2.2.

Figure 15.4 (bottom) is a box plot of reconstruction error $d_R(\mathcal{F}_\rho, \hat{\mathcal{F}}_n^k)$ associated with the k-truncated kernel-PCA estimate $\hat{\mathcal{F}}_n^k$ (the expected distance in $\mathcal{F}$ of samples to $\hat{\mathcal{F}}_n^k$), with $n = 1000$ and varying k. While d_R is computed analytically in this example, and $\mathcal{F}_\rho$ is fixed, the estimate $\hat{\mathcal{F}}_n^k$ is a random variable, and hence the variability in the graph. Notice from the figure that, as pointed out in [7] and discussed in Section 15.2.6, the reconstruction error $d_R(\mathcal{F}_\rho, \hat{\mathcal{F}}_n^k)$ is always a non-increasing function of k, due to the fact that the kernel-PCA estimates are nested: $\hat{\mathcal{F}}_n^k \subset \hat{\mathcal{F}}_n^{k'}$ for $k < k'$ (see Section 15.2.2). The graph is highly concentrated around a curve with a steep initial decay, until reaching some sufficiently high k, past which the reconstruction (pseudo) distance becomes stable, and does not vanish. In our experiments, this behavior is typical for the reconstruction distance and high-dimensional problems.

Due to the simple form of this example, we are able to compute analytically the spectrum of the true covariance C. In this case, the eigenvalues of C decay as $2\gamma/((k\pi)^2 + \gamma^2)$, with $k \in \mathbb{N}$, and therefore they have a polynomial decay

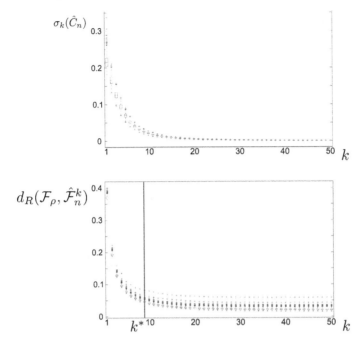

FIGURE 15.4: The spectrum of the empirical covariance (top), and the expected distance from a random sample to the empirical k-truncated kernel-PCA subspace estimate (bottom), as a function of k ($n = 1000$, 1000 trials shown in a boxplot). Our predicted plateau threshold k_n^* (Theorem 15.2) is a good estimate of the value k past which the distance stabilizes.

rate $r = 2$ (see Section 15.2.4). Given the known spectrum decay rate, we can estimate the plateau threshold $k = k_n^*$ in the bound of Theorem 15.2, which can be seen to be a good approximation of the observed start of a plateau in $d_R(\mathcal{F}_\rho, \hat{\mathcal{F}}_n^k)$ (Figure 15.4, bottom). Notice that our bound for this case (Corollary 15.2) similarly predicts a steep error decay until the threshold $k = k_n^*$, and a plateau afterwards.

15.5 Sketch of the Proofs

For the sake of completeness we sketch the main step in the proof of our main theoretical result, Theorem 15.1, with some details omitted in the interests of conciseness.

For each $\lambda > 0$, we denote by $r^\lambda(x) := \mathbf{1}\{x > \lambda\}$ the step function with a cut-off at λ. Given an empirical covariance operator C_n, we will consider the truncated version $r^\lambda(C_n)$ where, in this notation, r^λ is applied to the eigenvalues of C_n, that is, $r^\lambda(C_n)$ has the same eigen-structure as C_n, but its eigenvalues that are less than or equal to λ are clamped to zero.

In order to prove the bound of Equation (15.4), we begin by proving a more general upper bound of $d_{\alpha,p}(S_\rho, \hat{S}_n^k)$, which is split into a random ($\mathcal{A}$), and a deterministic part ($\mathcal{B}, \mathcal{C}$). The bound holds for all values of a free parameter $t > 0$, which is then constrained and optimized in order to find the (close to) tightest version of the bound.

Lemma 15.1. *Let $t > 0$, $0 \leq \alpha \leq 1/2$, and $\lambda = \sigma_k(C)$ be the k-th top eigenvalue of C, it is,*

$$d_{\alpha,p}(S_\rho, \hat{S}_n^k) \;\leq\; \underbrace{\|(C+tI)^{\frac{1}{2}}(C_n+tI)^{-\frac{1}{2}}\|_\infty^{2\alpha}}_{\mathcal{A}} \cdot \tag{15.9}$$

$$\cdot \; \underbrace{\{3/2(\lambda+t)\}^\alpha}_{\mathcal{B}} \cdot \underbrace{\|C^\alpha(C+tI)^{-\alpha}\|_p}_{\mathcal{C}}. \tag{15.10}$$

Note that the right-hand side of Equation (15.9) is the product of three terms, the left of which ($\mathcal{A}$) involves the empirical covariance operator C_n, which is a random variable, and the right two ($\mathcal{B}, \mathcal{C}$) are entirely deterministic. While the term $\mathcal{B}$ has already been reduced to the known quantities t, α, λ, the remaining terms are bounded next. We bound the random term $\mathcal{A}$ in the next Lemma, whose proof makes use of recent concentration results [52].

Lemma 15.2 (Term $\mathcal{A}$). *Let $0 \leq \alpha \leq 1/2$, for each $\frac{9}{n} \log \frac{n}{\delta} \leq t \leq \|C\|_\infty$, with probability $1 - \delta$ it is*

$$(2/3)^\alpha \leq \|(C+tI)^{\frac{1}{2}}(C_n+tI)^{-\frac{1}{2}}\|_\infty^{2\alpha} \leq 2^\alpha$$

Lemma 15.3 (Term $\mathcal{C}$). *Let C be a symmetric, bounded, positive semidefinite linear operator on $\mathcal{F}$. If $\sigma_k(C) \leq f(k)$ for $k \in \mathbb{N}$, where f is a decreasing function, then, for all $t > 0$ and $\alpha \geq 0$, it holds that*

$$\|C^\alpha(C+tI)^{-\alpha}\| p \leq \inf_{0 \leq u \leq 1} g_{u\alpha} t^{-u\alpha} \tag{15.11}$$

where $g_{u\alpha} = \left(f(1)^{u\alpha p} + \int_1^\infty f(x)^{u\alpha p} dx\right)^{1/p}$. Furthermore, if $f(k) = gk^{-1/\gamma}$, with $0 < \gamma < 1$ and $\alpha p > \gamma$, then it holds that

$$\|C^\alpha(C+tI)^{-\alpha}\| p \leq Q t^{-\gamma/p} \tag{15.12}$$

where $Q = (g^\gamma \Gamma(\alpha p - \gamma)\Gamma(1+\gamma)/\Gamma(\gamma))^{1/p}$.

The combination of Lemmas 15.1 and 15.2 leads to the main Theorem 15.1, which is a probabilistic bound, holding for every $k \in \{1, \ldots, n\}$, with a deterministic term $\|C^\alpha(C+tI)^{-\alpha}\| p$ that depends on knowledge of the covariance C. In cases in which some knowledge of the decay rate of C is available, Lemma 15.3 can be applied to obtain Theorem 15.2 and Corollary 15.1. Finally, Corollary 15.2 is simply a particular case for the reconstruction error $d_R(S_\rho, \cdot) = d_{\alpha,p}(S_\rho, \cdot)^2$, with $\alpha = 1/2, p = 2$.

As noted in Section 15.2.4, looser bounds would be obtained if classical Bernstein inequalities in Hilbert spaces [34] were used instead. In particular, Lemma 15.2 would result in a range for t of $qn^{-r/(r+1)} \le t \le \|C\|_\infty$, implying $k^* = O(n^{1/(r+1)})$ rather than $O(n^{1/r})$, and thus Theorem 15.2 would become (for $k \ge k^*$) $d_{\alpha,p}(S_\rho, S_n^k) = O(n^{-\alpha r/(r+1)+1/(p(r+1))})$ (compared with the sharper $O(n^{-\alpha+1/rp})$ of Theorem 15.2). For instance, for $p = 2$, $\alpha = 1/2$, and a decay rate $r = 2$ (as in the example of Section 15.4), it would be $d_{1/2,2}(S_\rho, S_n) = O(n^{-1/4})$ using Theorem 15.2, and $d_{1/2,2}(S_\rho, S_n) = O(n^{-1/6})$ using classical Bernstein inequalities.

15.6   Conclusions

The problem of set learning consists of estimating the smallest subset of the input space containing the data distribution. In this chapter the problem has been investigated by analyzing its relations with subspace learning, that consists of estimating the smallest linear subspace containing the distribution. In particular we showed that, given a suitable feature map, the set learning problem can be cast as a subspace learning problem in the associated feature space. In order to analyze the theoretical properties of the first problem, the statistical analysis for the second has been developed obtaining novel and sharper sample complexity upper bounds. Finally, by exploiting such results, the consistency of set learning has been established. The chapter is concluded by numerical examples that show the effectiveness of our analysis.

Acknowledgments

L. R. acknowledges the financial support of the Italian Ministry of Education University and Research FIRB project RBFR12M3AC.

Bibliography

[1] G. Beer. *Topologies on Closed and Closed Convex Sets*. Springer, 1993.

[2] M. Belkin and P. Niyogi. Laplacian eigenmaps for dimensionality reduction and data representation. *Neural computation*, 15(6):1373–1396, 2003.

[3] Y. Bengio, O. Delalleau, N.L. Roux, J.F. Paiement, P. Vincent, and M. Ouimet. Learning eigenfunctions links spectral embedding and kernel pca. *Neural Computation*, 16(10):2197–2219, 2004.

[4] Y. Bengio, J.-F. Paiement, P. Vincent, O. Delalleau, N. Le Roux, and M. Ouimet. Out-of-sample extensions for lle, isomap, mds, eigenmaps, and spectral clustering. *Advances in neural information processing systems*, 16:177–184, 2004.

[5] S. Bernstein. *The Theory of Probabilities*. Gastehizdat Publishing House, Moscow, 1946.

[6] G. Biau, B. Cadre, D. Mason, and B. Pelletier. Asymptotic normality in density support estimation. *Electron. J. Probab.*, 14:no. 91, 2617–2635, 2009.

[7] G. Blanchard, O. Bousquet, and L. Zwald. Statistical properties of kernel principal component analysis. *Machine Learning*, 66(2):259–294, 2007.

[8] I. Borg and P.J.F. Groenen. *Modern multidimensional scaling: Theory and applications*. Springer, 2005.

[9] V. Chandola, A. Banerjee, and V. Kumar. Anomaly detection: A survey. *ACM Computing Surveys (CSUR)*, 41(3):15, 2009.

[10] P. Cheng, W. Li, and P. Ogunbona. Kernel pca of hog features for posture detection. In *Image and Vision Computing New Zealand, 2009. IVCNZ'09. 24th International Conference*, pages 415–420. IEEE, 2009.

[11] A. Cuevas and R. Fraiman. A plug-in approach to support estimation. *Ann. Statist.*, 25(6):2300–2312, 1997.

[12] A. Cuevas and R. Fraiman. Set estimation. In *New perspectives in stochastic geometry*, pages 374–397. Oxford Univ. Press, Oxford, 2010.

[13] A. Cuevas and A. Rodríguez-Casal. Set estimation: an overview and some recent developments. In *Recent advances and trends in nonparametric statistics*, pages 251–264. Elsevier, Amsterdam, 2003.

[14] E. De Vito, L. Rosasco, and A. Toigo. A universally consistent spectral estimator for the support of a distribution. *Applied Computational Harmonic Analysis*, 2014. In press, DOI 10.1016/j.acha.2013.11.003.

[15] E. De Vito, L. Rosasco, and A. Toigo. Spectral regularization for support estimation. *Advances in Neural Information Processing Systems, NIPS Foundation*, pages 1–9, 2010.

[16] E. De Vito, L. Rosasco, and A. Toigo. Learning sets with separating kernels. *arXiv:1204.3573*, 2012.

[17] L. Devroye and G.L. Wise. Detection of abnormal behavior via nonparametric estimation of the support. *SIAM J. Appl. Math.*, 38(3):480–488, 1980.

[18] D.L. Donoho and C. Grimes. Hessian eigenmaps: Locally linear embedding techniques for high-dimensional data. *Proceedings of the National Academy of Sciences*, 100(10):5591–5596, 2003.

[19] R. Duda, P.E. Hart, and D.G. Stork. *Pattern classification*. John Wiley & Sons, 2012.

[20] L. Dümbgen and G. Walther. Rates of convergence for random approximations of convex sets. *Adv. in Appl. Probab.*, 28(2):384–393, 1996.

[21] J. Geffroy. Sur un probleme d'estimation géométrique. *Publ. Inst. Statist. Univ. Paris*, 13:191–210, 1964.

[22] J. Ham, D.D. Lee, S. Mika, and B. Schölkopf. A kernel view of the dimensionality reduction of manifolds. In *Proceedings of the Twenty-First International Conference on Machine Learning*, page 47. ACM, 2004.

[23] T. Hastie, R. Tibshirani, J. Friedman, and J. Franklin. The elements of statistical learning: data mining, inference and prediction. *The Mathematical Intelligencer*, 27(2):83–85, 2005.

[24] F. He, J.H. Yang, M. Li, and J.W. Xu. Research on nonlinear process monitoring and fault diagnosis based on kernel principal component analysis. *Key Engineering Materials*, 413:583–590, 2009.

[25] H. Hoffmann. Kernel pca for novelty detection. *Pattern Recognition*, 40(3):863–874, 2007.

[26] I. Jolliffe. *Principal component analysis*. Wiley Online Library, 2005.

[27] A.P. Korostelëv and A.B. Tsybakov. *Minimax theory of image reconstruction*. Springer-Verlag, New York, 1993.

[28] H.-J. Lee, S. Cho, and M.-S. Shin. Supporting diagnosis of attention-deficit hyperactive disorder with novelty detection. *Artificial intelligence in medicine*, 42(3):199–212, 2008.

[29] M. Loève. *Probability theory*, volume 45. Springer, 1963.

[30] M.L. Maestri, M.C. Cassanello, and G.I. Horowitz. Kernel pca performance in processes with multiple operation modes. *Chemical Product and Process Modeling*, 4(5), 2009.

[31] M. Markou and S. Singh. Novelty detection: a review–part 1: statistical approaches. *Signal Processing*, 83(12):2481–2497, 2003.

[32] A. Maurer and M. Pontil. K–dimensional coding schemes in hilbert spaces. *IEEE Transactions on Information Theory*, 56(11):5839–5846, 2010.

[33] S. Mika, B. Schölkopf, A.J. Smola, K.-R. Müller, M. Scholz, and G. Rätsch. Kernel pca and de-noising in feature spaces. In *NIPS*, volume 11, pages 536–542, 1998.

[34] I. Pinelis. Optimum bounds for the distributions of martingales in banach spaces. *The Annals of Probability*, pages 1679–1706, 1994.

[35] M. Reitzner. Random polytopes and the Efron-Stein jackknife inequality. *Ann. Probab.*, 31(4):2136–2166, 2003.

[36] A. Rényi and R. Sulanke. Über die konvexe hülle von n zufällig gewählten punkten. *Probability Theory and Related Fields*, 2(1):75–84, 1963.

[37] J.R. Retherford. *Hilbert Space: Compact Operators and the Trace Theorem*. London Mathematical Society Student Texts. Cambridge University Press, 1993.

[38] B. Ristic, B. La Scala, M. Morelande, and N. Gordon. Statistical analysis of motion patterns in ais data: Anomaly detection and motion prediction. In *Information Fusion, 2008 11th International Conference on*, pages 1–7. IEEE, 2008.

[39] S.T. Roweis and L.K. Saul. Nonlinear dimensionality reduction by locally linear embedding. *Science*, 290(5500):2323–2326, 2000.

[40] A. Rudi, G.D. Canas, and L. Rosasco. On the sample complexity of subspace learning. In *Advances in Neural Information Processing Systems*, pages 2067–2075, 2013.

[41] A. Rudi, F. Odone, and E. De Vito. Geometrical and computational aspects of spectral support estimation for novelty detection. *Pattern Recognition Letters*, 36:107–116, 2014.

[42] L.K. Saul and S.T. Roweis. Think globally, fit locally: unsupervised learning of low dimensional manifolds. *The Journal of Machine Learning Research*, 4:119–155, 2003.

[43] B. Schölkopf, J. Platt, J. Shawe-Taylor, A. Smola, and R. Williamson. Estimating the support of a high-dimensional distribution. *Neural Comput.*, 13(7):1443–1471, 2001.

[44] B. Schölkopf, A. Smola, and K.R. Müller. Kernel principal component analysis. *Artificial Neural Networks–ICANN'97*, pages 583–588, 1997.

[45] C.D. Scott and R.D. Nowak. Learning minimum volume sets. *J. Mach. Learn. Res.*, 7:665–704, 2006.

[46] J. Shawe-Taylor, C.K. Williams, N. Cristianini, and J. Kandola. On the eigenspectrum of the gram matrix and the generalization error of kernel-pca. *Information Theory, IEEE Transactions on*, 51(7), 2005.

[47] B. Sofman, J.A. Bagnell, and A. Stentz. Anytime online novelty detection for vehicle safeguarding. In *Robotics and Automation (ICRA), 2010 IEEE International Conference on*, pages 1247–1254. IEEE, 2010.

[48] I. Steinwart and A. Christmann. *Support vector machines*. Information science and statistics. Springer-Verlag. New York, 2008.

[49] I. Steinwart, D. Hush, and C. Scovel. A classification framework for anomaly detection. *J. Mach. Learn. Res.*, 6:211–232 (electronic), 2005.

[50] J. Sun, S. Boyd, L. Xiao, and P. Diaconis. The fastest mixing Markov process on a graph and a connection to a maximum variance unfolding problem. *SIAM review*, 48(4):681–699, 2006.

[51] J.B. Tenenbaum, V. De Silva, and J.C. Langford. A global geometric framework for nonlinear dimensionality reduction. *Science*, 290(5500):2319–2323, 2000.

[52] J.A. Tropp. User-friendly tools for random matrices: An introduction. 2012.

[53] A.B. Tsybakov. On nonparametric estimation of density level sets. *Ann. Statist.*, 25(3):948–969, 1997.

[54] F.J. Valero-Cuevas, H. Hoffmann, M.U. Kurse, J.J. Kutch, and E.A. Theodorou. Computational models for neuromuscular function. *Biomedical Engineering, IEEE Reviews in*, 2:110–135, 2009.

[55] R. Vert and J.-P. Vert. Consistency and convergence rates of one-class svms and related algorithms. *Journal of Machine Learning Research*, 7:817–854, 2006.

[56] E. De Vito, L. Rosasco, and A. Toigo. Learning sets with separating kernels. *Applied and Computational Harmonic Analysis*, 2013.

[57] K.Q. Weinberger and L.K. Saul. Unsupervised learning of image manifolds by semidefinite programming. In *Computer Vision and Pattern Recognition, 2004. CVPR 2004*, volume 2, pages II–988. IEEE, 2004.

[58] K.Q. Weinberger and L.K. Saul. Unsupervised learning of image manifolds by semidefinite programming. *International Journal of Computer Vision*, 70(1):77–90, 2006.

[59] C.K.I. Williams. On a connection between kernel pca and metric multidimensional scaling. *Machine Learning*, 46(1):11–19, 2002.

Chapter 16

Output Kernel Learning Methods

Francesco Dinuzzo[*]

IBM Research

Cheng Soon Ong[*]

NICTA

Kenji Fukumizu[*]

The Institute of Statistical Mathematics

Simultaneously solving multiple related estimation tasks is a problem known as *multi-task learning* in the machine learning literature. A rather flexible approach to multi-task learning consists of solving a regularization problem where a positive semidefinite *multi-task kernel* is used to model joint relationships between both inputs and tasks. Specifying an appropriate multi-task kernel in advance is not always possible, therefore it is often desirable to estimate one from the data. In this chapter, we overview a family of regularization techniques called output kernel learning (OKL), for learning a multi-task kernel that can be decomposed as the product of a kernel on the inputs and one on the task indices. The kernel on the task indices is optimized simultaneously with the predictive function by solving a joint two-level regularization problem.

---

[*]francesd@ie.ibm.com, cheng-soon.ong@nicta.com.au, fukumizu@ism.ac.jp

16.1   Learning Multi-Task Kernels

Supervised multi-task learning consists of estimating multiple functions $f_j : \mathcal{X}_j \to \mathcal{Y}$ from multiple datasets of input-output pairs

$$(x_{ij}, y_{ij}) \in \mathcal{X}_j \times \mathcal{Y}, \quad j = 1, \ldots, m, \quad i = 1, \ldots, \ell_j,$$

where m is the number of *tasks* and ℓ_j is the number of data pairs for the j-th task. In general, the input sets $\mathcal{X}_j$ and the output set $\mathcal{Y}$ can be arbitrary non-empty sets. If the input sets $\mathcal{X}_j$ are the same for all the tasks, i.e., $\mathcal{X}_j = \mathcal{X}$, and the power set $\mathcal{Y}^m$ can be given a vector space structure, one can equivalently think in terms of learning a single vector-valued function $f : \mathcal{X} \to \mathcal{Y}^m$ from a dataset of pairs with incomplete output data. The key point in multi-task learning is to exploit relationships between the different components f_j in order to improve performance with respect to solving each supervised learning problem independently.

For a broad class of multi-task (or multi-output) learning problems, a suitable positive semidefinite *multi-task kernel* can be used to specify the joint relationships between inputs and tasks [5]. The most general way to address this problem is to specify a similarity function of the form $K((x_1, i), (x_2, j))$ defined for every pair of input data (x_1, x_2) and every pair of task indices (i, j). In the context of a kernel-based regularization method, choosing a multi-task kernel amounts to designing a suitable reproducing kernel Hilbert space (RKHS) of vector-valued functions, over which the function f whose components are the different tasks f_j is searched. See [13] for details about the theory of RKHS of vector valued-functions.

Predictive performances of kernel-based regularization methods are highly influenced by the choice of the kernel function. Such influence is especially evident in the case of multi-task learning where, in addition to specifying input similarities, it is crucial to correctly model inter-task relationships. Designing the kernel allows us to incorporate domain knowledge by properly constraining the function class over which the solution is searched. Unfortunately, in many problems the available knowledge is not sufficient to uniquely determine a good kernel in advance, making it highly desirable to have data-driven automatic selection tools. This need has motivated a fruitful research stream that has led to the development of a variety of techniques for learning the kernel.

There is considerable flexibility in choosing the similarity function K, the only constraint being positive semidefiniteness of the resulting kernel. However, such flexibility may also be a problem in practice, since choosing a good multi-task kernel for a given problem may be difficult. A very common way to simplify such modeling is to utilize a multiplicative decomposition of the form

$$K((x_1, i), (x_2, j)) = K_X(x_1, x_2) K_Y(i, j),$$

where the *input kernel* K_X is decoupled from the *output kernel* K_Y. The same structure can be equivalently represented in terms of a matrix-valued kernel

$$H(x_1, x_2) = K_X(x_1, x_2) \cdot \mathbf{L}, \qquad (16.1)$$

where $\mathbf{L}$ is a positive semidefinite matrix with entries $\mathbf{L}_{ij} = K_Y(i, j)$. Since specifying the kernel function K_Y is completely equivalent to specifying the matrix $\mathbf{L}$, we will use the term *output kernel* to denote both of them, with a slight abuse of terminology.

Even after imposing such a simplified model, specifying the inter-task similarities in advance is typically impractical. Indeed, it is often the case that multiple learning tasks are known to be related, but no precise information about the structure or the intensity of such relationships is available. Simply fixing $\mathbf{L}$ to the identity, which amounts to sharing no information between the tasks, is clearly suboptimal in most of the cases. On the other hand, wrongly specifying the entries may lead to a severe performance degradation. It is therefore clear that, whenever the task relationships are subject to uncertainty, learning them from the data is the only meaningful way to proceed.

16.1.1 Multiple Kernel Learning

The most studied approach to automatic kernel selection, known as multiple kernel learning (MKL), consists of learning a conic combination of N basis kernels of the form

$$K = \sum_{k=1}^{N} d_k K_k, \qquad d_k \geq 0, \qquad k = 1, \ldots, N.$$

Appealing properties of MKL methods include the ability to perform selection of a subset of kernels via sparsity, and tractability of the associated optimization problem, typically (re)formulated as a convex program. Although most of the works on MKL focus on learning similarity measured between inputs, clearly the approach can also be used to learn a multi-task kernel of the form

$$K((x_1, i), (x_2, j)) = \sum_{k=1}^{N} d_k K_X^k(x_1, x_2) K_Y^k(i, j),$$

which includes the possibility of optimizing the matrix $\mathbf{L}$ in (16.1) as a conic combination of basis matrices, by simply choosing the input kernels K_X^k to be equal. In principle, proper complexity control allows us to combine an arbitrarily large, even infinite [1], number of kernels. However, computational and memory constraints force the user to specify a relatively small dictionary of basis kernels to be combined, which again calls for a certain amount of domain knowledge. Examples of works that employ an MKL approach to address multi-output or multi-task learning problems include [17, 11, 16].

16.1.2   Output Kernel Learning

A more direct approach to learning inter-task similarities from the data consists in searching the output kernel K_Y over the whole cone of positive semidefinite kernels, by optimizing a suitable objective functional. Equivalently, the corresponding matrix $\mathbf{L}$ can be searched over the cone of positive semidefinite matrices.

This can be accomplished by solving a two-level regularization problem of the form

$$\min_{\mathbf{L} \in \mathbb{S}_+} \min_{f \in \mathcal{H}_\mathbf{L}} \left(\sum_{j=1}^{m} \sum_{i=1}^{\ell_j} V(y_{ij}, f_j(x_{ij})) + \lambda \left(\|f\|_{\mathcal{H}_\mathbf{L}}^2 + \Omega(\mathbf{L}) \right) \right), \quad (16.2)$$

where (x_{ij}, y_{ij}) are input-output data pairs for the j-th task, V is a suitable loss function, $\mathcal{H}_\mathbf{L}$ is the RKHS of vector-valued functions associated with the reproducing kernel (16.1), Ω is a suitable matrix regularizer, and $\mathbb{S}_+$ is the cone of symmetric and positive semidefinite matrices. The regularization parameter $\lambda > 0$ should be properly selected in order to achieve a good trade-off between approximation of the training data and regularization. This can be achieved by hold-out validation, cross-validation, or other methods. We call such an approach output kernel learning (OKL). By virtue of a suitable representer theorem [13], the inner regularization problem in (16.2) can be shown to admit solutions of the form

$$\hat{f}_k(x) = \sum_{j=1}^{m} \mathbf{L}_{kj} \left(\sum_{i=1}^{\ell_j} c_{ij} K_X(x_{ij}, x) \right), \quad (16.3)$$

under mild hypothesis on V.

From the expression (16.3), we can clearly see that when $\mathbf{L}$ equals the identity, the external sum decouples and each optimal function $\hat{f}_k$ only depends on the corresponding dataset (*independent single task-learning*). On the other hand, when all the entries of the matrix $\mathbf{L}$ are equal, all the functions $\hat{f}_k$ are the same (*pooled single task-learning*). Finally, whenever $\mathbf{L}$ differs from the identity, the datasets from multiple tasks get mixed together and contribute to the estimates of other tasks.

16.1.2.1   Frobenius Norm Output Kernel Learning

A first OKL technique was introduced in [4] for the case where V is a square loss function, Ω is the squared Frobenius norm, and the input data x_{ij} is the same for all the output components f_j, leading to a problem of the form

$$\min_{\mathbf{L} \in \mathbb{S}_+} \min_{f \in \mathcal{H}_\mathbf{L}} \left(\sum_{i=1}^{\ell} (y_i - f_j(x_i))^2 + \lambda \left(\|f\|_{\mathcal{H}_\mathbf{L}}^2 + \|\mathbf{L}\|_F^2 \right) \right). \quad (16.4)$$

Such special structure of the objective functional allows us to develop an effective block coordinate descent strategy where each step involves the solution of a Sylvester linear matrix equation. A simple and effective computational scheme to solve (16.4) is described in [4]. Regularizing with the squared Frobenius norm ensures that the sub-problem with respect to $\mathbf{L}$ is well-posed. However, one may want to encourage different types of structures for the output kernel matrix, depending on the application.

16.1.2.2 Low-Rank Output Kernel Learning

When the output kernel is low-rank, the estimated vector-valued function maps into a low-dimensional subspace. Encouraging such low-rank structure is of interest in several problems. Along this line, [3, 2] introduce low-rank OKL, a method to discover relevant low dimensional subspaces of the output space by learning a low-rank kernel matrix. This method corresponds to regularizing the output kernel with a combination of the trace and a rank indicator function, namely

$$\Omega(\mathbf{L}) = \mathrm{tr}(\mathbf{L}) + I(\mathrm{rank}(\mathbf{L}) \leq p).$$

For $p = m$, the hard-rank constraint disappears and Ω reduces to the trace, which still encourages low-rank solutions. Setting $p < m$ gives up convexity of the regularizer but, on the other hand, allows to set a hard bound on the rank of the output kernel, which can be useful for both computational and interpretative reasons. The optimization problem associated with low-rank OKL is the following:

$$\min_{\mathbf{L} \in \mathbb{S}_+} \min_{f \in \mathcal{H}_{\mathbf{L}}} \left(\sum_{j=1}^{m} \sum_{i=1}^{\ell_j} (y_{ij} - f_j(x_{ij}))^2 + \lambda \left(\|f\|_{\mathcal{H}_{\mathbf{L}}}^2 + \mathrm{tr}(\mathbf{L}) \right) \right), \text{ s.t. } \mathrm{rank}(\mathbf{L}) \leq p.$$

(16.5)

The optimal output kernel matrix can be factorized as $\mathbf{L} = \mathbf{B}\mathbf{B}^T$, where the horizontal dimension of $\mathbf{B}$ is equal to the rank parameter p. Problem (16.5) exhibits several interesting properties and interpretations. Just as sparse MKL with a square loss can be seen as a nonlinear generalization of (grouped) Lasso, low-rank OKL is a natural kernel-based generalization of reduced-rank regression, a popular multivariate technique in statistics [9]. When $p = m$ and the input kernel is linear, low-rank OKL reduces to multiple least squares regression with nuclear norm regularization. Connections with reduced-rank regression and nuclear norm regularization are analyzed in [3].

For problems where the inputs x_{ij} are the same for all the tasks, optimization for low-rank OKL can be performed by means of a rather effective procedure that iteratively computes eigendecompositions; see Algorithm 1 in [3]. Importantly, the size of the involved matrices such as $\mathbf{B}$, the low rank factor of $\mathbf{L}$, can be controlled by selecting the parameter p. However, more general multi-task learning problems where each task is sampled in correspondence with different inputs require completely different methods. It turns out that an effective strategy to approach the problem consists of iteratively applying inexact preconditioned conjugate gradient (PCG) solvers to suitable linear operator equations that arise from the optimality conditions. Such linear operator equations are derived and analyzed in [2].

16.1.2.3 Sparse Output Kernel Learning

In many multitask learning problems it is known that some of the tasks might be related while some others are independent, but it is unknown in advance which of the tasks are related. In such cases, it may make sense to try and encourage sparsity in the output kernel by means of suitable regularization. For instance, by choosing an entry-wise ℓ_1 norm regularization $\Omega(\mathbf{L}) = \|\mathbf{L}\|_1$, one obtains the problem

$$\min_{\mathbf{L} \in \mathbb{S}_+} \min_{f \in \mathcal{H}_L} \left(\sum_{j=1}^{m} \sum_{i=1}^{\ell_j} (y_{ij} - f_j(x_{ij}))^2 + \lambda \left(\|f\|^2_{\mathcal{H}_L} + \|\mathbf{L}\|_1 \right) \right).$$

Encouraging a sparse output kernel may allow us to automatically discover clusters of related tasks. However, some of the tasks may already be known in advance to be unrelated. Such information can be encoded by also enforcing a hard constraint on the entries of the output kernel, for instance by means of the regularizer $\Omega(\mathbf{L}) = \|\mathbf{L}\|_1 + I(P_S(\mathbf{L}) = 0)$, where I is a indicator function, and P_S selects a subset S of the non-diagonal entries of $\mathbf{L}$ and projects them into a vector, yielding the additional constraint

$$\mathbf{L}_{ij} = 0, \qquad \forall(i,j) \in S.$$

The subproblem with respect to $\mathbf{L}$ is a convex nondifferentiable problem, also when hard sparsity constraints are present. Effective solvers for sparse output kernel learning problems are currently under investigation.

16.2 Applications

Multi-task learning problems where it is important to estimate the relationships between tasks are ubiquitous. In this section, we provide examples of such problems where OKL techniques have been applied successfully.

16.2.1 Collaborative Filtering and Preference Estimation

Estimating preferences of several users for a set of items is a typical instance of multi-task learning problem where each task is the preference function of one of the users, and exploiting similarities between the tasks matters. Preference estimation is a key problem addressed by collaborative filtering systems and recommender systems, which find wide applicability on the Web.

In the context of collaborative filtering, techniques such as low-rank matrix approximation are considered state of the art. In the following, we present some results from a study based on the MovieLens datasets (see Table 16.1), three popular collaborative filtering benchmarks containing collections of ratings in the range $\{1, \ldots, 5\}$ assigned by several users to a set of movies; for more details, see [2]. The study shows that, by exploiting additional information about the inputs (movies), OKL techniques are superior to plain low-rank matrix approximation.

TABLE 16.1: MovieLens datasets: total number of users, movies, and ratings.

Dataset	Users	Movies	Ratings
MovieLens100K	943	1682	10^5
MovieLens1M	6040	3706	10^6
MovieLens10M	69878	10677	10^7

The results reported in Table 16.2 correspond to a setup where a random test set is extracted, containing about the 50% of the ratings for each user, see also [15, 10]. Results under different test settings are also available, see [2]. The 25% of the remaining training data are used as a validation set to tune the regularization parameter. Performance is evaluated according to the root mean squared error (RMSE) on the test set. Regularized matrix factorization (RMF) corresponds to choosing the input kernel equal to $K_X(x_1, x_2) = \delta_K(x_1, x_2)$, where δ_K denotes the Kronecker delta (non-zero only when the two arguments are equal), so that no information other than the movie Id is exploited to express the similarity between the movies. The pooled and independent baselines correspond to choosing $\mathbf{L}_{ij} = 1$ and $\mathbf{L}_{ij} = \delta_K(i,j)$, respectively. The last method employed is low-rank OKL with rank parameter $p = 5$ fixed a priori for all three datasets, and input kernel designed as

$$K(x_1, x_2) = \delta_K(x_1^{id}, x_2^{id}) + \exp\left(-d_H(x_1^g, x_2^g)\right),$$

by taking into account movie Ids x_1^{id}, x_2^{id} and meta-data about genre categorization of the movies x_1^g, x_2^g available in all three datasets.

16.2.2 Structure Discovery in Multiclass Classification

Multi-class classification problems can also be seen as particular instances of multi-task learning where each real-valued task function f_j corresponds to a score for a given class. The training labels can be converted into sparse real vectors of length equal to the number of classes, with only one component different from zero. Employing an OKL method in this context allows not only training a multi-class classifier, but also learning the similarities between the classes.

TABLE 16.2: MovieLens datasets: test RMSE for low-rank OKL, RMF, pooled and independent single-task learning.

Dataset	RMF	Pooled	Independent	OKL
MovieLens100K	1.0300	1.0209	1.0445	**0.9557**
MovieLens1M	0.9023	0.9811	1.0297	**0.8945**
MovieLens10M	0.8627	0.9441	0.9721	**0.8501**

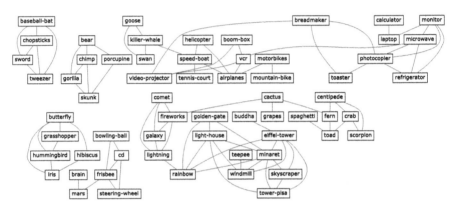

FIGURE 16.1: Caltech 256: learned similarities between classes. Only a subset of the classes is shown.

As an example, Figure 16.1 shows a visualization of the entries of output kernel matrix obtained by applying low-rank OKL to the popular Caltech 256 dataset [6, 7], containing images of several different categories of objects, including buildings, animals, tools, etc. By using 30 training examples for each class, the obtained classification accuracy on the test set (0.44) is close to state-of-the art results. At the same time, the graph obtained by thresholding the entries of the learned output kernel matrix with low absolute value, reveals clusters of classes that are meaningful and agree with common sense. Output kernel learning methods have also been applied in [8] to solve object recognition problems.

16.2.3   Pharmacological Problems

Multi-task learning problems are common in pharmacology, where data from multiple subjects are available. Due to the scarcity of data for each subject, it is often crucial to combine the information from different datasets in order to obtain a good estimation performance. Such combination needs to take into account the similarities between the subjects, while allowing for enough flexibility to estimate personalized models for each of them. Output

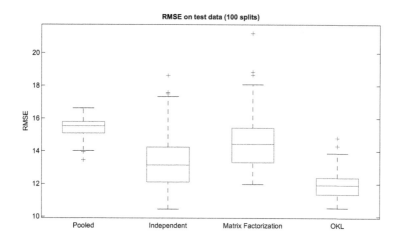

FIGURE 16.2: Experiment on pharmacokinetic data [2]. Root mean squared error averaged over 100 random splits for the 27 subject profiles in correspondence with different methods.

kernel learning methods have been successfully applied to pharmacological problems in [2], where two different problems are analyzed. Both problems can be seen as multi-task regression problems or matrix completion problems with side information.

The first problem consists of filling a matrix of drug concentration measurements for 27 subjects in correspondence with 8 different time instants after the drug administration, by having access to only 3 measurements per subject. Standard low-rank matrix completion techniques are not able to solve this problem satisfactorily, since they ignore the available knowledge about the temporal shape of the concentration curves. On the other hand, an OKL method allows us to easily incorporate such knowledge by designing a suitable input kernel that takes into account temporal correlation, as done in [2]. Figure 16.2 reports boxplots over the 27 subjects of the root mean squared error, averaged over 100 random selections of the three training measurements, showing a clear advantage of the OKL methodology with respect to both pooled and independent baselines, as well as a low-rank matrix completion technique that does not use side information.

The second problem analyzed in [2] has to do with completing a matrix of hamilton depression rating scale (HAMD) scores for 494 subjects in correspondence with 7 subsequent weeks, for which only a subset of 2855 entries is available [12]. Performance is evaluated by keeping 1012 properly selected entries for test purposes. In order to automatically select the regularization parameter λ, a further splitting of the remaining data is performed to obtain a validation set containing about 30% of the examples. Such splitting is performed randomly and repeated 50 times. By employing a low-rank OKL approach with a simple linear spline input kernel, one can observe significantly better results (Table 16.3) with respect to low-rank matrix completion and standard baselines; see [2] for further details.

TABLE 16.3: Drug efficacy assessment experiment [2]: best average RMSE on test data (and their standard deviation over 50 splits)

Pooled	Independent	RMF	OKL
6.86 (0.02)	6.72(0.16)	6.66(0.4)	**5.37**(0.2)

16.3 Concluding Remarks and Future Directions

Learning output kernels via regularization is an effective way to solve multi-task learning problems where the relationships between the tasks are uncertain or unknown. The OKL framework that we have discussed in this chapter is rather general and can be further developed in various directions. There are several practically meaningful constraints that could be imposed on the output kernel: sparsity patterns, hierarchies, groupings, etc. Effective optimization techniques for more general (non-quadratic) loss functions are still lacking and the use of a variety of matrix penalties for the output kernel matrix is yet to be explored. Extensions to semi-supervised and online problems are needed in order to broaden applicability of these techniques. Finally, some hybrid methods that combine learning of, possibly multiple, input and output kernels have been recently investigated [14] and are currently still under active investigation.

Bibliography

[1] A. Argyriou, C. A. Micchelli, and M. Pontil. Learning convex combinations of continuously parameterized basic kernels. In Peter Auer and Ron Meir, editors, *Learning Theory*, volume 3559 of *Lecture Notes in Computer Science*, pages 338–352. Springer Berlin/Heidelberg, 2005.

[2] F. Dinuzzo. Learning output kernels for multi-task problems. *Neurocomputing*, 118:119–126, 2013.

[3] F. Dinuzzo and K. Fukumizu. Learning low-rank output kernels. *Journal of Machine Learning Research - Proceedings Track*, 20:181–196, 2011.

[4] F. Dinuzzo, C. S. Ong, P. Gehler, and G. Pillonetto. Learning output kernels with block coordinate descent. In *Proceedings of the 28th Annual International Conference on Machine Learning*, Bellevue, WA, USA, 2011.

[5] T. Evgeniou, C. A. Micchelli, and M. Pontil. Learning multiple tasks with kernel methods. *Journal of Machine Learning Research*, 6:615–637, 2005.

[6] L. Fei-Fei, R. Fergus, and P. Perona. Learning generative visual models from few training examples: An incremental Bayesian approach tested on 101 object categories. In *CVPR*, page 178, 2004.

[7] G. Griffin, A. Holub, and P. Perona. Caltech-256 object category dataset. Technical Report 7694, Caltech, 2007.

[8] Z. Guo and Z. J. Wang. Cross-domain object recognition by output kernel learning. In *Multimedia Signal Processing (MMSP), 2012 IEEE 14th International Workshop on*, pages 372–377. IEEE, 2012.

[9] A. J. Izenman. Reduced-rank regression for the multivariate linear model. *Journal of Multivariate Analysis*, 5(2):248–264, 1975.

[10] M. Jaggi and M. Sulovský. A simple algorithm for nuclear norm regularized problems. In J. Fürnkranz and T. Joachims, editors, *Proceedings of the 27th International Conference on Machine Learning (ICML-10)*, pages 471–478, Haifa, Israel, June 2010. Omnipress.

[11] H. Kadri, A. Rakotomamonjy, F. Bach, and P. Preux. Multiple operator-valued kernel learning. In *NIPS*, pages 2438–2446, 2012.

[12] E. Merlo-Pich and R. Gomeni. Model-based approach and signal detection theory to evaluate the performance of recruitment centers in clinical trials with antidepressant drugs. *Clinical Pharmacology and Therapeutics*, 84:378–384, September 2008.

[13] C. A. Micchelli and M. Pontil. On learning vector-valued functions. *Neural Computation*, 17:177–204, 2005.

[14] V. Sindhwani, H.Q. Minh, and A. C. Lozano. Scalable matrix-valued kernel learning for high-dimensional nonlinear multivariate regression and granger causality. In *UAI*, 2013.

[15] K. Toh and S. Yun. An accelerated proximal gradient algorithm for nuclear norm regularized least squares problems. *Optimization Online*, 2009.

[16] C. Widmer, N. C. Toussaint, Y. Altun, and G. Rätsch. Inferring latent task structure for multitask learning by multiple kernel learning. *BMC Bioinformatics*, 11(Suppl 8):S5, 2010.

[17] A. Zien and C. S. Ong. Multiclass multiple kernel learning. In *Proceedings of the 24th International Conference on Machine Learning*, pages 1191–1198, 2007.

Chapter 17

Kernel-Based Identification of Systems with Multiple Outputs Using Nuclear Norm Regularization

Tillmann Falck

KU Leuven, ESAT-STADIUS

Bart De Moor

KU Leuven, ESAT-STADIUS and IMinds Medical IT

Johan A.K. Suykens

KU Leuven, ESAT-STADIUS

17.1   Introduction

System identification is a field that deals with modeling physical systems in terms of their input-output relationship. The goal is to estimate a model for the system from measurements of past inputs and outputs. Linear system identification [18] is a well developed field that is able to handle diverse system structures in efficient ways. In many modeling tasks it is straightforward to incorporate information from multiple input sources; however it is usually much more difficult to exploit the presence of multiple outputs. In an engineering context the notion of state-space descriptions popularized by Kalman [14] provides an intuitive method to characterize systems with multiple inputs and outputs. In a linear setting these can also be readily estimated from data, using, e.g., subspace identification techniques [29].

The class of nonlinear systems is much broader than that of linear systems. Hence, for the estimation of nonlinear systems a lot of system structures are still an active domain for research. Most established techniques for nonlinear system identification [26] formulate a nonlinear regression problem that is then solved by a variety of techniques, like wavelets, neural networks, or more recently, support vector machines and kernel-based methods. A common limitation for nonlinear regression techniques is that it is difficult to handle multiple outputs, and this is particularly true for most kernel-based estimation schemes. A frequently applied workaround is to estimate independent models for each individual output. This chapter introduces an advanced estimation scheme that is able to exploit dependencies between output variables. The technique is based on least-squares support vector machines [28] and a convex regularization term based on the nuclear norm [11].

Regularization schemes based on the nuclear norm have been successfully applied in different domains, such as multi-task learning [2], matrix completion [5, 21], tensor completion [25], and system identification [17, 16].

Kernel-based methods are a very popular choice for nonlinear regression due to their excellent modeling power. Support vector machines and in particular least-squares support vector machines are an attractive choice for several reasons. Firstly their foundation on convex optimization provides efficient means for their numerical solutions. Also the serious problem of local minima as occurring in many competing techniques is less pronounced as the main problems are convex and as such the global optimum can be attained. Secondly, in line with other kernel-based estimation techniques, support vector techniques are not as prone to the curse of dimensionality as other nonlinear estimation techniques. Thirdly, the optimization problem formulation makes it straightforward to introduce additional structure on the model and thereby incorporate prior information [27].

This chapter has two key contributions. On the one hand a new kernel-based model for nonlinear systems with multiple related outputs is proposed.

On the other hand a new primal-dual derivation of a kernel-based model with nuclear norm regularization is presented.

The structure is as follows. The first section motivates the problem setting and introduces the necessary background on the employed techniques. The section concludes with stating the model formulation for a nonlinear system with multiple related outputs. The next section characterizes some properties of the proposed model, in particular, uniqueness and the range of suitable regularization parameters. Section 17.4 derives the dual model. In a first step a kernel-based optimization problem is found and in a second step its solution is used to obtain a predictive equation. After establishing the primal as well as the dual formulation of the proposed model, some of its properties are illustrated using a numerical example. Finally the conclusions are given in the last section.

17.1.1   Notation

All vectors are column vectors and written as bold lowercase letters. Matrices are also bold but denoted by capital letters. Transposition is marked with a superscript T. Subscripts are used to denote elements of a set or components of a vector. It should be obvious from the context which one is applicable.

Furthermore a vector of all ones in dimension N is denoted by $\mathbf{1}_N$ and similarly a vector of all zeros by $\mathbf{0}_N$. The identity matrix of size N is written as $\boldsymbol{I}_N$.

The eigenvalues of a square matrix $\boldsymbol{X}$ are denoted by $0 \leq \lambda_1(\boldsymbol{X}) \leq \cdots \leq \lambda_{\max}(\boldsymbol{X})$. In the same way singular values of matrix $\boldsymbol{X}$ are denoted by $0 \leq \sigma_1(\boldsymbol{X}) \leq \cdots \leq \sigma_{\max}(\boldsymbol{X})$. Several matrix norms are used, the Frobenius norm $\|\boldsymbol{X}\|_F = \sqrt{\sum_{n=1}^{N} \sum_{m=1}^{M} X_{nm}^2} = \sqrt{\sum_{i=1}^{\min\{N,M\}} \sigma_i(\boldsymbol{X})^2}$ for $\boldsymbol{X} \in \mathbb{R}^{N \times M}$ and the matrix two-norm or operator or spectral norm $\|\boldsymbol{X}\|_2 = \sigma_{\max}(\boldsymbol{X})$. The notation $\|\boldsymbol{X}\|_*$ refers to the nuclear or trace norm and is defined in Subsection 17.2.5.

17.2   Problem Description and Motivation

Most real-world phenomena are nonlinear. Also, most real-world problems have more than one quantity of interest. These two statements, however, are not yet sufficient to motivate the central assumption exploited in this chapter. This assumption is that in many cases one has target variables that are somehow related.

Consider a power distribution network as depicted in Figure 17.1. Several consumers are connected to the electricity grid via power stations. An important task in the control of power grids to ensure their stability is the ability to

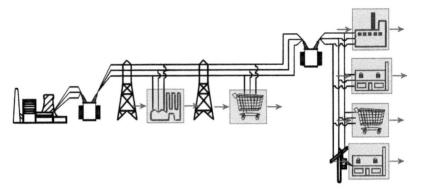

FIGURE 17.1: Power distribution network with different consumers connected to it. Shown are clusters of industrial customers, residential areas, and sections dominated by offices. (Illustration is based on public domain image by the United States Department of Energy.)

precisely forecast the load of the network [6]. One could approach this by using an independent model for each substation. This however does not need to be optimal. As shown in [1] on the example of the Belgian power distribution network there are only a small number of distinctive profiles. Some of these can be easily understood. There will be a behavior that will correspond to residential consumers, with peaks in the morning and the evening. Then there are power stations dominated by businesses that operate on a more or less 9-to-5 schedule. A distinctive third group are large industry consumers working in shifts 24/7. Within a fixed category of similar customers the behavior will be similar as it is driven by similar people in a similar environment. As the example is limited to Belgium, the weather conditions, holiday seasons, and habits will likely by similar all around. Therefore it is assumed that understanding the load at one power station will already reveal a lot of information relevant to power stations with similar customer profiles. The loads at different substations are related.

As a second motivational example consider sensors of an industrial process. There likely are large arrays of different sensors, such as temperature, pressure or flow rates, throughout the production environment. The process itself is however most likely governed by a complex but connected set of equations. Therefore all the measurements obtained from various parts of the process are related through the process itself.

17.2.1   Formal Problem Description

The problem addressed in this chapter is to estimate a mathematical model for the system depicted in Figure 17.2. Given input-output data

FIGURE 17.2: Nonlinear system with single input and multiple outputs.

FIGURE 17.3: Nonlinear system with single input and multiple outputs split into individual systems for every output.

$\mathcal{D} = \{(u_t, \boldsymbol{y}_t)\}_{t=1}^{N}$ with $\boldsymbol{y}_t \in \mathbb{R}^M$, estimate a model of the form

$$\hat{\boldsymbol{y}}_t = \hat{\boldsymbol{f}}(\boldsymbol{x}_t) \tag{17.1}$$

where $\boldsymbol{x}_t \in \mathbb{R}^D$ and $\hat{\boldsymbol{f}} : \mathbb{R}^D \to \mathbb{R}^M$. In nonlinear system identification the vector $\boldsymbol{x}_t$ is usually made up of past inputs and outputs to capture the memory of the system, i.e., $\boldsymbol{x}_t = [\boldsymbol{y}_{t-1}^T, \ldots, \boldsymbol{y}_{t-P}^T, u_t, \ldots, u_{t-Q}]^T$ where P and Q are the number of considered lags for past outputs and past inputs, respectively. To allow for a compact notation, a new data set $\mathcal{D}'$ can be defined in terms of $\boldsymbol{x}_t$ and $\boldsymbol{y}_t$ as $\mathcal{D}' = \{(\boldsymbol{x}_t, \boldsymbol{y}_t)\}_{t=R}^{N}$ where $R = \max(P, Q) + 1$. The size of this set is $N' = N - R + 1$.

17.2.2   Traditional vs. Proposed Handling of Multiple Outputs

The most common approach for dealing with multiple outputs is to split the system apart into individual systems with a single output each, as depicted in Figure 17.3. This allows to reuse all methods that are available for systems with a single output. For a system with an output vector given by $\boldsymbol{y}_t = [y_t^{(1)}, \ldots, y_t^{(M)}]^T$ this would, for example, correspond to M estimation problems of the form

$$\min_{f_m} \quad \eta\|f_m\| + \sum_{t=R}^{N} l(y_t^{(m)}, f_m(\boldsymbol{x}_t)) \tag{17.2}$$

where $m = 1, \ldots, M$, and $\eta > 0$ is a regularization parameter. The choice of the norm on f_m and the loss function l depend on the estimation technique and will be defined later. Besides being straightforward, this approach has the advantage that it offers the most flexibility for modeling an individual output variable. In this setting it is simple to employ different function bases for different outputs or to carry out the estimation using different loss functions. This might, for example, be advantageous in case the noise is very different across the outputs.

However, this approach also clearly has some drawbacks. As the estimation problems are completely decoupled, the information contained in one target variable cannot be used to improve the estimation of another. Therefore this chapter advocates leaving the system in the form shown in Figure 17.2. The corresponding estimation problem

$$\min_{\boldsymbol{f}} \quad \eta\|\boldsymbol{f}\| + \sum_{t=R}^{N} l(\boldsymbol{y}_t, \boldsymbol{f}(\boldsymbol{x}_t)) \tag{17.3}$$

is more complex, but at least in theory allows the use of all available information. For kernel-based models in a functional setting, models of this form have been described by [3]. In the following an alternative formulation in primal-dual framework will be derived.

17.2.3 Least Squares Support Vector Machines — A Primal-Dual Kernel-Based Model

A particular kind of kernel-based models are least-squares support vector machines (LS-SVMs). Like most nonlinear regression techniques, they address problems like those formulated in (17.2). Their name stems from the choice of the loss function l, which is the least squares loss $l(x, y) = (x - y)^2$. The regularization term is as in all support vector machines the squared ℓ_2-norm of the function f_m. In support vector techniques the model f is usually formulated in the so-called primal description as

$$\hat{y} = \hat{f}(\boldsymbol{x}_t) = \boldsymbol{w}^T \varphi(\boldsymbol{x}_t) + b. \tag{17.4}$$

Here $\boldsymbol{w} \in \mathbb{R}^{n_h}$ and $b \in \mathbb{R}$ are the model parameters, while $\varphi : \mathbb{R}^D \to \mathbb{R}^{n_h}$ is the feature map. The feature space is often high dimensional. In kernel-based techniques the feature map φ is most often not defined explicitly, but rather implicitly via a positive definite kernel function K. Based on Mercer's condition [19, 23] every positive definite function $K(\boldsymbol{x}, \boldsymbol{y})$ can be expressed as an inner product $K(\boldsymbol{x}, \boldsymbol{y}) = \varphi(\boldsymbol{x})^T \varphi(\boldsymbol{y})$. Therefore, by choosing a positive definite kernel K and reformulating the problem in terms of inner products of φ one can implicitly work in very high dimensional spaces without having to formulate an explicit parametrization of the basis. A popular choice for the kernel function is the Gaussian RBF kernel $K_{\mathrm{RBF}}(\boldsymbol{x}, \boldsymbol{y}) = \exp(-\|\boldsymbol{x} - \boldsymbol{y}\|_2^2/\sigma^2)$ with positive bandwidth $\sigma > 0$. The RBF kernel is associated with an infinite dimensional feature map.

The primal estimation problem for LS-SVM regression is

$$\min_{\boldsymbol{w}, b, e_t} \quad \frac{1}{2}\boldsymbol{w}^T\boldsymbol{w} + \frac{1}{2}\gamma\sum_{t=R}^{N} e_t^2 \tag{17.5}$$

$$\text{subject to} \quad y_t = \boldsymbol{w}^T\varphi(\boldsymbol{x}_t) + b + e_t, \quad t = R, \dots, N.$$

In the case of high dimensional or implicitly defined feature spaces it is not practical to work with this primal formulation. The kernel-based Lagrange dual formulation represents the solution in terms of the kernel function and changes the number of optimization variables from $n_h + 1$ to $N' + 1$. The derivation of the kernel-based model follows a scheme with five steps. As the proposed model formulation will require substantial effort to carry out some of these steps, the regular derivation will be briefly reproduced here for later reference.

Lemma 17.1 (Dual LS-SVM model [28]). *The dual model for (17.4) is given by*

$$\hat{y} = \hat{f}(z) = \sum_{t=R}^{N} \alpha_t K(x_t, z) + b. \tag{17.6}$$

The dual variables $\alpha = [\alpha_R, \ldots, \alpha_N]^T \in \mathbb{R}^{N'}$ and b are given as the solution of the linear system

$$\begin{bmatrix} \Omega + \gamma^{-1} I_{N'} & 1_{N'} \\ 1_{N'}^T & 0 \end{bmatrix} \begin{bmatrix} \alpha \\ b \end{bmatrix} = \begin{bmatrix} y \\ 0 \end{bmatrix}. \tag{17.7}$$

The kernel matrix $\Omega \in \mathbb{R}^{N' \times N'}$ is defined element-wise as $\Omega_{ij} = K(x_{R+i-1}, x_{R+j-1}) = \varphi(x_{R+i-1})^T \varphi(x_{R+j-1})$ for $i, j = 1, \ldots, N'$. The vector $y = [y_R, \ldots, y_N]^T \in \mathbb{R}^{N'}$ stacks all target values.

Proof.

1. Writing down the Lagrangian of the primal problem

$$\mathcal{L}(w, b, e_t, \alpha_t) = \frac{1}{2} w^T w + \frac{1}{2} \gamma \sum_{t=R}^{N} e_t^2 - \sum_{t=P}^{N} \alpha_t (w^T \varphi(x_t) + b + e_t - y_t). \tag{17.8}$$

2. Taking the derivatives of the Lagrangian with respect to the primal and dual variables, and

3. formulating the Karush-Kuhn-Tucker (KKT) conditions for optimality.

$$\frac{\partial \mathcal{L}}{\partial w} = 0_{n_h} : \quad w = \sum_{t=R}^{N} \alpha_t \varphi(x_t), \tag{17.9a}$$

$$\frac{\partial \mathcal{L}}{\partial b} = 0 : \quad \sum_{t=R}^{N} \alpha_t = 0, \tag{17.9b}$$

$$\frac{\partial \mathcal{L}}{\partial e_t} = 0 : \quad e_t = \gamma^{-1} \alpha_t, \quad t = R, \ldots, N, \tag{17.9c}$$

$$\frac{\partial \mathcal{L}}{\partial \alpha_t} = 0 : \quad y_t = w^T \varphi(x_t) + b + e_t, \quad t = R, \ldots, N. \tag{17.9d}$$

4. Writing down the dual optimization problem.

   The first line of the dual optimization problem (17.7) follows directly from substituting (17.9a) and (17.9c) into (17.9d) and applying the kernel trick, i.e. $\varphi(x)^T \varphi(y) = K(x, y)$. The second line is a reformulation of (17.9b) in vector notation.

5. Substitution of the dual solution into the primal model to obtain the dual model.

   Substituting (17.9a) into (17.4) and applying the kernel trick once more yields the dual model (17.6).

$\square$

17.2.4   Advantages of Primal-Dual Approach

LS-SVMs have several advantages. For instance, the basic core models almost solely rely on linear algebra and as such the mathematical tools are well understood by most. Also the solution is straightforward in any number of computing languages and can be achieved in just a handful of lines of code necessary to set up a linear system.

An advantage on a more fundamental modeling level, is the primal-dual formulation. The primal problem description (17.4) is parametric in nature and allows direct interpretation. Even more importantly, it allows us to formulate prior information on the modeled system in a straightforward fashion. This additional information can then be incorporated into the estimation problem as additional constraints. The strength of this approach is that it is on the one hand simple and on the other hand it is often possible to attach a straightforward physical interpretation to the introduced constraints. When deriving the dual model formulation, the information incorporated via constraints is embedded into the model itself. In some cases all information can be captured in an equivalent kernel function while in others the dual model obtains a special structure. For large scale problems, the primal-dual approach also offers a powerful approach to compute approximate solutions. Using a subsample of the full data set, one can estimate a finite dimensional approximation of the feature map. This approximation can then be utilized in the primal to solve the parametric problem using all available data. This is done in fixed-size kernel methods for dealing with large scale data [28].

Examples for prior information that can be incorporated into the prior model are symmetry $w^T \varphi(x) = \pm w^T \varphi(-x)$ or boundedness $w^T \varphi(x) \geq y_0$. A more thorough discussion on the primal-dual nature of LS-SVM-based models is given in [27, 7]. It is also not restricted to regression problems but discusses other applications of the support vector formalism as well. A more complex example from system identification is estimation of structured systems, like Hammerstein and Wiener-Hammerstein systems [12, 9]. In these one exploits a partitioning of the system in linear dynamical parts and a concentrated static nonlinearity.

17.2.5 Nuclear Norm

The trade-off between model fit and model complexity is governed by the regularization term, in the case of SVMs $w^T w$. Therefore by choosing another regularization term one can potentially introduce new and additional structure into the estimation problem. In convex optimization in general and especially in compressed sensing many new regularization schemes and their advantages have been studied. A particular one of those is the nuclear norm, a regularization term for matrices.

The nuclear norm or trace norm is the convex envelope of the rank function. As such it can be used as a surrogate to obtain convex approximations of otherwise NP-hard problems. For a rectangular matrix $X \in \mathbb{R}^{N \times M}$ it is defined as

$$\|X\|_* = \sum_{n=1}^{\min(M,N)} \sigma_n(X) \qquad (17.10)$$

where $\sigma_n(X)$ is the n-th singular value of the matrix X. The interpretation as convex envelope of the rank function holds for all $\|X\|_2 \leq 1$. Relating the nuclear norm to vector norms, it can be seen as a generalization of the ℓ_1-norm. In fact it is computed as the ℓ_1-norm of the singular values of a matrix. In similarity to its vector-cousin, it promotes sparsity. Whereas the ℓ_1-norm favors solutions with sparse vectors, the nuclear norm promotes solutions having a low rank. Conditions for sparse solutions are studied in [22].

To solve nuclear norm problems in polynomial time using general purpose convex optimization solvers the problem can be formulated as a semidefinite programming (SDP) problem. $\|X\|_*$ can be computed as [10, 13].

$$\max_{X, U = U^T, V = V^T} \frac{1}{2}\operatorname{tr}(U) + \frac{1}{2}\operatorname{tr}(V) \qquad (17.11)$$

$$\text{subject to} \quad \begin{bmatrix} U & X \\ X^T & V \end{bmatrix} \succeq 0.$$

The number of variables in this SDP embedding is large. Therefore only small scale problems can be solved using general purpose solvers. A brief overview of ongoing research on solvers is given in [22, Section 5]. Some examples are interior point solvers that exploit special problem structure [17] and first order methods based on subgradients [20, 15].

17.3   Primal Model for Multiple Outputs

17.3.1   Model Formulation

The LS-SVM model (17.5) can and has been used to identify systems with multiple outputs using the traditional decomposition formulation (17.2). Going to the joint estimation (17.3) is discussed next. The simplest extension from the regularization $\boldsymbol{w}^T\boldsymbol{w}$ for one model to the joint regularization is taking the sum over all models $\sum_{m=1}^{M} \boldsymbol{w}_m^T\boldsymbol{w}_m = \|\boldsymbol{W}\|_F^2$ where $\boldsymbol{W} = [\boldsymbol{w}_1, \ldots, \boldsymbol{w}_M]$ [28]. This approach has only limited appeal as it is purely of a cosmetic nature as the estimation problems are in essence still decoupled. An interesting property, expected at least in some multi-output systems, is that each output can be described as the combination of a small number of principal behaviors. In case of a linear model and linear combinations of principal models, this translates into $y_t^{(m)} = \sum_{l=1}^{L} \mu_{ml}\boldsymbol{w}_l^T\boldsymbol{x}_t$. Here $y_t^{(m)}$ denotes the m-th output of $\boldsymbol{y}_t$ and μ_{ml} the weighting coefficient for the contribution of the l-th principal model to the m-th output. This problem can be rewritten as $\boldsymbol{y}_t = \boldsymbol{W}^T\boldsymbol{x}_t$ subject to $\text{rank}(\boldsymbol{W}) \leq L$.

This relation is non-convex, but in the previous section a convex relaxation to this problem was already introduced. Hence, an approximate representation of the prior belief that the joint model should be a combination of a small number of principal models is given by a nuclear norm regularization. In this brief motivation, all arguments were presented in terms of linear models and linear combinations. The strength of kernel-based models and the primal-dual approach is that this transparently maps to nonlinear models. The proposed model formulation for a model with multiple related outputs is then

$$\min_{\boldsymbol{W},b,e_t} \quad \eta\|\boldsymbol{W}\|_* + \frac{1}{2}\sum_{t=R}^{N} \boldsymbol{e}_t^T \boldsymbol{T} \boldsymbol{e}_t \tag{17.12}$$

$$\text{subject to} \quad \boldsymbol{y}_t = \boldsymbol{W}^T\boldsymbol{\varphi}(\boldsymbol{x}_t) + \boldsymbol{b} + \boldsymbol{e}_t, \quad t = R, \ldots, N.$$

This description contains a few technical modifications with respect to the basic LS-SVM model (17.5), besides the transition to vector valued outputs and a nuclear norm-based regularization term. In the following derivations it turns out to be easier to work with a norm instead of a squared norm. Also it is convenient to have the regularization constant η next to the regularization term. A slight generalization to allow some of the flexibility of a decoupled formulation is the introduction of the positive definite weighting matrix $\boldsymbol{T} \in \mathbb{R}^{M \times M}$. It can be used to model correlation among the residuals of the different outputs and provides the means to have different regularization constants acting on each output. Note that the problem is still convex and is only subject to equality constraints. Therefore Slater's condition is trivially satisfied [4, pages 226–227] and as such strong duality holds for the problem. Thus the duality gap is zero.

17.3.2   Characterization of Proposed Multiple Output Model

In contrast to LS-SVMs and many other kernel-based estimation techniques, the regularization term in the case of (17.12) is not quadratic. This has several implications. The desired effect is that solutions with linearly dependent model parameters in the primal domain $\boldsymbol{w}_m$ are favored. Furthermore, the primal is defined in a typically high-dimensional feature space, therefore even a linear dependence of the parameters can represent complex interactions. On the technical side a property proved easily for a quadratic penalty term, uniqueness of the solution, is much less evident for the proposed scheme. Due to the similarity of the nuclear norm to ℓ_1-regularization in case of vectors, critical values for the regularization parameters exist from which the solution will remain constant. This section as well as the following two are based on [8, Chapter 6].

The following lemma establishes uniqueness of the solution.

Lemma 17.2. *For $\eta > 0$ the solution of (17.12) is unique in $\boldsymbol{W}$, $\boldsymbol{b}$ and $\boldsymbol{e}_t$.*

Proof. Considering the residuals $\boldsymbol{e}_t = \boldsymbol{y}_t - \boldsymbol{W}^T \boldsymbol{\varphi}(\boldsymbol{x}_t) - \boldsymbol{b}$, a sufficient condition for their uniqueness is the uniqueness of $\boldsymbol{W}$ and $\boldsymbol{b}$. By eliminating the residuals $\boldsymbol{e}_t$ from (17.12), the problem can be written in unconstrained form as $\min_{\boldsymbol{W},\boldsymbol{b}} \mathcal{J}(\boldsymbol{W}, \boldsymbol{b})$ with

$$\mathcal{J}(\boldsymbol{W}, \boldsymbol{b}) = \frac{1}{2}\|(\boldsymbol{\Phi}^T \boldsymbol{W} + \boldsymbol{1}_N \boldsymbol{b}^T - \boldsymbol{Y}^T)\boldsymbol{T}^{\frac{1}{2}}\|_F^2 + \eta\|\boldsymbol{W}\|_*. \qquad (17.13)$$

Then the lemma holds if the solution to this problem is unique.

First the solution for $\boldsymbol{b}$ will be derived. Note that $\boldsymbol{b}$ is not subject to regularization. Therefore, from $\boldsymbol{0}_M = \partial \mathcal{J}(\boldsymbol{W}, \boldsymbol{b})/\partial \boldsymbol{b}$, it follows that $\boldsymbol{b} = \frac{1}{N}(\boldsymbol{Y} - \boldsymbol{W}^T \boldsymbol{\Phi})\boldsymbol{1}_N$. Substitution of this relation into $\mathcal{J}$ yields

$$\tilde{\mathcal{J}}(\boldsymbol{W}) = \frac{1}{2}\|(\tilde{\boldsymbol{\Phi}}^T \boldsymbol{W} - \tilde{\boldsymbol{Y}}^T)\boldsymbol{T}^{\frac{1}{2}}\|_F^2 + \eta\|\boldsymbol{W}\|_*.$$

Here, $\tilde{\boldsymbol{Y}}^T = \boldsymbol{P}_1^\perp \boldsymbol{Y}^T$ and $\tilde{\boldsymbol{\Phi}}^T = \boldsymbol{P}_1^\perp \boldsymbol{\Phi}^T$ with $\boldsymbol{P}_1^\perp = \boldsymbol{I}_N - \frac{1}{N}\boldsymbol{1}_N \boldsymbol{1}_N^T$.

The uniqueness ot $\boldsymbol{W}$ in $\tilde{\mathcal{J}}$ is shown in the proof of Lemma 6.1 in [8, page 101]. The general idea of the proof is to split the optimization variable $\boldsymbol{W}$ into a part in the range of $\tilde{\boldsymbol{\Phi}}$ and a part in the null space of $\tilde{\boldsymbol{\Phi}}^T$. For the contribution in the range it is shown that the problem is strongly convex and thus unique. The part in null space is assumed to be nonzero and then it is shown that this leads to a contradiction, establishing the full proof. $\qquad \square$

The regularization parameter η needs to be chosen carefully to obtain a model with good generalization performance. In analogy to ℓ_1-regularization one can determine a critical value for η above which the solution remains constant.

Lemma 17.3. *For* $\eta \geq \eta_0 = \sigma_{\max}(\mathbf{\Phi} \mathbf{P}_1^{\perp} \mathbf{Y}^T \mathbf{T})$ *the solution of* (17.12) *is given by*

$$\mathbf{W}_0 = \mathbf{0} \qquad and \qquad b_0 = \frac{1}{N'} \mathbf{Y} \mathbf{1}_{N'}. \tag{17.14}$$

Here, $\mathbf{P}_1^{\perp} = \frac{1}{N'} \mathbf{1}_{N'} \mathbf{1}_{N'}^T$ *denotes the projector onto the null space of* $\mathbf{1}_{N'}$.

This result is proven as Lemma 6.2 in [8, page 102]. The proof relies on the necessary condition for optimality, that the subdifferential of (17.13) has to contain the element $(\mathbf{0}, \mathbf{0})$.

17.4   Dual Formulation of the Model

The model formulated in (17.12) is parametric and requires the selection of appropriate basis functions. However, one advantage of kernel-based modeling is that the choice of basis functions is simplified by reducing it to the choice of a kernel function. The kernel function often induces very large sets of basis functions, but the inherent regularization is an effective methodology to counter overfitting effects.

For LS-SVMs a kernel-based dual model was derived in Subsection 17.2.3. In analogy to that the derivation of a nonparametric model for (17.12) will be outlined in this section. The derivation is based on convex optimization and, as suggested in Subsection 17.2.3, the most challenging parts are stating the KKT conditions by differentiating the Lagrangian (Step 2 in proof of Lemma 17.1) and the formulation of a predictive model in terms of the dual solution (Step 5 in proof of Lemma 17.1).

Remark 17.1. *Some of the advantages of kernel-based modeling could also be exploited without explicitly deriving a dual-based model. The Nyström approximation [32] is able to compute a finite dimensional approximation of the feature map φ on a data set $\mathcal{D}'$. This results in a set of basis functions that approximately span the space induced by the kernel. The approximation is tailored to the distribution of the given data sample. A thorough description on how the Nyström approximation can be used to solve the primal estimation problem is given in [28] for fixed-size LS-SVMs. The same methodology could be employed to solve (17.12).*

17.4.1   Dual Optimization Problem

The kernel-based dual optimization problem corresponding to (17.12) is given by the following lemma.

Lemma 17.4. *The solution to* (17.12) *is equivalent to the solution of its Lagrange dual*

$$\max_{\boldsymbol{A} \in \mathbb{R}^{M \times N}} \operatorname{tr}(\boldsymbol{A}^T \boldsymbol{Y}) - \frac{1}{2} \operatorname{tr}(\boldsymbol{A}^T \boldsymbol{T}^{-1} \boldsymbol{A})$$

subject to (17.15)

$$\boldsymbol{A} \mathbf{1}_{N'} = \mathbf{0}_M, \quad \|\boldsymbol{G} \boldsymbol{A}^T\|_2 \le \eta$$

with $\boldsymbol{Y} = [\boldsymbol{y}_R, \ldots, \boldsymbol{y}_N] \in \mathbb{R}^{M \times N'}$. *The matrix* $\boldsymbol{G}$ *is defined as a matrix square root, such that* $\boldsymbol{G}^T \boldsymbol{G} = \boldsymbol{\Omega}$. *The elements of the kernel matrix* $\boldsymbol{\Omega}$ *can be computed using the kernel trick* $\Omega_{ij} = \varphi(\boldsymbol{x}_{R+i-1})^T \varphi(\boldsymbol{x}_{R+j-1}) = K(\boldsymbol{x}_{R+i-1}, \boldsymbol{x}_{R+j-1})$ *for* $i, j = 1, \ldots, N'$.

Here only the outline of the proof is given. For the full details please refer to [8, page 104 (Lemma 6.3)]. The key ingredient is to exploit the definition of the dual norm. This idea has also been applied in the context of robust support vector regression using an ℓ_2 regularization term in [24]. This allows us to restate the nondifferentiable term $\|\boldsymbol{W}\|_*$ as a linear function subject to additional constraints. The reformulated Lagrangian can then be used to compute a KKT system similar to the one derived in Subsection 17.2.3. An alternative approach would be to use conic duality; this eventually gives rise to the same dual problem stated above. A similarly structured problem has been derived in [8, pages 211 (Lemma 10.1)] using conic duality.

Proof.

1a. Write down the Lagrangian for (17.12)

$$\mathcal{L}(\boldsymbol{w}_m, b_m, \boldsymbol{e}_t, \boldsymbol{\alpha}_t) = \eta \|[\boldsymbol{w}_1, \ldots, \boldsymbol{w}_M]\|_* + \frac{1}{2} \sum_{t=R}^{N} \boldsymbol{e}_t^T \boldsymbol{T} \boldsymbol{e}_t$$

$$- \sum_{t=R}^{N} \boldsymbol{\alpha}_t^T (\boldsymbol{y}_t - [\boldsymbol{w}_1, \ldots, \boldsymbol{w}_M]^T \varphi(\boldsymbol{x}_t) - \boldsymbol{b} - \boldsymbol{e}_t). \quad (17.16)$$

1b. Replace $\|[\boldsymbol{w}_1, \ldots, \boldsymbol{w}_M]\|_*$ by $\max_{\boldsymbol{c}_m} \sum_{m=1}^{M} \boldsymbol{c}_m^T \boldsymbol{w}_m$ subject to $\|[\boldsymbol{c}_1, \ldots, \boldsymbol{c}_M]\|_2 \le 1$.

2a. The reformulated Lagrangian contains the maximization over a variable. Due to this it is still not differentiable. Therefore introduce a modified Lagrangian $\tilde{\mathcal{L}}_C$ such that $\mathcal{L}(\boldsymbol{w}_m, b_m, \boldsymbol{e}_t, \boldsymbol{\alpha}_t) = \max_{\|C\|_2 \le 1} \tilde{\mathcal{L}}_C(\boldsymbol{w}_m, b_m, \boldsymbol{e}_t, \boldsymbol{\alpha}_t)$. The dual function is defined as $g(\boldsymbol{\alpha}_t) = \inf_{\boldsymbol{\alpha}_t} \mathcal{L}(\boldsymbol{w}_m, b_m, \boldsymbol{e}_t, \boldsymbol{\alpha}_t) = \inf_{\boldsymbol{\alpha}_t} \max_{\|C\|_2 \le 1} \tilde{\mathcal{L}}_C(\boldsymbol{w}_m, b_m, \boldsymbol{e}_t, \boldsymbol{\alpha}_t)$. Using the saddle point property [4, Exercise 5.25] one can reverse the order of the inf and the max. For the saddle point property to hold $\xi(\boldsymbol{w}_m, b_m, \boldsymbol{e}_t, \boldsymbol{\alpha}_t) = \tilde{\mathcal{L}}_C(\boldsymbol{w}_m, b_m, \boldsymbol{e}_t, \boldsymbol{\alpha}_t)$ has to be closed and convex for every C in $\|C\|_2 \le 1$ and $\chi(C) = -\tilde{\mathcal{L}}_C(\boldsymbol{w}_m, b_m, \boldsymbol{e}_t, \boldsymbol{\alpha}_t)$ has to be closed and convex for every $(\boldsymbol{w}_m, b_m, \boldsymbol{e}_t, \boldsymbol{\alpha}_t)$. ξ is a convex quadratic function while χ is linear, hence the conditions are satisfied and the order of minimization and maximization can be reversed.

2b. Compute the derivatives of the modified Lagrangian $\tilde{\mathcal{L}}_C$ in all variables.

3. Formulate the KKT system:

$$\frac{\partial \tilde{\mathcal{L}}_C}{\partial \boldsymbol{w}_m} = \mathbf{0}_{n_h} : \quad \boldsymbol{c}_m = \sum_{t=R}^{N} \alpha_t^{(m)} \boldsymbol{\varphi}(\boldsymbol{x}_t), \quad m = 1, \dots, M, \quad (17.17\text{a})$$

$$\frac{\partial \tilde{\mathcal{L}}_C}{\partial b_m} = 0 : \quad \sum_{t=R}^{N} \alpha_t^{(m)} = 0, \quad m = 1, \dots, M, \quad (17.17\text{b})$$

$$\frac{\partial \tilde{\mathcal{L}}_C}{\partial \boldsymbol{e}_t} = \mathbf{0}_M : \quad \boldsymbol{e}_t = \boldsymbol{T}^{-1} \boldsymbol{\alpha}_t, \quad t = R, \dots, N, \quad (17.17\text{c})$$

$$\|[\boldsymbol{c}_1, \dots, \boldsymbol{c}_M]\|_2 \leq 1. \quad (17.17\text{d})$$

The last condition (17.17d) does not belong to the KKT system itself, but rather to the outer maximization problem. Note that the dual problem is defined as the maximization over the dual function. To allow for a simpler presentation, the maximization over $\boldsymbol{C}$ can be attributed to this final maximization step.

4. In contrast to the LS-SVM derivation in Subsection 17.2.3, (17.17a) does not provide an expansion for $\boldsymbol{w}_m$. Nevertheless, the expression can be substituted into (17.17d). Considering the square of this condition and applying the kernel trick, the inequality constraint in (17.15) can be derived. Substitution of all conditions into the Lagrangian and carrying out some simplifications yields the objective function of the dual problem.

□

Lemmas 17.2 and 17.3 established uniqueness for the primal problem and derived an upper bound for the regularization constant η. Similar results can be obtained for the dual. The corresponding relations can be found in Corollaries 6.4 and 6.5 in [8, pages 105-106]. The critical value for η in terms of the kernel is $\eta_0 = \sigma_{\max}(\boldsymbol{G} \boldsymbol{P}_1^{\perp} \boldsymbol{Y}^T \boldsymbol{T})$. For values larger than η_0 the solution of (17.15) is $\boldsymbol{A} = \boldsymbol{T} \boldsymbol{Y} \boldsymbol{P}_1^{\perp}$.

Remark 17.2. *Instead of relying on optimization theory and Lagrangian duality, SVM solutions can alternatively be derived using function estimation in reproducing kernel Hilbert spaces (RKHSs) [31] by proving a representer theorem [30]. For unitarily invariant matrix norms, of which the nuclear norm is a special case, this has been done in [3].*

The derivation outlined here has the advantage of being constructive and that additional constraints can be integrated straightforwardly. Also it yields a different optimization problem.

17.4.2 Predictive Model

Lemma 17.4 provides the optimal solution for (17.12) even if the feature map φ is not known explicitly. The predictive model (17.4) relies on the primal variables w_m, however. For problems with Tikhonov type regularization like (17.5), establishing a link between the primal and dual variables is immediately given by the KKT conditions, in this example (17.9a). This relation can directly be substituted into the primal model to obtain a predictive equation in terms of the dual variables; see Table 17.1. However, establishing an equivalent relation for the solutions of (17.12) is much more involved.

The link is given by studying the form of the dual norm $\|X\|_* = \max_{\|Z\|_2 \le 1} \operatorname{tr}(X^T Z)$. Theorem 6.6 in [8, p. 107] characterizes the set of matrices X that satisfy $\operatorname{tr}(X^T Z_0) = \xi_0$ and $\|X\|_* = \xi_0$ for given Z_0 and ξ_0. The elements of the resulting set are of the form $U_1 H V_1^T$ where H is positive semi-definite with $\operatorname{tr}(H) = \xi_0$ and U_1, V_1 contain the singular vectors corresponding to the largest singular value of Z_0. Thus the primal variables in the definition of the dual norm are not uniquely defined for a given dual matrix. Therefore, more information has to be used to recover the primal variables and derive a predictive model.

The approach followed in [8] is to substitute the set of solutions given above into the primal problem (17.12) and solve for H. This is relatively straightforward once the value of ξ_0 is recovered. Determining ξ_0 can be achieved by establishing that strong duality holds for the considered problem and thus the duality gap between the solutions of (17.12) and (17.15) is zero. Integrating all this information yields expansions of the primal variables in terms of the dual solution.

Theorem 17.1. *The optimal values for W and b in (17.12) in terms of the dual optimal solution A are given by*

$$W = \Phi A^T Q \qquad \text{and} \qquad b = \frac{1}{N}(Y - QA\Omega)1_N \qquad (17.18)$$

with

$$Q = \eta^{-2} P_\eta (Y A^T - T^{-1} A A^T) P_\eta \qquad (17.19)$$

where $P_\eta = V_\eta V_\eta^T$. The matrix V_η contains the eigenvectors of $A\Omega A^T$ corresponding to the eigenvalue η^2.

The derivation of this result is given in Section 6.5 of [8, pages 108–110]. Having found explicit parametrizations for the primal variables, formulating a predictive equation is straightforward. It directly follows from substitution of W and b given in (17.18) into the primal model $\widehat{y} = \widehat{f}(z) = W^T \varphi(z) + b$ and the application of the kernel trick.

Corollary 17.1. *With the definitions from Theorem 17.1 the predictive model for a new point z, in terms of the dual variables, is given by*

$$\widehat{y} = \widehat{f}(z) = \sum_{t=R}^{N} \widetilde{\alpha}_t K(x_t, z) + b. \qquad (17.20)$$

TABLE 17.1: Overview of parametric/primal and kernel-based/dual estimation problems and the corresponding models.

	PARAMETRIC MODEL	KERNEL-BASED MODEL
basis functions	choose φ	choose kernel K
model estimation	solve (17.12) for W and b	solve (17.15) for A
obtaining model representation	prespecified	obtain Q and b from Theorem 17.1
generating predictions	$\hat{f}(z) = W^T \varphi(z) + b$	$\hat{f}(z) = \sum_{t=R}^{N} \tilde{\alpha}_t K(x_t, z) + b$

The variables $\tilde{\alpha}_t$ form the matrix $\tilde{A} = [\tilde{\alpha}_1, \ldots, \tilde{\alpha}_{N'}]$, which is computed as $\tilde{A} = QA$.

In the following the essential steps of the proposed method are summarized to give a concise algorithm. It allows us to estimate a model from given data and to use this model to generate predictions at an unknown point z.

Algorithm 15

Given a kernel function $K(x, y)$, data $\{(x_t, y_t)\}_{t=R}^{N}$, and a regularization constant, proceed as follows:

1. Compute kernel matrix $\Omega_{ij} = K(x_{R+i-1}, x_{R+j-1})$ for $i, j = 1, \ldots, N'$.

2. Compute a matrix square root G such that $\Omega = G^T G$.

3. Solve dual problem (17.15) to obtain A.

4. Compute the compact eigenvalue decomposition of $A\Omega A^T$ and form the matrix V_η from the eigenvectors corresponding to the largest eigenvalue (η^2).

5. Evaluate (17.19) to obtain mixing matrix Q.

6. Generate predictions at a new point z by evaluating the model given by (17.20).

17.5   Extension to Variable Input Vectors

The last two sections considered a model of the form (17.4) for which the input vector x_t was the same for every output $y_t^{(m)}$. This section generalizes this model to situations in which the input vector is different for each output. This is a simplified version of the model discussed in Subsection 6.6.1 of [8, page 112], which additionally considers the possibility that the amount of data gathered for each output could be different.

The generalization to different input vectors is reflected in the modified primal estimation problem

$$\min_{W,b,e^{(m)}} \quad \eta\|W\|_* + \frac{1}{2}\sum_{m=1}^{M} t_m e^{(m)T} e^{(m)} \tag{17.21}$$

subject to

$$y^{(m)} = \Phi_m^T w_m + b_m \mathbf{1}_{N'} + e^{(m)}, \qquad m = 1, \ldots, M,$$

where $y^{(m)} = [y_R^{(m)}, \ldots, y_N^{(m)}]^T \in \mathbb{R}^{N'}$, $\Phi_m = [\varphi(x_R^{(m)}), \ldots, \varphi(x_N^{(m)})] \in \mathbb{R}^{n_h \times N'}$, $e^{(m)} = [e_R^{(m)}, \ldots, e_N^{(m)}]^T \in \mathbb{R}^{N'}$ and w_m is the m-th column of W.

Remark 17.3. *Note that the equality constraint in (17.21) is transposed with respect to the one in (17.12). It is also written in terms of the target variables (M constraints) instead of the samples (N' constraints). This is done to stay close to the notation of the corresponding section of [8] in which it is necessary to account for possible different dimensionalities of the considered quantities.*

The dual optimization problem for (17.21) is given in the following lemma, thus generalizing Lemma 17.4.

Lemma 17.5. *The solution to (17.21) is equivalent to the solution of its Lagrange dual*

$$\max_{\alpha^{(m)}} \quad \sum_{m=1}^{M} \alpha^{(m)T} y^{(m)} - \frac{1}{2}\sum_{m=1}^{M} \frac{1}{t_m} \alpha^{(m)T} \alpha^{(m)}$$

subject to

$$\mathbf{1}_{N'}^T \alpha^{(m)} = 0, \quad m = 1, \ldots, M, \tag{17.22}$$

$$\begin{bmatrix} \alpha^{(1)T}\Omega_{11}\alpha^{(1)} & \cdots & \alpha^{(1)T}\Omega_{1M}\alpha^{(M)} \\ \vdots & \ddots & \vdots \\ \alpha^{(M)T}\Omega_{M1}\alpha^{(1)} & \cdots & \alpha^{(M)T}\Omega_{MM}\alpha^{(M)} \end{bmatrix} \preceq \eta^2 I_M$$

with $\alpha^{(m)} \in \mathbb{R}^{N'}$. The $N' \times N'$ matrices Ω_{mn} are given by $\Phi_m^T\Phi_n$. They can be computed elementwise as $(\Omega_{mn})_{ij} = K(x_{R+i-1}^{(m)}, x_{R+j-1}^{(n)})$ for $i, j = 1, \ldots, N'$, and $m, n = 1, \ldots, M$.

The derivation of this dual formulation is almost a carbon copy of the proof of Lemma 17.4. Therefore only the differences will be highlighted.

Proof.

$1'$. Write down the Lagrangian for (17.21)

$$\mathcal{L}(\boldsymbol{w}_m, b_m, \boldsymbol{e}^{(m)}, \boldsymbol{\alpha}^{(m)}) = \eta\|[\boldsymbol{w}_1, \ldots, \boldsymbol{w}_M]\|_* + \frac{1}{2}\sum_{m=1}^{M} t_m \boldsymbol{e}^{(m)^T} \boldsymbol{e}^{(m)}$$

$$- \sum_{m=1}^{M} \boldsymbol{\alpha}^{(m)^T}(\boldsymbol{y}^{(m)} - \boldsymbol{\Phi}_m^T \boldsymbol{w}_m - b_m \mathbf{1}_{N'} - \boldsymbol{e}^{(m)}). \quad (17.23)$$

$3'$. Formulate the KKT system.

$$\frac{\partial \mathcal{L}}{\partial \boldsymbol{w}_m} = \mathbf{0}_{n_h} : \quad \boldsymbol{c}_m = \boldsymbol{\Phi}_m \boldsymbol{\alpha}^{(m)}, \quad m = 1, \ldots, M, \qquad (17.24\text{a})$$

$$\frac{\partial \mathcal{L}}{\partial b_m} = 0 : \quad \mathbf{1}_{N'}^T \boldsymbol{\alpha}^{(m)} = 0, \quad m = 1, \ldots, M, \qquad (17.24\text{b})$$

$$\frac{\partial \mathcal{L}}{\partial \boldsymbol{e}^{(m)}} = \mathbf{0}_{N'} : \quad \boldsymbol{e}^{(m)} = t_m^{-1} \boldsymbol{\alpha}^{(m)}, \quad m = 1, \ldots, M, \qquad (17.24\text{c})$$

$$\|[\boldsymbol{c}_1, \ldots, \boldsymbol{c}_M]\|_2 \le 1. \qquad (17.24\text{d})$$

$\square$

Lemma 17.6. *The quadratic matrix inequality in (17.22) can be reformulated as the following collection of smaller linear matrix inequalities (LMIs).*

$$\begin{bmatrix} \boldsymbol{I}_{N'} & \boldsymbol{G}_m \boldsymbol{\alpha}^{(m)} \\ \boldsymbol{\alpha}^{(m)^T} \boldsymbol{G}_m^T & r_m \end{bmatrix} \succeq \mathbf{0}, \quad m = 1, \ldots, M, \qquad (17.25\text{a})$$

$$\begin{bmatrix} \boldsymbol{I}_{2N'} & \boldsymbol{F}_{mn} \begin{bmatrix} \boldsymbol{\alpha}^{(m)} \\ \boldsymbol{\alpha}^{(n)} \end{bmatrix} \\ \begin{bmatrix} \boldsymbol{\alpha}^{(m)^T} & \boldsymbol{\alpha}^{(n)^T} \end{bmatrix} \boldsymbol{F}_{mn}^T & r_m + 2s_{mn} + r_n \end{bmatrix} \succeq \mathbf{0}, \qquad (17.25\text{b})$$

$$m = 1, \ldots, M, n = m+1, \ldots, M$$

$$\begin{bmatrix} r_1 & s_{12} & \cdots & s_{1M} \\ s_{12} & r_2 & \cdots & s_{2M} \\ \vdots & \vdots & \ddots & \vdots \\ s_{1M} & s_{2M} & \cdots & r_M \end{bmatrix} \preceq \eta^2 \boldsymbol{I}_M. \qquad (17.25\text{c})$$

Here $\boldsymbol{G}_m$ and $\boldsymbol{F}_{mn}$ are matrix factorizations such that $\boldsymbol{\Omega}_{mm} = \boldsymbol{G}_m^T \boldsymbol{G}_m$ and $\begin{bmatrix} \boldsymbol{\Omega}_{mm} & \boldsymbol{\Omega}_{mn} \\ \boldsymbol{\Omega}_{nm} & \boldsymbol{\Omega}_{nn} \end{bmatrix} = \boldsymbol{F}_{mn}^T \boldsymbol{F}_{mn}$ for $m = 1, \ldots, M$ and $n = m+1, \ldots, M$.

Proof. The proof is a repeated application of the Schur complement. Consider $\boldsymbol{\alpha}^{(m)^T}\boldsymbol{\Omega}_{mm}\boldsymbol{\alpha}^{(m)} \leq r_m$ and the factorization for $\boldsymbol{\Omega}_{mm}$ given above; then (17.25a) is a direct consequence of the Schur complement. Note that $\boldsymbol{\alpha}^{(m)^T}\boldsymbol{\Omega}_{mn}\boldsymbol{\alpha}^{(n)} = \boldsymbol{\alpha}^{(n)^T}\boldsymbol{\Omega}_{nm}\boldsymbol{\alpha}^{(m)}$. Then (17.25b) is, by its Schur complement, equivalent to $r_m + 2s_{mn} + r_n \geq \boldsymbol{\alpha}^{(m)^T}\boldsymbol{\Omega}_{mm}\boldsymbol{\alpha}^{(m)} + \boldsymbol{\alpha}^{(n)^T}\boldsymbol{\Omega}_{nn}\boldsymbol{\alpha}^{(n)} + 2\boldsymbol{\alpha}^{(m)^T}\boldsymbol{\Omega}_{mn}\boldsymbol{\alpha}^{(n)} \geq 0$. At the solution the inequalities will be tight, therefore (17.25) is a reformulation of the inequality in (17.22) as LMI. ☐

For the problems in (17.21) and (17.22) the relation between primal and dual variables is as follows.

Corollary 17.2. *Given the optimal dual solution* $\boldsymbol{\alpha}_1, \ldots, \boldsymbol{\alpha}_M$, *the primal optimal variables* $\boldsymbol{w}_1, \ldots, \boldsymbol{w}_M$ *and* $b_1, \ldots, b_M$ *can be represented as*

$$\boldsymbol{w}_m = \sum_{n=1}^{M} Q_{nm}\boldsymbol{\Phi}_n\boldsymbol{\alpha}_n \tag{17.26}$$

and

$$b_m = \frac{1}{N'}\left(\mathbf{1}_{N'}^T\boldsymbol{y}^{(m)} - \sum_{n=1}^{M}Q_{nm}\mathbf{1}_{N'}^T\boldsymbol{\Omega}_{mn}\boldsymbol{\alpha}_n\right), \tag{17.27}$$

for $m = 1, \ldots, M$ *and with* $(\boldsymbol{Q})_{mn} = Q_{mn}$ *and* $\boldsymbol{Q} = \boldsymbol{V}_\eta \boldsymbol{H}_\eta \boldsymbol{V}_\eta^T$.

The matrix $\boldsymbol{V}_\eta$ *contains the eigenvectors of* $\tilde{\boldsymbol{A}}^T\tilde{\boldsymbol{\Omega}}\tilde{\boldsymbol{A}}$ *corresponding to the largest eigenvalue* η^2. *Furthermore the matrix* $\boldsymbol{H}_\eta$ *is given as the solution to the auxiliary feasibility problem*

find $\quad \boldsymbol{H}_\eta$

subject to

$\boldsymbol{H}_\eta \succeq 0, \mathrm{tr}(\boldsymbol{H}_\eta) = \xi$

$\boldsymbol{y}^{(m)} = [\boldsymbol{\Omega}_{m,1}\boldsymbol{\alpha}_1, \ldots, \boldsymbol{\Omega}_{m,M}\boldsymbol{\alpha}_M]\boldsymbol{V}_\eta\boldsymbol{H}_\eta\boldsymbol{V}_\eta^T\boldsymbol{\epsilon}_m + b_m\mathbf{1}_{N'} + t_m^{-1}\boldsymbol{\alpha}_m,$

$\qquad\qquad\qquad\qquad\qquad\qquad\qquad m = 1, \ldots, M,$

where the $\boldsymbol{\epsilon}_i$'s *form the standard basis for* $\mathbb{R}^M$ *and* $\xi = \eta^{-2}\left(\sum_{m=1}^{M}\boldsymbol{\alpha}_m^T\boldsymbol{y}^{(m)} - \sum_{m=1}^{M}t_m^{-1}\boldsymbol{\alpha}_m^T\boldsymbol{\alpha}_m\right).$

The predictive model then follows immediately.

Corollary 17.3. *With the definitions from Corollary 17.2 the predictive model for a new point* $\boldsymbol{z}^{(m)}$, *in terms of the dual variables, is given by*

$$\widehat{y}^{(m)} = \hat{f}_m(\boldsymbol{z}^{(m)}) = \sum_{n=1}^{M}Q_{nm}\boldsymbol{k}_n(\boldsymbol{z}^{(m)})^T\boldsymbol{\alpha}_n + b_m, \tag{17.28}$$

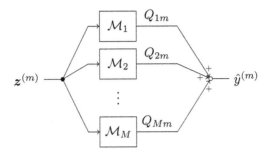

FIGURE 17.4: Conceptual visualization of the predictive Equation (17.28) for m-th output $\hat{y}^{(m)}$ given the corresponding input vector $z^{(m)}$. The individual models $\mathcal{M}_1, \ldots, \mathcal{M}_M$ are the same for each output. The weights $Q_{1m}, \ldots, Q_{Mm}$ specify how each of the models contributes to the prediction of an individual output. Each submodel $\mathcal{M}_n$ performs the mapping $z \to k_n(z)^T \alpha_n$.

with $k_n(\zeta) = [K(\zeta, x_R^{(n)}), \ldots, K(\zeta, x_N^{(n)})]^T$ for $m = 1, \ldots, M$. The predictive model is applicable to all outputs $\hat{y}^{(m)}$ with $m = 1, \ldots, M$. The matrix Q is given by the solution of the feasibility problem stated in Corollary 17.2.

The proofs for the results in this section can be found in [8, pp. 113–115] and in general are straightforward extensions of those in the previous section.

Remark 17.4 (Structure of predictive equation). *Note that the predictive Equation (17.28) is coupled by the matrix Q. The prediction for the m-th output is generated by feeding the corresponding input $z^{(m)}$ into each of the M submodels. Then the coefficients Q_{mn} determine the contribution of each of these submodels to the final prediction. This structure is also visualized in Figure 17.4.*

This is in contrast to the traditional modeling approach using a Frobenius norm regularization scheme. In such a formulation the predictive equations are decoupled from each other and only one model contributes to the prediction of each output. In the setting here this would correspond to the weighting matrix Q being the identity matrix I_M.

Remark 17.5 (Numerical complexity). *The extended model (17.21) presented in this section is a generalization of the basic model (17.12) discussed before. Hence, the question arises, how much more expensive is solving the more general, and consequently more powerful, problem.*

The relevant dimensionalities are summarized in Table 17.2. It can be seen that the number of unknowns is independent of the chosen formulation. The main difference is in the dimensionality of the data. As a different input vector is considered for each output this increase is inevitable. In the primal the dimensionality of the data matrix Φ grows linearly with the number of outputs

TABLE 17.2: Dimensionalities of primal and dual model formulations. n_h denotes the dimension of the feature space (number of basis functions), N' is the amount of data available for model estimation, and M is the number of model outputs.

	Identical inputs	Different input for each output
Primal		
Feature map $\boldsymbol{\Phi}$	$n_h \times N'$	$M \cdot (n_h \times N')$
Unknown $\boldsymbol{W}$	$n_h \times M$	$n_h \times M$
Dual		
Kernel matrix $\boldsymbol{\Omega}$	$N' \times N'$	$(M \cdot N') \times (M \cdot N')$
Unknowns α_{mt}	$N' \times M$	$N' \times M$

M. The dependency in the dual goes with M^2 as it compares each input sample to all the others.

Due to this significant increase of data to be processed, one should carefully consider whether the system can be modeled with a single consolidated input instead of individual ones.

17.6   Numerical Example

To illustrate several properties of the proposed model a numerical example is analyzed. In order to keep the analysis as simple as possible, an artificial problem is constructed. The algorithms are implemented using MATLAB and CVX, which severely limits the problem sizes that can be solved with moderate computational resources. Therefore the number of data is limited to 50 samples for training of the model, 100 samples to select hyper-parameters using cross validation, and, finally, 150 samples to evaluate the model performance on an independent test set. The model structure is chosen to closely match the assumptions embedded in the model,

$$Y = W_0^T \boldsymbol{\Phi} + \text{noise}. \tag{17.29}$$

Instead of generating nonlinear data and projecting it to a higher dimensional space, for the sake of simplicity, data for the evaluated feature map $\boldsymbol{\Phi}$ is generated directly. Data is drawn from a normal distribution to obtain $\boldsymbol{\Phi} \in \mathbb{R}^{50 \times 50}$. The core assumption of the model is that the true parameter matrix is low rank. Letting $\boldsymbol{W}_0 = \sum_{i=1}^{3} g_i r_i^T = \boldsymbol{G} \boldsymbol{R}^T$ with $\boldsymbol{G} \in \mathbb{R}^{50 \times 3}$ and $\boldsymbol{R} \in$

$\mathbb{R}^{20 \times 3}$, corresponds to a model with 20 outputs yet only three independent "behaviors".

To evaluate the predictive power of the proposed model it is compared to several other models. The considered models are as follows.

MIMO The proposed nuclear norm regularized model as given in Subsection 17.2.5.

RR The notable difference of the proposed model to classical approaches is the choice of the regularization term. The most common regularization would be a squared 2-norm type, corresponding to ridge regression. In this case the estimation problems for the different components are independent of each other. Each one can be seen as an individual LS-SVM model as in Subsection 17.2.3. Using this factored form has the advantage that different hyper-parameters can be chosen for every output. While in theory at least the regularization constant could also be chosen independently for each output in the proposed model, this would only be computationally tractable for very small numbers of outputs.

OLS To obtain an intuition on the upper and lower bounds for the predictive performance, two very simple models are considered. An upper bound is given by the model obtained using ordinary least squares, i.e., $\widehat{\boldsymbol{W}}_0 = \boldsymbol{\Phi}^{\dagger} \boldsymbol{Y}$.

OLS + oracle A lower bound is also generated based on ordinary least squares. However, in this case the true structure of the problem in the form of $\boldsymbol{R}$ is assumed given. The objective is therefore reduced to estimating $\boldsymbol{G}$, which in this example reduces the number of unknowns to be determined by a factor of 60.

Model performance is measured using the root mean squared error (RMSE), for a single output defined as

$$\text{RMSE}_m = \sqrt{\frac{1}{N'} \sum_{t=R}^{N} \left(y_t^{(m)} - \hat{f}_m(\boldsymbol{x}_t) \right)^2}.$$

The total RMSE is then given by

$$\text{RMSE} = \sqrt{\frac{1}{M} \sum_{m=1}^{M} \text{RMSE}_m^2}.$$

17.6.1   Agreement of Primal and Dual Solutions

To support that the derived dual model is equivalent to the proposed primal form, the problem is solved in both forms. This is easily possible as the

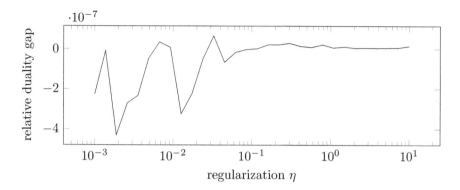

FIGURE 17.5: Relative duality gap $(p^* - d^*)/p^*$ between optimal value p^* of (17.12) and the optimal value d^* of (17.15) for toy example.

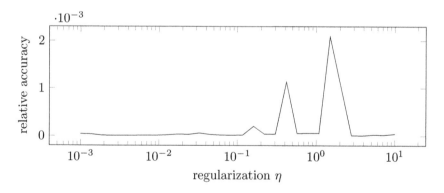

FIGURE 17.6: Relative accuracy of parameter estimate $\widehat{W}$ for toy example. For the primal problem in (17.12) the value W_{primal} is a direct result of the optimization problem. For the dual (17.15) the optimal parameter W_{dual} is recovered using the equation from Theorem 17.1. The reported quantity is $\|W_{\text{primal}} - W_{\text{dual}}\|_F / \|W_{\text{primal}}\|_F$.

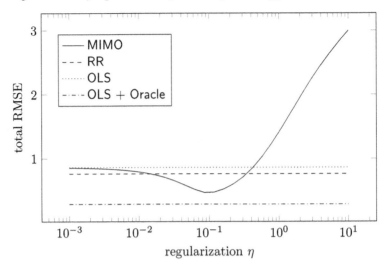

FIGURE 17.7: Model performance evaluated on a validation set for different values of the regularization parameter η. The OLS-based models are not subject to regularization and are thus constant. For RR the optimal regularization constant for each output is computed individually and only the best performance is indicated.

considered example has an explicitly given feature map. To check agreement of both model representations, the recovered parameter vector $\widehat{W}$ as well as the duality gap between both estimates are compared. From Figures 17.5 and 17.6 it can be seen that both solutions are in agreement up to numerical precision. This is expected as it has been shown that strong duality holds and therefore the duality gap has to be zero.

17.6.2   Cross-Validation Performance

Selecting good values for the model hyper-parameters is essential for most data-driven estimation schemes. A popular approach is to select the hyper-parameters based on cross-validation performance. This technique is also viable for the proposed model structure, as can be seen from Figure 17.7. The total RMSE on the validation set shows a distinct minimum. Also the proposed method MIMO yields a much lower error than the compared techniques RR and OLS, which have no information on the problem structure. Full knowledge of the structure, however, results in an even superior performance, as can be seen from the value of OLS + Oracle.

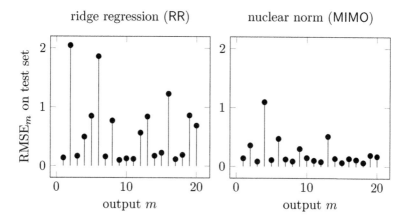

FIGURE 17.8: Performance of the models RR and MIMO on an independent test set. The accuracy of the predictions is shown for every output.

17.6.3 Predictive Performance

Figure 17.8 shows the predictive performance for RR and MIMO. It can be seen that for almost all outputs the prediction performance of the proposed model is improved with respect to the baseline. Table 17.3 shows the total RMSE for the different parts of the data. The results illustrate that in this example the proposed approach MIMO improves significantly over the traditional technique RR. It can also be seen that selecting the regularization parameter using cross validation works well as there is no evidence of overfitting.

TABLE 17.3: Predictive performance for multivariate toy dataset. All given quantities are RMSE values with respect to time and the output, for different partitions of the data.

	TRAINING	VALIDATION	TEST
OLS	0	0.8633	0.8459
OLS + Oracle	0.0920	0.2885	0.2981
RR	0.0175	0.7644	0.7867
MIMO	0.0184	0.4716	0.4974

17.7   Conclusions

This chapter introduced a novel identification scheme for nonlinear systems with multiple outputs. It is able to exploit relations between output variables. The benefits in predictive performance are illustrated on a small toy example. The model is derived in a primal-dual setting, which is new for models with nuclear norm regularization. Due to the primal-dual approach, it is straightforward to incorporate additional information in the estimation problem. The derivation for primal-dual models presented here can also be applied to other models with non-quadratic regularization terms. New algorithms and increasing computational power will allow the application of the presented technique on larger real world datasets in the future.

Acknowledgments

Research supported by: Research Council KUL: GOA/10/09 MaNet, PFV/10/002 (OPTEC), several PhD/postdoc and fellow grants; Flemish Government: IOF: IOF/KP/SCORES4CHEM, FWO: PhD/postdoc grants, projects: G.0377.12 (structured systems), G.083014N (block term decompositions), G.088114N (tensor-based data similarity), IWT: PhD Grants, projects: SBO POM, EUROSTARS SMART, iMinds 2013, Belgian Federal Science Policy Office: IUAP P7/19 (DYSCO, Dynamical systems, control and optimization, 2012-2017), EU: FP7-SADCO (MC ITN- 264735), ERC AdG A-DATADRIVE-B (290923), COST: Action ICO806: IntelliCIS. Johan A.K. Suykens is a professor and Bart De Moor is a full professor at KU Leuven.

Bibliography

[1] Carlos Alzate, Marcelo Espinoza, Bart De Moor, and Johan A.K. Suykens. Identifying customer profiles in power load time series using spectral clustering. In *Proceedings of the 19th International Conference on Artificial Neural Networks*, ICANN '09, pages 315–324, Berlin, Heidelberg, 2009. Springer-Verlag.

[2] Andreas Argyriou, Theodoros Evgeniou, and Massimiliano A. Pontil. Convex multi-task feature learning. *Machine Learning*, 73(3):243–272, January 2008.

[3] Andreas Argyriou, Charles A. Micchelli, and Massimiliano A. Pontil. When Is There a Representer Theorem? Vector Versus Matrix Regularizers. *Journal of Machine Learning Research*, 10:2507–2529, 2009.

[4] Stephen P. Boyd and Lieven Vandenberghe. *Convex Optimization*. Cambridge University Press, 2004.

[5] Emmanuel J. Candès and Yaniv Plan. Matrix completion with noise. *Proceedings of the IEEE*, 98(6):925–936, 2010.

[6] Marcelo Espinoza. *Structured Kernel Based Modeling and its Application to Electric Load Forecasting*. PhD thesis, Katholieke Universiteit Leuven, Belgium, June 2006.

[7] Marcelo Espinoza, Johan A. K. Suykens, Ronnie Belmans, and Bart De Moor. Electric Load Forecasting – Using kernel based modeling for nonlinear system identification. *IEEE Control Systems Magazine*, 27:43–57, October 2007.

[8] Tillmann Falck. *Nonlinear system identification using structured kernel based models*. PhD thesis, Katholieke Universiteit Leuven, Belgium, April 2013.

[9] Tillmann Falck, Philippe Dreesen, Kris De Brabanter, Kristiaan Pelckmans, Bart De Moor, and Johan A. K. Suykens. Least-Squares Support Vector Machines for the Identification of Wiener-Hammerstein Systems. *Control Engineering Practice*, 20(11):1165–1174, November 2012.

[10] Maryam Fazel. *Matrix Rank Minimization with Applications*. PhD thesis, Stanford, 2002.

[11] Maryam Fazel, Haitham A. Hindi, and Stephen P. Boyd. A rank minimization heuristic with application to minimum order system approximation. In *Proceedings of the American Control Conference*, Arlington, VA, USA, 2001.

[12] Ivan Goethals, Kristiaan Pelckmans, Johan A. K. Suykens, and Bart De Moor. Subspace identification of Hammerstein systems using least squares support vector machines. *IEEE Transactions on Automatic Control*, 50:1509–1519, October 2005.

[13] Martin Jaggi and Marek Sulovský. A simple algorithm for nuclear norm regularized problems. In *Proceedings of the 27th International Conference on Machine Learning (ICML-10)*, pages 471–478, Haifa, Israel, June 2010.

[14] Rudolph E. Kalman. Contributions to the theory of optimal control. *Boletín de la Sociedad Matemática Mexicana*, 5:102–119, 1960.

[15] Yong-Jin Liu, Defeng Sun, and Kim-Chuan Toh. An implementable proximal point algorithmic framework for nuclear norm minimization. *Mathematical programming*, 133(1-2):399–436, 2012.

[16] Zhang Liu, Anders Hansson, and Lieven Vandenberghe. Nuclear norm system identification with missing inputs and outputs. *Systems & Control Letters*, 62(8):605–612, 2013.

[17] Zhang Liu and Lieven Vandenberghe. Interior-Point Method for Nuclear Norm Approximation with Application to System Identification. *SIAM Journal on Matrix Analysis and Applications*, 31(3):1235–1256, January 2009.

[18] Lennart Ljung. *System identification: Theory for the User*. Prentice Hall PTR, Upper Saddle River, NJ, USA, 2nd edition, 1999.

[19] James Mercer. Functions of positive and negative type, and their connection with the theory of integral equations. *Philosophical Transactions of the Royal Society of London. Series A, Containing Papers of a Mathematical or Physical Character*, 209:415–446, 1909.

[20] Ting Kei Pong, Paul Tseng, Shuiwang Ji, and Jieping Ye. Trace norm regularization: Reformulations, algorithms, and multi-task learning. *SIAM Journal on Optimization*, 20(6):3465–3489, 2010.

[21] Benjamin Recht. A simpler approach to matrix completion. *Journal of Machine Learning Research*, 12:3413–3430, 2011.

[22] Benjamin Recht, Maryam Fazel, and Pablo A. Parrilo. Guaranteed Minimum-Rank Solutions of Linear Matrix Equations via Nuclear Norm Minimization. *SIAM Review*, 52(3):471–501, January 2010.

[23] Bernhard Schölkopf and Alexander J. Smola. *Learning with Kernels*. MIT Press, Cambridge, Mass., 2002.

[24] Pannagadatta K. Shivaswamy, Chiranjib Bhattacharyya, and Alexander J. Smola. Second Order Cone Programming Approaches for Handling Missing and Uncertain Data. *Journal of Machine Learning Research*, 7:1283–1314, 2006.

[25] Marco Signoretto, Quoc Tran Dinh, Lieven De Lathauwer, and Johan A. K. Suykens. Learning with tensors: a framework based on convex optimization and spectral regularization. *Machine Learning*, 94(3):303–351, 2014.

[26] Jonas Sjoberg, Qinghua Zhang, Lennart Ljung, Albert Benveniste, Bernard Delyon, Pierre-Yves Glorennec, Hakan Hjalmarsson, and Anatoli Juditsky. Nonlinear black-box modeling in system identification: a unified overview. *Automatica*, 31:1691–1724, December 1995.

[27] Johan A. K. Suykens, Carlos Alzate, and Kristiaan Pelckmans. Primal and dual model representations in kernel-based learning. *Statistics Surveys*, 4:148–183, August 2010.

[28] Johan A.K. Suykens, Tony Van Gestel, Jos De Brabanter, Bart De Moor, and Joos Vandewalle. *Least Squares Support Vector Machines*. World Scientific, 2002.

[29] Peter Van Overschee and Bart De Moor. *Subspace Identification for Linear Systems, Theory, Implementation, Applications*. Kluwer Academic Publishers, 1996.

[30] Grace Wahba. *Spline Models for Observational Data*. SIAM, 1990.

[31] Grace Wahba. Support Vector Machines, Reproducing Kernel Hilbert Spaces and the Randomized GACV. In Bernhard Schölkopf, Christoper J. C. Burges, and Alexander J. Smola, editors, *Advances in Kernel Methods - Support Vector Learning*, pages 69–88. MIT Press, Cambridge, MA, May 1998.

[32] Christopher K. I. Williams and Matthias Seeger. Using the Nyström Method to Speed Up Kernel Machines. In T. Leen, T. Dieterich, and V. Tresp, editors, *Neural Information Processing Systems 13*, pages 682–688. MIT Press, 2001.

Chapter 18

Kernel Methods for Image Denoising

Pantelis Bouboulis

Department of Informatics and Telecommunications, University of Athens

Sergios Theodoridis

Department of Informatics and Telecommunications, University of Athens

18.1 Introduction

One of the most commonly encountered problems in the area of image processing is that of noise removal. Any image taken by conventional film cameras or digital cameras can pick up noise from a variety of sources. This noise usually has a negative aesthetic effect on the human eye. Moreover, subsequent uses of the digitized image such as in applications of computer vision, classification, recognition, etc., require the noise to be removed in order to maximize the performance of the respective system. Hence, a large number of techniques have been proposed to address this problem, ranging from the typical low pass filters that convolve the original image with a predefined mask [11], to methods involving fractals, differential equations, and approximation theory. Among the most popular methodologies are the denoising methods based on wavelet theory (e.g., [19, 21, 26, 8, 9, 6]), where the coefficients of the image's wavelet expansion are manipulated so that the noise is removed, while the important features of the image are preserved. Other popular techniques include the denoising methods based on the theory of partial differential equations [27], methods for impulse detection ([1, 2, 22]), non-linear approaches that employ kernel regression and/or local expansion approximation techniques ([31]), and sparse approximation techniques ([10, 5, 15]. In most cases, the denoising techniques are developed assuming a particular noise model (Gaussian, impulse, etc.); thus, they are unable to effectively treat more complex models, which are often met in practical applications. In this chapter, we are interested in image denoising techniques that are based on the theory of reproducing kernel Hilbert spaces (RKHS). In general, these methods attempt to approximate the noise-free image as a linear combination of positive definite kernel functions, while all other components are considered as noise and hence they are discarded. The straightforward approach is to consider a kernel ridge regression rationale, minimizing the square error between the original image and the noise-free estimation, while, at the same time, constraining the norm of the expansion's coefficients. However, it turns out that although this simple method can remove noise components, it also discards fine details of the image. Hence, we will discuss more advanced techniques that exploit notions and ideas not only from functional analysis, but also from machine learning and the recently developed theory of sparse representations. Observe that this rationale makes no assumption for the noise distribution; hence it can be used to remove any type of noise.

In a relatively similar context, kernels have also been employed by other denoising methods that are not considered here. For example, in [14] and [35] a support vector regression approach is considered for the Gaussian noise case, while in [12] the kernel principal components of an image are extracted and this expansion is truncated to produce the denoising effect.

18.2   Preliminaries

Throughout this chapter, we use boldface letters for vectors and normal letters for scalars. Prior to delving into the main part of this chapter, i.e., the problem of denoising into RKHS, we devote some space to introducing the main notions and theories that are employed. We begin with a short discussion on the main properties of RKHS.

18.2.1   Reproducing Kernel Hilbert Spaces

In a nutshell, RKHS [29, 25] are inner product spaces of functions on $\mathcal{X}$, in which pointwise evaluation is a continuous (and hence bounded) linear evaluation functional. These spaces have been proved to be a very powerful tool in the context of statistics, complex and harmonic analysis, quantum mechanics, and, in particular, in machine learning applications [25, 28, 32]. It has been established that each such space is associated with a positive definite kernel, $\kappa(\mathbf{x}, \mathbf{y})$, i.e., a symmetric function of two variables of $\mathcal{X}$ that satisfies

$$\sum_{n,m=1}^{N} a_n a_m \kappa(\mathbf{x}_n, \mathbf{x}_m) \geq 0,$$

for all numbers a_n, a_m and points $\mathbf{x}_n, \mathbf{x}_m \in \mathcal{X}$, where $n, m = 1, 2, \ldots, N$. A Hilbert space, $\mathcal{H}$, is called an RKHS if there exists a positive definite kernel κ with the following properties:

- For every $x \in \mathcal{H}$, the function $\kappa(\cdot, \mathbf{x})$ belongs to $\mathcal{H}$.

- κ has the so-called *reproducing property*, i.e., $f(\mathbf{x}) = \langle f, \kappa(\cdot, \mathbf{x}) \rangle_{\mathcal{H}}$, for all $f \in \mathcal{H}, \mathbf{x} \in \mathcal{X}$.

A large variety of kernels can be found in the respective literature [33, 25, 7]. However, in this chapter, we focus our attention on one of the most widely used kernels, the Gaussian radial basis function:

$$\kappa_\sigma(\mathbf{x}, \mathbf{y}) = \exp\left(-\frac{\|\mathbf{x} - \mathbf{y}\|^2}{\sigma^2}\right),$$

due to its valuable characteristics (smoothness, universal approximation).

One of the most important properties of RKHS is that, although the space might be infinite-dimensional, the solution of any regularized risk minimization problem admits a finite representation. In fact, the optimal solution lies in the span of the kernels centered on the training points. This is ensured by the celebrated *Representer Theorem*.

Theorem 18.1 (Representer Theorem [13]). *Denote by $\Omega : [0, \infty) \to \mathbb{R}$ a strictly monotonic increasing function, by $\mathcal{X}$ a set and by $\ell : (\mathcal{X} \times \mathbb{R}^2)^N \to \mathbb{R} \cup \{\infty\}$ an arbitrary loss function. Then each minimizer $f \in \mathcal{H}$ of the regularized risk functional*

$$\ell((\mathbf{x}_1, z_1, f(\mathbf{x}_1)), \ldots, (\mathbf{x}_N, z_N, f(\mathbf{x}_N))) \ + \ \Omega(\|f\|_{\mathcal{H}}) \tag{18.1}$$

admits a representation of the form

$$f(\mathbf{x}) = \sum_{n=1}^{N} \alpha_n \kappa(\mathbf{x}_n, \mathbf{x}). \tag{18.2}$$

This theorem can be generalized to include the case where f has two components, one lying in $\mathcal{H}$ and the other in the span of a set of predefined independent functions, as follows:

Theorem 18.2 (Semi-Parametric Representer Theorem [25]). *Suppose that, in addition to the assumptions of the previous theorem, we are given $\Omega_2 : [0, \infty) \to \mathbb{R}$, another strictly monotonic increasing function, and a set* *of M real-valued functions $\{\psi_k\}_{k=1}^{M} : \mathcal{X} \to \mathbb{R}$, with the property that the $N \times M$ matrix $(\psi_p(\mathbf{x}_n))_{n,p}$ has rank M. Then any $\tilde{f} := f + \psi$, with $f \in \mathcal{H}$ and $\psi \in \mathfrak{H} = \overline{span\{\psi_k\}}$, where $\|\cdot\|$ is a norm defined in $\mathfrak{H}$, minimizing the regularized risk functional*

$$\ell\left((\mathbf{x}_1, z_1, f(\mathbf{x}_1)), \ldots, (\mathbf{x}_N, z_N, f(\mathbf{x}_N))\right) \quad + \quad \Omega\left(\|f\|_{\mathcal{H}}\right)$$
$$+ \quad \Omega_2\left(\|\psi\|\right) \qquad (18.3)$$

admits a representation of the form

$$\tilde{f}(\mathbf{x}) = \sum_{n=1}^{N} \alpha_n \kappa(\mathbf{x}_n, \mathbf{x}) + \sum_{k=1}^{M} \beta_k \psi_k(\mathbf{x}). \qquad (18.4)$$

In the denoising tasks that are dealt with in this chapter, ℓ takes the form

$$\ell\left((\mathbf{x}_1, z_1, f(\mathbf{x}_1)), \ldots, (\mathbf{x}_N, z_N, f(\mathbf{x}_N))\right) = \sum_{n=1}^{N} \mathcal{L}\left(z_n - f(\mathbf{x}_n)\right),$$

where z_n are the actual values of the (possibly corrupted) signal at $\mathbf{x}_n$ and $f(\mathbf{x}_n)$ are the reconstructed (noise free) values. The function $\mathcal{L}$ is either the ℓ_2, or the ℓ_1 norm. The regularization term $\Omega(f)$, on the other hand, takes the form $\Omega(f) = 1/2\|f\|_{\mathcal{H}}^2$, $\lambda > 0$. Recall that the RKHS considered here are associated with the Gaussian kernel. This implies an infinite dimensional space ([25]). Moreover, in the cases where $\mathcal{X} = \mathbb{R}^m$, $m > 0$, it can be shown that the induced norm of any function in this RKHS is given by

$$\|f\|_{\mathcal{H}} = \int_{\mathcal{X}} \sum_n \frac{\sigma^{2n}}{n!2^n} (O^n f(\mathbf{x}))^2 d\mathbf{x}, \qquad (18.5)$$

where $O^{2n} = \Delta^n$ and $O^{2n+1} = \nabla\Delta^n$, Δ being the Laplacian and ∇ the gradient operator (see [34]). The implication of this is that the regularization term "penalizes" the derivatives of the minimizer, resulting in a very smooth solution of the respective regularized risk minimization problem. In fact, this penalization has a stronger effect than the *total variation* scheme, which is often used in wavelet-based denoising (see for example [9, 6, 24, 16, 23]). Indeed, while the total variation penalizes only the first order derivatives, the term $\|f\|_{\mathcal{H}}^2$ penalizes derivatives of any order, resulting in very smooth estimates.

18.2.2   The Denoising Problem

In this chapter, we model the noisy image as

$$\hat{f}(x, y) = f(x, y) + \eta(x, y), \qquad (18.6)$$

for $x, y \in [0, 1]$, where f is the noise free image and η the additive noise [11]. Given the noisy image $\hat{f}$, the objective of any denoising method is to obtain an

FIGURE 18.1: A square $N \times N$ region.

estimate of the original image f. Usually, this task is carried out by exploiting some a priori knowledge concerning the noise distribution. For example, most wavelet-based methods assume the noise to be Gaussian. This is a reasonable assumption according to the central limit theorem, which states that the sum of different noise sources tends to approach a Gaussian distribution. In fact, some methods go as far as to require the actual variance of the noise distribution [8]. In contrast, the methods that will be presented in the following sections make no assumption regarding the underlying noise model. They use, however, some user-defined parameters that depend on the amount of the additive noise.

In the following, we will assume that the original "noisy" image is divided into smaller $N \times N$ square *regions of interest* (ROIs), as it is illustrated in Figure 18.1. Instead of applying the denoising process to the entire image, we will process each ROI sequentially. This is done for two reasons. Firstly, the time needed to solve the optimization tasks considered in the next sections increases polynomially with N. Secondly, working in each ROI separately enables us to change the parameters of the model in an adaptive manner, to account for the different levels of details in each ROI (Figure 18.2). Moreover, we will assume that each ROI represents the points on the surface of a function, $\hat{f}$, of two variables defined on $[0,1] \times [0,1]$. The pixel values of the digitized image are represented as $z_{n,m} = \hat{f}(x_n, y_m) \in [0, 255]$, where $x_n = n/(N-1)$, $y_m = m/(N-1)$ for $n, m = 0, 1, ..., N-1$. To simplify the notation, we will often rearrange the columns of each ROI so that it becomes a vector $\mathbf{z} \in \mathbb{R}^{N^2}$, as Figure 18.3 illustrates.

As this paper deals with kernel-based regression methods to address the denoising problem, we should also discuss another important issue. It is known that the accuracy of fit of these methods drops near the borders of the data set. There are several techniques to address this problem. For example, one could employ overlapping ROIs and finally compute the mean value over all

FIGURE 18.2: The two ROIs shown above differ in the level of details. The first one (A) belongs to a smooth area of the picture, while the second (B) contains edges.

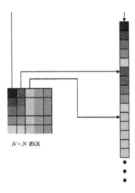

FIGURE 18.3: Rearranging the columns of a ROI so that it becomes a vector.

estimations, or discard the estimation of the pixels near the borders, etc. Another relative strategy is discussed in detail in Section 18.6.

18.3   Kernel Ridge Regression Modeling

This section provides two simple (non-adaptive) formulations of the denoising problem in RKHS. Motivated by these simple methods, we will present some more advanced techniques in the following sections.

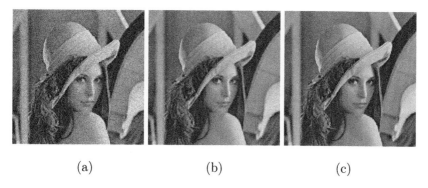

(a)   (b)   (c)

FIGURE 18.4: (a) The Lena image corrupted by 20 dB Gaussian noise (PSNR = 25.67 dB). (b) The reconstructed Lena image derived by the KRR-L_2 method with $\sigma = 0.5$, $\lambda = 4$, $N = 6$ (PSNR = 31.84 dB). (c) The reconstructed Lena image derived by the BM3D wavelet based method [8] with $\sigma = 15$ (PSNR = 34.6 dB).

18.3.1 KRR-L_2

A straightforward approach to the image denoising problem is to assume that the original noise free image, f, can be closely approximated by its projection to an RKHS induced by the Gaussian kernel. Hence, one may consider the so called *kernel ridge regression* (KRR) minimization task and solve:

$$\min_{g \in \mathcal{H}, c \in \mathbb{R}} \mathcal{C}(g, c) = \sum_{n,m=0}^{N-1} \left(g(x_n, y_m) + c - z_{n,m}\right)^2 + \lambda \|g\|_{\mathcal{H}}^2 + \lambda \cdot c^2, \quad (18.7)$$

where $f = g + c$. This is motivated by the universal approximation properties of the Gaussian kernel and by the fact that the RKHS associated to that kernel is comprised of smooth functions. Observe the bias factor, c, that has been explicitly used in (18.7). This is a common strategy in support vector regression (SVR) and kernel based regression tasks to improve their performance [29, 30, 25]. It turns out that the bias is important in order to counteract the effect of the regularizer, which affects the leveling of the solution. Recall that the semi-parametric representer theorem (Theorem 18.2) ensures that the solution of (18.7) takes the form $g_* = \sum_{n,m=0}^{N-1} a_n \kappa(\cdot, (x_n, y_m))$. Hence, the reconstructed denoised image is given by

$$f = \sum_{n,m=0}^{N-1} a_n \kappa(\cdot, (x_n, y_m)) + c. \quad (18.8)$$

To take the respective "noise free" pixels, one simply computes $f(x_n, y_m)$, for all $n, m = 0, 1, ..., N - 1$.

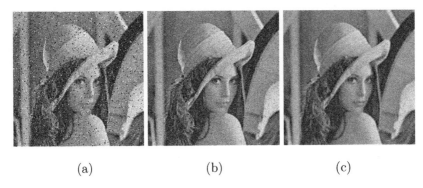

(a)   (b)   (c)

FIGURE 18.5: (a) The Lena image corrupted by 20 dB Gaussian noise and 5% salt and pepper noise (PSNR = 17.78 dB). (b) The reconstructed Lena image derived by the KRR-L_2 method with parameters $\sigma = 0.5$, $\lambda = 4$, $N = 12$ (PSNR = 26.89 dB). (c) The reconstructed Lena image derived by the KRR-L_2 method with parameters $\sigma = 0.5$, $\lambda = 50$, $N = 12$ (PSNR = 27.52 dB).

To solve (18.7), we recast it using matrix notation and replace g with its kernel expansion:

$$\mathcal{C}(a, c) = \left\| (K \quad 1) \begin{pmatrix} a \\ c \end{pmatrix} \right\|^2 + \lambda a^T K^T K a + \lambda \cdot c^2.$$

This is a strictly convex function with a unique minimum given by

$$\begin{pmatrix} a_* \\ c_* \end{pmatrix} = \left(A^T A + \lambda B \right)^{-1} A^T \mathbf{z}, \tag{18.9}$$

where $A = (K \quad c)$ and $B = \begin{pmatrix} K^T K & 0 \\ 0 & 1 \end{pmatrix}$. Besides the size of the ROIs, i.e., N, the task in (18.7) has two more user defined parameters: the kernel width, σ, and the regularization parameter, λ. Both can be used to control the denoising process. The larger the λ is, more influence is given to $\|g\|_{\mathcal{H}}^2$, resulting to a smoother picture. Similarly, large values of σ imply a smoother RKHS, also resulting in smoother estimates. Figures 18.4 and 18.5 demonstrate the KRR-L_2 denoising process described by (18.7) in two simple cases. In the first case, the Lena image is corrupted by 20 dB Gaussian noise. Figure 18.4 shows that KRR-L_2 can remove this type of noise relatively well (although not close enough to the state-of-the-art wavelet based methods). In the second case, the Lena image is corrupted by 20 dB Gaussian noise and 5% salt and pepper noise. In this situation, the KRR-L_2 rationale fails to effectively remove the outliers for all values of λ and σ (Figure 18.5). This is due to the square error loss function (L_2) employed by (18.7). For small values of λ and σ, the cost function forces f to fit the outlier data (i.e., the black and white pixels) as well as possible, resulting in small dark and light artifacts (Figure 18.5b). This

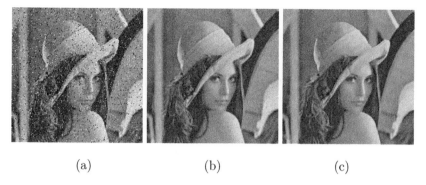

(a)   (b)   (c)

FIGURE 18.6: (a) The Lena image corrupted by 20 dB Gaussian noise and 5% salt and pepper noise (PSNR = 17.78 dB). (b) The reconstructed Lena image derived by the KRR-L_1 method with parameters $\sigma = 0.5$, $\lambda = 0.01$, $N = 5$ (PSNR = 30.13 dB). (c) The reconstructed Lena image derived by the BM3D [8] method with $\sigma = 45$ (PSNR = 29.7 dB)

effect is reduced as λ and/or σ are increased. However, in the latter case the resulting image is blurry (Figure 18.5c).

The KRR-L_2 denoising algorithm can be summarized in the following three steps:

- For each image pixel (i, j):

- Form the $N \times N$ ROI, $\mathbf{z}$ (so that the (i, j) pixel is at the top left corner of the ROI).

- Solve Problem (18.7) for that particular ROI.

- Move to the next pixel.

Note that each pixel is assigned to N^2 different values (since it belongs to each one of the N^2 regions of its neighboring pixels). The actual value that we assign to each pixel at the end of the algorithm is the mean of these values.

18.3.2   KRR-L_1

To enhance the performance of the KRR-L_2 denoising process in the case of impulse noise, one might consider replacing the L_2 (square) loss of (18.7) by an L_1 type loss function. This results in the KRR-L_1 method that can be cast as

$$\min_{g \in \mathcal{H}, c \in \mathbb{R}} \mathcal{C}(g, c) = \sum_{n,m=0}^{N-1} |g(x_n, y_m) + c - z_{n,m}| + \lambda \|g\|_{\mathcal{H}}^2. \quad (18.10)$$

Observe that in the KRR-L_1 method the bias is not penalized, in contrast to (18.7), where the penalization ensures the invertibility of the respective matrix. The task defined in (18.10) has the advantage that it relaxes the need to closely fit the noisy data, a property that can be advantageous in the case of impulse noise. Unfortunately, although its cost function is also a strictly convex function, the solution cannot be given in a closed form as in the case of the KRR-L_2 formulation. Moreover, note that (18.10) is not differentiable. Here, we mobilize the celebrated Polyak's projected subgradient method (see [20]). Polyak's Algorithm solves for the optimal value of $\mathbf{x} = \text{argmin}\{\mathcal{C}(\mathbf{x})\}$ iteratively and it can be summarized in the following recursion:

$$\mathbf{x}_{k+1} = \mathbf{x}_k - \gamma_k \cdot \frac{\nabla \mathcal{C}(\mathbf{x}_k)}{\|\nabla \mathcal{C}(\mathbf{x}_k)\|}$$

where γ_k is an arbitrary sequence such that $\sum_{k=1}^{\infty} \gamma_k = \infty$, $\sum_{k=1}^{\infty} \gamma_k^2 \in \mathbb{R}$ and $\nabla \mathcal{C}(\mathbf{x}_k)$ is any subgradient of the cost function $\mathcal{C}$ at $\mathbf{x}_k$. Hence, to implement the algorithm in the case of (18.10), we need to compute any of the subgradients $\mathcal{C}(g, c)$. Taking into account the reproducing property, after some algebra we can deduce that a suitable choice is:

$$\nabla \mathcal{C}(g, c) = \begin{pmatrix} \nabla_g \mathcal{C}(g, c) \\ \nabla_c \mathcal{C}(g, c) \end{pmatrix} =$$

$$= \begin{pmatrix} \sum_{n,m=0}^{N-1} \text{sign}\left(g(x_n, y_m) + c - z_{n,m}\right) \cdot \kappa\left(\cdot, (x_n, y_m)\right) + \lambda \cdot g \\ \sum_{n,m=0}^{N-1} \text{sign}\left(g(x_n, y_m) + c - z_{n,m}\right) \end{pmatrix}. \quad (18.11)$$

The KRR-L_1 denoising algorithm can be summarized in the following three steps:

- For each image pixel (i, j):

- Form the $N \times N$ "pixel centered" ROI, $\mathbf{z}$.

- Solve Problem (18.10) for that particular ROI using Polyak's method. (Here N is an odd number).

- Move to the next pixel.

Similar to the KRR-L_2 case, at the end of the algorithm the value of each image pixel is set as the mean of the N^2 values obtained by each ROI that contains it. Figure 18.6 compares the KRR-L_1 algorithm with the wavelet based method of [8]. It is shown that although the wavelet method provides a smoother picture (which is probably more eye pleasing), the KRR-L_1 method does a better job of preserving the edges and the details of the picture. KRR-L_1's main disadvantage is that it cannot effectively remove the "smaller" Gaussian noise resulting in obscure artifacts in the smooth areas of the picture. Although this effect can be reduced by increasing the regularization parameter λ of (18.10), it has the negative effect of producing a blurry image.

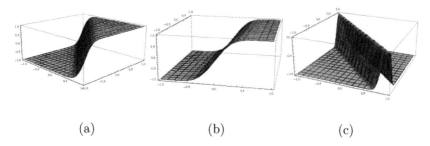

(a)	(b)	(c)

FIGURE 18.7: (a) Some of the functions that are used in the semi-parametric modeling. (a) $\mathrm{Erf}(4x + 4y)$, (b) $\mathrm{Erf}(3x)$, (c) $\mathrm{Erf}(-(4x + 4y)^2)$.

18.4 Semi-Parametric Modeling

Motivated by the KRR-L_1 formulation, in this section we describe a more advanced denoising method, by adopting the semi-parametric modeling (Theorem 18.2), as the means to remedy the smoothing effects associated with the problem formulation of the previous section. To this end, we consider a set of real valued two dimensional functions $\{\psi_k,\ k = 1, \ldots, K\}$, suitably selected to model edges. Although there can be numerous choices for this purpose, here we consider three types of functions: a) bivariate polynomials of order 1, b) functions of the form $\mathrm{Erf}(a \cdot x + b \cdot y + c)$, where Erf is the error function, i.e.,

$$\mathrm{Erf}(x) = \frac{2}{\sqrt{\pi}} \int_0^x e^{-t^2}\, dt,$$

(which can approximate ridges - see Figure 18.7(a), (b)) and c) functions of the form $\mathrm{Exp}(-(a \cdot x + b \cdot y + c)^2)$ (see Figure 18.7(c)) for several suitable predefined choices of a, b and c.

The regularized risk minimization problem is now reformulated as follows:

$$\underset{g \in \mathcal{H},\ \beta \in \mathbb{R}^K, h \in \mathbb{R}^4}{\text{minimize}}\quad \mathcal{C}(g, \boldsymbol{h}, \boldsymbol{\beta}) =$$

$$\frac{1}{N^2} \sum_{n=0}^{N-1} \sum_{m=0}^{N-1} \left| g(x_n, y_m) + h_0 + h_1 x_n + h_2 y_m + h_3 x_n y_m \right. \tag{18.12}$$

$$\left. + \sum_{k=1}^{K} \beta_k \psi_k(x_n, y_m) - z_{n,m} \right| + \frac{\lambda}{2N^2}\|f\|_{\mathcal{H}}^2 + \frac{\mu}{2K}\sum_{k=1}^{K}\beta_k^2 + \frac{\mu_1}{2}\sum_{l=1}^{3}h_l^2,$$

where $\boldsymbol{\beta} = (\beta_1, \ldots, \beta_K)$, $\boldsymbol{h} = (h_0, h_1, h_2, h_3)$. In this case, we assume that f_* belongs to the space $\mathcal{H} + \Psi + \mathcal{P}$, where $\Psi = \mathrm{span}\{\psi_k,\ k = 1, \ldots, K\}$ and $\mathcal{P}$ is the space of the bivariate polynomials of order 1. In other words, we recast Problem (18.10), to account for some extra parameters, i.e., β_k,

$k = 1, \ldots, K$, h_i, $i = 0, \ldots, 3$ (that contribute to the preservation of the fine details of the image), which are also regularized. In all the examples presented below $K = 32$. This approach (learning edge models from a rich set of basis functions) bears some similarities to the modeling taken by the K-SVD algorithm [10]. According to Theorem 18.2, f_* will have a finite representation of the form:

$$f_*(x, y) = \sum_{n=0}^{N-1} \sum_{m=0}^{M-1} \alpha_{n,m} \kappa((x_n, y_m), (x, y))$$

$$+ \sum_{k=1}^{K} \beta_k \psi_k(x, y) + h_0 + h_1 x + h_2 y + h_3 xy. \tag{18.13}$$

To solve (18.12) the Polyak's projected subgradient method is employed once more. After some algebra we can deduce that the required subgradients are:

$$\nabla \mathcal{C}(g, \boldsymbol{h}, \boldsymbol{\beta}) = (\nabla \mathcal{C}_g(g, \boldsymbol{h}, \boldsymbol{\beta}), \nabla \mathcal{C}_{h_0}(f, \boldsymbol{h}, \boldsymbol{\beta}), \ldots, \nabla \mathcal{C}_{h_3}(g, \boldsymbol{h}, \boldsymbol{\beta}),$$
$$\nabla \mathcal{C}_{\beta_1}(g, \boldsymbol{h}, \boldsymbol{\beta}), \ldots, \nabla \mathcal{C}_{\beta_K}(g, \boldsymbol{h}, \boldsymbol{\beta}))^T, \tag{18.14}$$

where

$$\nabla \mathcal{C}_g(g, \boldsymbol{h}, \boldsymbol{\beta}) = \frac{1}{N^2} \Big(\sum_{n=0}^{N-1} \sum_{m=0}^{N-1} sign\,(e_{n,m}(g, \boldsymbol{h}, \boldsymbol{\beta})) \cdot$$
$$\cdot \, \kappa\,((x_n, y_m), (\cdot, \cdot)) + \lambda \cdot f \Big), \tag{18.15}$$

$$\nabla \mathcal{C}_{h_0}(g, \boldsymbol{h}, \boldsymbol{\beta}) = \frac{1}{N^2} \left(\sum_{n=0}^{N-1} \sum_{m=0}^{N-1} sign\,(e_{n,m}(g, \boldsymbol{h}, \boldsymbol{\beta})) \right), \tag{18.16}$$

$$\nabla \mathcal{C}_{h_1}(g, \boldsymbol{h}, \boldsymbol{\beta}) = \frac{1}{N^2} \left(\sum_{n=0}^{N-1} \sum_{m=0}^{N-1} sign\,(e_{n,m}(g, \boldsymbol{h}, \boldsymbol{\beta})) \cdot x_n \right)$$
$$+ \mu_1 \cdot h_1, \tag{18.17}$$

$$\nabla \mathcal{C}_{h_2}(g, \boldsymbol{h}, \boldsymbol{\beta}) = \frac{1}{N^2} \left(\sum_{n=0}^{N-1} \sum_{m=0}^{N-1} sign\,(e_{n,m}(g, \boldsymbol{h}, \boldsymbol{\beta})) \cdot y_m \right)$$
$$+ \mu_1 \cdot h_2, \tag{18.18}$$

$$\nabla \mathcal{C}_{h_3}(g, \boldsymbol{h}, \boldsymbol{\beta}) = \frac{1}{N^2} \left(\sum_{n=0}^{N-1} \sum_{m=0}^{N-1} sign\,(e_{n,m}(g, \boldsymbol{h}, \boldsymbol{\beta})) \cdot x_n y_m \right)$$
$$+ \mu_1 \cdot h_3, \tag{18.19}$$

and

$$\nabla \mathcal{C}_{\beta_k}(g, \boldsymbol{h}, \boldsymbol{\beta}) = \frac{1}{N^2} \Big(\sum_{n=0}^{N-1} \sum_{m=0}^{N-1} sign\,(e_{n,m}(g, \boldsymbol{h}, \boldsymbol{\beta})) \cdot$$

$$\cdot \psi_k(x_n, y_m) \Big) + \frac{\mu}{K} \cdot \beta_k, \qquad (18.20)$$

for $k = 1, \ldots, K$, where the $e_{n,m}(g, \boldsymbol{h}, \boldsymbol{\beta})$ term is given by:

$$e_{n,m}(g, \boldsymbol{h}, \boldsymbol{\beta}) = g(x_n, y_m) + h_0 + h_1 x_n + h_2 y_m +$$

$$h_3 x_n y_m + \sum_{k=1}^{K} \beta_k \psi_k(x_n, y_m) - z_{n,m}.$$

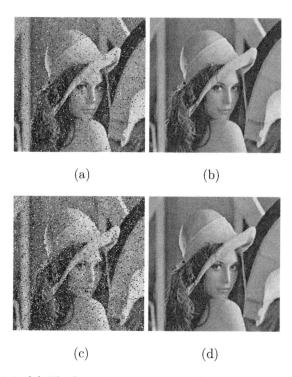

(a)   (b)

(c)   (d)

FIGURE 18.8: (a) The Lena image corrupted by 20 dB Gaussian noise and 5% salt and pepper noise (PSNR = 17.78 dB). (b) The reconstructed Lena image derived by the semi-parametric method with $N = 5$ (PSNR = 32.34 dB). (c) The Lena image corrupted by 20 dB Gaussian noise and 10% salt and pepper noise (PSNR = 15.10 dB). (d) The reconstructed Lena image derived by the semi-parametric method with $N = 5$ (PSNR = 31.69 dB).

The regularization parameters μ, μ_1 play an important role in the edge-preservation properties of the algorithm. The proposed algorithm adjusts μ and μ_1 in an adaptive manner, so that they take small values in ROIs that contain a lot of edges and large values in ROIs of smoother areas (the user defined regularization parameter λ is kept fixed). The reason for this approach is that small values for the regularization parameters, μ and μ_1, enhance the contribution of the semi-parametric part, which is desirable around edges, while in smoother regions, this effect needs to be suppressed. As the algorithm moves from one ROI to the next, it solves Problem (18.12) using suitably selected regularization parameters. More specifically, we consider six different types of ROIs, depending on how large the respective mean gradient of their pixels is. Once the type of region has been decided, we assign values to μ and μ_1 accordingly. This is accomplished using the values of the vectors $\boldsymbol{\mu}$ and $\boldsymbol{\mu}_1$ (6 elements each). The elements of those vectors, i.e., μ_i and $\mu_{1,i}$, $i = 1, 2 \ldots, 6$, contain the regularization values associated with any region of type i. For example, if the algorithm decides that the current ROI belongs to a very smooth area, it associates that ROI to the number 6 and sets $\mu = \boldsymbol{\mu}_6$ and $\mu_1 = \boldsymbol{\mu}_{1,6}$. In effect, the proposed method gives higher weight to kernel smoothing in smooth regions and performs sparse modeling for edge regions. A more thorough analysis of this algorithm with all the implementation details can be found in [4] and the respective code can be found in bouboulis.mysch.gr/kernels.html. A simpler version of the algorithm can be summarized in the following steps:

1. Input: The noisy image, the size of the ROIs, N, and the regularization parameters λ, $\boldsymbol{\mu}$, $\boldsymbol{\mu}_1$.

   (a) Compute the mean gradient of all pixels in each ROI.

   (b) Assign each ROI to a specific type, say i, according to the mean gradient of its pixels.

   (c) Set $\mu = \boldsymbol{\mu}_i$ and $\mu_1 = \boldsymbol{\mu}_{1,i}$.

   (d) Solve Problem 18.12 using the regularization parameters λ, μ, μ_1 using the Polyak's projected subgradient method.

   (e) Move to the next ROI.

2. Output: The denoised image.

As is shown in Figure 18.8, the semi-parametric denoising method based on the task given in (18.12) and Polyak's subgradient method demonstrates significantly superior behavior compared to the state-of-the-art wavelet based denoising methods (Figure 18.6), when both Gaussian and impulse noise are present. The main drawback of this approach is an increased need for computational resources. In fact, the algorithm (implemented in C) may require more than 5 minutes to remove the noise from a typical 512×512 grayscale picture. Moreover, in the case of images corrupted by Gaussian noise only, wavelet-based methods are more efficient (Figure 18.9).

(a)   (b)   (c)

FIGURE 18.9: (a) The Lena image corrupted by 20 dB Gaussian noise (PSNR = 25.67 dB). (b) The reconstructed Lena image derived by the semiparametric method (PSNR = 32.66 dB). (c) The reconstructed Lena image derived by the BM3D [8] method with $\sigma = 15$ (PSNR = 34.6 dB).

18.5   A Robust Approach

Most image denoising methods (including the ones presented in the previous sections) adopt Equation (18.6) as a means to model the relation between the noisy and the original image and minimize for different kinds of empirical loss, without explicitly taking into account the outliers. In this context, these techniques are suitable for removing white Gaussian noise, as any other type of noise that follows a heavy tailed distribution (e.g., Gamma distribution, impulses, etc.), usually causes the model to overfit (especially when the model parameters are chosen so that the fine details of the image are preserved). On the other hand, if the priority is to enhance the smoothing effect and remove the complete set of outliers, then the fine details will probably be lost. To resolve these issues, this section considers a more robust approach. The outliers are explicitly modeled as a sparse vector, $\boldsymbol{u}$, and the popular *orthogonal matching pursuit* (OMP) algorithm is employed to identify its support. Hence, the relation between the noisy and the original image is modeled by

$$z_i = f(\mathbf{x}_i) + \eta_i + \boldsymbol{u}_i, \tag{18.21}$$

where z_i, $i = 1, \ldots, N^2$ are the pixel values of the $N \times N$ digitized image (rearranged to become a vector, as Section 18.2 outlines), $f(\mathbf{x}) = \sum_{j=1}^{N^2} a_j \kappa(\cdot, \mathbf{x}_j) + c$, is the nonlinear function representing the noisy-free image, η is an unobservable bounded noise sequence, and $\boldsymbol{u}$ is a sparse vector representing the possible outliers. Note that the image can be corrupted by any type of noise (e.g., Gaussian) plus some outliers. In this case, we consider that η is the noise contribution within some bounds, while beyond those bounds the noise

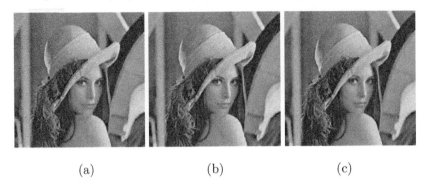

(a)   (b)   (c)

FIGURE 18.10: (a) The Lena image has been corrupted by 20 dB Gaussian noise and 5% salt and pepper noise (PSNR = 17.78 dB) and then reconstructed using sparse modeling (KROMP). Model parameters were set to $\lambda = 1$, $\sigma = 0.3$, and $\epsilon = 45$ (PSNR = 32.14 dB). (b) The Lena image has been corrupted by 20 dB Gaussian noise and 10% salt and pepper noise (PSNR = 15.10 dB) and then reconstructed via KROMP. Model parameters: $\lambda = 1$, $\sigma = 0.35$, and $\epsilon = 45$ (PSNR = 31.3 dB). (c) The Lena image has been corrupted by 20 dB Gaussian noise and then reconstructed using KROMP. Model parameters were set to $\lambda = 1$, $\sigma = 0.3$, and $\epsilon = 45$ (PSNR = 32.68 dB).

components are considered to be outliers. In this context, our objective is to solve

$$
\begin{aligned}
&\underset{a,u\in\mathbb{R}^{N^2} c\in\mathbb{R}}{\text{minimize}} && \|u\|_0 \\
&\text{subject to} && \sum_{i=1}^{N^2}\left(z_i - \sum_{j=1}^{N^2} a_j \kappa(\mathbf{x}_i, \mathbf{x}_j) - c - u_j\right)^2 + \lambda(c^2 + \|a\|^2) \le \epsilon,
\end{aligned}
\tag{18.22}
$$

where $\lambda > 0$ is the regularization parameter. Observe that the regularization of the constraint in (18.22) does not employ the standard RKHS norm (i.e., $\|f\|_{\mathcal{H}}^2 = a^T K a$), which excels in terms of generalization performance. Instead, as our primary goal is to reduce the square error (and we do not care about the generalization performance), we choose to directly penalize the elements of a. Adopting matrix notation, (18.22) can be shown to take the form:

$$
\begin{aligned}
&\underset{a,u\in\mathbb{R}^{N^2} c\in\mathbb{R}}{\text{minimize}} && \|u\|_0 \\
&\text{subject to} && \|\mathbf{z} - A\mathbf{w}\|^2 + \lambda \mathbf{w}^T B \mathbf{w} \le \epsilon,
\end{aligned}
\tag{18.23}
$$

where

$$
A = \begin{bmatrix} K & 1 & I_n \end{bmatrix}, \mathbf{w} = \begin{bmatrix} \alpha \\ c \\ u \end{bmatrix}, B = \begin{bmatrix} I_{N^2} & 0 & O_{N^2} \\ 0^T & 1 & 0^T \\ O_{N^2} & 0 & O_{N^2} \end{bmatrix},
$$

Algorithm 16: Kernel Regularized OMP (KROMP)

Input: K, $\boldsymbol{y}$, λ, ϵ

  Initialization: $k := 0$

  $S_{ac} = \{1, 2, ..., n+1\}$, $S_{inac} = \{n+2, ..., 2n+1\}$

  $A_{ac} = [K\ \mathbf{1}]$, $A_{inac} = I_n = [\boldsymbol{e}_1 \cdots \boldsymbol{e}_n]$

  Solve: $\boldsymbol{w}^{(0)} := \operatorname{argmin}_{\boldsymbol{w}} ||A_{ac}\boldsymbol{w} - \mathbf{z}||_2^2 + \lambda ||\boldsymbol{\alpha}||_2^2 + \lambda c^2$

  Initial Residual: $\boldsymbol{r}^{(0)} = A_{ac}\boldsymbol{w}^{(0)} - \mathbf{z}$

  while $||\boldsymbol{r}^{(k)}||_2 > \epsilon$ **do**

    $k := k + 1$

    Find: $j_k := \operatorname{argmax}_{j \in S_{inac}} |r_j^{(k-1)}|$

    Update Support:

    $S_{ac} = S_{ac} \cup \{j_k\}$, $S_{inac} = S_{inac} - \{j_k\}$

    $A_{ac} = [A_{ac}\ \boldsymbol{e}_{j_k}]$

    Update Current solution:

    $\mathbf{z}^{(k)} := \operatorname{argmin}_{\mathbf{z}} ||A_{ac}\boldsymbol{w} - \mathbf{z}||_2^2 + \lambda ||\boldsymbol{\alpha}||_2^2 + \lambda c^2$

    Update Residual: $\boldsymbol{r}^{(k)} = A_{ac}\boldsymbol{w}^{(k)} - \mathbf{z}$

  end while

Output: $\boldsymbol{w} = (\boldsymbol{\alpha}, c, \boldsymbol{u})^T$ after k iterations

---

while I_{N^2} denotes the unitary matrix, $\mathbf{1}$ and $\mathbf{0}$ the vectors of ones and zeros, respectively, and O_{N^2} the all zero square matrix. Problem 18.23 aims at finding the sparsest vector of outliers, $\boldsymbol{u}$, so that the mean square error between the actual noisy data and the estimated noise-free data plus the outliers remains below a certain tolerance, ϵ, while, at the same time, the solution remains smooth. The latter is ensured by the regularization part of (18.23) as in KRR-L_2. The solution of (18.23), determines the noise-free image, i.e., $\boldsymbol{y}_{est} = K\boldsymbol{a} + c\mathbf{1}$.

To solve (18.23), we employ a two-step iterative approach. Firstly, A is initialized as $A^{(0)} = [K\ \ \mathbf{1}]$. The first step adopts the OMP rationale to compute the most correlated column, j_k, of I_n with respect to the latest residual, i.e., $\boldsymbol{r}^{(k)} = A\boldsymbol{w}^{(k)} - \mathbf{z}$, and then A is augmented to include the $\boldsymbol{e}_{j_k}$ column. The second step attempts to solve $\min_{\boldsymbol{w}} J(\boldsymbol{w})$. It takes a few lines of elementary algebra to conclude that $J(\boldsymbol{w})$ attains a unique minimum, which can be updated iteratively using Cholesky decomposition. Similar approaches, which are based on the OMP rationale to identify the support of sparse vectors, have already been used in denoising tasks [5]. The procedure proposed here is called *Kernel Regularized OMP* (KROMP) and it is summarized in Algorithm 16. More details regarding KROMP, including convergence properties, recovery of the support, and the solution of $\min_{\boldsymbol{w}} J(\boldsymbol{w})$ using Cholesky decomposition (which is not shown here) can be found in [17, 3, 18]. In the following, we give only the major results, omitting the respective proofs.

Theorem 18.3. *The norm of the residual vector,* $r^{(k)} = A_{ac}w^{(k)} - z$, *in Algorithm 16 is strictly decreasing. Moreover, the algorithm will always converge, i.e.,* $\|r^{(k)}\|_2$ *will eventually drop below* ϵ.

Theorem 18.4. *Assuming that* z *admits a representation of the form* $z = Ka_0 + c_0 \mathbf{1} + u_0 + \eta$, *where* $\|u\|_0 = S$ *and* $\|\eta\|_2 \leq \epsilon$, *then Algorithm 16 recovers the exact sparsity pattern of* u, *after* N *steps, if*

$$\epsilon \leq \frac{\left| \min_{k=1,2,\ldots,N^2} \{u_k, u_k \neq 0\} \right|}{2} - \|f_0 - f^{(0)}\|_2,$$

with $f_0 = \begin{pmatrix} K & 1 \end{pmatrix} \cdot \begin{pmatrix} a_0 \\ c_0 \end{pmatrix}$ *and* $f^{(0)} = \begin{pmatrix} K & 1 \end{pmatrix} \cdot \begin{pmatrix} a^{(0)} \\ c^{(0)} \end{pmatrix}$.

Theorem 18.5. *Assuming that* z *admits a representation as in Theorem 18.4, then the approximation error of Algorithm 16, after* S *steps, is bounded by*

$$\|w^{(N)} - w_0\| \leq \frac{|\lambda| \|(a_0 \quad c_0)\| + \sqrt{N}\left(1 + \|(K \quad 1)\|_2\right)\epsilon}{\lambda_{min}(\Omega^T\Omega + \lambda I_0)},$$

where $\lambda_{min}(\Omega^T\Omega + \lambda I_0)$ *is the minimum eigenvalue of the matrix* $\Omega^T\Omega + \lambda I_0$, *where*

$$\Omega = \begin{pmatrix} K & 1 & I \end{pmatrix}, \quad I_0 = \begin{pmatrix} 0 & 0 \\ 0 & I \end{pmatrix}.$$

Similar to the case of KRR-L_2, the user-selected parameter λ, controls the quality of the reconstruction regarding the bounded noise. The parameter ϵ on the other hand, controls the recovery of the outliers' pattern. Small values of ϵ lead to the recovery of all noise samples (even those originating from a Gaussian source) as impulses, filling up the vector u, which will no longer be sparse. Similarly, if λ is very small, f will closely fit the noisy data including some possible outliers (overfitting). In contrast, if ϵ is very large, only a handful of outliers (possibly none) will be detected. If λ is large, then f will be smooth resulting in a blurry picture.

Another important issue that needs to be addressed is that the accuracy of fit of any kernel-based regression technique drops near the borders of the data set. Hence we need to take special care of these points. The proposed denoising algorithmic scheme takes this important fact into consideration, removing these points from the reconstructed image. The procedure splits the noise image into small square ROIs, with dimensions $N_1 \times N_1$. Then, it applies the KROMP algorithm (see Algorithm 16) sequentially to each ROI. However, only the centered $N_2 \times N_2$ pixels are used to reconstruct the noise free image. Hence, in order to avoid gaps in the final reconstructed image, the ROIs contain overlapping sections (Figure 18.11).

It is evident that this procedure requires three user defined parameters: (a) the regularization parameter, λ, (b) the parameter ϵ that controls the outlier

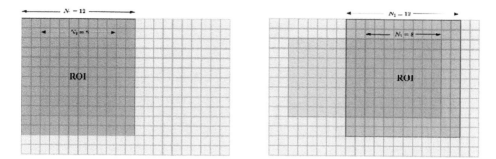

FIGURE 18.11: Although each ROI contains $N_1 \times N_1$ pixels (dark gray), only $N_2 \times N_2$ of them are used for the reconstruction of the image. This is to account for the reduced accuracy of the reconstruction at the pixels lying close to the borders of the respective ROI. Hence, each ROI has overlapping sections with its neighbors.

selection of the OMP mechanism, and (c) the Gaussian kernel parameter σ. The size of the ROIs can also be adjusted by the user, but in the following we will consider that $N_1 = 12$ and $N_2 = 8$. As λ regulates the smoothness of the estimation, it is controlled in an adaptive manner, so that smaller values of λ are used in areas of the picture that contain edges. Hence, while the user defines a single value for λ, say λ_0, the denoising scheme computes the mean gradient magnitude of each ROI and classifies them into three categories. The ROIs of the first class, i.e., those with the largest values of gradient magnitude, solve (18.23) using $\lambda = \lambda_0$, while for the other two classes the algorithm employs $\lambda = 5 * \lambda_0$ and $\lambda = 15 * \lambda_0$, respectively. Such choices came out after extensive experimentation, and the algorithm is not sensitive to them. The denoising scheme can be summarized in the following steps:

- Compute the mean gradient magnitude of each $N_2 \times N_2$ ROI.

- Classify them into three categories.

- For each $N_2 \times N_2$ ROI, say $\hat{R}$:

  1. Extend the ROI to size $N_1 \times N_1$. Let $\hat{R}_e$ be the extended ROI.
  2. Apply the KROMP algorithm to the extended ROI using the respective λ. Let R_e be the output of the algorithm.
  3. Reconstruct the noise-free ROI, R, taking only the centered $N_2 \times N_2$ points of R_e.

Compared to the semi-parametric case (Section 18.4), the present denoising approach offers comparable performance (see Figures 18.9 and 18.10), while at the same time it admits significantly reduced complexity. While a typical semi-parametric denoising task requires several minutes to complete,

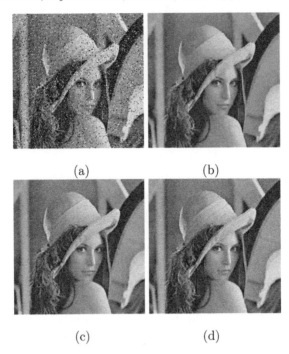

(a)   (b)

(c)   (d)

FIGURE 18.12: (a) The Lena image corrupted by 20 dB Gaussian noise and 5% impulses generated by a uniform distribution in the interval $[-300, 300]$ (PSNR = 19.23 dB). (b) The reconstructed Lena image derived by BM3D [8] with $\sigma = 40$ (PSNR = 30.6 dB). (c) The reconstructed Lena image derived by the semi-parametric model with $N = 5$ (PSNR = 32.39 dB). (d) The reconstructed Lena image derived by KROMP with $\sigma = 0.3$, $\lambda = 1$, and $\epsilon = 40$ (PSNR = 32.14 dB).

the denoising approach offered by KROMP takes only a few seconds. This is due to slow convergence of the subgradient technique adopted in the semi-parametric scenario. Table 18.1 compares the performance of the denoising method based on KROMP with the wavelet based denoising method BM3D presented in [8]. It is shown that KROMP offers superior results when the larger portion of the noise is due to outliers. In cases where the Gaussian noise is dominant, the BM3D offers comparable or even better performance (Table 18.1).

TABLE 18.1: Comparing the wavelet-based BM3D method of [8] with KROMP. In the case of the impulse noise the term "sp" stands for "salt and pepper" noise, while the term "u" stands for uniform impulses in the intraval $[-300, 300]$.

| Image | Noise | | PSNR | Denoising Method | | | |
| | Gaussian | Impulse | | BM3D | | KROMP | |
				PSNR	parameters	PSNR	parameters
Lena	20 dB	5% sp	17.78 dB	29.70 dB	$\sigma = 45$	32.14 dB	$\sigma = 0.3, \lambda = 1, \epsilon = 45$
Lena	20 dB	10% sp	15.10 dB	28.30 dB	$\sigma = 45$	31.31 dB	$\sigma = 0.35, \lambda = 1, \epsilon = 45$
Lena	20 dB	5% u	19.23 dB	30.60 dB	$\sigma = 45$	32.14 dB	$\sigma = 0.3, \lambda = 1, \epsilon = 45$
Lena	20 dB	10% u	16.80 dB	29.43 dB	$\sigma = 45$	31.55 dB	$\sigma = 0.35, \lambda = 1, \epsilon = 45$
Lena	15 dB	5% sp	16.47 dB	29.30 dB	$\sigma = 45$	29.71 dB	$\sigma = 0.4, \lambda = 2, \epsilon = 70$
Lena	15 dB	10% sp	14.42 dB	27.90 dB	$\sigma = 45$	29.11 dB	$\sigma = 0.4, \lambda = 2, \epsilon = 70$
Lena	15 dB	5% u	17.63 dB	30.31 dB	$\sigma = 40$	29.80 dB	$\sigma = 0.35, \lambda = 1.5, \epsilon = 70$
Lena	15 dB	10% u	15.84 dB	29.02 dB	$\sigma = 45$	29.18 dB	$\sigma = 0.4, \lambda = 2, \epsilon = 70$
Boat	20 dB	5% sp	17.52 dB	27.82 dB	$\sigma = 45$	29.74 dB	$\sigma = 0.3, \lambda = 1, \epsilon = 45$
Boat	20 dB	10% sp	14.92 dB	26.59 dB	$\sigma = 45$	29.10 dB	$\sigma = 0.3, \lambda = 1, \epsilon = 45$
Boat	20 dB	5% u	19.08 dB	28.81 dB	$\sigma = 40$	29.79 dB	$\sigma = 0.3, \lambda = 1, \epsilon = 45$
Boat	20 dB	10% u	16.65 dB	27.76 dB	$\sigma = 40$	29.39 dB	$\sigma = 0.3, \lambda = 1, \epsilon = 45$
Boat	15 dB	5% sp	16.16 dB	27.41 dB	$\sigma = 45$	27.60 dB	$\sigma = 0.35, \lambda = 1.5, \epsilon = 70$
Boat	15 dB	10% sp	14.16 dB	26.21 dB	$\sigma = 45$	27.08 dB	$\sigma = 0.35, \lambda = 1.5, \epsilon = 70$
Boat	15 dB	5% u	17.25 dB	28.48 dB	$\sigma = 35$	27.68 dB	$\sigma = 0.35, \lambda = 1.5, \epsilon = 70$
Boat	15 dB	10% u	15.57 dB	27.25 dB	$\sigma = 45$	27.12 dB	$\sigma = 0.35, \lambda = 1.5, \epsilon = 70$

18.6   Conclusions

Motivated by the Kernel Ridge Regression scheme and the theory of sparse representations, we presented two methods suitable for image denoising tasks. Both techniques employ the popular Gaussian kernel and change their parameters adaptively in each image region according to the amount of fine details contained. Besides the RKHS modeling (based on the semi-parametric representer theorem), the first method employs a rich basis of functions to model image edges and minimizes a regularized L_1 error function to account for outliers. In contrast, the second method models the outliers explicitly using a sparse vector, $\boldsymbol{u}$, and minimizes the L_0 norm of $\boldsymbol{u}$, over the set where the L_2 error function remains small. Each outlier is selected iteratively using an orthogonal matching pursuit rationale and the procedure is terminated when the L_2 error drops below a predefined threshold. Both methods excel, compared to the most up-to-date wavelet methods, when the underlying noise has strong heavy tailed components (e.g., impulses), as it is demonstrated through experiments. Although the first method offers slightly better performance, it has significantly increased complexity, due to slow convergence (a common disadvantage of Polyak's method). On the other hand, the second method offers comparable complexity with the latest wavelet based methods.

Bibliography

[1] E. Abreu, M. Lightstone, S. Mitra, and K. Arakawa. A new efficient approach for the removal of impulse noise from highly corrupted images. *IEEE Trans. Im. Proc.*, 5:1012–1025, 1996.

[2] E. Besdok. Impulsive noise suppression from images by using anfis interpolant and lillietest. *EURASIP J. Appl. Si. Pr.*, 16:2423–2433, 2004.

[3] P. Bouboulis, G. Papageorgiou, and S. Theodoridis. Robust image denoising in rkhs via orthogonal matching pursuit. In *ICASSP*, 2014 to appear.

[4] P. Bouboulis, K. Slavakis, and S. Theodoridis. Adaptive kernel-based image denoising employing semi-parametric regularization. *IEEE Transactions on Image Processing*, 19(6):1465–1479, 2010.

[5] Alfred M. Bruckstein, David L. Donoho, and Michael Elad. From sparse solutions of systems of equations to sparse modeling of signals and images. *SIAM review*, 51(1):34–81, 2009.

[6] P. L. Combettes and J.-C. Pesquet. Image restoration subject to a total variation constraint. *IEEE Trans. Im. Proc.*, 13(9):1213–1222, 2004.

[7] N. Cristianini and J. Shawe-Taylor. *An introduction to support vector machines and other kernel-based learning methods.* Cambridge University Press, 2000.

[8] K. Dabov, A. Foi, V. Katkovnik, and K. Egiazarian. Image denoising by sparse 3d transform-domain collaborative filtering. *IEEE Transactions on Image Processing*, 16(8):2080–2095, 2007.

[9] S. Durand and J. Froment. Reconstruction of wavelet coefficients using total variation minimization. *SIAM J. Sci. Comput.*, 24:1754–1767, 2003.

[10] M. Elad and M. Aharon. Image denoising via sparse and redundant representations over learned dictionaries. *IEEE Tran. Im. Proc.*, 15(12):3736–3745, 2006.

[11] R. C. Gonzalez and R.E. Woods. *Digital Image Processing.* Prentice Hall, 2002.

[12] K. Kim, M. O. Franz, and B. Scholkopf. Iterative kernel principal component analysis for image modeling. *IEEE Trans. Pattern Anal. Mach. Intell.*, 27(9):1351–1366, 2005.

[13] G. S. Kimeldorf and G. Wahba. Some results on Tchebycheffian spline functions. *J. Math. Anal. Applic.*, 33:82–95, 1971.

[14] Dalong Li. Support vector regression based image denoising. *Image Vision Comput.*, 27:623–627, 2009.

[15] J. Liu, X.-C. Tai, H. Huang, and Z. Huan. A weighted dictionary learning model for denoising images corrupted by mixed noise. *Image Processing, IEEE Transactions on*, 22(3):1108–1120, 2013.

[16] S. Osher and L. I. Rudin. Feature-oriented image enhancement using shock filters. *SIAM J. Numer. Anal.*, 27:919–940, 1990.

[17] G. Papageorgiou, P. Bouboulis, and S. Theodoridis. Robust kernel-based regression using orthogonal matching pursuit. In *MLSP*, September 2013.

[18] G. Papageorgiou, P. Bouboulis, and S. Theodoridis. Robust kernel-based regression using orthogonal matching pursuit with applications to image denoising. to appear.

[19] M. Petrou and C. Petrou. *Image Processing: The Fundamentals,.* Wiley, 2nd edition, 2010.

[20] B. T. Polyak. *Introduction to Optimization.* New York: Optimization Software, 1987.

[21] J. Portilla, V. Strela, M. Wainwright, and E. P. Simoncelli. Image denoising using scale mixtures of gaussians in the wavelet domain. *IEEE Transactions on Image Processing,* 12(11):1338–1351, 2003.

[22] R. Garnett, T. Huegerich, and C. Chui. A universal noise removal algorithm with an impulse detector. *IEEE Trans. Im. Proc.,* 14(11):1747–1754, 2005.

[23] L. I. Rudin and S. Osher. Total variation based image restoration with free local constraints. *Proc. IEEE Int. Conf. Image Processing,* 1:31–35, 1994.

[24] L. I. Rudin, S. Osher, and E. Fatemi. Nonlinear total variation based noise removal algorithms. *Physica D.,* 60:259–268, 1992.

[25] B. Scholkopf and A.J. Smola. *Learning with Kernels: Support Vector Machines, Regularization, Optimization and Beyond.* MIT Press, 2002.

[26] L. Sendur and I.W. Selesnick. Bivariate shrinkage functions for wavelet-based denoising exploiting interscale dependency. *IEEE Trans. Signal Process.,* 50(11):2744–2756, 2002.

[27] K. Seongjai. PDE-based image restoration: A hybrid model and color image denoising. *IEEE Trans. Im. Proc.,* 15(5):1163–1170, 2006.

[28] J. Shawe-Taylor and N. Cristianini. *Kernel Methods for Pattern Analysis.* Cambridge University Press, 2004.

[29] K. Slavakis, P. Bouboulis, and S. Theodoridis. Online learning in Reproducing Kernel Hilbert Spaces. In Rama Chellappa and Sergios Theodoridis, editors, *Academic Press Library in Signal Processing,* volume 1, Signal Processing Theory and Machine Learning, Chapter 17, pages 883–987. Academic Press, 2014.

[30] J.A.K. Suykens, T. Van Gestel, J. De Brabanter, B. De Moor, and J. Vandewalle. *Least Squares Support Vector Machines.* World Scientific, Singapore, 2002.

[31] H. Takeda, S. Farsiu, and P. Milanfar. Kernel regression for image processing and reconstruction. *IEEE Tran. Im. Proc.,* 16(2):349–366, 2007.

[32] S. Theodoridis and K. Koutroumbas. *Pattern Recognition.* Academic Press, 4th edition, Nov. 2008.

[33] V.N. Vapnik. *The Nature of Statistical Learning Theory.* Springer Verlag, 1999.

[34] Alan L. Yuille and Norberto M. Grzywacz. A mathematical analysis of the motion coherence theory. *International Journal of Computer Vision*, 3(2):155–175, 1989.

[35] S. Zhang and Y. Chen. Image denoising based on wavelet support vector machine. *IICCIAS 2006*.

Chapter 19

Single-Source Domain Adaptation with Target and Conditional Shift

Kun Zhang

Max Planck Institute for Intelligent Systems

Bernhard Schölkopf

Max Planck Institute for Intelligent Systems

Krikamol Muandet

Max Planck Institute for Intelligent Systems

Zhikun Wang

Max Planck Institute for Intelligent Systems

Zhi-Hua Zhou

National Key Laboratory for Novel Software Technology, Nanjing University

Claudio Persello

Max Planck Institute for Intelligent Systems and University of Trento

19.1   Introduction

The goal of supervised learning is to infer a function f from a training set $\mathbf{D}^{tr} = \{(x_1^{tr}, y_1^{tr}), ..., (x_m^{tr}, y_m^{tr})\} \subseteq \mathcal{X} \times \mathcal{Y}$, where $\mathcal{X}$ and $\mathcal{Y}$ denote the domains of predictors X and target Y, respectively. The estimated f is expected to generalize well on the test set $\mathbf{D}^{te} = \{(x_1^{te}, y_1^{te}), ..., (x_n^{te}, y_n^{te})\} \subseteq \mathcal{X} \times \mathcal{Y}$, where y_i^{te} are unknown. Traditionally, the training set and test set are assumed to follow the same distribution. However, in many real world problems, the training data and test data have different distributions, i.e., $P_{XY}^{tr} \neq P_{XY}^{te}$,[1] and the goal is to find a learning machine that performs well on the test domain. This problem is known as *domain adaptation* in machine learning.

If the data distribution changes arbitrarily, training data would be of no use to make predictions on the test domain. To perform domain adaptation successfully, relevant knowledge in the training (or source) domain should be transferred to the test (or target) domain. For instance, the situation where P_{XY}^{tr} and P_{XY}^{te} only differ in the marginal distribution of the covariate (i.e., $P_X^{tr} \neq P_X^{te}$, while $P_{Y|X}^{tr} = P_{Y|X}^{te}$) is termed covariate shift [25, 33, 10] or sample selection bias [37], and has been well studied. For surveys on domain adaptation for classification, see, e.g., [13, 17, 1].

In particular, we address the situation where both the marginal distribution P_X and the conditional distribution $P_{Y|X}$ may change across the domains. Clearly, we need to make certain assumptions for the training domain to be adaptable to the test domain. We first consider the case where $P_{X|Y}$ is the same on both domains. As a consequence of Bayes' rule, the changes in P_X and $P_{Y|X}$ are caused by the change in P_Y, the marginal distribution of the target variable. We term this situation *target shift* (TarS) which is frequently encountered in practice; for instance, it is known as choice-based or endogenous stratified sampling [15] in econometrics, and is sometimes called prior probability shift [31].

We further discuss the situation where P_Y remains the same, while $P_{X|Y}$ changes, as termed *conditional shift* (ConS). Estimation of $P_{X|Y}^{te}$ under ConS is in general ill-posed; we consider a rather practical yet identifiable case where $P_{X|Y}$ changes under location-scale (LS) transformations on X. We show how to transform the training points to mimic the distribution of test data and facilitate learning on the test domain. Finally, the situation in which both P_Y and $P_{X|Y}$ change across domains is termed *generalized target shift* (GeTarS);

---

[1]We use P to denote the probability density or mass function.

we focus on LS-GeTarS, i.e., GeTarS with $P_{X|Y}$ changes under LS transformations, and propose practical methods to estimate both changes, making domain adaptation possible.

Recently there have been major advances in discovering causal information from purely observational data [29, 18, 24, 9, 38, 39, 16, 11, 40]. In particular, it has been shown that causal information can be derived from changes in data distributions [34]; on the other hand, knowledge of the data generating process, or causal knowledge, would imply how the data distribution changes across domains and helps in domain adaptation. It has been demonstrated that a number of learning tasks, especially semi-supervised learning, can be understood from the causal point of view [22]. The problems studied here, TarS, ConS, and GeTarS, have clear causal interpretations. Throughout this chapter, we assume that Y is a cause of X.[2] If we further know that X depends on the domain (or selection variable) only via Y, we have the *TarS* situation: the marginal distribution of the cause, P_Y, describes the process that generates Y in the domain, and $P_{X|Y}$ describes the data generating mechanism for X from the cause Y, which is independent of the domain. According to [35], the invariance of $P_{X|Y}$ with respect to the change in P_Y is one of the features of the causal system $Y \to X$. Consider the clinical diagnosis as an example. The disease is naturally considered as the cause of symptoms; moreover, the marginal distribution of the disease could change across different regions, but the conditional distribution of the symptoms given the disease is expected to be invariant. Furthermore, if both Y and the domain are causes of X while Y is independent of the domain, we have the ConS situation. More generally, the situation where Y is a cause of X and both P_Y and $P_{X|Y}$ depend on the domain corresponds to GeTarS.

In the classification scenario, target shift was referred to the class imbalance problem by [12]. To solve it, sometimes it is assumed that P_Y^{te} is known *a priori* [14], or that some knowledge about the change in P_Y is known [36]. However, this is usually not the case in practice. In [2], the authors proposed to estimate P_Y^{te} with an expectation-maximization (EM) algorithm. Unfortunately, this approach has to estimate $P_{X|Y}^{tr}$, which is a difficult task if the dimensionality of X is high; moreover, it does not apply to regression problems. In fact, lack of information on P_Y^{te} causes the main difficulty in domain adaptation under TarS.

In this chapter we provide practical approaches for domain adaptation under TarS, LS-ConS, and LS-GeTarS, by sample importance reweighting or sample transformation.[3] The approach for TarS also applies to regression. Kernel embedding of both conditional and marginal distributions provides a

---

[2]This is usually the case, especially for classification: In many cases features were generated from classes, while the latter are hidden and to be estimated. For instance, in handwriting digit recognition, the written digit image was generated from the digit one intended to write, which is to be estimated from the image.

[3]A preliminary version was presented at the 30th International Conference on Machine Learning [41].

convenient tool to estimate the importance weights and the sample transformations. With it, we are able to avoid estimating any distribution explicitly, and the proposed approaches apply to high-dimensional problems without any difficulty. We note that kernel distribution embedding has been used to correct for covariate shift in [10, 6], but the studied problems are inherently different: they used the kernel mean matching to estimate the ratio P_X^{te}/P_X^{tr}, avoiding estimating P_X^{te} and P_X^{tr} explicitly from data; in our problems we are interested in how P_Y^{te} is different from P_Y^{tr} (for TarS and GeTarS) or how $P_{X|Y}^{tr}$ changes to $P_{X|Y}^{te}$ (for ConS and GeTarS), but there are no data points available to estimate P_Y^{te} or $P_{X|Y}^{te}$, making the problems much more difficult to solve.

19.2   Distribution Shift Correction

In this section, we outline two different frameworks for distribution shift correction, namely, *importance reweighting* and *sample transformation*, and briefly introduce the classification and regression machines used in this chapter.

19.2.1   Importance Reweighting

We aim to find the function $f(x)$ that minimizes the expected loss on test data. Assume *the support of P_{XY}^{te} is contained by that of P_{XY}^{tr}*. The expected loss is

$$R[P^{te}, \theta, L(x, y; \theta)] = \mathbb{E}_{(X,Y) \sim P^{te}}[L(x, y; \theta)]$$

$$= \int P_{XY}^{tr} \cdot \frac{P_{XY}^{te}}{P_{XY}^{tr}} \cdot L(x, y, \theta) dx dy$$

$$= \mathbb{E}_{(X,Y) \sim P^{tr}}[\beta^*(y) \cdot \gamma^*(x, y) \cdot L(x, y; \theta)], \qquad (19.1)$$

where θ denotes the parameters in the loss function $L(x, y; \theta)$, $\beta^*(y) \triangleq P_Y^{te}/P_Y^{tr}$ and $\gamma^*(x, y) \triangleq P_{X|Y}^{te}/P_{X|Y}^{tr}$. Here we *factorize P_{XY} as $P_Y P_{X|Y}$ instead of $P_X P_{Y|X}$* because it provides a more convenient way to handle the change in P_{XY}, according to our assumptions given later. In practice, we minimize the empirical loss:

$$\widehat{R}[P^{te}, \theta, L(x, y; \theta)] = \frac{1}{m} \sum_{i=1}^{m} \beta^*(y_i^{tr}) \gamma^*(x_i^{tr}, y_i^{tr}) L(x_i^{tr}, y_i^{tr}; \theta), \qquad (19.2)$$

to find the supervised learning machine that is expected to work well on test data, if $\beta^*(y_i^{tr}) \gamma^*(x_i^{tr}, y_i^{tr})$ are given.

Note that although (19.2) converges almost surely to (19.1), it could have a very high or even infinite variance if in some regions P_{XY}^{tr} is very small while P_{XY}^{te} is large—in this situation, a small number of data points in such regions would have high weights, so that the learning machine is very sensitive to the locations of such points, i.e., it has high random errors [20]. Readers who are interested in how to reduce the variance of the empirical expected loss and the corresponding learning machine may refer to, e.g., [25].

19.2.2   Sample Transformation and Reweighting

Sample reweighting only applies when the support of P_{XY}^{te} is contained in that of P_{XY}^{tr}; even under this condition, it is usually very difficult to estimate $\gamma^*(x, y)$ without prior knowledge on how $P_{X|Y}$ changes. Therefore, in the case where both P_Y and $P_{X|Y}$ change, the application of the sample reweighting scheme is rather limited. Instead, if we can find the transformation from $P_{X|Y}^{tr}$ to $P_{X|Y}^{te}$, i.e., find the transformation $\mathcal{T}$ such that the conditional distribution of $X^{new} = \mathcal{T}(X^{tr}, Y^{tr})$ satisfies $P_{X|Y}^{new} = P_{X|Y}^{te}$, we can calculate the expected loss on the test domain:

$$R[P^{te}, \theta, L(x, y; \theta)] = \mathbb{E}_{(X,Y) \sim P^{te}}[L(x, y; \theta)]$$

$$= \int P_Y^{tr} \cdot \beta^*(y) \cdot P_{X|Y}^{te} \cdot L(x, y; \theta) dx dy$$

$$= \int P_Y^{tr} \cdot \beta^*(y) \cdot P_{X|Y}^{new} \cdot L(x, y; \theta) dx dy,$$

$$= \mathbb{E}_{(X,Y) \sim P_{XY}^{new}}[\beta^*(y) \cdot L(x, y; \theta)].$$

Note that Y^{tr} is an argument of the transformation $\mathcal{T}$, i.e., $\mathcal{T}$ might be different at different Y values. This empirical loss can be calculated on the transformed training points $(\mathbf{x}^{new}, \mathbf{y}^{tr})$ with weights β^*:

$$\widehat{R}[P^{te}, \theta, L(x, y; \theta)] = \frac{1}{m} \sum_{i=1}^{m} \beta^*(y_i^{tr}) L(x_i^{new}, y_i^{tr}; \theta). \qquad (19.3)$$

19.2.3   Classification and Regression Machines

We consider both classification and regression problems. For the former problem, we adopt the support vector machine (SVM), and for the latter we use the kernel ridge regression (KRR). The standard formulation of both SVM and KRR can be straightforwardly modified to incorporate the importance weights according to (19.2) and (19.3). All parameters in the learning machines (e.g., the kernel width and regularization parameter) are selected by cross-validation.

Reweighted support vector classification: Support vector classifiers can be extended to incorporate non-uniform importance weights of the training instances. Associated to each training instance is the importance weight $\beta^*(y_i)\gamma^*(x_i, y_i)$, which can be incorporated into (19.2) via the following minimization problem:

$$\underset{\theta,b,\xi}{\text{minimize}} \quad \frac{1}{2}\|\theta\|^2 + C\sum_{i=1}^{n}\beta^*(y_i)\gamma^*(x_i, y_i)\xi_i \tag{19.4a}$$

$$\text{subject to} \quad y_i(\langle\theta, \phi(x_i)\rangle + b) \geq 1 - \xi_i, \xi_i \geq 0 \tag{19.4b}$$

where $\phi(x)$ is a feature map from $\mathcal{X}$ to a feature space $\mathcal{F}$. The dual of the above problem is

$$\underset{\alpha}{\text{minimize}} \quad \frac{1}{2}\sum_{i=1}^{n}\sum_{j=1}^{n}\alpha_i\alpha_j k(x_i, x_j) - \sum_{i=1}^{n}\alpha_i \tag{19.5a}$$

$$\text{subject to} \quad 0 \leq \alpha_i \leq \beta^*(y_i)\gamma^*(x_i, y_i)C, \tag{19.5b}$$

$$\sum_{i=1}^{n}\alpha_i y_i = 0 \tag{19.5c}$$

Here $k(x, x') \triangleq \langle\phi(x), \phi(x')\rangle_{\mathcal{F}}$ denotes the inner product between the feature maps. We have modified the LIBSVM implementation[4] for reweighted instances.

Reweighted kernel ridge regression (KRR): The original kernel ridge regression [21] represents the vector of fitted target values as $\mathbf{f} = Kc$, where K is the kernel matrix of $\mathbf{x}^{tr}$, and find the estimate of c by minimizing $(\mathbf{y}^{tr} - Kc)^{\mathsf{T}}(\mathbf{y}^{tr} - Kc) + \lambda_x c^{\mathsf{T}} Kc$. The estimate is $\hat{c} = (K + \lambda_x I)^{-1}\mathbf{y}^{tr}$ and consequently, the fitted target values are $\hat{\mathbf{f}} = K\hat{c} = K(K + \lambda_c I)^{-1}\mathbf{y}^{tr}$. Similarly, the reweighted kernel ridge regression minimizes $(\mathbf{y}^{tr} - Kc)^{\mathsf{T}} \cdot \text{diag}\{\beta^*(\mathbf{y}^{tr}) \odot \gamma^*(\mathbf{x}^{tr}, \mathbf{y}^{tr})\} \cdot (\mathbf{y}^{tr} - Kc) + \lambda_x c^{\mathsf{T}} Kc$, where $\odot$ denotes the Hadamard (or entrywise) product. This gives $\hat{c} = [K + \lambda_x \text{diag}^{-1}\{\beta^*(\mathbf{y}^{tr}) \odot \gamma^*(\mathbf{x}^{tr}, \mathbf{y}^{tr})\}]^{-1}\mathbf{y}^{tr}$ and hence, the fitted values are $\mathbf{f} = K[K + \lambda_x \cdot \text{diag}^{-1}\{\beta^*(\mathbf{y}^{tr}) \odot \gamma^*(\mathbf{x}^{tr}, \mathbf{y}^{tr})\}]^{-1}\mathbf{y}^{tr}$.

19.3   Correction for Target Shift

Unfortunately, unlike the covariate shift, the weights $\beta^*(y_i)\gamma^*(x_i, y_i)$ cannot be directly estimated because P_Y^{te} and $P_{X|Y}^{te}$ are unknown on the test data. Below we first consider the situation where $P_{X|Y}^{te} = P_{X|Y}^{tr}$, i.e., $\gamma^*(x, y) \equiv 1$, and propose a practical method to estimate $\beta^*(\mathbf{y}^{tr})$ as well as P_Y^{te} based on kernel embedding of conditional and marginal distributions. In subsequent sections, we further extend the results to the case where not only does P_Y change, but $P_{X|Y}$ is also allowed to change.

---

[4]http://www.csie.ntu.edu.tw/~cjlin/libsvm/ .

19.3.1 Assumptions

We first consider Target Shift (TarS):

$\mathbf{A}_1^{\mathbf{TarS}}$: $P_{X|Y}^{te} = P_{X|Y}^{tr}$ and $P_Y^{te} \neq P_Y^{tr}$.

That is, the difference between P_{XY}^{tr} and P_{XY}^{te} is caused by a shift in target distribution P_Y. Figure 19.1 shows a causal interpretation of TarS. For classification

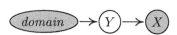

FIGURE 19.1: A causal model for TarS.

problems, it is possible to estimate P_Y^{te} in an iterative way by maximizing the likelihood on $\mathbf{x}^{te}$, for instance, with the EM algorithm [2]; however, such approaches involve estimation of $P_{X|Y}^{tr}$ explicitly, which is difficult for high-dimensional problems. They are also not practical for regression.

We make the following assumptions on P_Y^{te} and $P_{X|Y}^{tr}$:

$\mathbf{A}_2^{\mathbf{TarS}}$: The support of P_Y^{te} is contained in the support of P_Y^{tr} (i.e., roughly speaking, the training set is richer than the test set).

$\mathbf{A}_3^{\mathbf{TarS}}$: There exists only one possible distribution of Y that, together with $P_{X|Y}^{tr}$, leads to P_X^{te}.

Imagine that we can draw a biased sample from the training data; here the selection variable depends only on Y, i.e., it is independent of X given Y. Denote by $P^{new}(\cdot)$ the distribution on this sample. Note that $P_{X|Y}^{new} = P_{X|Y}^{tr} = P_{X|Y}^{te}$. Thus, we can make P_X^{new} identical to P_X^{te} by adjusting P_Y^{new}.

Let $\beta(y)$ be the ratio of the P_Y^{new} to P_Y^{tr}, i.e., $P_Y^{new} = \beta(y) \cdot P_Y^{tr}$. To make P_X^{new} identical to P_X^{te}, we can adjust $\beta(y)$ to minimize $\mathcal{D}(P_X^{te}, P_X^{new}) = \mathcal{D}\left(P_X^{te}, \int P_X^{tr} \beta(y) P_{X|Y}^{tr} dy\right)$, where $\mathcal{D}$ measures the difference between two distributions; it can be the mean square error or the Kullback-Leibler distance. To solve this problem, we have to estimate $P_{X|Y}^{tr}$ and P_X^{tr} from the training set, and moreover, the integral makes optimization very difficult.

19.3.2 A Kernel Mean Matching Approach

Instead, we solve this problem by making use of the kernel mean embedding of the marginal and conditional distributions; see Table 19.1 for the notation we use. The kernel mean embedding of P_X [26, 5] is a point in the reproducing kernel Hilbert space (RKHS) given by $\mu[P_X] = \mathbb{E}_{X \sim P_X}[\psi(X)]$, and its empirical estimate is $\hat{\mu}[P_X] = \frac{1}{m} \sum_{i=1}^m \psi(x_i)$. The embedding of the conditional distribution has been studied by [28, 27]. The embedding of $P_{X|Y}$ can be considered as an operator mapping from $\mathcal{G}$ to $\mathcal{F}$, defined as $\mathcal{U}[P_{X|Y}] = \mathcal{C}_{XY} \mathcal{C}_{YY}^{-1}$, where $\mathcal{C}_{XY}$ and $\mathcal{C}_{YY}$ denote the (uncentered) cross-covariance and covariance operators, respectively [3]. Furthermore, we have

$$\mu[P_X] = \mathcal{U}[P_{X|Y}]\mu[P_Y].$$

We make the following assumption on the kernel k for X and l for Y:

TABLE 19.1: Notation used in this chapter.

random variable	X	Y
domain	$\mathcal{X}$	$\mathcal{Y}$
observation	x	y
data matrix	$\mathbf{x}$	$\mathbf{y}$
kernel	$k(x, x')$	$l(y, y')$
kernel matrix on training set	K	L
feature map	$\psi(x)$	$\phi(y)$
feature matrix on training set	Ψ	Φ
RKHS	$\mathcal{F}$	$\mathcal{G}$

$\mathbf{A_4^{TarS}}$: Product kernel kl on $\mathcal{X} \times \mathcal{Y}$ is characteristic.

For characteristic kernels, the kernel mean map μ from the space of the distribution to the RKHS is injective, meaning that all information of the distribution is preserved [4, 30]. In this chapter we use the Gaussian kernel, i.e., $k(x_i, x_j) = \exp\left(-\frac{\|x_i - x_j\|^2}{2\sigma^2}\right)$, where σ is the kernel width. Note that under assumptions A_3^{TarS} and A_4^{TarS}, for the embedding $\mathcal{U}[P_{X|Y}^{tr}]$, which is a mapping from $\mathcal{G}$ to $\mathcal{F}$, the pre-image of $\mu[P_X^{te}]$ is unique.

The kernel mean embedding of P_Y^{new} is

$$\mu[P_Y^{new}] = \mathbb{E}_{Y \sim P_Y^{new}}[\phi(Y)] = \mathbb{E}_{Y \sim P_Y^{tr}}[\beta(y)\phi(Y)]. \tag{19.6}$$

The embedding of P_X^{new} is then given by $\mu[P_X^{new}] = \mathcal{U}[P_{X|Y}^{tr}]\mu[P_Y^{new}]$. Consequently, in the population version, we can find $\beta(y)$ by minimizing the maximum mean discrepancy (MMD):

$$\left\|\mu[P_X^{new}] - \mu[P_X^{te}]\right\| = \left\|\mathcal{U}[P_{X|Y}^{tr}]\mu[P_Y^{new}] - \mu[P_X^{te}]\right\|$$
$$= \left\|\mathcal{U}[P_{X|Y}^{tr}]\mathbb{E}_{Y \sim P_Y^{tr}}[\beta(y)\phi(y)] - \mu[P_X^{te}]\right\|, \tag{19.7}$$

subject to $\beta(y) \geq 0$ and $\mathbb{E}_{P_Y^{tr}}[\beta(y)] = 1$, which guarantees that $P_Y^{new} = \beta(y)P_Y^{tr}$ is a valid distribution.

Theorem 19.1. *Under assumptions A_2^{TarS}, A_3^{TarS}, and A_4^{TarS}, the minimization problem (19.7) is convex in β. Further suppose A_1^{TarS} holds. Then the solution to (19.7) is $\beta(y) = \frac{P_Y^{te}(y)}{P_Y^{tr}(y)}$.*

Proof. In (19.7), $\mathcal{U}[P_{X|Y}^{tr}]$ is a linear operator, $\mathbb{E}_{Y \sim P_Y^{tr}}[\beta(y)\phi(y)]$ is linear in β. Further note that the constraints are convex. We can see that the optimization problem (19.7) is convex in β.

According to assumption A_1^{TarS}, we have $\mu[P_X^{te}] = \mathcal{U}[P_{X|Y}^{tr}]\mu[P_Y^{te}]$, and the function in (19.7) reduces to

$$\left\|\mathcal{U}[P_{X|Y}^{tr}] \cdot \left\{\mathbb{E}_{Y \sim P_Y^{tr}}[\beta(y)\phi(y)] - \mu[P_X^{te}]\right\}\right\|.$$

It achieves zero, which is clearly a minimum, when $\mathbb{E}_{Y \sim P_Y^{tr}}[\beta(y)\phi(y)] = \mu[P_Y^{te}]$. It is equivalent to $\beta(y)P_Y^{tr}(y) = P_Y^{te}(y)$, since the kernel l is characteristic. Moreover, combining assumptions A_1^{TarS} and A_4^{TarS} implies that there is no other solution of $\beta(y)$ to (19.7). $\qquad\qquad\qquad\qquad\qquad\qquad\qquad\qquad\square$

In practice we have to use an empirical version. The empirical estimate of $\mathcal{U}_{X|Y}$ is $\hat{\mathcal{U}}_{X|Y} = \Psi(L+\lambda I)^{-1}\Phi^\intercal$. Recall that m and n are the sizes of the training and test sets. Denote by $\mathbf{1}_n$ the vector of 1's of length n, and by K^c the "cross" kernel matrix between $\mathbf{x}^{te}$ and $\mathbf{x}^{tr}$, i.e., $K_{ij}^c = k(x_i^{te}, x_j^{tr})$. Let β stand for $\beta(\mathbf{y}^{tr})$ and β_i for $\beta(y_i^{tr})$. The empirical version of the square of (19.7) is

$$
\left\| \hat{\mathcal{U}}_{X|Y} \cdot \frac{1}{m}\sum_{i=1}^{m} \beta_i \phi(y_i^{tr}) - \frac{1}{n}\sum_{i=1}^{n} \psi(x_i^{te}) \right\|^2
$$

$$
= \left\| \frac{1}{m}\hat{\mathcal{U}}_{X|Y}\phi(\mathbf{y}^{tr})\beta - \frac{1}{n}\psi(\mathbf{x}^{te})\mathbf{1}_n \right\|^2
$$

$$
= \frac{1}{m^2}\beta^\intercal\phi^\intercal(\mathbf{y}^{tr})\hat{\mathcal{U}}_{X|Y}^\intercal\hat{\mathcal{U}}_{X|Y}\phi(\mathbf{y}^{tr})\beta - \frac{2}{mn}\mathbf{1}_n^\intercal\psi^\intercal(\mathbf{x}^{te})\hat{\mathcal{U}}_{X|Y}\phi(\mathbf{y}^{tr})\beta + \text{const}
$$

$$
= \frac{1}{m^2}\beta^\intercal L(L+\lambda I)^{-1}K(L+\lambda I)^{-1}L\beta - \frac{2}{mn}\mathbf{1}_n^\intercal K^c(L+\lambda I)^{-1}L\beta + \text{const}
$$

$$
= \frac{1}{m^2}\beta^\intercal \underbrace{\Omega K\Omega^T}_{\triangleq A} \beta - \frac{2}{mn}\mathbf{1}_n^\intercal \underbrace{K^c\Omega^T}_{\triangleq M} \beta + \text{const}, \qquad (19.8)
$$

where we use short-hand notation $\Omega \triangleq L(L+\lambda I)^{-1}$. As shown by [10, Lemma 3], if $\beta_i \in [0, B_\beta]$, i.e., B_β is the upper bound of β, given that β_i has finite mean and non-zero variance, the sample mean $\frac{1}{m}\sum_{i=1}^{m}\beta_i$ converges in distribution to a Gaussian variable with mean $\mathbb{E}_{P_Y^{tr}}[\beta(y)]$ and standard deviation bounded by $\frac{B_\beta}{2\sqrt{m}}$. As $\mathbb{E}_{P_Y^{tr}}[\beta(y)] = 1$, we have the following constrained quadratic programming (QP) problem:

$$
\underset{\beta}{\text{minimize}} \qquad \frac{1}{2}\beta^\intercal A\beta - \frac{m}{n}M\beta,
$$

$$
\text{s.t.} \qquad \beta_i \in [0, B_\beta] \text{ and } \left| \sum_{i=1}^{m} \beta_i - m \right| \le m\epsilon,
$$

where a good choice of ϵ is $\mathcal{O}\left(\frac{B}{2\sqrt{m}}\right)$.

Note that β estimated this way is not necessarily a function of y: different data points in the training set with the same y value could correspond to different β values. We also found that the β values estimated by solving the above optimization problem usually change dramatically along with y. We can improve the estimation quality of β by making use of reparameterization. First consider the case where Y is discrete. Let C be the cardinality of Y and denote by $v_1, ..., v_C$ its possible values. We can define a matrix $R^{(d)}$ where

$R_{iq}^{(d)}$ is 1 if $y_i = v_q$ and is zero everywhere else. β can then be reparameterized as

$$\beta = R^{(d)}\alpha, \tag{19.9}$$

where the C-dimensional vector α is the new parameter.

We then consider the case where Y is continuous. Usually both distributions P_Y^{tr} and P_Y^{te} are smooth, and so is $\beta(y)$. Therefore, we would like to enforce the smoothness of $\beta(y)$ with respect to y. Let $R^{(c)} \triangleq L_\beta(L_\beta + \lambda_\beta I)^{-1}$, where L_β is a kernel matrix of $\mathbf{y}$ with the Gaussian kernel and λ_β is the regularization parameter.[5] Inspired by KRR [21], we parameterize $\beta(\mathbf{y}^{tr})$ as

$$\beta = R^{(c)}\alpha, \tag{19.10}$$

with new parameter α.[6] One can consider β as a smoothed version of α.

Finally, we find α (and β) in both cases by solving:

$$
\begin{aligned}
\underset{\alpha}{\text{minimize}} \quad & \frac{1}{2}\alpha^{\mathsf{T}}\big[R^{\mathsf{T}}AR\big]\alpha - \frac{m}{n}\big[MR\big]\alpha, \\
\text{s.t.} \quad & 0 \leq \big[R\alpha\big]_i \leq B_\beta \text{ and } \big|\mathbf{1}_m^{\mathsf{T}}R\alpha - m\big| \leq m\epsilon,
\end{aligned} \tag{19.11}
$$

where R stands for $R^{(d)}$ or $R^{(c)}$, depending on whether Y is discrete or continuous. In all our experiments, we set $B_\beta = 10$ and $\epsilon = \frac{B_\beta}{4\sqrt{m}}$. We then set β^* in (19.2) to the estimated β and $\gamma^*(x_i, y_i) \equiv 1$. Minimizing (19.2) produces the classifier or regression model after correction for TarS.

19.4   Location-Scale Conditional Shift

In practice $P_{X|Y}^{tr}$ and $P_{X|Y}^{te}$ might differ to some extent. It is certainly not possible to transfer useful knowledge from the training domain to the test domain if $P_{X|Y}$ changes arbitrarily. However, under certain assumptions on the change in $P_{X|Y}$, one could estimate $P_{X|Y}^{te}$ without knowing Y on test data. In this section we assume that $P_{X|Y}$ changes across domains and that

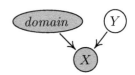

FIGURE 19.2: A causal model for ConS.

---

[5]Note that although L_β and L are both kernel matrices of $\mathbf{y}$, they have different purposes and might have different hyperparameters, so we use different notations.

[6]When m is large, say, bigger than 1000, we parameterize β as a linear combination of the nonlinear principal components of $\mathbf{y}$ with the weight vector α. The nonlinear principal components were extracted by kernel principal component analysis (kPCA) [23]. In this way the dimensionality of α is much less than m and the optimization procedure is very efficient.

$P_Y^{tr} = P_Y^{te}$. We term this situation Conditional Shift (ConS); Figure 19.2 gives its causal interpretation. This situation might be less realistic in practice and will not be considered in our experiments; however, it serves as a foundation of a more general situation, GeTarS, which will be studied in Section 19.5. When considering ConS and GeTarS, we focus on classification problems.

19.4.1   Assumptions and Identifiability

In some situations, we can formulate how the conditional distribution changes. For instance, for the same image, features such as intensities and colors are influenced by illumination, viewing angles, etc., which might change across domains. Modeling such a change enables distribution matching between the training domain and test domain, and consequently improves the performance on the test domain. Here we use the approach of *transforming training data to reproduce the covariate distribution on the test domain*; see Section 19.2. Since we can model the transformation from $P_{X|Y}^{tr}$ to $P_{X|Y}^{te}$, we do not need the condition that the support of $P_{X|Y}^{te}$ is contained in that of $P_{X|Y}^{tr}$, making the approach more practical.

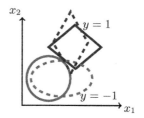

FIGURE 19.3: An illustration of LS-ConS where Y is binary and X is two-dimensional. Lines are contours of $P_{X|Y}(x|y = -1)$ and $P_{X|Y}(x|y = 1)$. Solid and dashed lines represent the contours on the training and test domains, respectively.

We assume that the shape of the distribution of each feature X_i, as well as the dependence structure between features, is preserved across the domains. More precisely, we assume that given any y value, $P_{X_i|Y}^{te}$ and $P_{X_i|Y}^{tr}$ only differs in the location and scale:

$\mathbf{A^{ConS}}$:   There exists $\mathbf{w}(Y^{tr}) = \text{diag}[w_1(Y^{tr}), ..., w_d(Y^{tr})]$ and $\mathbf{b}(Y^{tr}) = [b_1(Y^{tr}), ..., b_d(Y^{tr})]^\intercal$, where d is the dimensionality of X, such that the conditional distribution of $X^{new} \triangleq \mathbf{w}(Y^{tr})X^{tr} + \mathbf{b}(Y^{tr})$ given Y^{tr} is the same as that of X^{te} given Y^{te}.

We term this situation location-scale ConS (LS-ConS). In matrix form, the transformed training points

$$\mathbf{x}^{new} \triangleq \mathbf{x}^{tr} \odot \mathbf{W} + \mathbf{B}, \tag{19.12}$$

where the ith columns of $\mathbf{W}$ and $\mathbf{B}$ are $[w_1(y_i), ..., w_d(y_i)]^\intercal$ and $[b_1(y_i), ..., b_d(y_i)]^\intercal$, respectively, are expected to have the same distribution as the test data. Figure 19.3 illustrates how the contours of $P_{X|Y}$ change across domains under LS-ConS.

Let $\mathbf{w}_q$ and $\mathbf{b}_q$ be the transformation to be applied to the data points with

$y = v_q$, $q = 1, ..., C$. The following theorem states that $P_{X|Y}^{new}$ is identifiable under some conditions on $P_{X|Y}^{tr}(x|y = v_q)$.

Theorem 19.2. *Let $P_{X|Y}^{(\mathbf{w}_q, \mathbf{b}_q)}(x|y = v_q)$ be the LS transformed version of $P_{X|Y}^{tr}(x|y = v_q)$ with parameters $(\mathbf{w}_q, \mathbf{b}_q)$ and $P_Y^{te} = P_Y^{tr}$. Suppose A^{ConS} holds, i.e., $\forall q$, $\exists(\mathbf{w}_q^*, \mathbf{b}_q^*)$ such that $P_{X|Y}^{(\mathbf{w}_q^*, \mathbf{b}_q^*)}(x|y = v_q) = P_{X|Y}^{te}(x|y = v_q)$. Further assume*

A_2^{ConS}: *Set $\{c_{q1}P_{X|Y}^{(\mathbf{w}_q, \mathbf{b}_q)}(x|y = v_q) + c_{q2}P_{X|Y}^{(\mathbf{w}_q', \mathbf{b}_q')}(x|y = c_q); q = 1, ..., C\}$ is linearly independent $\forall$ c_{q1}, c_{q2} $(c_{q1}^2 + c_{q2}^2 \neq 0)$, $\mathbf{w}_q$, $\mathbf{w}_q'$ $(||\mathbf{w}_q||_F^2 + ||\mathbf{w}_q'||_F^2 \neq 0)$, and $\mathbf{b}_q$, $\mathbf{b}_q'$.*

If $\exists$ $(\mathbf{w}_q, \mathbf{b}_q)$ such that $P_X^{te} = \sum_q P_Y^{tr}(y_q)P_{X|Y}^{(\mathbf{w}_q, \mathbf{b}_q)}(x|y = v_q)$, then we have $\forall$ q, $P_{X|Y}^{(\mathbf{w}_q, \mathbf{b}_q)}(x|y = v_q) = P_{X|Y}^{te}(x|y = v_q)$.

Proof. This theorem is a special case of Theorem 19.3, which shall be presented in Section 19.5: In Theorem 19.3, setting $P_Y^{new} = P_Y^{tr} = P_Y^{te}$ gives this theorem.   □

A necessary condition for A_2^{ConS} is that $P_{X|Y}^{tr}(x|y = v_q)$, $q = 1, ..., C$, are linearly independent after any LS transformations. Roughly speaking, the higher d, the less likely for this assumption to be violated.

19.4.2   A Kernel Approach

As in Section 19.3.2, we parameterize $\mathbf{W}$ and $\mathbf{B}$ as $\mathbf{W} = RG$ and $\mathbf{B} = RH$, where $\mathbf{G}$ and $\mathbf{H}$ are the parameters to be estimated, and R is $R^{(c)}$ or $R^{(d)}$, depending on whether Y is discrete or continuous. In this way $\mathbf{W}$ and $\mathbf{B}$ are guaranteed to be functions of y, and the number of parameters is greatly reduced.

Noting the relationship between X^{new} and X^{tr}, and using the substitution rule, we have

$$\mathcal{U}[P_{X|Y}^{new}] = \mathcal{C}_{X^{new}Y} C_{YY}^{-1}$$
$$= \mathbb{E}_{(X^{new}, Y) \sim P_{XY}^{new}}[\psi(X^{new}) \otimes \phi^\mathsf{T}(Y)] \, \mathbb{E}_{Y \sim P_Y^{tr}}^{-1}[\phi(Y) \otimes \phi^\mathsf{T}(Y)]$$
$$= \mathbb{E}_{(X^{tr}, Y) \sim P_{XY}^{tr}}[\psi(X^{new}) \otimes \phi^\mathsf{T}(Y)] \cdot \mathbb{E}_{Y \sim P_Y^{tr}}^{-1}[\phi(Y) \otimes \phi^\mathsf{T}(Y)].$$

The empirical estimate of $\mathcal{U}[P_{X|Y}^{new}]$ is consequently

$$\hat{\mathcal{U}}[P_{X|Y}^{new}] = \frac{1}{m}\psi(\mathbf{x}^{new}) \cdot \phi^\mathsf{T}(\mathbf{y}^{tr}) \cdot \left[\frac{1}{m}\phi(\mathbf{y}^{tr})\phi^\mathsf{T}(\mathbf{y}^{tr}) + \tilde{\lambda}I\right]^{-1}$$
$$= \tilde{\Psi}(L + \lambda I)^{-1}\Phi^\mathsf{T}, \tag{19.13}$$

where $\tilde{\Psi} = \psi(\mathbf{x}^{new})$.

Let $\tilde{K}$ be the kernel matrix corresponding to the feature matrix $\tilde{\Psi}$, i.e., $\tilde{K}_{i,j} = k(x_i^{new}, x_j^{new})$, and $\tilde{K}^c$ the cross kernel matrix between $\mathbf{x}^{te}$ and $\mathbf{x}^{new}$, i.e., $\tilde{K}_{ij}^c = k(x_i^{te}, x_j^{new})$. We aim to minimize $||\mu[P_X^{new}] - \mu[P_X^{te}]||^2$, whose empirical version is

$$
\begin{aligned}
J^{ConS} &\triangleq \left|\left|\hat{\mu}[P_X^{new}] - \hat{\mu}[P_X^{te}]\right|\right|^2 = \left|\left|\hat{\mathcal{U}}[P_{X|Y}^{new}]\hat{\mu}[P_Y^{tr}] - \hat{\mu}[P_X^{te}]\right|\right|^2 \\
&= \frac{1}{m^2}\mathbf{1}_m^{\mathsf{T}}\phi^{\mathsf{T}}(\mathbf{y}^{tr})\hat{\mathcal{U}}^{\mathsf{T}}[P_{X|Y}^{new}]\hat{\mathcal{U}}[P_{X|Y}^{new}]\phi(\mathbf{y}^{tr})\mathbf{1}_m \\
&\quad - \frac{2}{mn}\mathbf{1}_n^{\mathsf{T}}\psi^{\mathsf{T}}(\mathbf{x}^{te})\hat{\mathcal{U}}[P_{X|Y}^{new}]\phi(\mathbf{y}^{tr})\mathbf{1}_m + \text{const} \\
&= \frac{1}{m^2}\mathbf{1}_m^{\mathsf{T}}\Omega\tilde{K}\Omega^T\mathbf{1}_m - \frac{2}{mn}\mathbf{1}_n^{\mathsf{T}}\tilde{K}^c\Omega^T\mathbf{1}_m + \text{const}. \quad (19.14)
\end{aligned}
$$

We then estimate $\mathbf{W}$ (or $\mathbf{G}$) together with $\mathbf{B}$ (or $\mathbf{H}$) by minimizing J^{ConS}. The gradient of J^{ConS} with respect to $\tilde{K}$ and $\tilde{K}^c$ is

$$
\frac{\partial J^{ConS}}{\partial \tilde{K}} = \frac{1}{m^2}(L + \lambda I)^{-1}L\mathbf{1}_m \cdot \mathbf{1}_m^{\mathsf{T}}L(L + \lambda I)^{-1}, \quad \text{and}
$$

$$
\frac{\partial J^{ConS}}{\partial \tilde{K}^c} = -\frac{2}{mn}\mathbf{1}_n\mathbf{1}_m^{\mathsf{T}}L(L + \lambda I)^{-1}.
$$

Using the chain rule, we further have the gradient of J^{ConS} with respect to the entries of $\mathbf{G}$ and $\mathbf{H}$:

$$
\frac{\partial J^{ConS}}{\partial G_{pq}} = \text{Tr}\left[\left(\frac{\partial J^{ConS}}{\partial \tilde{K}}\right)^{\mathsf{T}} \cdot (\mathbf{D}_{pq} \odot \tilde{K})\right] - \text{Tr}\left[\left(\frac{\partial J^{ConS}}{\partial \tilde{K}^c}\right)^{\mathsf{T}} \cdot (\mathbf{E}_{pq} \odot \tilde{K}^c)\right],
$$

$$
\frac{\partial J^{ConS}}{\partial H_{pq}} = \text{Tr}\left[\left(\frac{\partial J^{ConS}}{\partial \tilde{K}}\right)^{\mathsf{T}} \cdot (\tilde{\mathbf{D}}_{pq} \odot \tilde{K})\right] - \text{Tr}\left[\left(\frac{\partial J^{ConS}}{\partial \tilde{K}^c}\right)^{\mathsf{T}} \cdot (\tilde{\mathbf{E}}_{pq} \odot \tilde{K}^c)\right],
$$

where

$$
[\mathbf{D}_{pq}]_{ij} = -\frac{1}{l^2}(x_{jq}^{new} - x_{iq}^{new})(x_{jq}^{tr}R_{jp} - x_{iq}^{tr}R_{ip}),
$$

$$
[\mathbf{E}_{pq}]_{ij} = -\frac{1}{l^2}x_{jq}^{tr}R_{jp}(x_{jq}^{new} - x_{iq}^{te}),
$$

$$
[\tilde{\mathbf{D}}_{pq}]_{ij} = -\frac{1}{l^2}(x_{jq}^{new} - x_{iq}^{new})(R_{jp} - R_{ip}),
$$

$$
[\tilde{\mathbf{E}}_{pq}]_{ij} = -\frac{1}{l^2}R_{jp}(x_{jq}^{new} - x_{iq}^{te}).
$$

In practice we also regularize (19.14) to prefer the change in $P_{X|Y}$ to be as little as possible, i.e., to make entries of $\mathbf{W}$ close to one and those of $\mathbf{B}$ close to zero. This is particularly useful in case assumption A_2^{ConS} is violated; we then prefer the slightest change in the conditional, among all possibilities. The regularization term is

$$
J^{reg} = \frac{\lambda_{LS}}{m} \cdot ||\mathbf{W} - \mathbf{1}_m\mathbf{1}_d^{\mathsf{T}}||_F^2 + \frac{\lambda_{LS}}{m} \cdot ||\mathbf{B}||_F^2. \quad (19.15)
$$

whose gradient with respect to $\mathbf{G}$ and $\mathbf{H}$ is $\frac{\partial J^{reg}}{\partial \mathbf{G}} = \frac{2\lambda_{LS}}{m} R^{\mathsf{T}}(\mathbf{W} - \mathbf{1}_m \mathbf{1}_d^{\mathsf{T}})$ and $\frac{\partial J^{reg}}{\partial \mathbf{H}} = \frac{2\lambda_{LS}}{m} R^{\mathsf{T}} \mathbf{B}$. In our experiments we fix λ_{LS} to 0.001.

We use the scaled conjugate gradient (SCG) to minimize $J^{ConS} + J^{reg}$. After estimating $\mathbf{W}$ and $\mathbf{B}$, we transform $\mathbf{x}^{tr}$ to $\mathbf{x}^{new}$ according to (19.12), and $(\mathbf{x}^{new}, \mathbf{y}^{tr})$ would have the same distribution as the test data, under assumption A^{ConS}. Consequently, the classifier or regressor trained on $(\mathbf{x}^{new}, \mathbf{y}^{tr})$ is expected to generalize well to the test domain.

19.5   Location-Scale Generalized Target Shift

We then consider a more general situation where both P_Y and $P_{X|Y}$ change, called Generalized Target Shift (GeTarS). Figure 19.4 gives the causal model underlying the GeTarS situation.

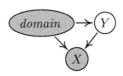

FIGURE 19.4: Causal model for GeTarS.

In this setting, we assume that $P_Y^{te} \neq P_Y^{tr}$ and that assumption A^{ConS} holds, i.e., we consider LS-GeTarS, and aim to estimate the importance weights $\beta^*(y_i) \triangleq \frac{P_Y^{te}(y_i)}{P_Y^{tr}(y_i)}$ and the matrices $\mathbf{W}$ and $\mathbf{B}$ in (19.12). They would transform the training data to mimic the distribution of the test data, and the learning machine learned on the reweighted transformed data is expected to work well on the test data. Parameters can be estimated by reweighting and transforming the training data to reproduce P_X^{te}, i.e., by minimizing $||\mu[P_X^{new}] - \mu[P_X^{te}]||$, where $P_X^{new} = \int P_Y^{new} P_{X|Y}^{new} dy$, $P_Y^{new} = \beta P_Y^{tr}$, and $P_{X|Y}^{new}(x|y = v_q) = P_{X|Y}^{(\mathbf{w}_q, \mathbf{b}_q)}(x|y = v_q)$. The following theorem provides the identifiability of p_Y^{new} and $P_{X|Y}^{new}$.

Theorem 19.3. *Suppose A^{ConS} holds. Under assumption A_2^{ConS}, if there exist $(\mathbf{w}_q, \mathbf{b}_q)$ such that $P_X^{te} = \sum_{q=1}^{C} P_Y^{new}(y = v_q) P_{X|Y}^{(\mathbf{w}_q, \mathbf{b}_q)}(x|y = v_q)$, then we have $P_Y^{new} = P_Y^{te}$, and $\forall\, q$, $P_{X|Y}^{(\mathbf{w}_q, \mathbf{b}_q)}(x|y = v_q) = P_{X|Y}^{te}(x|y = v_q)$.*

Proof. Combining assumption A^{ConS}, i.e., $P_X^{te} = \sum_q P_Y^{te}(v_q) P_{X|Y}^{(\mathbf{w}_q^*, \mathbf{b}_q^*)}(x|y = v_q)$, and the condition in Theorem 19.3, we have

$$\sum_q P_Y^{te}(y_q) P_{X|Y}^{(\mathbf{w}_q^*, \mathbf{b}_q^*)}(x|y = v_q) = \sum_q P_Y^{new}(y_q) P_{X|Y}^{(\mathbf{w}_q, \mathbf{b}_q)}(x|y = v_q)$$

$$\Rightarrow \sum_i \left[P_Y^{te}(v_q) P_{X|Y}^{(\mathbf{w}_q^*, \mathbf{b}_q^*)}(x|y_q) - P_Y^{new}(y_q) P_{X|Y}^{(\mathbf{w}_q, \mathbf{b}_q)}(x|y = v_q) \right] = 0.$$

Because of assumption A_2^{ConS}, we know that $\forall\ q$,

$$P_Y^{te}(y = v_q)P_{X|Y}^{(\mathbf{w}_q^*, \mathbf{b}_q^*)}(x|y = v_q) - P_Y^{new}(y_q)P_{X|Y}^{(\mathbf{w}_q, \mathbf{b}_q)}(x|y = v_q) = 0.$$

Taking the integral of the above equation gives $P_Y^{new}(v_q) = P_Y^{te}(v_q)$. This further implies $P_{X|Y}^{(\mathbf{w}_q, \mathbf{b}_q)}(x|y_q) = P_{X|Y}^{(\mathbf{w}_q^*, \mathbf{b}_q^*)}(x|y_q) = P_{X|Y}^{te}(x|y = v_q)$.

$\square$

Combining (19.6) and (19.13), we can find the empirical version of $||\mu[P_X^{new}] - \mu[P_X^{te}]||^2$:

$$
\begin{aligned}
J &= \left\|\hat{\mu}[P_X^{new}] - \hat{\mu}[P_X^{te}]\right\|^2 = \left\|\hat{\mathcal{U}}[P_{X|Y}^{new}]\hat{\mu}[P_Y^{te}] - \hat{\mu}[P_X^{te}]\right\|^2 \\
&= \left\|\frac{1}{m}\hat{\mathcal{U}}[P_{X|Y}^{new}]\phi(\mathbf{y}^{tr})\beta - \frac{1}{n}\psi(\mathbf{x}^{te})\mathbf{1}_n\right\|^2 \\
&= \frac{1}{m^2}\beta^{\mathsf{T}}\phi^{\mathsf{T}}(\mathbf{y}^{tr})\hat{\mathcal{U}}^{\mathsf{T}}[P_{X|Y}^{new}]\hat{\mathcal{U}}[P_{X|Y}^{new}]\phi(\mathbf{y}^{tr})\beta - \frac{2}{mn}\mathbf{1}_n^{\mathsf{T}}\psi^{\mathsf{T}}(\mathbf{x}^{te})\hat{\mathcal{U}}[P_{X|Y}^{new}]\phi(\mathbf{y}^{tr})\beta \\
&\qquad\qquad + \text{const} \\
&= \frac{1}{m^2}\beta^{\mathsf{T}}\Omega\tilde{K}\Omega^T\beta - \frac{2}{mn}\mathbf{1}_n^{\mathsf{T}}\tilde{K}^c\Omega^T\beta.
\end{aligned}
\tag{19.16}
$$

When minimizing J, we would also like the difference between $P_{X|Y}^{te}$ and $P_{X|Y}^{tr}$, as measured by J^{reg} given in (19.15), to be as little as possible. Combining both constraints, we estimate the involved parameters β, $\mathbf{W}$, and $\mathbf{B}$ by minimizing

$$J^{GeTarS} = J + \lambda_{LS}J^{reg}. \tag{19.17}$$

Finally, for parameter estimation, we iteratively alternate between the QP to minimize (19.16) w.r.t β and the SCG optimization procedure with respect to $\{\mathbf{W}, \mathbf{B}\}$. Algorithm 17 summarizes this procedure. For details of the two optimization sub-procedures, see Sections 19.3 and 19.4, respectively. After estimating the parameters, we train the learning machine by minimizing the weighted loss (19.3) on $(\mathbf{x}^{new}, \mathbf{y}^{tr})$.

The MATLAB source code for correcting TarS and LS-GeTarS is available at http://people.tuebingen.mpg.de/kzhang/Code-TarS.zip.

19.6   Determination of Hyperparameters

As discussed in Section 19.2, all hyperparameters in the subsequent learning machines reweighted SVM and KRR are selected by importance weighted cross-validation [32]. In addition, there are three types of hyperparameters. One is the kernel width of X to construct the kernel matrix K. In our experiments we normalize all variables in X to unit variance, and use some empirical

---

Algorithm 17 Estimating weights β^*, $\mathbf{W}$, and $\mathbf{B}$ under LS-GeTarS

    Input: training data $(\mathbf{x}^{tr}, \mathbf{y}^{tr})$ and test data $\mathbf{x}^{te}$

    Output: weights β and $\mathbf{x}^{new}$ corresponding to the training data points

    $\beta \leftarrow \mathbf{1}_m$, $\mathbf{W} \leftarrow \mathbf{1}_m \mathbf{1}_d^\mathsf{T}$, $\mathbf{B} \leftarrow \mathbf{0}$

    repeat

        fix $\mathbf{W}$ and $\mathbf{B}$ and estimate β by minimizing (19.17) with QP, under the constraint on β given in Section 19.3;

        fix β and estimate $\mathbf{W}$ and $\mathbf{B}$ by minimizing (19.17) with SCG;

    until convergence

    $\beta^* \leftarrow \beta$, $\mathbf{x}^{new} = \mathbf{x}^{tr} \odot \mathbf{W} + \mathbf{B}$.

---

values for those kernel widths: they are set to $0.8\sqrt{d}$ if the sample size $m \leq 200$, to $0.3\sqrt{d}$ if $m > 1200$, or to $0.5\sqrt{d}$ otherwise, where d is the dimensionality of X. This simple setting always works well in all our experiments; for a more principled strategy, one might refer to [7].

The second type of hyperparameters are involved in the parameterization of β for regression under TarS (the kernel width for L_β and regularization parameter λ_β) and λ_{LS} for LS-GeTarS in (19.17). We set these parameters by cross-validation. (On some large data sets we simply set λ_{LS} to 0.0001 to save computational load.) Although the objective functions (Equation 19.8 for TarS, and Equation 19.16 for LS-GeTarS) are the sum of squared errors, the corresponding problems are considered unsupervised, or in particular, as density estimation problems, rather than supervised. We treat P_X^{new} as the distribution given by the model, and $\mathbf{x}^{te}$ as the corresponding observed data points. They are different from the classical density estimation problem in that here we use the MMD between P_X^{new} and P_X^{te} as the loss function. We divide $\mathbf{x}^{te}$ into five equal size subsamples, use four of them to estimate β or $\mathbf{W}$ and $\mathbf{B}$, and the remaining one for testing. Finally we find the values of these hyperparameters that give the smallest cross-validated loss, which is (19.8) for regression under TarS or (19.16) for LS-GeTarS. The last type of hyperparameters, including hyperparameters in L and the regularization parameter λ, are learned by extending Gaussian process regression to the multi-output case [40].

19.7   Simulations

We use simulations to study the performance of the proposed approach for TarS and LS-GeTarS in four scenarios. Their settings are

(a) a *nonlinear regression problem under TarS*, where $X = Y + 3\tanh(Y) +$

E, $E \sim \mathcal{N}(0, 1.5^2)$; $Y^{tr} \sim N(0, 2^2)$, and $Y^{te} \sim 0.8\mathcal{N}(1, 1) + 0.2\mathcal{N}(0.2, 0.5^2)$,

(b) a *classification problem under TarS*, where

$$X|_{Y=-1} \sim \mathcal{N}\left(\mathbf{0}, \begin{bmatrix} 0.21 & 0.09 \\ 0.09 & 0.21 \end{bmatrix}\right),$$

$$X|_{Y=1} \sim \mathcal{N}\left(\begin{bmatrix} 1 \\ 1 \end{bmatrix}, \begin{bmatrix} 0.31 & -0.06 \\ -0.06 & 0.31 \end{bmatrix}\right),$$

$P_Y^{tr}(y = -1) = 0.6$, and $P_Y^{te}(y = -1) = 0.2$,

(c) a *classification problem approximately following LS-GeTarS*, where

$$X^{tr}|_{Y^{tr}=-1} \sim \mathcal{N}\left(\mathbf{0}, \begin{bmatrix} 0.24 & 0.22 \\ 0.22 & 0.24 \end{bmatrix}\right),$$

$$X^{tr}|_{Y^{tr}=1} \sim \mathcal{N}\left(\begin{bmatrix} 1 \\ 1 \end{bmatrix}, \begin{bmatrix} 0.16 & -0.03 \\ -0.03 & 0.16 \end{bmatrix}\right),$$

$$X^{te}|_{Y^{te}=-1} \sim \mathcal{N}\left(\begin{bmatrix} 0.5 \\ 0 \end{bmatrix}, \begin{bmatrix} 0.12 & 0.11 \\ 0.11 & 0.12 \end{bmatrix}\right),$$

$$X^{te}|_{Y^{te}=1} \sim \mathcal{N}\left(\begin{bmatrix} 2 \\ 1.3 \end{bmatrix}, \begin{bmatrix} 0.27 & -0.04 \\ -0.04 & 0.27 \end{bmatrix}\right),$$

$P_Y^{tr}(y = -1) = 0.6$, and $P_Y^{te}(y = -1) = 0.3$, and

(d) a *classification problem under non-LS-GeTarS with slight change in the conditional*, where

$$X^{tr}|_{Y^{tr}=-1} \sim \mathcal{N}\left(\mathbf{0}, \begin{bmatrix} 0.16 & 0 \\ 0 & 0.16 \end{bmatrix}\right),$$

$$X^{tr}|_{Y^{tr}=1} \sim \mathcal{N}\left(\begin{bmatrix} 0.9 \\ 0.9 \end{bmatrix}, \begin{bmatrix} 0.23 & 0 \\ 0 & 0.23 \end{bmatrix}\right),$$

$$X^{te}|_{Y^{te}=-1} \sim \mathcal{N}\left(\begin{bmatrix} -0.1 \\ 0 \end{bmatrix}, \begin{bmatrix} 0.10 & -0.03 \\ -0.03 & 0.10 \end{bmatrix}\right),$$

$$X^{te}|_{Y^{te}=1} \sim \mathcal{N}\left(\begin{bmatrix} 0.9 \\ 0.8 \end{bmatrix}, \begin{bmatrix} 0.11 & 0.05 \\ 0.05 & 0.11 \end{bmatrix}\right),$$

$P_Y^{tr}(y = -1) = 0.6$, and $P_Y^{te}(y = -1) = 0.2$.

See Figure 19.5 (left) for the training and test points generated according to the four settings in one random replication. The training and test sets consist of 500 and 400 data points, respectively.

We compare our approaches to correction for TarS (Section 19.3) and for LS-GeTarS (Section 19.5) with the baseline (unweighted) least squares KRR or SVM, the importance weighting approach to correction for covariate shift (CovS) proposed in [10, 6], as well as two "oracle" approaches: one uses the theoretical values of $\beta^*(y) = P_Y^{te}/P_Y^{te}$, and the other trains the learning machine directly on the test set. Note that the result learned on the test set certainly has the best performance, but in practice it cannot be applied; it is given to show the limit of the performance that any domain-adaptation approach can achieve. Since in the considered classification problems X is low-dimensional, it is possible to apply the EM algorithm proposed by [2] to estimate P_Y^{te}, so it is also included for comparison. We repeated the simulations 100 times.

Figure 19.5 (right) shows the boxplot of the performances of all approaches, measured by the mean square error (MSE) or classification error on the test set; for illustrative purposes, the left panels show the data points generated in one replication as well as the regression lines or decision boundaries learned by selected approaches. Under TarS, *(a, b)*, and non-LS-GeTarS with slightly changing conditionals, *(d)*, compared to the baseline unweighted method, clearly our approaches for TarS and LS-GeTarS improve the performance significantly. For regression under TarS, the estimated β values are very close to the theoretical ones, as seen from the lower-right corner of Figure 19.5 (a, left). EM achieves a similar performance as TarS, since $P_{X|Y}$ can be modeled well in this simple case. In *(c)* the conditional $P_{X|Y}$ changes significantly, such that none of the approaches correcting for CovS or TarS helps, but since the change approximately follows LS-GeTarS, our approach for LS-GeTarS greatly improves the classification performance. Compared to the unweighted method, the important reweighting approach for CovS slightly improves the performance in settings *(b)* and *(d)*, and make it worse in *(a)* and *(c)*.

19.8   Experiments on Pseudo Real-World Data

Table 19.2 reports the results on pseudo real-world data sets. In these experiments, we split each data set into training set and test set. The percentage of training samples ranges from 60% to 80%. Then, we perform the biased sampling on the training data to obtain the shifted training set. Letting $P(s = 1|y)$ be the probability of sample x being selected given that its true output value is y, we consider the following two biased sampling schemes for selecting training data: (1) **Weighted Label** uses $P(s = 1|y) = \exp(a + by)/(1 + \exp(a + by))$ denoted by `label(a,b)`, and (2) **PCA**. In this case, we generate biased sampling schemes over the features. Firstly, a kernel PCA is performed on the data. We select the first principal component and the corresponding projection values. The biased sampling scheme is then a normal distribution with mean $m + (\bar{m} - m)/a$ and variance $(\bar{m} - m)/b$ where m and $\bar{m}$ are the minimum value of the projection and the mean of the projection, respectively. We denote this sampling scheme by `PCA(a,b,`σ`)`, where σ is the bandwidth of the Gaussian RBF kernel. In summary, the LS-GeTarS outperforms unweight,

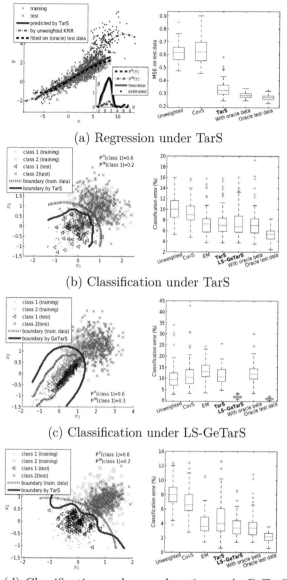

(a) Regression under TarS

(b) Classification under TarS

(c) Classification under LS-GeTarS

(d) Classification under non-location-scale GeTarS

FIGURE 19.5: Four simulation settings together with the performances of different approaches. Left panels show the data points together with the decision boundaries (or regression lines) obtained by selected approaches in one replication, and right panels give the boxplot of the performances of different approaches for 100 random replications. (a) For a regression problem with X depending on Y nonlinearly. (b) For a classification problem under TarS. (c) For a classification problem under shape-preserving GeTarS. (d) For a classification problem under GeTarS but the shape of the conditional distribution changes. Note that y-values of the test data were not given in the training phase, and they are plotted for illustrative purposes.

CovS, and TarS on 5 out of 6 data sets for a classification problem. The TarS outperforms all other approaches on one of these data sets. For regression problem, TarS outperforms the unweight and CovS on 7 out of 12 data sets.

19.9   Experiments on Real-World Data

We evaluate the performance of the proposed approaches for regression and classification on real data. We first consider prediction of nonstationary processes, and then tackle the remote sensing image classification problem, with images acquired on different areas.

19.9.1   Regression under TarS

We first applied our approach for prediction on suitable data selected from the cause-effect pairs.[7] We selected Data Set No. 68, since 1) the data are nonstationary time series, 2) there is a strong dependence between the two variables so that one can be predicted non-trivially by the other, and 3) the variables are believed to have a direct causal relation, so that the invariance of the conditional distribution of one variable (effect) given the other (cause) is likely to hold approximately. Figure 19.6 (top) shows the time series as well as the joint distribution. Here X and Y stand for the number of bytes sent by a computer at the tth minute and the number of open http connections at the same time, respectively. It is natural to have the causal relation $Y \to X$, and we aim to predict Y from X without making use of temporal dependence in the data. One subsample was always used for training, because on it Y has large values. The remaining data were divided into four subsets, and each time one of them was used for testing and the others included for training.

Figure 19.6 (bottom) shows the estimated β^* values on the four test sets; they match P_Y^{te} well. Table 19.3 gives the MSE on the four test sets produced by different approaches. Note that to achieve robustness of the prediction result, we incorporated an exponent q for β^* as the importance weights, as in correction for CovS with importance re-weighting [25]. $q = 1$ (i.e., the proposed standard approach) and $q = 0.5$ were used. From Table 19.3 one can see TarS gives the best results on all four test sets.

19.9.2   Remote Sensing Image Classification

Hyperspectral remote sensing images are characterized by a dense sampling of the spectral signature of different land-cover types. We used a benchmark data set in the literature that consists of data acquired by the Hyperion sensor

---

[7]http://webdav.tuebingen.mpg.de/cause-effect/

TABLE 19.2: The results of different distribution shift correction schemes. The results are averaged over 10 trials for regression problems (marked *) and 30 trials for classification problems. We report the normalized mean squared error (NMSE) for regression problem and test error for classification problem.

Data Set	Sampling Scheme	NMSE/test error ± std. error				
		Unweight	CovS	TarS	LS-GeTarS
1. Abalone*	label(1,10)	0.4447±0.0223	0.4497±0.0125	**0.4430±0.0208**	–
2. CA Housing*	PCA(10,5,0.1)	0.4075±0.0298	**0.3944±0.0346**	0.4565±0.0422	–
3. Delta Alilerons (1)*	label(1,10)	**0.3120±0.0040**	0.3408±0.0278	0.3451±0.0280	–
4. Ailerons*	PCA(1e3,4,0.1)	0.1360±0.0350	0.1486±0.0264	**0.1329±0.0174**	–
5. haberman (1)	label(0.2,0.8)	0.2699±0.0304	0.2699±0.0315	**0.2676±0.0287**	0.2619±0.0352
6. Bank8FM*	PCA(3,6,0.1)	0.0477±0.0014	0.0590±0.0117	**0.0452±0.0070**	–
7. Bank32nh*	PCA(3,6,0.01)	0.5210±0.0318	**0.5171±0.0131**	0.5483±0.0455	–
8. cpu-act*	PCA(4,2,1e-12)	0.2026±0.0382	0.2042±0.0316	**0.2000±0.0474**	–
9. cpu-small*	PCA(4,2,1e-12)	0.1314±0.0347	0.2009±0.0849	**0.0769±0.0100**	–
10. Delta Ailerons(2)*	PCA(1e3,4,0.1)	0.4496±0.0236	0.3373±0.0596	**0.3258±0.0274**	–
11. Boston House*	PCA(2,4,1e-4)	0.5128±0.1269	**0.4966±0.0970**	0.5342±0.0777	–
12. kin8nm*	PCA(8,5,0.1)	0.5382±0.0425	**0.5266±0.1248**	0.6079±0.0976	–
13. puma8nh*	PCA(4,4,0.1)	0.6093±0.0629	0.5894±0.0361	**0.5595±0.0297**	–
14. haberman(2)	PCA(2,2,0.01)	0.2736±0.0374	0.2725±0.0422	0.2724±0.0367	**0.2579±0.0241**
15. Breast Cancer	label(0.3,0.7)	0.2699±0.0304	0.3196±0.1468	0.2670±0.0319	**0.2609±0.0510**
16. India Diabetes	label(0.3,0.7)	0.2742±0.0268	0.2797±0.0354	0.2846±0.0364	**0.2700±0.0599**
17. Ionosphere	label(0.3,0.7)	0.0865±0.0294	0.1079±0.0563	**0.0846±0.0559**	0.0938±0.0294
18. German Credit	label(0.2,0.8)	0.3000±0.0284	0.2802±0.0354	0.2846±0.0364	**0.2596±0.0368**

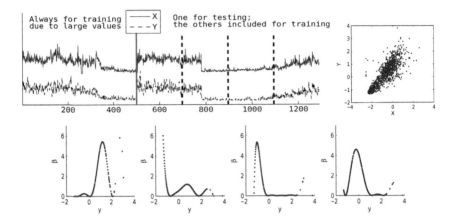

FIGURE 19.6: Prediction results on Pair 68 of the cause-effect pairs. Top: time series data of X and Y (left, shifted apart for clarity) and the joint distribution (right). Bottom: estimated β^* values on the four test sets.

TABLE 19.3: Prediction performance (MSE) on test sets.

Test set	Unweight.	CovS	CovS ($q = 0.5$)	TarS	TarS ($q=0.5$)
1	0.3789	0.3844	0.3802	0.3310	**0.3229**
2	0.0969	0.1126	0.1071	0.0937	**0.0887**
3	0.0578	0.0673	0.0659	**0.0466**	0.0489
4	0.2054	0.2126	0.2136	0.2008	**0.1630**

of the Earth Observing 1 (EO-1) satellite in an area of the Okavango Delta, Botswana, with 145 features; for details of this data set, see [8]. The labeled reference samples were collected on two different and spatially disjoint areas (Area 1 and Area 2), thus representing possible spatial variabilities of the spectral signatures of classes. The samples taken on each area were partitioned into a training set TR and a test set TS by random sampling.

The numbers of labeled reference samples for each set and class are reported in Table 19.4. TR_1, TS_1, TR_2, and TS_2 have sample sizes 1242, 1252, 2621, and 627, respectively. One would expect that not only the prior probabilities of the classes Y, but also the conditional distribution of X given Y would change across them, due to physical factors related to ground (e.g., different soil moisture or composition), vegetation, and atmospheric conditions. Our target is to do domain adaptation from TR_1 to TS_2 and from TR_2 to TS_1, as in [19].

TABLE 19.4: Number of training (TR_1 and TR_2) and test (TS_1 and TS_2) patterns acquired in the two spatially disjoint areas for the experiment on remote sensing image classification.

| Class | Number of patterns | | | |
| | Area 1 | | Area 2 | |
	TR_1	TS_1	TR_2	TS_2
Water	69	57	213	57
Hippo grass	81	81	83	18
Floodplain grasses1	83	75	199	52
Floodplain grasses2	74	91	169	46
Reeds1	80	88	219	50
Riparian	102	109	221	48
Firescar2	93	83	215	44
Island interior	77	77	166	37
Acacia woodlands	84	67	253	61
Acacia shrublands	101	89	202	46
Acacia grasslands	184	174	243	62
Short mopane	68	85	154	27
Mixed mopane	105	128	203	65
Exposed soil	41	48	81	14
Total	1242	1252	2621	627

After estimating the weights and/or the transformed training points (with $\lambda_{LS} = 10^{-4}$), we applied the multi-class classifier with an RBF kernel, provided by LIBSVM, on the weighted or transformed data. Each time, the kernel size and parameter C were chosen by five-fold cross-validation over the sets $\{2^{5/2}, 2^{3/2}, 2^{1/2}, 2^{-1/2}, 2^{-3/2}, 2^{-5/2}\} \cdot \sqrt{d}$ and $\{2^6, 2^8, 2^{10}, 2^{12}, 2^{14}, 2^{16}, 2^{18}\}$, respectively. (We found that the selected values always belonged to the interior of the sets.)

Table 19.5 shows the overall classification error (i.e., the fraction of misclassified points) obtained by different approaches for each domain adaptation problem. We can see that in this experiment, correction for target shift does not significantly improve the performance; in fact, the β values for most classes are rather close to one. However, correction for conditional shift with LS-GeTarS reduces the overall classification error from 20.73% to 11.96% for domain adaptation from TR_1 to TS_2, and from 25.32% to 14.54% for that from TR_2 to TS_1. Covariate shift helps slightly for $TR_2 \rightarrow TS_1$, probably because our classifier is rather simple in that all dimensions have the same kernel size.

TABLE 19.5: The misclassification rate on remote sensing data set under different distribution shift correction schemes.

Problem	Unweight	CovS	TarS	LS-GeTarS
$TR_1 \rightarrow TS_2$	20.73%	20.73%	20.41%	**11.96%**
$TR_2 \rightarrow TS_1$	26.36%	25.32%	26.28%	**14.54%**

Correction for conditional shift with LS-GeTarS reduces the overall classification error. In addition to the overall classification error, we also report the number of correctly classified points from each class; see Figure 19.7. One can see that for both domain adaptation problems, LS-GeTarS improves the classification accuracy on classes 11, 9, and 3. It also leads to significant improvement on class 13 for the problem $TR_1 \rightarrow TS_2$, and on class 2 for $TR_1 \rightarrow TS_2$. Note that this is a multi-class classification problem and we aim to improve the overall classification accuracy; to achieve that, the accuracy of some particular classes, such as classes 10 and 6, may become worse. Figure 19.8 plots some of the estimated scale transformation coefficients $\mathbf{w}(y^{tr})$ and location transformations $\mathbf{b}(y^{tr})$ that are significant (i.e., $\mathbf{w}(y^{tr})$ is significantly different from one, and $\mathbf{b}(y^{tr})$ different from zero). Roughly speaking, the transformation learned for the domain adaptation problem $TR_2 \rightarrow TS_1$ is the inverse of that for the problem $TR_1 \rightarrow TS_1$.

19.10   Conclusion and Discussions

We have considered domain adaptation where both the distribution of the covariate and the conditional distribution of the target given the covariate change across domains. From the causal point of view, we assume the target causes the covariate, such that the change in the data distribution can be modeled easily. In particular, we studied three situations, target shift, con-

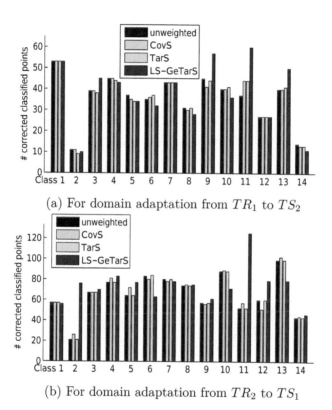

(a) For domain adaptation from TR_1 to TS_2

(b) For domain adaptation from TR_2 to TS_1

FIGURE 19.7: The number of correctly classified data points for each class and each approach. (a) $TR1$ as training set and $TS2$ as test set. (b) $TR2$ as training set and TS_1 as test set.

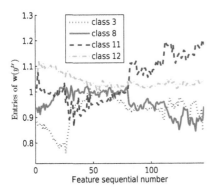

(a) Estimated scale transformation coefficient for selected classes for domain adaptation $TR_1 \rightarrow TS_2$.

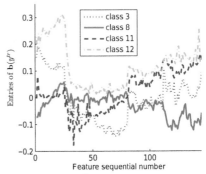

(b) Estimated location transformation for selected classes for domain adaptation $TR_1 \rightarrow TS_2$.

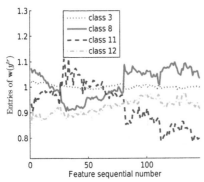

(c) Estimated scale transformation coefficient for selected classes for domain adaptation $TR_2 \rightarrow TS_1$.

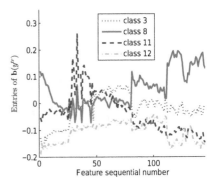

(d) Estimated location transformation for selected classes for domain adaptation $TR_2 \rightarrow TS_1$.

FIGURE 19.8: Estimated scale transformation coefficient $\mathbf{w}(y^{tr})$ and location transformation $\mathbf{b}(y^{tr})$ for selected classes by correction for LS-GeTarS. (a, b) For domain adaptation from TR_1 to TS_2. (c, d) For domain adaptation from TR_2 to TS_1.

ditional shift, and generalized target shift, which combines the above two situations. We presented practical approaches to handling them based on the kernel mean embedding of conditional and marginal distributions. Simulations were conducted to verify our theoretical claims. Experimental results on diverse real-world problems, including remote sensing image classification, showed that (generalized) target shift often happens in domain adaptation and that the proposed approaches could substantially improve the classification or regression performance accordingly.

Acknowledgments

The authors would like to thank D. Janzing for helpful discussions and M. Crawford for kindly providing the hyperspectral data set. Z.-H. Zhou was supported by the National Science Foundation of China (61333014) and the National Key Basic Research Program of China (2014CB340501). C. Persello has been supported by the Autonomous Province of Trento and the European Community in the framework of the project "Trentino - PCOFUND-GA-2008-226070 (call 3 - post-doc 2010 Outgoing)".

Bibliography

[1] J. Candela, M. Sugiyama, A. Schwaighofer, and N. Lawrence, editors. *Dataset Shift in Machine Learning.* MIT Press, 2009.

[2] Y. S. Chan and H. T Ng. Word sense disambiguation with distribution estimation. In *Proceedings of the 19th International Joint Conference on Artificial Intelligence*, pages 1010–1015, Scotland, 2005.

[3] K. Fukumizu, F. R. Bach, M. I. Jordan, and C. Williams. Dimensionality reduction for supervised learning with reproducing kernel Hilbert spaces. *Journal of Machine Learning Research*, 5:73–99, 2004.

[4] K. Fukumizu, A. Gretton, X. Sun, and B. Schölkopf. Kernel measures of conditional dependence. In J.C. Platt, D. Koller, Y. Singer, and S. Roweis, editors, *Advances in Neural Information Processing Systems 20*, pages 489–496, Cambridge, MA, 2008. MIT Press.

[5] A. Gretton, K. Borgwardt, M. Rasch, B. Schölkopf, and A. Smola. A kernel method for the two-sample-problem. In *NIPS 19*, pages 513–520, Cambridge, MA, 2007. MIT Press.

[6] A. Gretton, A. Smola, J. Huang, M. Schmittfull, K. Borgwardt, and B. Schölkopf. Covariate shift and local learning by distribution matching. In J. Quiñonero-Candela, M. Sugiyama, A. Schwaighofer, and N. Lawrence, editors, *Dataset shift in machine learning*, pages 131–160. MIT Press, Cambridge, MA, 2008.

[7] A. Gretton, B. Sriperumbudur, D. Sejdinovic, H. Strathmann, S. Balakrishnan, M. Pontil, and K. Fukumizu. Optimal kernel choice for large-scale two-sample tests. In *NIPS 25*. MIT Press, 2012.

[8] J. Ham, Y. Chen, M. M. Crawford, and J. Ghosh. Investigation of the random forest framework for classification of hyperspectral data. *IEEE Trans. Geosci. Remote Sens.*, 43(3):492–501, 2005.

[9] P.O. Hoyer, D. Janzing, J. Mooji, J. Peters, and B. Schölkopf. Nonlinear causal discovery with additive noise models. In *Advances in Neural Information Processing Systems 21*, Vancouver, B.C., Canada, 2009.

[10] J. Huang, A. Smola, A. Gretton, K. Borgwardt, and B. Schölkopf. Correcting sample selection bias by unlabeled data. In *NIPS 19*, pages 601–608, 2007.

[11] D. Janzing, J. Mooij, K. Zhang, J. Lemeire, J. Zscheischler, P. Daniusvis, B. Steudel, and B. Schölkopf. Information-geometric approach to inferring causal directions. *Artificial Intelligence*, pages 1–31, 2012.

[12] N. Japkowicz and S. Stephen. The class imbalance problem: A systematic study. *Intelligent Data Analysis*, 6:429–450, 2002.

[13] J. Jiang. *A literature survey on domain adaptation of statistical classifiers*, 2008. http://sifaka.cs.uiuc.edu/jiang4/domain_adaptation/survey.

[14] Y. Lin, Y. Lee, and G. Wahba. Support vector machines for classification in nonstandard situations. *Machine Learning*, 46:191–202, 2002.

[15] C. Manski and S. Lerman. The estimation of choice probabilities from choice-based samples. *Econometrica*, 45:1977–1988, 1977.

[16] J. Mooij, O. Stegle, D. Janzing, K. Zhang, and B. Schölkopf. Probabilistic latent variable models for distinguishing between cause and effect. In *Advances in Neural Information Processing Systems 23 (NIPS 2010)*, Curran, NY, USA, 2010.

[17] S. J. Pan and Q. Yang. A survey on transfer learning. *IEEE Transactions on Knowledge and Data Engineering*, 22:1345–1359, 2010.

[18] J. Pearl. *Causality: Models, Reasoning, and Inference*. Cambridge University Press, Cambridge, 2000.

[19] C. Persello. Interactive domain adaptation for the classification of remote sensing images using active learning. *IEEE Geoscience and Remote Sensing Letters*, 10:736–740, 2013.

[20] C. P. Robert and G. Casella. *Monte Carlo Statistical Methods*. Springer Press, New York, 2nd edition, 2004.

[21] C. Saunders, A. Gammerman, and V. Vovk. Ridge regression learning algorithm in dual variables. In *15th International Conference on Machine Learning*, pages 515–521, Madison, WI, 1998.

[22] B. Schölkopf, D. Janzing, J. Peters, E. Sgouritsa, K. Zhang, and J. Mooij. On causal and anticausal learning. In *Proc. 29th International Conference on Machine Learning (ICML 2012)*, Edinburgh, Scotland, 2012.

[23] B. Schölkopf, A. Smola, and K. Muller. Nonlinear component analysis as a kernel eigenvalue problem. *Neural Computation*, 10:1299–1319, 1998.

[24] S. Shimizu, P.O. Hoyer, A. Hyvärinen, and A.J. Kerminen. A linear non-Gaussian acyclic model for causal discovery. *Journal of Machine Learning Research*, 7:2003–2030, 2006.

[25] H. Shimodaira. Improving predictive inference under covariate shift by weighting the log-likelihood function. *Journal of Statistical Planning and Inference*, 90:227–244, 2000.

[26] A. Smola, A. Gretton, L. Song, and B. Schölkopf. A Hilbert space embedding for distributions. In *Proceedings of the 18th International Conference on Algorithmic Learning Theory*, pages 13–31. Springer-Verlag, 2007.

[27] L. Song, B. Boots, S. Siddiqi, G. Gordon, and A. Smola. Hilbert space embeddings of hidden Markov models. In *Proceedings of the 26th International Conference on Machine Learning*, Haifa, Israel, 2010.

[28] L. Song, J. Huang, A. Smola, and K. Fukumizu. Hilbert space embeddings of conditional distributions with applications to dynamical systems. In *International Conference on Machine Learning (ICML 2009)*, June 2009.

[29] P. Spirtes, C. Glymour, and R. Scheines. *Causation, Prediction, and Search*. MIT Press, Cambridge, MA, 2nd edition, 2001.

[30] B. Sriperumbudur, K. Fukumizu, and G. Lanckriet. Universality, characteristic kernels and RKHS embedding of measures. *Journal of Machine Learning Research*, 12:2389–2410, 2011.

[31] A. Storkey. When training and test sets are different: Characterizing learning transfer. In J. Candela, M. Sugiyama, A. Schwaighofer, and N. Lawrence, editors, *Dataset Shift in Machine Learning*, pages 3–28. MIT Press, 2009.

[32] M. Sugiyama, M. Krauledat, and K. R. Müller. Covariate shift adaptation by importance weighted cross validation. *Journal of Machine Learning Research*, 8:985–1005, December 2007.

[33] M. Sugiyama, T. Suzuki, S. Nakajima, H. Kashima, P. von Bünau, and M. Kawanabe. Direct importance estimation for covariate shift adaptation. *Annals of the Institute of Statistical Mathematics*, 60:699–746, 2008.

[34] J. Tian and J. Pearl. Causal discovery from changes: a bayesian approach. In *Proceedings of the 17th Conference on Uncertainty in Artificial Intelligence (UAI2001)*, pages 512–521, 2001.

[35] J. Woodward. *Making things happen: A theory of causal explanation.* Oxford University Press, New York, 2003.

[36] Y. Yu and Z. H. Zhou. A framework for modeling positive class expansion with single snapshot. In *PAKDD 2008*, 2008.

[37] B. Zadrozny. Learning and evaluating classifiers under sample selection bias. In *21st International Conference on Machine Learning*, pages 114–121, Banff, Canada, 2004.

[38] K. Zhang and A. Hyvärinen. Acyclic causality discovery with additive noise: An information-theoretical perspective. In *Proc. European Conference on Machine Learning and Principles and Practice of Knowledge Discovery in Databases (ECML PKDD) 2009*, Bled, Slovenia, 2009.

[39] K. Zhang and A. Hyvärinen. On the identifiability of the post-nonlinear causal model. In *Proceedings of the 25th Conference on Uncertainty in Artificial Intelligence*, Montreal, Canada, 2009.

[40] K. Zhang, J. Peters, D. Janzing, and B. Schölkopf. Kernel-based conditional independence test and application in causal discovery. In *Proceedings of the 27th Conference on Uncertainty in Artificial Intelligence (UAI 2011)*, Barcelona, Spain, 2011.

[41] K. Zhang, B. Schölkopf, K. Muandet, and Z. Wang. Domain adaptation under target and conditional shift. In *Proceedings of the 30th International Conference on Machine Learning, JMLR: W&CP Vol. 28*, 2013.

Chapter 20

Multi-Layer Support Vector Machines

Marco A. Wiering

Institute of Artificial Intelligence and Cognitive Engineering, University of Groningen

Lambert R.B. Schomaker

Institute of Artificial Intelligence and Cognitive Engineering, University of Groningen

20.1 Introduction

Support vector machines (SVMs) [24, 8, 20, 22] and other learning algorithms based on kernels have been shown to obtain very good results on many different classification and regression datasets. SVMs have the advantage of generalizing very well, but the standard SVM is limited in several ways. First, the SVM uses a single layer of support vector coefficients and is therefore a shallow model. Deep architectures [17, 14, 13, 4, 25, 6] have been shown to be very promising alternatives to these shallow models. Second, the results of the SVM rely heavily on the selected kernel function, but most kernel functions have limited flexibility in the sense they they are not trainable on a dataset. Therefore, it is a natural step to go from the standard single-layer SVM to the multi-layer SVM (ML-SVM). Just like the invention of the backpropagation algorithm [26, 19] allowed to construct multi-layer perceptrons from perceptrons, this chapter describes techniques for constructing and training multi-layer SVMs consisting only of SVMs.

There is a lot of related work in multiple kernel learning (MKL) [16, 3, 21, 18, 31, 10]. In these approaches, some combination functions of a set of fixed kernels are adapted to the dataset. As has been shown by a number of experiments, linear combinations of base kernels do not often help to get significantly better performance levels. Therefore, in [7] the authors describe the use of non-linear (polynomial) combinations of kernels and their results show that this technique is more effective. An even more recent trend in MKL is the use of multi-layer MKL. In [9], a general framework for two-layer kernel machines is described, but unlike in the current study, no experimental results were reported in which both layers used non-linear kernels. In [32], multi-layer MKL is described where mixture coefficients of different kernels are stored in an exponential function kernel. These coefficients in the second layer of the two-layer MKL algorithm are trained using a min-max objective function. In [5] a new type of kernel is described, which is useful for mimicking a deep learning architecture. The neural support vector machine (NSVM) [28] is also related to the multi-layer SVM. The NSVM is a novel algorithm that uses neural networks to extract features that are given to a support vector machine for giving the final output of the architecture. Finally, the current chapter extends the ideas in [27] by describing a classification and autoencoder method using multi-layer support vector machines.

Contributions. We describe a simple method for constructing and training multi-layer SVMs. The hidden-layer SVMs in the architecture learn to extract relevant features or latent variables from the inputs and the output-layer SVMs learn to approximate the target function using the extracted features from the hidden-layer SVMs. We can easily make the association with multi-layer perceptrons (MLPs) by letting a complete SVM replace each individual neuron. However, in contrast to the MLP, the ML-SVM algorithm is trained using a min-max objective function: the hidden-layer SVMs are trained to minimize the dual-objective function of the output-layer SVMs and the output-layer SVMs are trained to maximize their dual-objective functions. This min-max optimization problem is a result of going from the primal objective to the dual objective. Therefore, the learning dynamics of the ML-SVM are entirely different compared to the MLP in which all model parameters are trained to minimize the same error function. When compared to other multi-layer MKL approaches, the ML-SVM does not make use of any combination weights, but trains support vector coefficients and the biases of all SVMs in the architecture. Our experimental results show that the ML-SVM significantly outperforms state-of-the-art machine learning techniques on regression, classification and dimensionality reduction problems.

We have organized the rest of this chapter as follows. Section 20.2 describes the ML-SVM algorithm for regression problems. In Section 20.3, the ML-SVM algorithm is introduced for classification problems. In Section 20.4, the autoencoding ML-SVM is described. In Section 20.5, experimental results on 10 regression datasets, 8 classification datasets, and a dimensionality reduction problem are presented. Finally, Section 20.6 discusses the findings and describes future work.

20.2 Multi-Layer Support Vector Machines for Regression Problems

We will first describe the multi-layer SVM for regression problems. We use a regression dataset: $\{(\mathbf{x}_1, y_1), \ldots, (\mathbf{x}_\ell, y_\ell)\}$, where $\mathbf{x}_i$ are input vectors and y_i are the scalar target outputs. The architecture of a two-layer SVM is shown in Figure 20.1.

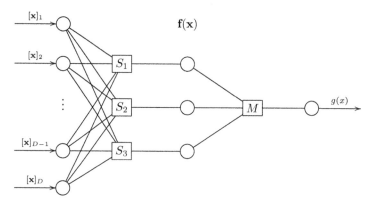

FIGURE 20.1: Architecture of a two-layer SVM. In this example, the hidden layer consists of three SVMs S_a.

The two-layer architecture contains an input layer of D inputs. Then, there are a total of d SVMs S_a, each one learning to extract one latent variable $\mathbf{f}(\mathbf{x}|\theta)_a$ from an input pattern $\mathbf{x}$. Here θ denotes the trainable parameters in the hidden-layer SVMs (which are the support vector coefficients and the biases). Finally, there is the main support vector machine M that learns to approximate the target function using the extracted feature vector as input. For computing the hidden-layer representation $\mathbf{f}(\mathbf{x}|\theta)$ of input vector $\mathbf{x}$, we use:

$$\mathbf{f}(\mathbf{x}|\theta)_a = \sum_{i=1}^{\ell}(\alpha_i^*(a) - \alpha_i(a))K_1(\mathbf{x}_i, \mathbf{x}) + b_a, \tag{20.1}$$

which is iteratively used by each SVM S_a to compute the element $\mathbf{f}(\mathbf{x}|\theta)_a$. In this equation, $\alpha_i^*(a)$ and $\alpha_i(a)$ are support vector coefficients for SVM S_a, b_a is its bias, and $K_1(\cdot, \cdot)$ is a kernel function for the hidden-layer SVMs. For computing the output of the whole ML-SVM, the main SVM maps the extracted hidden-layer representation to an output:

$$g(\mathbf{f}(\mathbf{x}|\theta)) = \sum_{i=1}^{\ell}(\alpha_i^* - \alpha_i)K_2(\mathbf{f}(\mathbf{x}_i|\theta), \mathbf{f}(\mathbf{x}|\theta)) + b. \tag{20.2}$$

Here, $K_2(\cdot, \cdot)$ is the kernel function in the output layer of the multi-layer SVM. The primal objective for a linear regression SVM M can be written as:

$$\min_{\mathbf{w}, \theta, \boldsymbol{\xi}, \boldsymbol{\xi}^*, b} J(\mathbf{w}, \theta, \boldsymbol{\xi}, \boldsymbol{\xi}^*, b) = \frac{1}{2}\|\mathbf{w}\|^2 + C\sum_{i=1}^{\ell}(\xi_i + \xi_i^*) \tag{20.3}$$

subject to constraints:

$$y_i - \mathbf{w} \cdot \mathbf{f}(\mathbf{x}_i|\theta) - b \le \varepsilon + \xi_i \quad ; \quad \mathbf{w} \cdot \mathbf{f}(\mathbf{x}_i|\theta) + b - y_i \le \varepsilon + \xi_i^* \tag{20.4}$$

and $\xi_i, \xi_i^* \ge 0$. Here C is a metaparameter, ϵ is an error tolerance value used in the Hinge (ϵ-insensitive) loss function, and ξ_i and ξ_i^* are slack variables that tolerate errors larger than ϵ, but which should be minimized. The dual-objective function for the regression problem for the main SVM M is:

$$\min_{\theta} \max_{\boldsymbol{\alpha}, \boldsymbol{\alpha}^*} J(\theta, \boldsymbol{\alpha}, \boldsymbol{\alpha}^*) = -\varepsilon \sum_{i=1}^{\ell}(\alpha_i^* + \alpha_i) + \sum_{i=1}^{\ell}(\alpha_i^* - \alpha_i)y_i$$

$$-\frac{1}{2}\sum_{i,j=1}^{\ell}(\alpha_i^* - \alpha_i)(\alpha_j^* - \alpha_j)K_2(\mathbf{f}(\mathbf{x}_i|\theta), \mathbf{f}(\mathbf{x}_j|\theta)) \tag{20.5}$$

subject to: $0 \le \alpha_i^*, \alpha_i \le C$ and $\sum_{i=1}^{\ell}(\alpha_i - \alpha_i^*) = 0$. The second constraint in generally known as the bias constraint.

Our learning algorithm adjusts the SVM coefficients of all SVMs through the min-max formulation of the dual-objective function $J(\cdot)$ of the main SVM. Note that the min-max optimization problem is a result of going from the primal objective to the dual objective. In the primal objective, it is a joint minimization with respect to θ and the α coefficients. However, by dualizing the primal objective of the main SVM, it is turned into a min-max problem.

We have implemented a simple gradient ascent algorithm to train the SVMs. The method adapts all SVM coefficients α_i^* and α_i toward a (local) maximum of $J(\cdot)$, where λ is the learning rate. The resulting gradient ascent learning rule for α_i is:

$$\alpha_i \leftarrow \alpha_i + \lambda(-\epsilon - y_i + \sum_{j=1}^{\ell}(\alpha_j^* - \alpha_j)K_2(\mathbf{f}(\mathbf{x}_i|\theta), \mathbf{f}(\mathbf{x}_j|\theta))). \qquad (20.6)$$

The resulting gradient ascent learning rule for α_i^* is:

$$\alpha_i^* \leftarrow \alpha_i^* + \lambda(-\epsilon + y_i - \sum_{j=1}^{\ell}(\alpha_j^* - \alpha_j)K_2(\mathbf{f}(\mathbf{x}_i|\theta), \mathbf{f}(\mathbf{x}_j|\theta))) \qquad (20.7)$$

The support vector coefficients are set to 0 if they become less than 0, and set to C if they become larger than C. We also added a penalty term to respect the bias constraint, so actually the gradient ascent algorithm trains the support vector coefficients to maximize the objective $J'(\cdot) = J(\cdot) - c_1 \cdot (\sum_i(\alpha_i - \alpha_i^*))^2$, with c_1 some metaparameter. Although this simple strategy works well, this ad-hoc optimization strategy could also be replaced by a gradient projection method for which convergence properties are better understood.

In the experiments we will make use of radial basis function (RBF) kernels in both layers of a two-layer SVM. Preliminary results with other often used kernels were somewhat worse. For the main SVM and hidden-layer SVMs the RBF kernel is defined respectively by:

$$K_2(\mathbf{f}(\mathbf{x}_i|\theta), \mathbf{f}(\mathbf{x}|\theta)) = \exp(-\sum_{a=1}^{d} \frac{(\mathbf{f}(\mathbf{x}_i|\theta)_a - \mathbf{f}(\mathbf{x}|\theta)_a)^2}{\sigma_2}) \qquad (20.8)$$

$$K_1(\mathbf{x}_i, \mathbf{x}) = \exp(-\sum_{a=1}^{D} \frac{(\mathbf{x}_i^a - \mathbf{x}^a)^2}{\sigma_1}) \qquad (20.9)$$

where σ_2 and σ_1 determine the widths of the RBF kernels in the output and hidden layers. The ML-SVM constructs a new dataset for each hidden-layer SVM S_a with a backpropagation-like technique for making examples: $(\mathbf{x}_i, \mathbf{f}(\mathbf{x}_i|\theta)_a - \mu \cdot \partial J(\cdot)/\partial \mathbf{f}(\mathbf{x}_i|\theta)_a)$, where μ is some metaparameter, and $\partial J(\cdot)/\partial \mathbf{f}(\mathbf{x}_i|\theta)_a$ for the RBF kernel is given by:

$$\frac{\partial J(\cdot)}{\partial \mathbf{f}(\mathbf{x}_i|\theta)_a} = (\alpha_i^* - \alpha_i)\sum_{j=1}^{\ell}(\alpha_j^* - \alpha_j)\frac{\mathbf{f}(\mathbf{x}_i|\theta)_a - \mathbf{f}(\mathbf{x}_j|\theta)_a}{\sigma_2} \cdot K_2(\mathbf{f}(\mathbf{x}_i|\theta), \mathbf{f}(\mathbf{x}_j|\theta)).$$

$$(20.10)$$

We constrain the target values for hidden-layer features between -1 and 1, so if some target output is larger than 1 for a feature we simply set the target value to 1. To allow the hidden-layer SVMs to extract different features, symmetry breaking is necessary. For this, we could randomly initialize the trainable parameters in each hidden-layer SVM. However, we discovered that a better way to initialize the hidden-layer SVMs is to let them train on different perturbed versions of the target outputs. Therefore we initially construct a dataset $(\mathbf{x}_i, y_i + \gamma_i^a)$, with γ_i^a some random value $\in [-\gamma, \gamma]$ for the hidden-layer SVM S_a, where γ is another metaparameter. In this way, the ML-SVM resembles a stacking ensemble approach [30], but due to the further training with the min-max optimization process, these approaches are still very different. The complete algorithm is given in Algorithm 18.

Algorithm 18 The multi-layer SVM algorithm

    Initialize output SVM
    Initialize hidden-layer SVMs
    Compute kernel matrix for hidden-layer SVMs
    Train hidden-layer SVMs on perturbed dataset
    repeat
        Compute kernel matrix for output-layer SVM
        Train output-layer SVM
        Use backpropagation to create training sets for hidden-layer SVMs
        Train hidden-layer SVMs
    until maximum number of epochs is reached

In the algorithm alternated training of the main SVM and hidden-layer SVMs a number of epochs are executed. An epoch here is defined as training the main SVM and the hidden-layer SVM a single time on their respective datasets with our gradient ascent technique that uses a small learning rate and a fixed number of iterations. The bias values of all SVMs are set by averaging over the errors on all examples.

Theoretical insight. Due to the min-max optimization problem and the two layers with non-linear kernel functions, the ML-SVM loses the property that the optimization problem is convex. However, similar to multiple-kernel learning, training the output-layer SVM given the outputs of the hidden layer remains a convex learning problem. Furthermore, the datasets generated with the backpropagation technique explained above are like normal training datasets. Since training an SVM on a dataset is a convex learning problem, these newly created datasets are also convex learning problems for the hidden-layer SVMs. By using the pre-training of hidden-layer SVMs on perturbed versions of the target outputs, the learning problem of the output-layer SVM becomes much simpler. In fact, this resembles a stacking ensemble approach [30], but unlike any other ensemble approach, the ML-SVM is further optimized using the min-max optimization process. This is interesting, because it is different from other approaches in which the same error function is minimized by all model parameters. Still, it could also be seen as a disadvantage, because min-max learning is not yet well understood in the machine learning community.

20.3   Multi-Layer Support Vector Machines for Classification Problems

In the multi-layer SVM classifier, the architecture contains multiple support vector classifiers in the output layer. To deal with multiple classes, we use a binary one vs. all classifier M_c for each class c. We do this even with 2 classes for convenience. We use a classification dataset for each classifier M_c: $\{(\mathbf{x}_1, y_1^c), \ldots, (\mathbf{x}_\ell, y_\ell^c)\}$, where $\mathbf{x}_i$ are input vectors and $y_i^c \in \{-1, 1\}$ are the target outputs that denote if the example $\mathbf{x}_i$ belongs to class c or not. All classifiers M_c share the same hidden-layer of regression SVMs. M_c determines its output on an example $\mathbf{x}$ as follows:

$$g_c(\mathbf{f}(\mathbf{x}|\theta)) = \sum_{i=1}^{\ell} y_i^c \alpha_i^c K_2(\mathbf{f}(\mathbf{x}_i|\theta), \mathbf{f}(\mathbf{x}|\theta)) + b_c. \qquad (20.11)$$

Here $\mathbf{f}(\mathbf{x}_i|\theta)$ is computed with the hidden-layer SVMs as before. The values α_i^c are the support vector coefficients for classifier M_c. The value b_c is its bias. After computing all output values for all classifiers, the class with the highest output is assumed to be the correct class label (with ties being broken randomly). The primal objective for a linear support vector classifier M_c can be written as:

$$\min_{\mathbf{w}^c, \boldsymbol{\xi}, b, \theta} J_c(\mathbf{w}^c, \boldsymbol{\xi}, b, \theta) = \frac{1}{2}||\mathbf{w}^c||^2 + C \sum_{i=1}^{\ell} \xi_i \qquad (20.12)$$

subject to: $y_i^c(\mathbf{w}^c \cdot \mathbf{f}(\mathbf{x}_i|\theta) + b_c) \geq 1 - \xi_i$, and $\xi_i \geq 0$. Here C is a metaparameter and ξ_i are slack variables that tolerate errors, but which should be minimized. The dual-objective function for the classification problem for classifier M_c is:

$$\min_{\theta} \max_{\boldsymbol{\alpha}^c} J_c(\theta, \boldsymbol{\alpha}^c) = \sum_{i=1}^{\ell} \alpha_i^c - \frac{1}{2} \sum_{i,j=1}^{\ell} \alpha_i^c \alpha_j^c y_i^c y_j^c K_2(\mathbf{f}(\mathbf{x}_i|\theta), \mathbf{f}(\mathbf{x}_j|\theta)) \quad (20.13)$$

subject to: $0 \leq \alpha_i^c \leq C$, and $\sum_{i=1}^{\ell} \alpha_i^c y_i^c = 0$. Whenever the ML-SVM is presented a training pattern $\mathbf{x}_i$, each classifier in the multi-layer SVM uses gradient ascent to adapt its α_i^c values towards a local maximum of $J_c(\cdot)$ by:

$$\alpha_i^c \leftarrow \alpha_i^c + \lambda(1 - \sum_{j=1}^{\ell} \alpha_j^c y_j^c y_i^c K_2(\mathbf{f}(\mathbf{x}_i|\theta), \mathbf{f}(\mathbf{x}_j|\theta))) \qquad (20.14)$$

where λ is a metaparameter controlling the learning rate of the values α_i^c. As before, the support vector coefficients are kept between 0 and C. Because we use a gradient ascent update rule, we use an additional penalty term $c_1(\sum_{j=1}^{\ell} \alpha_j^c y_j^c)^2$ with metaparameter c_1 so that the bias constraint is respected.

As in the regression ML-SVM, the classification ML-SVM constructs a new dataset for each hidden-layer SVM S_a with a backpropagation-like technique for making examples. However, in this case the aim of the hidden-layer SVMs is to minimize the sum of objectives $\sum_c J_c(\cdot)$. Therefore, the algorithm constructs a new dataset using: $(\mathbf{x}_i, \mathbf{f}(\mathbf{x}_i|\theta)_a - \mu \sum_c \partial J_c(\cdot)/\partial \mathbf{f}(\mathbf{x}_i|\theta)_a)$, where μ is some metaparameter, and $\partial J_c(\cdot)/\partial \mathbf{f}(\mathbf{x}_i|\theta)_a$ for the RBF kernel is:

$$\frac{\partial J_c(\cdot)}{\partial \mathbf{f}(\mathbf{x}_i|\theta)_a} = \alpha_i^c y_i^c \sum_{j=1}^{\ell} \alpha_j^c y_j^c \frac{\mathbf{f}(\mathbf{x}_i|\theta)_a - \mathbf{f}(\mathbf{x}_j|\theta)_a}{\sigma_2} \cdot K_2(\mathbf{f}(\mathbf{x}_i|\theta), \mathbf{f}(\mathbf{x}_j|\theta)) \quad (20.15)$$

The target outputs for hidden-layer features are again kept between -1 and 1. The datasets for hidden-layer SVMs are made so that the sum of the dual-objective functions of the output SVMs is minimized. All SVMs are trained with the gradient ascent algorithm on their constructed datasets. Note that the hidden-layer SVMs are still regression SVMs, since they need to output continuous values. For the ML-SVM classifier, we use a different initialization procedure for the hidden-layer SVMs. Suppose there are d hidden-layer SVMs and a total of c_{tot} classes. The first hidden-layer SVM is first pre-trained on inputs and perturbed target outputs for class 0, the second on the perturbed target outputs for class 1, and the kth hidden-layer SVM is pre-trained on the perturbed target outputs for class k modulo c_{tot}. The bias values are computed in a similar way as in the regression ML-SVM, but for the output SVMs only examples with non-bound support vector coefficients (which are not 0 or C) are used.

20.4   Multi-Layer Support Vector Machines for Dimensionality Reduction

The architecture of the ML-SVM autoencoder differs from the single-output regression ML-SVM in two respects: (1) The output layer consists of D nodes, the same number of nodes the input layer has. (2) It utilizes a total of D support vector regression machines M_c, which each take the entire hidden-layer output as input and determine the value of one of the outputs.

The forward propagation of a pattern $\mathbf{x}$ of dimension D determines the representation in the hidden layer. The hidden layer is then used as input for each support vector machine M_c that determines its output with:

$$g_c(\mathbf{f}(\mathbf{x}|\theta)) = \sum_{i=1}^{\ell} (\alpha_i^{c*} - \alpha_i^c) K_2(\mathbf{f}(\mathbf{x}_i|\theta), \mathbf{f}(\mathbf{x}|\theta)) + b_c. \quad (20.16)$$

Again we make use of RBF kernels in both layers. The aim of the ML-SVM autoencoder is to reconstruct the inputs in the output layer using a bottleneck of hidden-layer SVMs, where the number of hidden-layer SVMs is in general much smaller than the number of inputs. The ML-SVM autoencoder tries to find the SVM coefficients θ such that the hidden-layer representation $\mathbf{f}(\cdot)$ is most useful for accurately reconstructing the inputs, and thereby codes the features most relevant to the input distribution. This is similar to neural network autoencoders [23, 12]. Currently popular deep architectures [14, 4, 25] stack these autoencoders one by one, which is also possible for the ML-SVM autoencoder.

The dual objective of each support vector machine M_c is:

$$\min_{\theta} \max_{\alpha^{c*}, \alpha^c} J_c(\theta, \alpha_i^{c(*)}) = -\varepsilon \sum_{i=1}^{\ell} (\alpha_i^{c*} + \alpha_i^c) + \sum_{i=1}^{\ell} (\alpha_i^{c*} - \alpha_i^c) y_i^c$$

$$-\frac{1}{2} \sum_{i,j=1}^{\ell} (\alpha_i^{c*} - \alpha_i^c)(\alpha_j^{c*} - \alpha_j^c) K_2(\mathbf{f}(\mathbf{x}_i|\theta), \mathbf{f}(\mathbf{x}|\theta)) \qquad (20.17)$$

subject to: $0 \leq \alpha_i^c, \alpha_i^{c*} \leq C$, and $\sum_{i=1}^{\ell}(\alpha_i^{c*} - \alpha_i^c) = 0$. The minimization of this equation with respect to θ is a bit different from the single-node ML-SVM. Since all SVMs share the same hidden layer, we cannot just minimize $J(\cdot)$ for every SVM separately. It is actually this shared nature of the hidden layer that enables the ML-SVM to perform autoencoding. Therefore the algorithm creates new datasets for the hidden-layer SVMs by backpropagating the sum of the derivatives of all dual objectives $J_c(\cdot)$. Thus, the ML-SVM autoencoder uses: $(\mathbf{x}, \mathbf{f}(\mathbf{x}|\theta)_a - \mu \sum_{c=1}^{D} \frac{\partial J_c(\cdot)}{\partial \mathbf{f}(\mathbf{x}|\theta)_a})$ to create new datasets for the hidden-layer SVMs.

20.5 Experiments and Results

We first performed experiments on regression and classification problems to compare the multi-layer SVM (we used two layers) to the standard SVM and also to a multi-layer perceptron. Furthermore, we performed experiments with an image dataset where it was the goal to obtain the smallest reconstruction error with a limited number of hidden components.

20.5.1 Experiments on Regression Problems

We experimented with 10 regression datasets to compare the multi-layer SVM to an SVM, both using RBF kernels. We note that both methods are trained with the simple gradient ascent learning rule, adapted to also consider the penalty for obeying the bias constraint, although standard algorithms for the SVM could also be used. The first 8 datasets are described in [11] and the other 2 datasets are taken from the UCI repository [1]. The number of examples per dataset ranges from 43 to 1049, and the number of input features is between 2 and 13. The datasets are split into 90% training data and 10% test data. For optimizing the metaparameters we have used particle swarm optimization (PSO) [15]. There are in total around 15 metaparameters for the ML-SVM such as the learning rates for the two layers, the values for the error tolerance ϵ, the values for C, the number of gradient ascent iterations in the gradient ascent algorithm, the values for respecting the bias constraint c_1, the RBF kernel widths σ_1 and σ_2, the number of hidden-layer SVMs, the value for the perturbation value γ used for pre-training the hidden-layer SVMs, and the maximal number of epochs. PSO saved us from laborious manual tuning of these metaparameters. We made an effective implementation of PSO that also makes use of the UCB bandit algorithm [2] to eliminate unpromising sets of metaparameters. We always performed 100,000 single training-runs to obtain the best metaparameters that took at most 2 days on a 32-CPU machine on the largest dataset. For the gradient ascent SVM algorithm we also used 100,000 evaluations with PSO to find the best metaparameters, although our implementation of the gradient ascent SVM has 7 metaparameters, which makes it easier to find the best ones. Finally, we used 1000 or 4000 new cross validation runs with the best found metaparameters to compute the mean squared error and its standard error of the different methods for each dataset.

TABLE 20.1: The mean squared errors and standard errors of the gradient ascent SVM, the two-layer SVM, and results published in [11] for an MLP on 10 regression datasets. N/A means not available.

Dataset	Gradient ascent SVM	ML-SVM	MLP
Baseball	0.02413 ± 0.00011	$\mathbf{0.02294 \pm 0.00010}$	0.02825
Boston Housing	0.006838 ± 0.000095	$\mathbf{0.006381 \pm 0.000091}$	0.007809
Concrete Strength	0.00706 ± 0.00007	$\mathbf{0.00621 \pm 0.00005}$	0.00837
Diabetes	0.02719 ± 0.00026	$\mathbf{0.02327 \pm 0.00022}$	0.04008
Electrical Length	0.006382 ± 0.000066	0.006411 ± 0.000070	0.006417
Machine-CPU	0.00805 ± 0.00018	$\mathbf{0.00638 \pm 0.00012}$	0.00800
Mortgage	0.000080 ± 0.000001	0.000080 ± 0.000001	0.000144
Stock	0.000862 ± 0.000006	$\mathbf{0.000757 \pm 0.000005}$	0.002406
Auto-MPG	6.852 ± 0.091	6.715 ± 0.092	N/A
Housing	$\mathbf{8.71 \pm 0.14}$	9.30 ± 0.15	N/A

In Table 20.1 we show the results of the standard SVM trained with gradient ascent and the results of the two-layer SVM. The table also shows the results for a multi-layer perceptron (MLP) reported in [11] on the first 8 datasets. The MLP used sigmoidal hidden units and was trained with backpropagation. We note that Graczyk et al. [11] only performed 10-fold cross validation and did not report any standard errors.

The results show that the two-layer SVM significantly outperforms the other methods on 6 datasets ($p < 0.001$) and only performs worse than the standard SVM on the Housing dataset from the UCI repository. The average gain over all datasets is 6.5% error reduction. The standard errors are very small because we performed 1000 or 4000 times cross validation. We did this because we observed that with less cross validation runs the results were less trustworthy due to their stochastic nature caused by the randomized splits into different test sets. We also note that the results of the gradient ascent SVM are a bit better than the results obtained with an SVM in [11]. We think that the PSO method is more capable in optimizing the metaparameters than the grid search employed in [11]. Finally, we want to remark that the results of the MLP are worse than those of the two other approaches.

20.5.2   Experiments on Classification Problems

We compare the multi-layer classification SVM to the standard SVM and a multi-layer perceptron trained with backpropagation with one hidden layer with sigmoid activation functions. Early stopping was implemented in the MLP by optimizing the number of training epochs. For the comparison we use 8 datasets from the UCI repository. In these experiments we have used SVMLight as standard SVM and optimized the metaparameters (σ and C) with grid search (also with around 100,000 evaluations). We also optimized the metaparameters (number of hidden units, learning rate, number of epochs) for the multi-layer perceptron. The metaparameters for the multi-layer SVM are again optimized with PSO.

TABLE 20.2: The accuracies and standard errors on the 8 UCI classification datasets. Shown are the results of an MLP, a support vector machine (SVM), and the two-layer SVM.

Dataset	MLP	SVM	ML-SVM
Hepatitis	84.3 ± 0.3	81.9 ± 0.3	$\mathbf{85.1} \pm 0.1$
Breast Cancer W.	97.0 ± 0.1	96.9 ± 0.1	97.0 ± 0.1
Ionosphere	91.1 ± 0.1	94.0 ± 0.1	$\mathbf{95.5} \pm 0.1$
Ecoli	87.6 ± 0.2	87.0 ± 0.2	87.3 ± 0.2
Glass	64.5 ± 0.4	70.1 ± 0.3	$\mathbf{74.0} \pm 0.3$
Pima Indians	77.4 ± 0.1	77.1 ± 0.1	77.2 ± 0.2
Votes	96.6 ± 0.1	96.5 ± 0.1	96.8 ± 0.1
Iris	97.8 ± 0.1	96.5 ± 0.2	$\mathbf{98.4} \pm 0.1$
Average	87.0	87.5	88.9

We report the results on the 8 datasets with average accuracies and standard errors. We use 90% of the data for training data and 10% for test data.

We have performed 1000 new random cross validation experiments per method with the best found metaparameters (and 4000 times for Iris and Hepatitis, since these are smaller datasets). The results are shown in Table 20.2. The multi-layer SVM significantly ($p < 0.05$) outperforms the other methods on 4 out of 8 classification datasets. On the other problems the multi-layer SVM performs equally well as the other methods. We also performed experiments with the gradient ascent SVM on these datasets, but its results are very similar to those obtained with SVMLight, so we do not show them here. On some datasets such as Breast Cancer Wisconsin and Votes, all methods perform equally well. On some other datasets, the multi-layer SVM reduces the error of the SVM a lot. For example, the error on Iris is 1.6% for the multi-layer SVM compared to 3.5% for the standard SVM. The MLP obtained 2.2% error on this dataset. Finally, we also optimized and tested a stacking ensemble SVM method, which uses an SVM to directly map the outputs of the pretrained hidden-layer SVMs to the desired output without further min-max optimization. This approach obtained 2.3% error on Iris and is therefore significantly outperformed by the multi-layer SVM.

20.5.3   Experiments on Dimensionality Reduction Problems

The used dataset in the dimensionality reduction experiment contains a total of 1300 instances of gray-scaled images of the left eyes manually cropped from pictures in the "labeled faces in the wild" dataset. The images, shown in Figure 20.2, are normalized and have a resolution of 20 by 20 pixels, and thus have 400 values per image. The aim of this experiment is to see how well the autoencoder ML-SVM performs compared to some state-of-the-art methods. The goal of the used dimensionality reduction algorithms is to accurately encode the input data using fewer dimensions than the number of inputs. A well known, but suboptimal technique for doing this is the use of principal component analysis.

FIGURE 20.2: Examples of some of the cropped gray-scaled images of left eyes that are used in the dimensionality reduction experiment.

We compared the ML-SVM to principal component analysis (PCA) and a neural network autoencoding method. We used a state-of-the-art neural network autoencoding method, named a denoising autoencoder [25], for which we optimized the metaparameters. The autoencoders were trained using stochastic gradient descent with a decreasing learning rate. In each epoch, all samples in the training set were presented to the network in a random order. To improve generalization performance of the standard neural network autoencoder [23], in the denoising autoencoder each input sample is augmented with Gaussian noise, while the target stayed unaltered. We also added l_1 regularization on the hidden layer of the network to increase sparsity. These additions improved the performance of this non-linear autoencoder.

We also compared the ML-SVM to principal component analysis using a multi-variate partial-least squares (PLS) regression model with standardized inputs and outputs [29]. It can easily be shown that the standard PLS algorithm in autoencoder mode is actually equivalent to a principal component projection (with symmetric weights in the layer from the latent variable bottleneck layer to the output layer). The attractiveness of applying the PLS autoencoder in this case is the elegant and efficient implementation of the standard PLS algorithm to compute the principal components.

For these experiments, random cross validation is used to divide the data in a training set containing two thirds (867 examples) of the dataset, and a test set containing one third. The methods are compared by measuring the reconstruction error for different numbers of (non-linear) principal components: we used 10, 20, and 50 dimensions to encode the eye images. The root mean square error of 10 runs and standard errors are computed for the comparison.

TABLE 20.3: The RMSE and standard errors for different numbers of principal components for principal component analysis, a denoising autoencoder (DAE), and a multi-layer support vector machine (ML-SVM)

#dim	PCA	DAE	ML-SVM
10	0.1242 ± 0.0004	0.1211 ± 0.0002	$\mathbf{0.1202 \pm 0.0003}$
20	0.0903 ± 0.0003	0.0890 ± 0.0002	$\mathbf{0.0875 \pm 0.0003}$
50	0.0519 ± 0.0002	0.0537 ± 0.0001	$\mathbf{0.0513 \pm 0.0002}$

The results of these experiments can be found in Table 20.3. These results show a significantly better ($p<0.05$) performance for autoencoding with the use of a multi-layer support vector machine compared to the denoising autoencoder and PCA. As known from the literature, the difference to PCA decreases when more principal components are used.

20.5.4   Experimental Analysis of the Multi-Layer SVM

We also studied why the multi-layer SVM outperforms the SVM in many cases. For this we will examine the Iris dataset again, but in more detail. For this dataset the multi-layer SVM and the MLP perform much better than the standard SVM with an RBF kernel (see Table 20.2). We performed the experiments again with the previous best found metaparameters, but set the C-values to 3.0 for all methods so that the dual-objectives of different methods can be easily compared. This did not significantly change the performances. Furthermore, we set the number of epochs to 14.

Figure 20.3 shows the evolution of the training and test errors for three methods. The reported errors are averaged over 1000 simulations. We compare the standard SVM trained with gradient ascent, the multi-layer SVM, and a multi-layer SVM in which the hidden-layer SVMs were not pre-trained on perturbed class labels, but completely randomly initialized. In the case of the standard SVM, the epochs refer to the number of repetitions of the gradient ascent algorithm. For the multi-layer SVMs, the epoch counter is increased after only training the output SVMs or after only training the hidden-layer SVMs. The training times on a single training set are less than one second for the Iris dataset for all methods. In epoch 0, the output layer SVMs were initialized with constant positive support vector coefficients and by PSO optimized kernel widths. Therefore, they immediately work quite well since the SVM and the ML-SVM behave like a k-nearest neighbor or locally weighted learning method in this case.

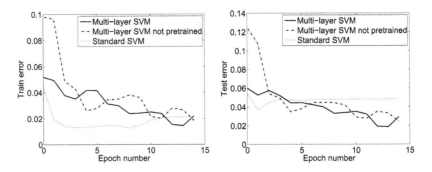

FIGURE 20.3: (A) Training error results on the Iris dataset. (B) Test error results on the Iris dataset.

The results show that the standard SVM obtains a low training error quicker than the other methods, but that its test error is higher (its best test error is 3.7%). The best test error is obtained by the (pre-trained) multi-layer SVM after 13 epochs, when it obtains a test error of 1.9%. The error of the multi-layer SVM that is not pre-trained starts much higher than with the other methods, but this method is still able to obtain a test error of 2.8% and significantly outperforms the standard SVM. The standard SVM obtains its best performance after a single training epoch with the gradient ascent algorithm during which 10 training iterations of the support vector coefficients were executed. Figures 20.3(A) and 20.3(B) show that for the multi-layer SVMs the test errors are very close to the training errors, except for the beginning. This behavior is due to the strong regularization power of the output-layer SVMs. Even with many hidden-layer SVMs, generalization performance can be made excellent by setting the regularization parameter C to a small value.

We also plotted the evolution of the average values of the dual-objective function that correspond to the evolution of the training and test errors shown before. Again this plot shows averages of 1000 simulations. Figure 20.4 shows that the gradient ascent SVM monotonically increases the dual-objective function (between epochs 1 and 14, the dual-objective value increases from 33 to 70). As can be seen in Figure 20.3(B), this does not lead to always improving test errors. This may have to do with not exactly fulfilling the bias constraint. However, when PSO is used it optimizes the number of epochs to overcome this problem (it found the best value of 1 for the number of epochs). The multi-layer SVMs alternate between minimizing and maximizing the dual-objective function. The min-max optimization process is quite complex, because multiple metaparameters influence the learning updates. Therefore, the dual objective does not just increase, then decrease in the next epoch, etc. Instead, the dual-objective function increases for some epochs, then decreases, etc., without any signs of convergence. The three figures show that the dual-objective should be minimized to obtain the lowest test errors. However, standard SVMs can only maximize the dual-objective function. Therefore, the flexibility of the hidden layer in the ML-SVM is especially fruitful to minimize the dual-objective function and thereby obtain lower test errors.

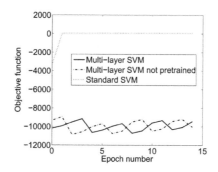

FIGURE 20.4: The evolution of the dual-objective value on the Iris dataset.

20.6 Discussion and Future Work

The multi-layer SVM consists of a hidden layer of SVMs and an output layer of SVMs that learn to approximate the target function using the outputs of the hidden-layer SVMs. The results show that the ML-SVM can outperform other state-of-the-art machine learning algorithms. By going from a single SVM to the multi-layer SVM, we have made the SVM a deeper architecture. Compared to other deep neural network architectures, the ML-SVM has the advantage that due to the strong regularization power of the output-layer SVMs, the system does not easily overfit the data. Therefore, the ML-SVM could potentially perform very well with very large input vectors and few training examples. On the other hand, training an SVM with many examples is more computationally demanding than training a deep neural network architecture.

There are several advantages of the ML-SVM algorithm. First, the method is very flexible in adapting the kernel functions compared to other multiple-kernel learning algorithms. Second, the algorithm is straightforward to implement by using the gradient ascent algorithm and the backpropagation technique. Finally, the training method uses a min-max optimization process, which is interesting and not (yet) applicable to neural network training.

There remains future work to be done in order to increase the power of the ML-SVM. First of all, the current implementation uses many metaparameters. Instead of using PSO to optimize the metaparameters, many different real-coded optimization algorithms can be employed. Second, the ML-SVM becomes very large for large datasets and then needs a lot of training time. To deal with large datasets, we want to explore stochastic gradient ascent techniques instead of the batch gradient ascent method we used in this chapter. We can also include more diversity in the hidden-layer SVMs, for example by letting them use different subsets of inputs, different examples, or different kernels. Finally, we want to develop a more rigorous theory to explain why the ML-SVM performs so well and test the ML-SVM on challenging handwriting and image recognition datasets.

Acknowledgments

Thanks to Mark Embrechts, Adrian Millea, Arnold Meijster, Aleke Nolte, Egbert van der Wal, Marten Schutten, Rick van der Mark, Michiel van der Ree, and Marijn Stollenga for helping with the experiments.

Bibliography

[1] A. Asuncion and D.J. Newman. UCI machine learning repository, 2007.

[2] P. Auer, N. Cesa-Bianchi, and P. Fischer. Finite-time analysis of the

multiarmed bandit problem. *Machine Learning*, 47(2-3):235–256, May 2002.

[3] F.R. Bach, G.R.G. Lanckriet, and M.I. Jordan. Multiple kernel learning, conic duality, and the SMO algorithm. In *Proceedings of the Twenty-First International Conference on Machine Learning*, ICML '04, pages 6–15, 2004.

[4] Y. Bengio, P. Lamblin, D. Popovici, and H. Larochelle. Greedy layer-wise training of deep networks. In *Advances in Neural Information Processing Systems*, volume 13, pages 153–160, 2007.

[5] Y. Cho and L. K. Saul. Kernel methods for deep learning. *Advances in Neural Information Processing Systems*, 22:342–350, 2009.

[6] D.C. Ciresan, U. Meier, L.M. Gambardella, and J. Schmidhuber. Deep big simple neural nets excel on handwritten digit recognition. *Neural Computation*, 22(12):3207–3220, 2010.

[7] C. Cortes, M. Mohri, and A. Rostamizadeh. Learning non-linear combinations of kernels. *Advances in Neural Information Processing Systems*, 22:396–404, 2009.

[8] N. Cristianini and J. Shawe-Taylor. *Support Vector Machines and Other Kernel-Based Learning Methods*. Cambridge University Press, 2000.

[9] F. Dinuzzo. Kernel machines with two layers and multiple kernel learning. *CoRR*, 2010.

[10] M. Gönen and E. Alpaydin. Multiple kernel learning algorithms. *Journal of Machine Learning Research*, pages 2211–2268, July 2011.

[11] M. Graczyk, T. Lasota, Z. Telec, and B. Trawinski. Nonparametric statistical analysis of machine learning algorithms for regression problems. In *Knowledge-Based and Intelligent Information and Engineering Systems*, pages 111–120. 2010.

[12] G.E. Hinton. Training product of experts by minimizing constrastive divergence. *Neural Computation*, 14:1771–1800, 2002.

[13] G.E. Hinton, S. Osindero, and Y.W. Teh. A fast learning algorithm for deep belief nets. *Neural Computation*, pages 1527–1554, 2006.

[14] G.E. Hinton and R.R. Salakhutdinov. Reducing the dimensionality of data with neural networks. *Science*, 313:504–507, 2006.

[15] J. Kennedy and R. Eberhart. Particle swarm optimization. In *Proceedings of the IEEE International Conference on Neural Networks*, volume 4, pages 1942–1948, 1995.

[16] G.R.G. Lanckriet, N. Cristianini, P. Bartlett, L. El Ghaoui, and M.I. Jordan. Learning the kernel matrix with semidefinite programming. *Journal of Machine Learnine Research*, 5:27–72, 2004.

[17] Y. LeCun, B. Boser, J. S. Denker, D. Henderson, R. E. Howard, W. Hubbard, and L. D. Jackel. Back-propagation applied to handwritten zip code recognition. *Neural Computation*, 1(4):541–551, 1989.

[18] A. Rakotomamonjy, F. Bach, S. Canu, and Y. Grandvalet. More efficiency in multiple kernel learning. In *Proceedings of the 24th international conference on machine learning*, pages 775–782, 2007.

[19] D. E. Rumelhart, G. E. Hinton, and R. J. Williams. Learning internal representations by error propagation. In *Parallel Distributed Processing*, volume 1, pages 318–362. MIT Press, 1986.

[20] B. Schölkopf and A. Smola. *Learning with Kernels: Support Vector Machines, Regularization, Optimization, and Beyond*. MIT Press, 2002.

[21] S. Sonnenburg, G. Rätsch, and C. Schäfer. A general and efficient multiple kernel learning algorithm. In *Advances in Neural Information Processing Systems 18*, pages 1273–1280, Cambridge, MA, 2006.

[22] J.A.K. Suykens, T. Van Gestel, J. De Brabanter, B. De Moor, and J. Vandewalle. *Least Squares Support Vector Machines*. World Scientific Pub, 2002.

[23] M. Turk and A. Pentland. Eigenfaces for recognition. *Cognitive Neuroscience*, 3(1):71–86, 1991.

[24] V. Vapnik. *The Nature of Statistical Learning Theory*. Springer-Verlag, 1995.

[25] P. Vincent, H. Larochelle, Y. Bengio, and P-A. Manzagol. Extracting and composing robust features with denoising autoencoders. In *Proceedings of the 25th International Conference on Machine Learning*, pages 1096–1103, 2008.

[26] P. J. Werbos. Advanced forecasting methods for global crisis warning and models of intelligence. In *General Systems*, volume XXII, pages 25–38, 1977.

[27] M.A. Wiering, M. Schutten, A. Millea, A. Meijster, and L.R.B. Schomaker. Deep support vector machines for regression problems. In *Proceedings of the International Workshop on Advances in Regularization, Optimization, Kernel Methods, and Support Vector Machines: Theory and Applications*, 2013.

[28] M.A. Wiering, M.H. van der Ree, M.J. Embrechts, M.F. Stollenga, A. Meijster, A. Nolte, and L.R.B. Schomaker. The neural support vector machine. In *Proceedings of the 25th Benelux Artificial Intelligence Conference (BNAIC)*, 2013.

[29] S. Wold, M. Sjöström, and L. Eriksson. PLS-regression: a basic tool of chemometrics. *Chemometrics and Intelligent Laboratory Systems*, 58:109–130, 2001.

[30] D.H. Wolpert. Stacked generalization. *Neural Networks*, 5:241–259, 1992.

[31] Z. Xu, R. Jin, I. King, and M.R. Lyu. An extended level method for efficient multiple kernel learning. In *Advances in Neural Information Processing Systems 20*, pages 1825–1832. Curran Associates, Inc., 2008.

[32] J. Zhuang, I.W. Tsang, and S.C.H. Hoi. Two-layer multiple kernel learning. In *AISTATS*, pages 909–917, 2011.

Chapter 21

Online Regression with Kernels

Steven Van Vaerenbergh

Department of Communications Engineering, University of Cantabria

Ignacio Santamaría

Department of Communications Engineering, University of Cantabria

Online machine learning algorithms are designed to learn from one data instance at a time. They are typically used in real-time scenarios, such as prediction or tracking problems, where data arrive sequentially and instant decisions must be made. The real-time nature of these settings implies that shortly after the decision is made, the true label will be made available, which allows the learning algorithm to adjust its solution before a new datum is received.

Online kernel methods extend the nonlinear learning capabilities of standard batch kernel methods to online environments. Especially important for these techniques is that they maintain their computational load moderate during each iteration, in order to perform fast updates in real time. Ideally, they should not only be able to learn in a stationary environment but also in nonstationary settings, where they must forget outdated information and adapt their solution to respond to changes in time. Online kernel methods also find use in batch scenarios where the amount of data is too high to fit in the machine's memory, and one or several passes over the data are to be performed.

In this chapter we focus on the problem of online regression. We will give an overview of the most important kernel-based methods for this problem, which have been developed over the span of the last decade. We start by formulating the online solution to the kernel ridge regression problem, and we point out different strategies to overcome the bottlenecks associated with using kernels in online methods. The discussed techniques are often referred to as *kernel adaptive filtering* algorithms, due to their close relationship with classical adaptive filters from the signal processing literature. After reviewing the most relevant algorithms in this area, we introduce an evaluation framework that allows us to compare their performance. We finish the discussion with a brief overview of the recent and future research directions.

21.1   Basic Principles of Kernel Machines

Kernel-based methods have had considerable success in a wide range of areas over the past decade, since they allow us to reformulate many nonlinear problems as convex optimization problems. In this section, we briefly outline the basic principles behind kernel methods, and we introduce the main ideas behind constructing online kernel-based algorithms.

21.1.1   Kernel Methods

Kernel methods rely on a nonlinear transformation of the input data $\mathbf{x}$ into a high-dimensional reproducing kernel Hilbert space (RKHS), in which it is more likely that the transformed data $\phi(\mathbf{x})$ are linearly separable. In this space, denoted the *feature space*, inner products can be calculated by using a positive definite kernel function $\kappa(\cdot, \cdot)$ satisfying Mercer's condition:

$$\kappa(\mathbf{x}, \mathbf{x}') = \langle \phi(\mathbf{x}), \phi(\mathbf{x}') \rangle, \tag{21.1}$$

where $\mathbf{x}$ and $\mathbf{x}'$ represent two different data points. This duality between positive definite kernels and feature spaces allows us to transform any inner-product based linear algorithm to an alternative, nonlinear algorithm by replacing the inner products with kernels [1, 15]. The solution, obtained as a linear functional in the feature space, then corresponds to the solution of a nonlinear problem in the input space.

Thanks to the Representer Theorem [14], a large class of optimization problems in RKHS have solutions that can be expressed as kernel expansions in terms of the training data only. Specifically, kernel-based learning aims at finding a nonlinear relationship $f : \mathcal{X} \to \mathbb{R}$ that can be expressed as the kernel expansion

$$f(\mathbf{x}) = \sum_{n=1}^{N} \alpha(n) \kappa(\mathbf{x}_n, \mathbf{x}). \tag{21.2}$$

In this relationship, N is the number of training data, and $\alpha(n) \in \mathbb{R}$ are denoted the *expansion coefficients*. The training data $\mathbf{x}_n$ used in this expansion are sometimes referred to as *bases*.

Kernel methods are non-parametric techniques, since they do not assume any specific model. Indeed, their solution is expressed as the functional representation (21.2) that relies explicitly on the training data.

As an introductory example algorithm, we will now describe the batch approach to the standard kernel regression problem, known as kernel ridge regression. Its formulation will lie at the core of the online algorithms we will review later in this chapter.

21.1.2 Kernel Ridge Regression

Assume we are given N input data points $\mathbf{x}_n$, $n = 1, \ldots, N$, in a D-dimensional space, and the corresponding outputs y_n. In order to adopt a matrix-based formulation, we stack the input data into an $N \times D$ matrix $\mathbf{X}$ and the output data into the vector $\mathbf{y}$. In linear regression, the regularized least-squares problem consists of seeking a vector $\mathbf{w} \in \mathbb{R}^{D \times 1}$ that solves

$$\min_{\mathbf{w}} \|\mathbf{y} - \mathbf{X}\mathbf{w}\|^2 + c\|\mathbf{w}\|^2, \tag{21.3}$$

where c is a positive Tikhonov regularization constant. The solution to this problem is given by

$$\hat{\mathbf{w}} = (\mathbf{X}^\top \mathbf{X} + c\mathbf{I})^{-1} \mathbf{X}^\top \mathbf{y}. \tag{21.4}$$

In the absence of regularization, i.e., $c = 0$, the solution is only well defined when $\mathbf{X}$ has full column rank, and thus $(\mathbf{X}^\top \mathbf{X})^{-1}$ exists.

In order to obtain the kernel-based version of Equation (21.3), we first transform the data and the solution into the feature space,

$$\min_{\phi(\mathbf{w})} \|\mathbf{y} - \phi(\mathbf{X})\phi(\mathbf{w})\|^2 + c\|\phi(\mathbf{w})\|^2. \tag{21.5}$$

Here, the shorthand notation $\phi(\mathbf{X})$ refers to the data matrix that contains the transformed data, stacked as rows, $\phi(\mathbf{X}) = [\phi(\mathbf{x}_1), \ldots, \phi(\mathbf{x}_N)]^\top$. The Representer Theorem [14] states that the solution $\phi(\mathbf{w})$ of this problem can be expressed as a linear combination of the training data, in particular

$$\phi(\mathbf{w}) = \sum_{n=1}^{N} \alpha(n)\phi(\mathbf{x}_n) = \phi(\mathbf{X})^{\top}\alpha, \qquad (21.6)$$

where $\alpha = [\alpha(1), \ldots, \alpha(N)]^{\top}$. After substituting Equation (21.6) into Equation (21.5), and defining the matrix $\mathbf{K} = \phi(\mathbf{X})\phi(\mathbf{X})^{\top}$, we obtain

$$\min_{\alpha} \|\mathbf{y} - \mathbf{K}\alpha\|^2 + c\alpha^{\top}\mathbf{K}\alpha. \qquad (21.7)$$

The matrix $\mathbf{K}$ is denoted as the *kernel matrix*, and its elements represent the inner products in the feature space, calculated as kernels $k_{ij} = \kappa(\mathbf{x}_i, \mathbf{x}_j)$.

Equation (21.7) represents the kernel ridge regression problem [12], and its solution is given by

$$\hat{\alpha} = (\mathbf{K} + c\mathbf{I})^{-1}\mathbf{y}. \qquad (21.8)$$

21.2   Framework for Online Kernel Methods

Online learning methods update their solution iteratively. In the standard online learning framework, each iteration consists of several steps, as outlined in Algorithm 19. During the n-th iteration, the algorithm first receives an input datum, $\mathbf{x}_n$. Then, it calculates the estimated output $\hat{y}_n$ corresponding to this datum. The learning setup is typically supervised, in that the true outcome y_n is made available next, which enables the algorithm to calculate the loss $L(\cdot)$ incurred on the data pair $(\mathbf{x}_n, y_n)$, and, finally, to update its solution. Initialization is typically performed by setting the involved weights to zero.

Algorithm 19 Protocol for online, supervised learning.

    Initialize variables as empty or zero.
    for $n = 1, 2, \ldots$ **do**
        Observe input $\mathbf{x}_n$.
        Predict output $\hat{y}_n$.
        Observe true output y_n.
        Update solution based on $L(e_n)$, with $e_n = y_n - \hat{y}_n$.
    end for

A typical setup for online system identification with a kernel-based method is depicted in Figure 21.1. It represents an unknown nonlinear system, whose input data $\mathbf{x}_n$ and response y_n (including additive noise r_n) can be measured at different time steps, and an adaptive kernel-based algorithm, which is used to identify the system's response.

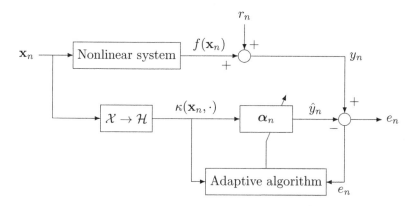

FIGURE 21.1: Kernel-based adaptive system identification. (Figure adapted from [8].)

Online algorithms should be capable of operating during extended periods of time, processing large amounts of data. Kernel methods rely on the functional representation (21.2), which grows as the amount of observations increases. A naive implementation of an online kernel method will therefore require growing computational resources during operation, leading to performance issues once the available memory is insufficient to store the training data or once the computations for one update take more time than the interval between incoming data [6].

The standard approach to overcoming this issue is to limit the number of terms in Equation (21.2) to a representative subset of observations, called a *dictionary*. In the sequel, we will review several dictionary-learning methods that can be carried out online. They will serve as building blocks for online kernel methods.

21.2.1 Online Dictionary Learning

The dictionary learning process aims to identify the bases in the expansion (21.2) that can be discarded without incurring a significant performance loss. After discarding these bases, a new, approximate expansion is obtained as

$$\hat{f}(\mathbf{x}) = \sum_{i=1}^{m} \alpha(i)\kappa(\mathbf{u}_i, \mathbf{x}), \tag{21.9}$$

where $m < N$. The dictionary $\mathcal{D} = \{\mathbf{u}_1, \ldots, \mathbf{u}_m\}$ consists of several data vectors $\mathbf{u}_i$ which are selected from the input data $\mathbf{x}_n$. They should be chosen carefully so that they represent the entire input data set sufficiently well.

In the online setting, the classical way of constructing a dictionary is by *growing*, i.e., by starting from an empty dictionary and gradually adding those bases that fulfill a certain criterion. If the dictionary is not allowed to grow beyond a specified maximum size, it may be necessary to discard bases at some point. This process is referred to as *pruning*. In Table 21.1 we have listed several criteria to grow dictionaries, in the top half, and several criteria to prune them, in the lower half. We will review these criteria here briefly, and revisit them in detail when discussing the learning algorithms that use them.

TABLE 21.1: Standard criteria for deciding whether to include new data when growing a dictionary, and which data to prune.

criterion	type	complexity
all	growing	—
coherence	growing	$\mathcal{O}(m)$
ALD	growing	$\mathcal{O}(m^2)$
oldest	pruning	—
least weight	pruning	$\mathcal{O}(m)$
least a posteriori SE	pruning	$\mathcal{O}(m^2)$

A simple criterion to check whether the newly arriving datum is sufficiently informative is called the *coherence criterion* [11]. Given the dictionary $\mathcal{D}$ at some iteration and the newly arriving datum $\mathbf{x}$, this criterion defines the *coherence* of the datum as the quantity

$$\mu = \max_{\mathbf{u}_i \in \mathcal{D}} \kappa(\mathbf{u}_i, \mathbf{x}). \tag{21.10}$$

In essence, the coherence criterion checks the similarity, as measured by the kernel function, between the new datum and the most similar dictionary point. Only if the coherence is below a certain predefined threshold, $\mu \leq \mu_0$, the datum is inserted into the dictionary. The higher the threshold μ_0 is chosen, the more data will be accepted in the dictionary. It is a simple and effective criterion that has low computational complexity: it only requires calculating m kernel functions.

A more sophisticated dictionary growth criterion was introduced in [5]. Whenever a new datum is observed, this criterion measures how well the datum can be approximated in the feature space as a linear combination of the dictionary bases in that space. It does so by checking if the following approximate linear dependence (ALD) condition holds:

$$\delta := \min_{\mathbf{a}} \left\| \sum_{i=1}^{m} a(i)\phi(\mathbf{u}_i) - \phi(\mathbf{x}) \right\|^2 \leq \nu, \tag{21.11}$$

where ν is a precision threshold. The lower ν is chosen, the more data will be accepted into the dictionary. In contrast to the coherence criterion, which only compares the new datum to one dictionary basis at a time, the ALD criterion looks for the best combination of all bases. This search corresponds to a least-squares problem, which, if solved iteratively, requires quadratic complexity in terms of M, per iteration. In comparison to the coherence criterion, ALD is computationally more complex. In return, it is able to construct more compact dictionaries that represent the same information with fewer bases.

In practice, it is often necessary to specify a maximum dictionary size M, or *budget*, that may not be exceeded. In order to avoid exceeding the budget, one could simply stop allowing any inclusions in the dictionary once the budget is reached, hence locking the dictionary. Nevertheless, it is not unimaginable that after reaching the budget some new datum may still be observed that is more informative than a currently stored dictionary basis. In this case, the quality of the algorithm's solution will improve by pruning the said dictionary basis and by adding the new, more informative datum.

At this point it is interesting to remark that there exists a conceptual difference between growing and pruning strategies: While growth criteria are concerned with determining *whether or not* to include a new datum, pruning criteria deal with determining *which* datum to discard.

In time-varying environments, it may be useful to simply discard the oldest bases, as these were observed when the underlying model was possibly most different from the current model. This strategy is at the core of sliding-window algorithms, which, in every iteration, accept the new datum and discard the oldest basis, thereby maintaining a dictionary of fixed size [22].

A different pruning strategy is obtained by observing that the solution takes the form of the functional representation (21.9). In this kernel expansion, each dictionary element $\mathbf{u}_i$ has an associated weight $\alpha(i)$. Hence, a low-complexity pruning strategy simply consists of discarding the dictionary element that has the least weight $|\alpha(i)|$ associated to it, as proposed in [16].

A more sophisticated pruning strategy was introduced in [3] and [4]. It consists of selecting the element that causes the least squared error (SE) to the solution after being pruned. The relevance of a dictionary basis, according to this criterion, can be calculated as

$$\frac{|\alpha(i)|}{[\mathbf{K}^{-1}]_{ii}}, \tag{21.12}$$

where $[\mathbf{K}^{-1}]_{ii}$ is the i-th element on the diagonal of the inverse kernel matrix, calculated for the dictionary bases. We will refer to this criterion as the *least a posteriori SE criterion*. Similarly to the ALD criterion, $\mathbf{K}^{-1}$ and $\boldsymbol{\alpha}$ can be updated iteratively, yielding a complexity per iteration of $\mathcal{O}(m^2)$.

21.3 Kernel Recursive Least-Squares Regression Methods

We now formulate the recursive update for the kernel ridge regression solution (21.8), known as kernel recursive least-squares (KRLS). First, we describe the evergrowing approach, in which all training data appear in the solution. By introducing dictionary strategies into this technique, we then obtain several different practical algorithms that limit their solution's growth.

21.3.1   Recursive Updates of the Least-Squares Solution

Assume an online scenario in which $n-1$ data have been processed at the $n-1$-th iteration. The regression solution from Equation (21.8) reads

$$\boldsymbol{\alpha}_{n-1} = \dot{\mathbf{K}}_{n-1}^{-1} \mathbf{y}_{n-1}, \tag{21.13}$$

where we denote the regularized kernel matrix with a dot, $\dot{\mathbf{K}} = \mathbf{K} + c\mathbf{I}$, in order to simplify the notation. In the next iteration, n, a new data pair $(\mathbf{x}_n, y_n)$ is received and we wish to update the solution (21.13) recursively. Following the online protocol from Algorithm 19, we first calculate the predicted output

$$\hat{y}_n = \mathbf{k}_n^\top \boldsymbol{\alpha}_{n-1}, \tag{21.14}$$

in which $\mathbf{k}_n = [\kappa(\mathbf{x}_n, \mathbf{x}_1), \dots, \kappa(\mathbf{x}_n, \mathbf{x}_{n-1})]^\top$, and we obtain the a priori error for this datum, $e_n = y_n - \hat{y}_n$. The updated kernel matrix is

$$\dot{\mathbf{K}}_n = \begin{bmatrix} \dot{\mathbf{K}}_{n-1} & \mathbf{k}_n \\ \mathbf{k}_n^\top & k_{nn} + c \end{bmatrix}, \tag{21.15}$$

with $k_{nn} = \kappa(\mathbf{x}_n, \mathbf{x}_n)$. By introducing the variables

$$\mathbf{a}_n = \dot{\mathbf{K}}_{n-1}^{-1} \mathbf{k}_n, \tag{21.16}$$

and

$$\gamma_n = k_{nn} + c - \mathbf{k}_n^\top \mathbf{a}_n, \tag{21.17}$$

the new inverse kernel matrix is calculated as

$$\dot{\mathbf{K}}_n^{-1} = \frac{1}{\gamma_n} \begin{bmatrix} \gamma_n \dot{\mathbf{K}}_{n-1}^{-1} + \mathbf{a}_n \mathbf{a}_n^\top & -\mathbf{a}_n \\ -\mathbf{a}_n & 1 \end{bmatrix}. \tag{21.18}$$

Finally, the updated solution $\boldsymbol{\alpha}_n$ is obtained as

$$\boldsymbol{\alpha}_n = \begin{bmatrix} \boldsymbol{\alpha}_{n-1} - \mathbf{a}_n e_n / \gamma_n \\ e_n / \gamma_n \end{bmatrix}. \tag{21.19}$$

Equations (21.18) and (21.19) are efficient updates that allow us to obtain the new solution in $\mathcal{O}(n^2)$ time and memory, based on the previous solution. Directly applying Equation (21.13) at iteration n would require $\mathcal{O}(n^3)$ cost, so the recursive procedure is preferred in online scenarios. For a detailed derivation of this evergrowing formulation we refer to [5, 18].

21.3.2 Approximate Linear Dependency KRLS Algorithm

The KRLS algorithm from [5] uses the recursive solution we described previously, and it introduces the ALD criterion in order to reduce the growth of the functional representation. While the name KRLS was coined for the algorithm proposed in [5], this algorithm is only one of the many possible implementations of the KRLS principle. We will therefore refer to it as ALD-KRLS. An outline of this algorithm is given in Algorithm 20.

At every iteration, ALD-KRLS decides whether or not to increase its order, based on its dictionary growth criterion. If the criterion is fulfilled, it performs a *full update*, consisting of an order increase of the dictionary and the algorithm variables. If the criterion is not fulfilled, the dictionary is maintained, but instead of simply discarding the data pair $(\mathbf{x}_n, y_n)$ altogether, the solution coefficients are updated (though not expanded) with the information contained in this datum. We denote this type of update as a *reduced update*.

Algorithm 20 KRLS algorithm with reduced growth.

Initialize variables as empty or zero.
  for $n = 1, 2, \ldots$ **do**
      Observe input $\mathbf{x}_n$.
      Predict output: Equation (21.14)
      Observe true output y_n.
      if dictionary growth criterion is fulfilled **then**
          Expand dictionary: $\mathcal{D}_n = \mathcal{D}_{n-1} \cup \{\mathbf{x}_n\}$
          Update inverse kernel matrix: Equation (21.18)
          Update expansion coefficients: Equation (21.19)
      else
          Maintain dictionary: $\mathcal{D}_n = \mathcal{D}_{n-1}$
          Perform reduced update of expansion coefficients.
      end if
  end for

ALD-KRLS does not include regularization, $c = 0$. In this case, the coefficients $\mathbf{a}_n = [a_n(1), \ldots, a_n(m)]^\top$ that minimize the ALD condition (21.11) are given by Equation (21.16), and the norm of the best linear combination is given by $\delta_n = \gamma_n$. In order to perform a reduced update, ALD-KRLS keeps track of a projection matrix that is used to project the information of redundant bases onto the current solution, before discarding them. For detail, refer to [5].

21.3.3 Sliding-Window KRLS Algorithm

ALD-KRLS assumes a stationary model, and, therefore, it is not suitable as a tracking algorithm. In addition, there does not seem to be an obvious extension that allows for tracking, mainly because it summarizes past information into a compact formulation that allows for little manipulation. This is a somewhat surprising fact, taking into account that it is derived from the linear recursive least-squares (RLS) algorithm, which is easily extendible into several tracking formulations, for instance by considering a forgetting factor.

In order to address this issue, a sliding-window KRLS (SW-KRLS) algorithm was proposed in [22]. Instead of summarizing previous data, it simply stores a window of the last M data as its dictionary. In each step it adds the new datum and discards the oldest datum, leading to a sliding-window approach. In order to expand the inverse kernel matrix with a new datum it uses Equation (21.18). To discard the oldest datum it relies on the following relationship. By breaking up the kernel matrix and its inverse as

$$\dot{\mathbf{K}}_{n-1} = \begin{bmatrix} a & \mathbf{b}^T \\ \mathbf{b} & \mathbf{D} \end{bmatrix}, \quad \dot{\mathbf{K}}_{n-1}^{-1} = \begin{bmatrix} e & \mathbf{f}^T \\ \mathbf{f} & \mathbf{G} \end{bmatrix}, \tag{21.20}$$

the inverse of the reduced kernel matrix is found as

$$\mathbf{D}^{-1} = \mathbf{G} - \mathbf{f}\mathbf{f}^T/e. \tag{21.21}$$

Finally, the coefficients $\boldsymbol{\alpha}_n$ are obtained through Equation (21.13), in which the vector $\mathbf{y}_n$ now contains only the M last outputs. This vector is stored along with the dictionary.

SW-KRLS is a conceptually very simple algorithm that obtains reasonable performance in a wide range of scenarios, including non-stationary environments. Nevertheless, its performance is limited by the quality of the bases in its dictionary, over which it has no control. In particular, it has no means to avoid redundancy in its dictionary or to maintain older bases that are relevant to its kernel expansion. In order to improve this performance, a fixed-budget KRLS (FB-KRLS) algorithm was proposed in [21]. Instead of discarding the oldest data point in each iteration, it discards the data point that causes the least error upon being discarded, using the least a posteriori SE pruning criterion from Table 21.1. In stationary scenarios, this extension obtains significantly better results. In non-stationary cases, however, it does not offer any advantage, and a different approach is required, as we discuss in the sequel.

21.3.4   Bayesian Interpretation

The standard KRLS equations, as described above, can also be derived from a Bayesian perspective. As we will see, the obtained solution is equivalent to the KRLS update, though the Bayesian approach does not only provide the mean value for the predicted solution but also its entire posterior distribution. The full derivation, which is based on the framework of Gaussian Processes (GPs) [10], can be found in [18, 9].

A Bayesian setting requires a model that describes the observations, and priors on the parameters of this model. The GP regression model assumes that the outputs can be modeled as some noiseless latent function of the inputs plus an independent noise component

$$y = f(\mathbf{x}) + r, \tag{21.22}$$

and then sets a zero mean[1] GP prior on $f(\mathbf{x})$ and a Gaussian prior on r:

$$f(\mathbf{x}) \sim \mathcal{GP}(\mathbf{0}, \kappa(\mathbf{x}, \mathbf{x}')), \qquad r \sim \mathcal{N}(0, \sigma^2), \tag{21.23}$$

where σ^2 is a hyperparameter that specifies the noise power. The notation $\mathcal{GP}(m(\mathbf{x}), \kappa(\mathbf{x}, \mathbf{x}'))$ refers to a GP distribution over functions in terms of a mean function $m(\mathbf{x})$ (zero in this case) and a *covariance* function $\kappa(\mathbf{x}, \mathbf{x}')$, equivalent to a kernel function. The covariance function specifies the a priori relationship between values $f(\mathbf{x})$ and $f(\mathbf{x}')$ in terms of their respective locations, and it is parameterized by a small set of hyperparameters, grouped in vector $\boldsymbol{\theta}$.

By definition, the marginal distribution of a GP at a finite set of points is a joint Gaussian distribution, with its mean and covariance being specified by the homonymous functions evaluated at those points. Thus, the joint distribution of outputs $\mathbf{y} = [y_1, \ldots, y_N]^\top$ and the corresponding latent vector $\mathbf{f} = [f(\mathbf{x}_1), \ldots, f(\mathbf{x}_N)]^\top$ is

$$\begin{bmatrix} \mathbf{y} \\ \mathbf{f} \end{bmatrix} \sim \mathcal{N}\left(\mathbf{0}, \begin{bmatrix} \mathbf{K} + \sigma^2\mathbf{I} & \mathbf{K} \\ \mathbf{K} & \mathbf{K} \end{bmatrix}\right). \tag{21.24}$$

By conditioning on the observed outputs $\mathbf{y}$, the posterior over the latent vector can be inferred

$$\begin{aligned} p(\mathbf{f}|\mathbf{y}) &= \mathcal{N}(\mathbf{f}|\mathbf{K}(\mathbf{K} + \sigma^2\mathbf{I})^{-1}\mathbf{y}, \mathbf{K} - \mathbf{K}(\mathbf{K} + \sigma^2\mathbf{I})^{-1}\mathbf{K}) \\ &= \mathcal{N}(\mathbf{f}|\boldsymbol{\mu}, \boldsymbol{\Sigma}). \end{aligned} \tag{21.25}$$

Assuming this posterior is obtained for the data up till time instant $n-1$, the predictive distribution of a new output y_n at location $\mathbf{x}_n$ is computed as

$$p(y_n|\mathbf{x}_n, \mathbf{y}_{n-1}) = \mathcal{N}(y_n|\mu_{\mathrm{GP},n}, \sigma^2_{\mathrm{GP},n}) \tag{21.26a}$$

$$\mu_{\mathrm{GP},n} = \mathbf{k}_n^\top(\mathbf{K}_{n-1} + \sigma^2\mathbf{I})^{-1}\mathbf{y}_{n-1} \tag{21.26b}$$

$$\sigma^2_{\mathrm{GP},n} = \sigma^2 + k_{nn} - \mathbf{k}_n^\top(\mathbf{K}_{n-1} + \sigma^2\mathbf{I})^{-1}\mathbf{k}_n. \tag{21.26c}$$

The mode of the predictive distribution, given by $\mu_{\mathrm{GP},n}$ in Equation (21.26b), coincides with the solution of KRLS, given by Equation (21.18), showing that the regularization in KRLS can be interpreted as a noise power σ^2. Furthermore, the variance of the predictive distribution, given by $\sigma^2_{\mathrm{GP},n}$ in Equation (21.26c), coincides with Equation (21.17), which is used by the dictionary criterion for ALD-KRLS.

---

[1] It is customary to subtract the sample mean from the data $\{y_n\}_{n=1}^N$, and then to assume a zero-mean model.

Using Equations (21.26), a recursive update of the complete GP can be found as

$$p(\mathbf{f}_n|\mathbf{X}_n, \mathbf{y}_n) = \mathcal{N}(\mathbf{f}_n|\boldsymbol{\mu}_n, \boldsymbol{\Sigma}_n) \tag{21.27a}$$

$$\boldsymbol{\mu}_n = \begin{bmatrix} \boldsymbol{\mu}_{n-1} \\ \hat{y}_n \end{bmatrix} + \frac{e_n}{\hat{\sigma}_{yn}^2} \begin{bmatrix} \mathbf{h}_n \\ \hat{\sigma}_{fn}^2 \end{bmatrix} \tag{21.27b}$$

$$\boldsymbol{\Sigma}_n = \begin{bmatrix} \boldsymbol{\Sigma}_{n-1} & \mathbf{h}_n \\ \mathbf{h}_n^\top & \hat{\sigma}_{fn}^2 \end{bmatrix} - \frac{1}{\hat{\sigma}_{yn}^2} \begin{bmatrix} \mathbf{h}_n \\ \hat{\sigma}_{fn}^2 \end{bmatrix} \begin{bmatrix} \mathbf{h}_n \\ \hat{\sigma}_{fn}^2 \end{bmatrix}^\top, \tag{21.27c}$$

where $\mathbf{h}_n = \boldsymbol{\Sigma}_{n-1}\mathbf{K}_{n-1}^{-1}\mathbf{k}_n$, and $\hat{\sigma}_{fn}^2$ and $\hat{\sigma}_{yn}^2$ are the predictive variances of the latent function and the new output, respectively, calculated at the new input. Details can be found in [18].

The update Equations (21.27) are formulated in terms of the predictive mean and covariance, $\boldsymbol{\mu}_n$ and $\boldsymbol{\Sigma}_n$, which allows us to interpret them directly in terms of the underlying GP. They can be reformulated in terms of $\boldsymbol{\alpha}_n$ and a corresponding matrix, as shown in [3]. Specifically, the relationship between $\boldsymbol{\mu}_n$ and $\boldsymbol{\alpha}_n$ is (see [18])

$$\boldsymbol{\alpha}_n = \mathbf{K}_n^{-1}\boldsymbol{\mu}_n \tag{21.28a}$$

$$= \dot{\mathbf{K}}_n^{-1}\mathbf{y}_n. \tag{21.28b}$$

Interestingly, while standard KRLS obtains $\boldsymbol{\alpha}_n$ based on the *noisy* observations $\mathbf{y}_n$ through Equation (21.28b), the GP-based formulation shows that $\boldsymbol{\alpha}_n$ can be obtained equivalently using the values of the *noiseless* function evaluated at the inputs $\boldsymbol{\mu}_n$, through Equation (21.28a).

The advantage of using a full GP model is that not only does it allow us to update the predictive mean, as does KRLS, but it keeps track of the entire predictive distribution of the solution [9]. This allows, for instance, establishing confidence intervals when predicting new outputs. And, more importantly in adaptive contexts, it allows us to explicitly handle the uncertainty about all learned data. In the sequel, we will review a recursive algorithm that exploits this knowledge to perform tracking.

21.3.5   KRLS Tracker Algorithm

In [18], a KRLS tracker (KRLS-T) algorithm was devised that explicitly handles uncertainty about the data, based on the above discussed probabilistic Bayesian framework. While in stationary environments it operates identically to the earlier proposed Sparse Online GP algorithm (SOGP) from [3], it includes a *forgetting mechanism* that enables it to handle non-stationary scenarios as well.

In non-stationary scenarios, adaptive online algorithms should be capable of tracking the changes of the observed model. This is possible by weighting past data less heavily than more recent data. A quite radical example of forgetting is provided by the SW-KRLS algorithm, which either assigns full validity to the data (those in its window), or discards them entirely.

KRLS-T includes a framework that permits several forms of forgetting. We focus on the forgetting strategy called "back to the prior" (B2P), in which the mean and covariance are replaced through

$$\boldsymbol{\mu} \leftarrow \sqrt{\lambda}\boldsymbol{\mu} \tag{21.29a}$$
$$\boldsymbol{\Sigma} \leftarrow \lambda\boldsymbol{\Sigma} + (1-\lambda)\mathbf{K}. \tag{21.29b}$$

As shown in [18], this particular form of forgetting corresponds to blending the informative posterior with a "noise" distribution that uses the same color as the prior. In other words, forgetting occurs by taking a step back towards the prior knowledge. Since the prior has zero mean, the mean is simply scaled by the square root of the forgetting factor λ. The covariance, which represents the posterior uncertainty on the data, is pulled towards the covariance of the prior. Interestingly, a regularized version of RLS (known as *extended RLS*) can be obtained by using a linear kernel with the B2P forgetting procedure. Standard RLS can be obtained by using a different forgetting rule (see [18]).

Equations (21.29) may seem like an *ad-hoc* step to enable forgetting. However, it can be shown that the whole learning procedure — including the mentioned controlled forgetting step — corresponds exactly to a principled non-stationary scheme within the GP framework, as described in [20]. It is sufficient to consider an augmented input space that includes the time stamp t of each sample and define a *spatio-temporal* covariance function:

$$\kappa_{\mathrm{st}}([t \ \mathbf{x}^\top]^\top, [t' \ \mathbf{x}'^\top]^\top) = \kappa_{\mathrm{t}}(t, t')\kappa_{\mathrm{s}}(\mathbf{x}, \mathbf{x}'), \tag{21.30}$$

where $\kappa_{\mathrm{s}}(\mathbf{x}, \mathbf{x}')$ is the already-known spatial covariance function and $\kappa_{\mathrm{t}}(t, t')$ is a temporal covariance function giving more weight to samples that are closer in time. Inference on this augmented model effectively accounts for non-stationarity in $f(\cdot)$ and recent samples have more impact in predictions for the current time instant. It is fairly simple to include this augmented model in the online learning process described in the previous section. When the temporal covariance is set to $k_{\mathrm{t}}(t, t') = \lambda^{\frac{|t-t'|}{2}}$, $\lambda \in (0, 1]$, inference in the augmented spatio-temporal GP model is exactly equivalent to using (21.29) after each update (21.27).

This equivalence has interesting consequences. Most importantly, it implies that the optimal hyperparameters for the recursive problem can be determined by performing standard GP hyperparameter estimation techniques, such as Type-II maximum likelihood estimation, on the equivalent spatio-temporal batch problem. This is an important accomplishment in kernel adaptive filtering theory, as it allows us to determine the hyperparameters in a principled manner, including kernel parameters, the noise level, and the forgetting factor λ. See [20] for further details.

The KRLS-T algorithm is summarized in Algorithm 21. Its first step in each iteration consists of applying a forgetting strategy, which takes away

---

Algorithm 21 KRLS Tracker (KRLS-T) algorithm.

Initialize variables as empty or zero.

for $n = 1, 2, \ldots$ **do**

    Forget: replace $\boldsymbol{\mu}_{n-1}$ and $\boldsymbol{\Sigma}_{n-1}$ through Equations 21.29

    Observe input $\mathbf{x}_n$.

    Calculate predictive mean: $\hat{y}_n = \mathbf{k}_n \mathbf{K}_{n-1}^{-1} \boldsymbol{\mu}_{n-1}$

    Calculate predictive variance $\hat{\sigma}_{yn}^2$.

    Observe true output y_n.

    Compute $\boldsymbol{\mu}_n$, $\boldsymbol{\Sigma}_n$, $\mathbf{K}_n^{-1}$.

    Add basis $\mathbf{x}_n$ to the dictionary.

    if number of bases in the dictionary $> M$ **then**

        Determine the least relevant basis, $\mathbf{u}_m$.

        Remove basis $\mathbf{u}_m$ from $\boldsymbol{\mu}_n$, $\boldsymbol{\Sigma}_n$, $\mathbf{K}_n^{-1}$

        Remove basis $\mathbf{u}_m$ from the dictionary.

    end if

end for

---

some of the weight of older information. KRLS-T accepts every datum into its dictionary (as long as this does not render $\mathbf{K}_n^{-1}$ rank-deficient), and at the end of each iteration it prunes the least relevant basis, after projecting its information onto the remaining bases. For pruning, it uses the least a posteriori SE (see Table 21.1). Additional details can be found in [18].

21.4   Stochastic Gradient Descent with Kernels

Up till this point we have reviewed several online algorithms that recursively estimate the batch solution to the kernel ridge regression problem. These algorithms have quadratic complexity in terms of the number of data that they store in their dictionary, $\mathcal{O}(m^2)$, which may be excessive in certain scenarios. It is possible to obtain algorithms with lower complexity, typically $\mathcal{O}(m)$, by performing approximations to the optimal recursive updates, as we will describe here.

The starting point is, again, the kernel ridge regression problem, which we repeat for convenience:

$$\min_{\phi(\mathbf{w})} J = \|\mathbf{y} - \phi(\mathbf{X})\phi(\mathbf{w})\|^2 + c\|\phi(\mathbf{w})\|^2.$$

Earlier, we dealt with techniques that focus on the batch solution to this problem. The same solution can be obtained through an iterative procedure, called the steepest-descent method [13]. It consists in iteratively applying the rule

$$\phi(\mathbf{w}) \leftarrow \phi(\mathbf{w}) - \frac{\eta}{2}\frac{\partial J}{\partial\phi(\mathbf{w})}, \tag{21.31}$$

where η represents a learning rate. After replacing the derivative in Equation (21.31) by its instantaneous estimate, and, omitting regularization momentarily, we obtain a low-cost online algorithm with the following stochastic gradient descent update rule

$$\phi(\mathbf{w}_n) = \phi(\mathbf{w}_{n-1}) + \eta\left(y_n\phi(\mathbf{x}_n) - \phi(\mathbf{x}_n)\phi(\mathbf{w}_{n-1})^\top\phi(\mathbf{x}_n)\right)$$
$$= \phi(\mathbf{w}_{n-1}) + \eta e_n\phi(\mathbf{x}_n).$$

By relying on the representer theorem [14], $\phi(\mathbf{w}_n)$ and $\phi(\mathbf{w}_{n-1})$ are expressed as linear combinations of the transformed data, yielding

$$\sum_{i=1}^{n}\alpha_n(i)\phi(\mathbf{x}_i) = \sum_{i=1}^{n-1}\alpha_{n-1}(i)\phi(\mathbf{x}_i) + \eta e_n\phi(\mathbf{x}_n), \tag{21.32}$$

where $\alpha_n(i)$ denotes the i-th element of the vector $\boldsymbol{\alpha}_n$. If regularization is not omitted, the update rule reads

$$\sum_{i=1}^{n}\alpha_n(i)\phi(\mathbf{x}_i) = (1 - \eta c)\sum_{i=1}^{n-1}\alpha_{n-1}(i)\phi(\mathbf{x}_i) + \eta e_n\phi(\mathbf{x}_n). \tag{21.33}$$

The update (21.32) is the core equation used to derive kernel least mean square (KLMS) algorithms. In essence, these algorithms are kernelized versions of the classical least-mean-squares (LMS) algorithm [13]. Similar to the previously discussed online kernel algorithms, KLMS algorithms usually also build an online dictionary, but since their core update is of linear complexity in term of the number of points in the dictionary, $\mathcal{O}(M)$, their dictionary update should not exceed this complexity.

KLMS algorithms possess the interesting property that their learning rule also provides them with a tracking mechanism, at no additional cost. KRLS algorithms, on the other hand, require their standard design to be specifically extended in order to obtain this property, as we discussed. In what follows we will discuss the mechanics of the three most popular KLMS algorithms, highlighting their similarities and differences.

Several forms exist to obtain the new weights $\boldsymbol{\alpha}_n$ in such a way that Equation (21.32) holds. The simplest form to obtain the weights at step n consists of maintaining the previous weights, i.e., $\alpha_n(i) = \alpha_{n-1}(i)$, for $i = 1, \ldots, n-1$ and adding a new weight $\alpha_n(n)$ that accounts for the term $\eta e_n\phi(\mathbf{x}_n)$. This is the update mechanism behind NORMA [6] and Q-KLMS [2].

21.4.1 Naive Online Regularized Risk Minimization Algorithm

Naive online regularized risk minimization algorithm (NORMA) is a family of stochastic-gradient online kernel-based algorithms [6]. It includes regularization and thus uses Equation (21.33) as it basic update. We will focus on its standard form for regression with a squared loss function. By concentrating all the novelty in the new coefficients α_n, the update for NORMA at time step n reads

$$\alpha_n = \begin{bmatrix} (1 - \eta c)\alpha_{n-1} \\ \eta e_n \end{bmatrix}.$$ (21.34)

Note that the coefficients shrink as n grows. Therefore, after a certain amount of iterations, the oldest coefficient can be discarded without affecting the solution's quality. Hence, a sliding-window dictionary mechanism is obtained that prevents the functional representation from growing too large during online operation.

21.4.2   Quantized KLMS

A second algorithm that concentrates all the novelty in one coefficient is quantized KLMS (Q-KLMS) [2]. The main characteristic of Q-KLMS is that it slows down its growth by constructing a dictionary through a *quantization* process. For each new datum, Q-KLMS uses the coherence criterion from [11] to check whether or not to add the datum to the dictionary. Q-KLMS was proposed based on a specific version of the coherence criterion that uses the Euclidean distance, though any kernel could be used in its criterion. Note that the calculation of the coherence criterion has linear cost in terms of the current dictionary size, M, making it especially useful for KLMS algorithms.

If the coherence condition is not fulfilled, $\mu \leq \mu_0$, Q-KLMS adds the new datum to its dictionary and updates the coefficients as follows

$$\alpha_n = \begin{bmatrix} \alpha_{n-1} \\ \eta e_n \end{bmatrix}.$$ (21.35)

If the coherence condition is fulfilled, $\mu > \mu_0$, Q-KLMS only updates the coefficient of the dictionary element that is closest to the new datum. If we denote the index of the closest dictionary element as j, the update rule for this case reads

$$\alpha_n(j) = \begin{bmatrix} \alpha_{n-1}(j) + \eta e_n \end{bmatrix},$$ (21.36)

Note that Q-KLMS does not include regularization. As shown in [7], it is a member of a class of KLMS algorithms that possess a self-regularizing property.

21.4.3   Kernel Normalized LMS

Instead of concentrating all the novelty in one coefficient, a different approach is followed in [11]. Specifically, the new coefficient vector α_n is obtained by projecting the previous vector, α_{n-1}, onto the line defined by $\mathbf{k}_n^\top \alpha - y_n = 0$.

As shown in [11], this yields the following normalized KLMS update

$$\boldsymbol{\alpha}_n = \begin{bmatrix} \boldsymbol{\alpha}_{n-1} \\ 0 \end{bmatrix} + \frac{\eta e_n}{\epsilon + \|\mathbf{k}_n\|^2} \begin{bmatrix} \mathbf{k}_n \\ k_{nn} \end{bmatrix}, \tag{21.37}$$

when the dictionary is to be expanded, where ϵ is a regularization constant. This algorithm, denoted kernel normalized LMS (KNLMS) [11], uses the online sparsification based on the coherence criterion to determine whether or not to include new data. In particular, when the new datum does not meet the coherence condition, KNLMS updates its coefficients without increasing the order, following the rule

$$\boldsymbol{\alpha}_n = \boldsymbol{\alpha}_{n-1} + \frac{\eta e_n}{\epsilon + \|\mathbf{k}_n\|^2} \mathbf{k}_n. \tag{21.38}$$

21.5   Performance Comparisons

The regression performance of online kernel methods has been studied in several ways. The standard manner to compare the performance is to analyze their learning curves, which depict their regression error over time. This, however, requires us to choose the parameters for each algorithm that are optimal in some sense. We will analyze some learning curves first, on a standard data set, and then we will show how a more global comparison can be obtained that encompasses multiple parameter configurations for each algorithm. Unless stated otherwise, we will use a Gaussian kernel of the form

$$\kappa(\mathbf{x}, \mathbf{x}') = \sigma_0^2 \exp\left(-\frac{\|\mathbf{x} - \mathbf{x}'\|^2}{2\ell^2}\right), \tag{21.39}$$

in which σ_0^2 is the signal power and ℓ is the length scale.

The results from this section can be reproduced by the code included in an open-source toolbox, available at http://sourceforge.net/projects/kafbox/, which we have developed specifically for this purpose. It contains implementations of the most popular kernel adaptive filtering algorithms.

21.5.1   Online Regression on the KIN40K Data Set

In the first experiment we train the online algorithms to perform regression of the KIN40K data set[2]. This data set is obtained from the forward kinematics of an 8-link all-revolute robot arm, and it represents a stationary regression problem. It contains 40000 examples, each consisting of an 8-dimensional input

---

[2]Available at http://www.cs.toronto.edu/~delve/data/datasets.html.

vector and a scalar output. We randomly select 10000 data points for training and use the remaining 30000 points for testing the regression.

For all algorithms we use an anisotropic Gaussian kernel in which the hyperparameters were determined offline by standard GP regression. In particular, the noise-to-signal ratio was $\sigma_n^2/\sigma_0^2 = 0.0021$. Each algorithm performs a single run over the data. The performance is measured as the normalized mean-square error (NMSE) on the test data set at different points throughout the training run.

The results are displayed in Fig. 21.2. The algorithm-specific parameters were set as follows: ALD-KRLS uses $\nu = 0.1$, KRLS-T uses $M = 500$, SW-KRLS uses $M = 500$, Q-KLMS uses $\eta = 0.5$ and $\epsilon_u = 1$, and NORMA uses $\eta = 0.5$ and $M = 1500$. Note that apart from the method mentioned in [20] and the typical cross-validation, parameters for these algorithms are typically determined by heuristics.

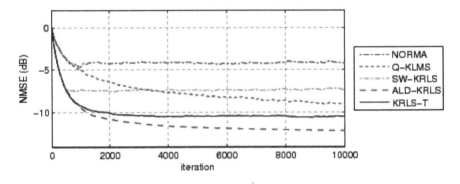

FIGURE 21.2: Learning curves on the KIN40K data set, with specific parameters per algorithm.

While Figure 21.2 shows some interesting results, one may wonder if it is possible to improve the performance of a specific algorithm by tweaking its parameters. Indeed, algorithms that use a budget parameter M to determine their maximum dictionary size will typically obtain lower NMSE values if the budget is raised. Nevertheless, this will increase their computational complexity. A similar phenomenon is observed for all algorithms. Algorithms that use a threshold to determine their budget, such as ALD-KRLS and Q-KLMS, obtain a better steady-state NMSE at the cost of higher complexity. There is thus a trade-off between the cost of an algorithm, in terms of computation and memory required, and its performance, in terms of the error it obtains and how fast it converges to its steady-state. In the following, we will analyze these trade-offs instead of the learning curves, as they may provide us with a more global picture of the performance of the algorithms.

21.5.2 Cost-versus-Performance Trade-Offs

The computational cost of an algorithm is often measured as the CPU time. Nevertheless, this measure depends on the machine and the particular implementation of the algorithm. For a fairer comparison, we use a count of the floating point operations (FLOPS) instead, which are measured explicitly by the toolbox used for the experiments. We also measure the used memory, in terms of the number of bytes necessary to store the variables. In the results, we report the maximum FLOPS and maximum bytes per iteration, as these are the values that impose limits on the hardware.

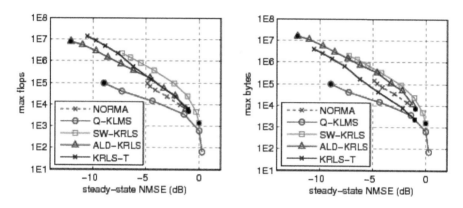

FIGURE 21.3: Maximum FLOPS used per iteration (left) and maximum bytes used per iteration (right), in the online regression experiment on the KIN40K data set, for different parameter configurations and different algorithms. Parameter values are represented in Table 21.2 and the black dots represent the first configuration, for each algorithm.

Figure 21.3 illustrates the trade-offs obtained by each algorithm. The markers represent the results obtained for different sets of algorithm parameters. Each algorithm has one budget-related parameter, which also determines the computational and memory complexities. We fix every other parameter to a value close to its optimum and vary the budget parameter over a wide range. The full parameter values are displayed in Table 21.2. The steady-state NMSE is measured as the average NMSE over the last 1000 iterations.

The best-performing algorithm configurations are located to the left in both plots of Figure 21.3, corresponding to low steady-state errors, and to the bottom, corresponding to low algorithmic complexities. The black dots show the first configuration of each algorithm, and typically they also represent an initial parameter setting beyond which it is difficult to move, for instance due to numerical limits.

By leaving the NMSE results out of Figure 21.3, we obtain the plot of Figure 21.4. It shows that for all algorithms there is an approximately linear relationship between the number of bytes stored in memory and the number

TABLE 21.2: Parameters used in the KIN40K online regression.

Method	Fixed parameter	Varying parameters
NORMA	$\eta = 0.5$	$\tau = 100, 200, 500, 1000, 1500, 2000$
Q-KLMS	$\eta = 0.5$	$\epsilon_u = 1, 1.2, 1.5, 2, 3, 5, 10, 12, 15$
SW-KRLS	—	$M = 10, 20, 50, 100, 200, 300, 400, 500$
ALD-KRLS	—	$\nu = .1, .2, .3, .4, .5, .6, .7, .8, .9, .95, .99$
KRLS-T	$\lambda = 1$	$M = 10, 20, 50, 100, 200, 300, 400, 500$

of FLOPS required per iteration. Notice the logarithmic scales used. Algorithms that lie below the diagonal are more efficient with computation, i.e., when using the same amount of memory they require less computations, for a given NMSE. Algorithms that lie above the diagonal are more efficient with memory, i.e., when performing the same amount of computation they require less memory, for a given NMSE.

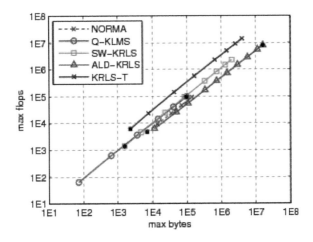

FIGURE 21.4: FLOPS per iteration versus bytes stored, in the KIN40K experiment.

21.5.3   Empirical Convergence Analysis

Apart from the steady-state error, an important measure for online and adaptive algorithms is the speed at which they converge to this error, called the *convergence rate*. Some theoretical convergence analyses have been carried out on specific algorithms, for instance the KNLMS algorithm in [8]. Here, in line with the previous experiments, we will perform an empirical convergence analysis that compares several algorithms.

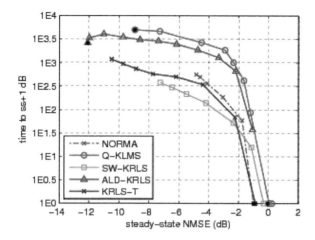

FIGURE 21.5: Trade-off between steady-state error and convergence rate in the KIN40K experiment.

In order to estimate the convergence rate, we measure the number of iterations it takes an algorithm to get within 1 dB of its steady-state error. The trade-off between this measure and the steady-state NMSE is shown in Figure 21.5. Again, results that are most to the left or to the bottom of the plot represent the most interesting algorithm configurations, as they reach the lowest steady-state error or have the fastest convergence, respectively.

21.5.4 Experiments with Real-World Data

Figure 21.6 shows the convergence results obtained on two real-world data sets. The first data set is obtained by measuring the response of a wireless communication channel. The online algorithm requires us to learn the nonlinear channel response and track its changes in time. The second data set is a recording of a patient's body surface during robotic radiosurgery. The online algorithm is used to predict the position of several markers on the body as the patient moves, so that the robot can use this prediction to compensate for its mechanical delay in positioning itself. More detailed descriptions of both experiments can be found in [19]. The parameters used in both experiments can be found in Tables 21.3 and 21.4, respectively.

In the first case, as the solution is to be implemented for real-time operation on a compact, low-power device, the maximum amount of FLOPS is typically fixed and the algorithm that obtains the lowest error under this condition is chosen for implementation. In the second case there is usually no such restriction on complexity, as large surgical robots can be equipped with sufficient resources. Instead, the main requirement is to maintain the prediction error as low as possible.

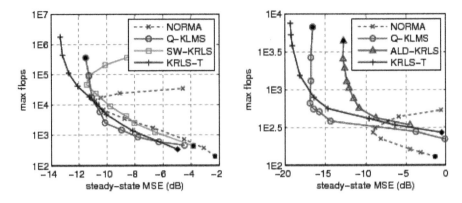

FIGURE 21.6: Results obtained for identifying the wireless communication channel (left) in the identification problem and for predicting a patient's position in the prediction problem.

TABLE 21.3: Parameters used in the online channel identification.

Method	Fixed	Varying parameters
NORMA	$\eta = 0.5$	$\tau = 5, 10, 20, 50, 100, 200, 310, 400, 500, 700, 1000$
Q-KLMS	$\eta = 0.6$	$\epsilon_u = 0.1, 1, 2, 3, 4, 5, 6, 7, 8$
SW-KRLS	—	$M = 5, 10, 15, 20, 30, 50, 70, 100, 150, 200, 300$
KRLS-T	$\lambda = 0.995$	$M = 2, 5, 10, 15, 20, 30, 50, 100, 200$

TABLE 21.4: Parameters used in the motion prediction.

Method	Fixed	Varying parameters
NORMA	$\eta = 0.99$	$\tau = 3, 4, 5, 10, 15, 20, 30, 40, 60$
Q-KLMS	$\eta = 0.99$	$\epsilon_u = .2, .5, 1, 2, 2.5, 2.75, 4, 6, 7$
ALD-KRLS	—	$\nu = .0001, .001, .003, .01, .02, .05, .1, .3, .5$
KRLS-T	$\lambda = 0.999$	$M = 3, 4, 5, 7, 10, 20, 50, 70, 100$

21.6   Further Reading

We have given an overview of one decade of research in the field of online regression with kernels. The standard approach followed in this field, which consists of constructing kernel-based versions of classical adaptive filtering algorithms such as LMS and RLS, has produced several efficient state-of-the-art algorithms. Some other, related classes of algorithms that we have not discussed here are kernel affine projection algorithms (KAPA) [11, 7], which occupy the middle ground between KLMS and KRLS, and projection-based subgradient methods [17].

Several new directions are also being explored to improve the learning capabilities of these algorithms. A major focus of new algorithms is on the automatic learning of hyperparameters. Some algorithms pursue this by performing stochastic natural gradient descent in an online manner, in order to approximately maximize the marginal likelihood [10]. Other algorithms follow approaches inspired by the recent advances in neural networks, and they focus on online multi-kernel learning, where the parameter learning consists of assigning weights to several different kernels that are applied in parallel [23]. Many approaches in this direction seek sparse solutions by performing L_1-norm based learning.

Bibliography

[1] Nachman Aronszajn. Theory of reproducing kernels. *Transactions of the American mathematical society*, 68(3):337–404, 1950.

[2] Badong Chen, Songlin Zhao, Pingping Zhu, and José C. Príncipe. Quantized kernel least mean square algorithm. *IEEE Transactions on Neural Networks and Learning Systems*, 23(1):22–32, January 2012.

[3] Lehel Csató and Manfred Opper. Sparse online Gaussian processes. *Neural Computation*, 14(3):641–668, 2002.

[4] Bas J. De Kruif and Theo J. A. De Vries. Pruning error minimization in least squares support vector machines. *IEEE Transactions on Neural Networks*, 14(3):696–702, 2003.

[5] Yaakov Engel, Shie Mannor, and Ron Meir. The kernel recursive least squares algorithm. *IEEE Transactions on Signal Processing*, 52(8):2275–2285, August 2004.

[6] Jyrki Kivinen, Alexander J. Smola, and Robert C. Williamson. Online learning with kernels. *IEEE Transactions on Signal Processing*, 52(8):2165–2176, August 2004.

[7] Weifeng Liu, José C. Príncipe, and Simon Haykin. *Kernel Adaptive Filtering: A Comprehensive Introduction*. Wiley, 2010.

[8] Wemerson D. Parreira, Jose Carlos M. Bermudez, Cédric Richard, and

Jean-Yves Tourneret. Stochastic behavior analysis of the Gaussian kernel least-mean-square algorithm. *IEEE Transactions on Signal Processing*, 60(5):2208–2222, 2012.

[9] Fernando Pérez-Cruz, Steven Van Vaerenbergh, Juan José Murillo-Fuentes, Miguel Lázaro-Gredilla, and Ignacio Santamaría. Gaussian processes for nonlinear signal processing: An overview of recent advances. *IEEE Signal Processing Magazine*, 30:40–50, July 2013.

[10] Carl Edward Rasmussen and Christopher K. I. Williams. *Gaussian Processes for Machine Learning*. MIT Press, 2006.

[11] Cédric Richard, José Carlos M. Bermudez, and Paul Honeine. Online prediction of time series data with kernels. *IEEE Transactions on Signal Processing*, 57(3):1058–1067, March 2009.

[12] Craig Saunders, Alexander Gammerman, and Volodya Vovk. Ridge regression learning algorithm in dual variables. In *Proceedings of the 15th International Conference on Machine Learning (ICML)*, pages 515–521, Madison, WI, USA, July 1998.

[13] Ali H. Sayed. *Fundamentals of Adaptive Filtering*. Wiley-IEEE Press, 2003.

[14] Bernhard Schölkopf, Ralf Herbrich, and Alexander J. Smola. A generalized representer theorem. In *Computational learning theory*, pages 416–426. Springer, 2001.

[15] Bernhard Schölkopf and Alexander J. Smola. *Learning with Kernels*. The MIT Press, Cambridge, MA, USA, 2002.

[16] Johan A.K. Suykens, Jos De Brabanter, Lukas Lukas, and Joos Vandewalle. Weighted least squares support vector machines: robustness and sparse approximation. *Neurocomputing*, 48(1):85–105, 2002.

[17] Sergios Theodoridis, Konstantinos Slavakis, and Isao Yamada. Adaptive learning in a world of projections. *IEEE Signal Processing Magazine*, 28(1):97 –123, January 2011.

[18] Steven Van Vaerenbergh, Miguel Lázaro-Gredilla, and Ignacio Santamaría. Kernel recursive least-squares tracker for time-varying regression. *IEEE Transactions on Neural Networks and Learning Systems*, 23(8):1313–1326, August 2012.

[19] Steven Van Vaerenbergh and Ignacio Santamaría. A comparative study of kernel adaptive filtering algorithms. In *2013 IEEE Digital Signal Processing (DSP) Workshop and IEEE Signal Processing Education (SPE)*, 2013.

[20] Steven Van Vaerenbergh, Ignacio Santamaría, and Miguel Lázaro-Gredilla. Estimation of the forgetting factor in kernel recursive least squares. In *2012 IEEE International Workshop on Machine Learning for Signal Processing (MLSP)*, September 2012.

[21] Steven Van Vaerenbergh, Ignacio Santamaría, Weifeng Liu, and José C. Príncipe. Fixed-budget kernel recursive least-squares. In *2010 IEEE Int. Conf. on Acoustics, Speech, and Signal Processing (ICASSP)*, Dallas, USA, April 2010.

[22] Steven Van Vaerenbergh, Javier Vía, and Ignacio Santamaría. A sliding-window kernel RLS algorithm and its application to nonlinear channel identification. In *2006 IEEE Int. Conf. on Acoustics, Speech, and Signal Processing (ICASSP)*, pages 789–792, Toulouse, France, May 2006.

[23] Masahiro Yukawa. Multikernel adaptive filtering. *IEEE Transactions on Signal Processing*, 60(9):4672–4682, 2012.

Index

Printed and bound by CPI Group (UK) Ltd, Croydon, CR0 4YY

23/10/2024

01777693-0015